The Practical Handbook of Well Control

The Practical Handbook of Well Control teaches readers to safeguard well safety and integrity in drilling well engineering. Offering an applied and scientific point of view, it covers fundamental aspects of well control and includes practical procedures and well control methods for land and offshore operations. It features a wealth of questions to commonly encountered problems and comprehensive answers at the end of each topical discussion to test reader comprehension.

- Written in a concise, accessible way by experienced oilfield and academic experts
- Covers all related technical subjects of well control
- Describes modern aspects of well control, including automatic well control, advances in outflow measurement, applied data analytics artificial intelligence and IoT techniques for early kick detection, mud gas separators, and riser gas modeling
- Includes case studies to familiarize readers with real-world problems and solutions
- Offers a full explanation of each problem to familiarize readers with commonly faced issues
- Features sample exercises with comprehensive answers useful for IWCF and IADC exams

This handbook serves as a valuable reference and workbook for field drilling and workover engineers in the energy sector.

Rahman Ashena has over a decade of experience in industrial field work as a senior well completion/workover and drilling supervisor engineer. He is an expert in data analytics / ML techniques and well engineering with TTT degree and IWCF Level IV certificates. Ashena is the director of smart energy and data analytics (SEDA) group operating in Houston and Kuala Lumpur, specialized in providing consultancy and engineering services for oil and gas, renewable energy including geothermal, and data analytics sectors with current projects in SE Asia and the Middle East. Ashena is a grid energy planning engineer contractor at Oklahoma department of commerce. He completed his PhD, Master's, and bachelor's degrees in petroleum Well Engineering respectively at Montanuniversität Leoben, Curtin University, and Petroleum University of Technology. Ashena is a Chartered Engineer recognized by IMechE and a Professional Engineer by the board of Engineers Australia (EA).

The Practical Handbook of Well Control

Edited by
Rahman Ashena

CRC Press is an imprint of the
Taylor & Francis Group, an informa business

Designed cover image: shutterstock

First edition published 2025
by CRC Press
2385 NW Executive Center Drive, Suite 320, Boca Raton FL 33431

and by CRC Press
4 Park Square, Milton Park, Abingdon, Oxon, OX14 4RN

CRC Press is an imprint of Taylor & Francis Group, LLC

Library of Congress Cataloging-in-Publication Data
Names: Ashena, Rahman, editor.
Title: The practical handbook of well control / edited by Rahman Ashena.
Description: First edition. | Boca Raton : CRC Press, 2026. | Includes bibliographical references and index.
Summary: "The Practical Handbook of Well Control teaches readers to safeguard well safety and integrity in drilling well engineering. Offering an applied and scientific point of view, it covers fundamental aspects of well control and includes practical procedures and well control methods for land and offshore operations. It features a wealth of questions to commonly encountered problems and comprehensive answers at the end of each topical discussion to test reader comprehension. Written in a concise, accessible way by experienced oilfield and academic experts. Covers all related technical subjects of well control. Describes modern aspects of well control including automatic well control, advances in outflow measurement, artificial intelligence and IoT techniques for early kick detection, mud gas separators and riser gas modelling. Includes case studies to familiarize readers with real-world problems and solutions. Offers a full explanation of each problem to familiarize readers with commonly faced issues. Features sample exercises with comprehensive answers useful for IWCF and IADC exams. This handbook serves as a valuable reference and workbook for field drilling and workover engineers in the energy sector"—Provided by publisher.
Identifiers: LCCN 2024029123 (print) | LCCN 2024029124 (ebook) | ISBN 9781032753850 (hbk) | ISBN 9781032753874 (pbk) | ISBN 9781003473770 (ebk)
Subjects: LCSH: Oil well drilling—Handbooks, manuals, etc. | Oil well completion—Handbooks, manuals, etc. | Oil fields—Safety measures—Handbooks, manuals, etc. | Oil wells—Blowouts—Handbooks, manuals, etc.
Classification: LCC TN871.2 .P6293 2026 (print) | LCC TN871.2 (ebook) | DDC 622/.3381—dc23/eng/20241007
LC record available at https://lccn.loc.gov/2024029123
LC ebook record available at https://lccn.loc.gov/2024029124

ISBN: 978-1-032-75385-0 (hbk)
ISBN: 978-1-032-75387-4 (pbk)
ISBN: 978-1-003-47377-0 (ebk)

DOI: 10.1201/9781003473770

Typeset in Times LT Std
by Apex CoVantage, LLC

Contents

Preface

For safety concerns, well control is of particular importance for drilling or other well operations in today's petroleum and geothermal industries. The consequences of losing safety by loss of well control (LOWC) incidents or blowouts are severe, including fatalities, environmental contamination including oil spills, damage to total loss of equipment, and loss of reputation.

Considering its importance, most regulatory agencies in the world require that all well operations personnel be trained in well control. Despite such requirements, studies on well control statistics show that LOWC incidents have not reduced considerably compared with the past or have even remained almost the same in many regions. It is thus as vital as ever has been to continue training well operations crew on the principles of well control. With the potential for well control training being still present, it should be emphasized using different educational sources. Books are considered an essential source for training. The basic premise of well control books is that they should present not only the fundamentals in a more straightforward way but also the current techniques and practices in such a way as to be useful for the industry practitioners and researchers. Well control is multidisciplinary, including a broad spectrum of topics requiring basics in physics, geomechanics, etc. For a stepwise approach, this book first covers fundamental principles and then delivers advanced concepts such as innovative data analytics approaches, providing readers with a comprehensive understanding of the subject matter. Hence, the total number of pages in this book has exceeded 500, reflecting the extensive coverage of topics.

This handbook is intended to be not only a well control training aid but also a source for presenting some modern well control techniques to readers. The industry, and at times academia, commonly face the challenge of losing valuable experiential insights and technological advancements over time, or those who know may not have time or resources to present them to others. This book strives to serve as a conduit through which the wealth of industry experiences is gracefully transmitted to the awaiting students or learners. Therefore, for modern well control topics including automated well control, industry internet of things (IoT), artificial intelligence (AI) and machine learning (ML) approaches, automatic well control, and more engineering look into mud gas separators due their critical role in well control events, some experts of considerable knowledge and experience were sought as chapter authors. Bringing up such innovative topics can offer valuable insights and inspire creative thinking in well control.

Last but not least, the book presents some loss of well control (LOWC) case studies to transfer valuable lessons, equipping readers with the knowledge to avert failures in well control in the future endeavors.

Rahman Ashena
Houston, US

Authors

Rahman Ashena has around ten years of field work experience as a drilling supervisor and senior well completion/workover. Ashena is specialized in oil and gas and geothermal well operations and engineering, including well control, geomechanics, wellbore fluid flow modeling, well integrity, cementing, and fluids, and has been a consultant in different well engineering projects in the Middle East and SE Asia. He is the founder of Smart Energy and Data Analytics (SEDA)-Group in Houston, USA, providing consultation in oil and gas and renewable and data analytics. Ashena completed his PhD, master's, and bachelor's degrees in petroleum well engineering, respectively, at Montanuniversität Leoben in 2017, Curtin University in 2009, and Petroleum University of Technology in 2007. Ashena also holds a master's in data science and analytics. His PhD dissertation was on an innovative geomechanics approach in drilling. Ashena has more than 10 years' experience of training in drilling and production for university and industry. Ashena has published more than 70 technical peer-reviewed papers with international publishers, and is the assistant editor-in-chief of Journal of Petroleum Exploration and Production Technology in Springer. Ashena is the corresponding author of the book "Coring Methods and Systems" published by Springer and a book chapter in "Artificial Intelligent Approaches in Petroleum Geosciences" published by Springer. His next book is a handbook on cement well integrity.

Bryan Atchison graduated with an honors degree in mechanical engineering in 1988. Following two years working with a drilling contractor, he returned to university, graduating with a master's in petroleum engineering. He has held drilling supervisor, engineering, superintendent, and drilling manager positions in operators. Atchison then joined Robert Gordon University in 2017 in a business-facing role, delivering and creating courses that optimize the skill sets of well engineering personnel for well control and high-risk critical operations using numerical drilling simulators. The university environment coupled with a career in well engineering enabled the research and development for automated well control. He has written several papers associated with automated well control for the Society of Petroleum Engineering, the Institute of Chemical Engineering, and the European Safety and Reliability Conference. Bryan is a founder, investor, and managing director of Safe Influx.

Muhammad Abbas Bhatti holds a MSc in Oil and Gas Engineering with Distinction from Robert Gordon University, in Aberdeen, and has over nine years of experience in the upstream oil and gas industry. Bhatti worked as a drilling engineer at Safe Influx Ltd. Prior to Safe Influx, Muhammad worked on various onshore drilling projects as a driller in HPHT regions in Pakistan, where he has developed a diverse set of skills and attributes.

Asad Elmgerbi is a distinguished figure in the field of petroleum engineering, renowned for his academic excellence and groundbreaking contributions to drilling operations. He completed his master and PhD degrees at Montanuniversität Leoben (Austria) in 2012 and 2023. His doctoral research focused on developing integrated system capable of automatically detecting and classifying most of downhole problems including losses, kicks, stuck-pipe, string failure, and wellbore instability, efficiently and flawlessly. Elmgerbi's illustrious career extends beyond academia, as he has also amassed a wealth of practical experience in the industry. His tenure as a production supervisor with Wintershall Libya from 2002 to 2005 provided invaluable insights into the operational challenges and intricacies of on-site management. Subsequently, Elmgerbi assumed roles as a drilling engineer from 2005 to 2009, both globally and in Libya with Wintershall, where he demonstrated his expertise in optimizing drilling processes and maximizing operational efficiency. In 2014, Elmgerbi made a seamless transition into academia, joining Montanuniversität Leoben as a senior lecturer. During his academic journey, Elmgerbi has supervised more than 23 master theses, and he has published more than 20 papers.

Stefan Erakovic is a skilled petroleum engineer with a passion for drilling engineering. He began his academic journey at the University of Belgrade, where he obtained his bachelor's degree in petroleum engineering in 2021. Building upon his foundational knowledge, Stefan pursued his master's education at Montanuniversitaet Leoben, Austria, specializing in petroleum engineering with a focus on drilling engineering. Following the completion of his studies, Stefan ventured into the professional realm, joining Equinor ASA as a drilling and well operations engineer in Norway. In this role, Stefan plays a pivotal part in the development and execution of well programs, collaborating closely with multidisciplinary teams comprising geologists, geophysicists, and engineers. His responsibilities encompass the preparation of comprehensive operational procedures and the facilitation of various meetings to address risks, optimize costs, and disseminate best practices.

Ali Ghalambor is currently an international consultant with more than 46 years of industrial and academic experience. He is internationally recognized for his technical contributions on fundamental and applied research on formation damage control, well drilling, well completions, and production operations. Dr. Ghalambor is the recipient of numerous international and domestic awards for his technical and service contributions to the profession. He is a registered professional engineer in the U.S. State of Texas and the District of Columbia. Dr. Ghalambor held engineering and supervisory positions at Marlin Drilling, Tenneco Oil, Amerada Hess Corporation, and Occidental Research Corporation. He previously served as the American Petroleum Institute (API) endowed professor, Head of the Petroleum Engineering Department, and Director of the Energy Institute at the University of Louisiana at Lafayette. He was also the technical director and a program manager at the Qatar National Research Fund of Qatar Foundation. He has authored or coauthored 17 books and manuals and more than 250 technical papers. Dr. Ghalambor holds BS and MS degrees in petroleum engineering from the

University of Southwestern Louisiana and a PhD from Virginia Polytechnic Institute and State University.

Farzad Ghorbani is an experienced drilling and well control engineer with about 30 years experience in the Middle East, and has a master's in petroleum engineering. Ghorbani has several publications on drilling and completion engineering. Ghorbani is extremely skilled in field work and has been an experienced drilling and well control instructor and has trained many drillers and drilling supervisors.

Terrence Josiah is a master student in data science at Leeds Beckett University, United Kingdom, with a deep passion for data analysis and machine learning. His expertise in data science positions me as a forward-thinking individual capable of effective collaboration with technical teams. His experience includes conducting some research with the help of his strong understanding in Python, SQL, and Tableau. Additionally, his experiences in several data analyst positions coupled with some well-done data science and machine learning projects enable him to become an innovative thinker and contribute to future technological advancements.

Nabe Konate is a PhD candidate in petroleum engineering at the University of Oklahoma, specializing in the development of smart drilling fluids to address thermal short-circuiting in enhanced geothermal systems (EGS). From 2021 to 2023, he served as a data analyst for Salt and Light Energy Equipment, where he created a performance tracking tool to evaluate the efficiency of fracking fluid ends in daily operations. This innovative tool also enabled partner companies and clients to monitor the performance of various crews and technicians. From 2018 to 2021, Konate worked as a research associate at the University of Oklahoma to conduct significant research in unconventional drilling, focusing on developing and testing inhibitive mud systems to enhance the efficiency of drilling in unconventional shale formations, while minimizing drilling-related issues. Konate's expertise spans wellbore stability, drilling performance, wellbore integrity, and well control, making him a valuable contributor to advancements in both the oil and gas and geothermal energy sectors.

Elliot H. Kurnia is a petroleum engineer with a wide interest in subjects drilling and completion engineering and an excellent performance during his bachelor studies in petroleum engineering. Having completed his internship on well control, he is passionate in well engineering and machine learning methods to potentially help advance the industry even further technologically.

Saeed Salehi is currently an endowed professor of petroleum engineering at Texas A&M International University. Previously worked as drilling and well engineer in industry, he has more than ten years of academic experience teaching and delivering university and industry customized courses. While in academia, he has managed more than $10 million in academic R&D expenditure. Dr. Salehi has more than 200 articles published in scholarly oil and gas journals and refereed conferences. Dr. Salehi is recipient of prestigious awards from SPE such as SPE Drilling Engineering

Award, SPE Health, Safety, Security, Environment, and Social Responsibility, and SPE Distinguished Achievement for Petroleum Engineering Faculty Award. He has leadership recognition as past chairman of SPE Evangeline section in Louisiana. He currently serves as Co-Chair, Steering and Finance Committee Chair for SPE International Conference and Exhibition on Formation Damage Control. His research interest includes subsurface energy both fossil fuels and renewables such as geothermal energy. Other areas of interest are geological carbon storage, mitigation of surface and subsurface contamination, issues of process safety and environmental protection when developing subsurface energy resources through a combination of experimentation, finite element and fluid dynamics modeling, data analytics, and analytical and computational tool development. The long-term goal of his research program is to provide viable and sustainable solutions for the protection of environment and enable a sustainable, affordable, and secure energy supply.

Daniel A. Tetteh is a research scientist at Leidos Inc., supporting the U.S. National Energy Technology Laboratory (NETL). He holds dual master's degrees in petroleum engineering and data science and analytics from the University of Oklahoma. During his master's studies, Tetteh contributed to significant projects, including mud gas separators analysis and investigations into renewable energy sources like geological hydrogen storage and geothermal energy. Tetteh's research has been showcased at conferences such as the World Hydrogen Energy Conference and the SPE/IADC Geothermal Rising Conference. He received the Livermore Lab Foundation Carbon Fellowship in 2022 and interned at Lawrence Livermore National Laboratory. With multiple publications and active membership in professional societies, Daniel is committed to advancing sustainable energy. Currently, he works on developing innovative tools for sustainable energy, including web applications for oil and gas data visualization, analysis, and decision support in carbon capture and storage and geological hydrogen storage.

Contributors

Rahman Ashena
Smart Energy and Data Analytics (SEDA) Group
Houston, Texas

Bryan Atchison
Safe Influx Ltd
Aberdeen, Scotland

Muhammad Abbas Bhatti
Safe Influx Ltd
Aberdeen, Scotland

Asad Elmgerbi
Montanuniversitaet Leoben
Leoben, Austria

Stefan Erakovic
Equinor
Stavanger, Norway

Ali Ghalambor
Oil Center Research International LLC
Lafayette, Louisiana

Farzad Ghorbani
Independent Consultant
National Iranian Oil Company
Ahvaz, Iran

Terrence Josiah
Leeds Beckett University
Leeds, United Kingdom

Nabe Konate
University of Oklahoma
Oklahoma City, Oklahoma

Elliot H. Kurnia
SEDA-Group
Kuala Lumpur, Malaysia

Saeed Salehi
University of Oklahoma
Oklahoma City, Oklahoma

Daniel A. Tetteh
University of Oklahoma
Oklahoma City, Oklahoma

Acknowledgments

The author wishes to thank all the authors who contributed to this book, as well as their employers who allowed them to participate in this venture. I also appreciate my drilling managers who taught me drilling, including S. Christman, A. Hekmatinia, H. Safaei, A. Shirani, M. Moghadasi, A. Iravani, and H. Bahraini for their valuable technical discussions.

Acknowledgments

1 Introduction

Rahman Ashena

1.1 DEFINITIONS

Control of a machine, process, or system is defined as the process of making it work in a required way.

Well control simply means to keep our well, rig, and drilling crew safe and away from any danger. Therefore, wells cannot be allowed to flow in an uncontrolled manner. In this book, it is important to distinguish between two terms: "well control event" and "loss of well control (LOWC) incident or event". In other words, well control events which consist of "*kick*" and "*blowout*" should be differentiated. A well control event or situation occurs when a well might kick/flow, and this would not follow standard well operations expectations or would be unsafe. A kick is defined as a sudden undesirable fluid influx from the formation into the wellbore. It occurs when the primary hydrostatic barrier is lost. During drilling operations, the primary barrier is the hydrostatic mud pressure. Some examples of well control events consist of monitored or measured additional surface volume of drilling fluid at the mud tanks (called "pit gain") or rushing fluid flow out of the well. If the kick influx becomes uncontrolled and unmanageable, either by reaching the surface or entering another underground formation, then it would be also called a serious well control event or an LOWC incident. Some sources (e.g., Holand, 2017) call such an incident an "LOWC incident", whereas some other sources simply call it a blowout.

There are numerous LOWC incidents worldwide in onshore and offshore regions. Several historical LOWC incidents remain unrecorded or are solely documented within the confines of in-house company records. From 2000 to 2015, Holand (2017) found 156 recorded offshore LOWC incidents during well operations in the US Gulf of Mexico (GoM) and Pacific Outer Continental Shelf (OCS), Canada East Coat, Norway, the UK, Denmark, Brazil, and Australia.

SINTEF, which is one of Europe's largest independent research organizations in Norway, categorizes and defines LOWC incidents as follows (see Table 1.1).

- *Blowout:* A blowout is an incident where formation fluid flows out of the well (surface flow) or between formation layers (underground blowout) after all the predefined technical well barriers or the activation of the same have failed. Figure 1.1 shows a schematic of an underground blowout.
- *Well release:* The reported incident is a well release if oil or gas flowed from the well from some point where flow was not intended, and the flow was stopped by the use of the barrier system that was available on the well at the time the incident started.

DOI: 10.1201/9781003473770-1

TABLE 1.1
Categories of Loss of Well Control (LOWC) Incidents/Events by SINTEF (Holand, 2017)

Main	Category	Sub category	Comments/Example
Blowout and well release	Blowout (surface flow)	Totally uncontrolled flow, from a deep zone	Totally uncontrolled incidents with surface/subsea flow.
		Totally uncontrolled flow, from a shallow zone	Typically the diverter system fails.
		Shallow gas "controlled" subsea release only	Typical incident for, for example, riser-less drilling is performed when the well starts to flow. The rig is pulled away.
	Blowout (underground flow)	Underground flow only	
		Underground flow mainly, limited surface flow	The limited surface flow will be incidents where a minor flow has appeared, and typically the BOP has been activated to shut the surface flow.
	Well release	Limited surface flow before the secondary barrier was activated	Typical incident will be with flow through the drill-pipe, and the shear ram is activated.
		Tubing blown out of well, then the secondary barrier is activated	Typical incident occurring during completion or workover. Shear ram is used to close the well after the tubing has been blown out of the well.
	Diverted well release	Shallow gas controlled flow (diverted)	All incidents where the diverter system functioned as intended.
	Unknown	Unknown	Unknown may be selected for both the category and the subcategory.

- *Shallow gas:* Any gas zone penetrated before the BOP has been installed. Any zone penetrated after the BOP is installed is not shallow gas (typical Norwegian definition of shallow gas).

Technically, well control is the technical methodology of first preventing formation fluids from entering the wellbore, controlling them (kick influx) in case they occur, and extracting them safely from the well to prevent conversion to loss of well control or a blowout. Well control is important for all phases of drilling, completion, production, and workover operations. Kick prevention is also called primary well control which is met by maintaining mud pressure greater than the formation pressure. Controlling the kick influx downhole by well shut-in is called secondary well control which is done by using equipment particularly the Blowout Preventers (BOPs) to shut in the well and restrict the influx. In case none of the above worked out and a blowout gets in progress,

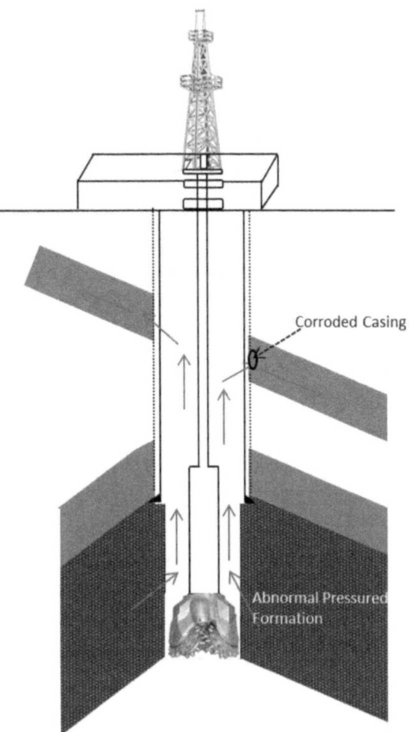

FIGURE 1.1 A typical underground blowout in a well with lost well integrity (due to corrosion and leakage in casing).

tertiary well control is applied, which is out of scope of this manual and any typical well control training. Because of criticality of well control issues, the responsibility of the well control event/situation is both on the drilling crew from the contractor as well as on the operator; thus, both sides should be aware and well trained.

1.2 CONSEQUENCES OF LOWC

Loss of well control (LOWC) incidents usually cause harm to many important assets of companies and societies consisting of people, environment, equipment/infrastructure or economic damage, and reputation. Different types of harm are as follows:

- Physical and psychological harm to the personnel or drilling crew and to the people living in the environment in the region. For example, on April 20, 2010, the Deepwater Horizon oil rig exploded off the Gulf Coast, with 11 people fatalities and 17 people injured.
- Environmental contamination of water, soil, and air leaving irreversible damage to the environment where human inhabitants and animals live. The largest marine oil spill in the history is the Macondo's blowout's spill where during

the 87-day period, 3.19 million barrels were spilled into the Gulf of Mexico (Noaa, 2017). Some other sources like Holand (2017) estimated the spill to be even larger at 4.25 million barrels. It fouled the coasts of Florida, Alabama, Mississippi, Louisiana, and Texas and launched a six-year long environmental and legal battle. It left the region reeling from a disrupted coastal economy; a devastated ecosystem; the deaths of as many as 105,400 sea birds and 7,600 adult and 160,000 juvenile sea turtles; and up to a 51% decrease in dolphins in Louisiana's Barataria Bay (Noaa, 2017). Figure 1.5 shows a photo of the huge size of the oil spill spread in the Gulf of Mexico (GoM). The next, largest spill incidents offshore occurred in 2009 in Montara, Australia, with 29,600 barrels and in 2010 in Brazil with 3,700 barrels.

- Economic damage due to the total loss of the equipment and infrastructure located on the well, compensation cost, and the cost associated with controlling the incident well. Figure 1.2 shows the Transocean's rig prior to and following the Macondo's blowout in 2010, leading to the total loss of the rig and infrastructure. Figure 1.3 and Figure 1.4 show two views of the Deepwater Horizon Macondo's Blowout that occurred in Gulf of Mexico in 2010. For the compensation cost, a Federal District judge approved the largest environmental damage settlement in US history – $20.8 billion – on April 4, 2016. The settlement ended all civil and criminal penalty claims against the owners and operators of the rig – BP, Anadarko, TransOcean, and Halliburton – under the Clean Water Act and the Oil Pollution Act. It also included economic damage claims submitted by the five Gulf states and their local governments (Noaa, 2017).
- Damage to the reputation of the companies involved and political party running the state/country. This damage can cause loss of future projects for the company and in a larger scale for the whole oil and gas sector. Some companies may collapse due to such incidents. As for the Macondo's blowout, the associated economic costs did not cause the collapse of BP due to its large capacity, finances, and size. Huge environmental contamination due to such blowouts can make the oil and gas sector notorious to be named as

a)

b)

FIGURE 1.2 Transocean's rig before (a) and after (b) Macondo's blowout in 2010.

FIGURE 1.3 An example of surface blowout in Deepwater Horizon Macondo Blowout in 2010; www.awesomestories.com/asset/view/158535 (last accessed in February 15, 2020).

FIGURE 1.4 Another view of the surface blowout in Deepwater Horizon Macondo Blowout in 2010; https://slate.com/technology/2016/09/bp-is-to-blame-for-deep-water-horizon-but-its-mistake-was-actually-years-of-small-mistakes.html (last accessed in February 15, 2020).

FIGURE 1.5 Oil spill in the Gulf of Mexico (*The New York Times*, 2024).

a main source of global pollution. Thus, the public would suspect economic advantages of oil and gas activities creating resistance against oil and gas and therefore less projects for oil and gas and an increased tendency for growth in energy transition activities.

Causes of loss of well control (LOWC) incidents can be external or nonexternal. External are imposed from outside such as natural disasters. Normal causes are due to the loss of primary and secondary barriers, which will be covered in Chapter 2, Chapter 4, and Chapter 5.

1.3 HUMAN ERROR AND DEFENSES

An important reason for the occurrence of well control events is the human factor or error. Humans may make errors. Let us imagine a situation where an individual is doing something in the expectation of a certain outcome which does not happen. Human error is an imbalance between what the situation requires, what the individual intends to do, and what he/she does. Human errors happen when people:

- Intend to do the right thing but with the wrong outcome (e.g., give the correct instruction to the wrong person).
- Do the wrong thing for the situation (e.g., turn off the pit gain alarm).
- Fail to do anything when action is required (e.g., fail to report faulty equipment or repair it if possible).

The reasons for human errors are not only the imperfect nature and limitations of human abilities but also the system in which the people work. As people are one part of the system that includes all the other parts of the organization (including organization, training, policies, procedures, equipment, technology, and environment),

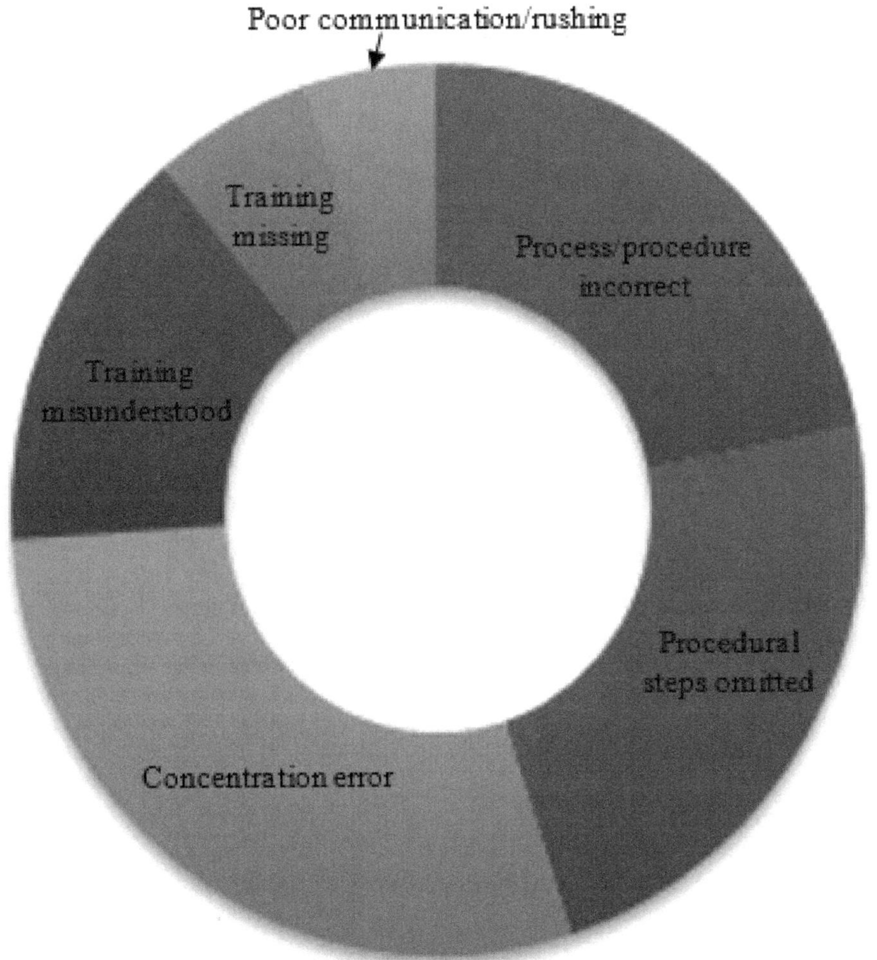

FIGURE 1.6 Root causes of human error which indicate that "human error" is indeed "system error". www.questionmark.com/to-your-health-to-err-is-human-but-assessments-can-help/ (last accessed in February 2, 2020).

"human error" is indeed the "system error" (refer to Figure 1.6). Therefore, this is where poor training may have caused human error. Thus, competent training and assessment contribute to the mitigation or elimination of human error.

1.3.1 Defenses against Human Error

To analyze and investigate human error, there is a concept called *defenses against human error*. In drilling and well control, some of these defenses are:

- Appropriate and correct training
- Checking the mud weights periodically (following the policy/procedures for that)

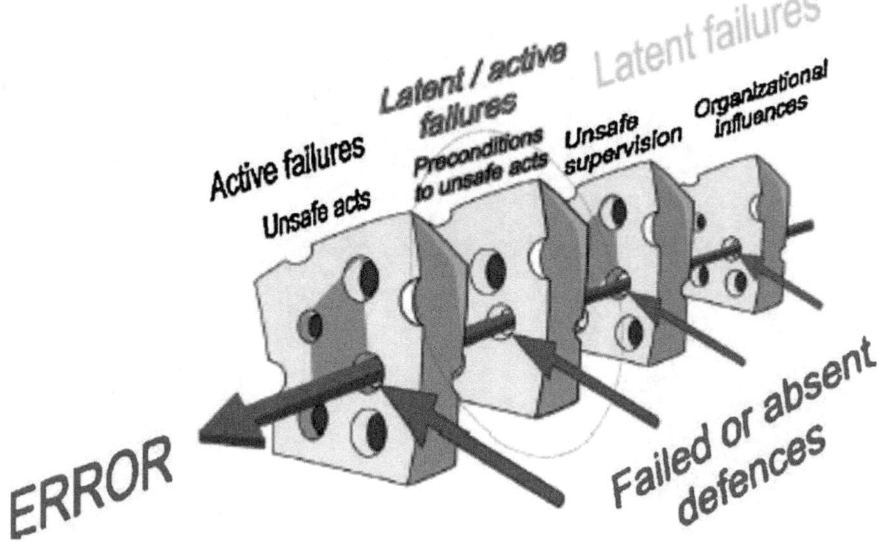

FIGURE 1.7 Reason's Swiss Cheese Model for how human errors occur. The holes in each slice show the failures in each defense. Some failures are latent (they were made in the past at some point and lay dormant), and some are active. The more the holes in the slices, the greater the possibility that the human error results in incidents or accidents (www.researchgate.net/figure/Diagram-of-Reasons-Swiss-cheese-model-of-human-error-15-Each-of-the-cheese-slices_fig1_283260847, last accessed in February 2, 2020).

- Setting alarms correctly
- Following correct testing procedures
- Challenging response procedures (challenge something you are told to do, but you know it is wrong)

Defenses or barriers against human error can be illustrated graphically as slices of a "Swiss Cheese Model". This model which was raised by Prof. James Reason is shown in Figure 1.7. As shown in the figure, each layer of defense is shown as a cheese slice in which there are some holes or failures. Some failures are latent, meaning they were made at some point in the past and lay dormant. These may be introduced at the time a well barrier was designed or may be associated with poor management decisions and policies. On the other hand, some failures are active which are made by frontline personnel such as supervisors and drillers. The more holes are in a system's defenses, the more likely that the errors result in incidents or accidents. In certain circumstances, when all holes line up, and a blowout occurs (Figure 1.8).

In case of an incident in the workplace/rig site, a system of recording and reporting incidents and/or accidents is often found. These systems of recording and reporting are greatly prominent as unsafe behaviors or acts, unsafe conditions, or unsafe equipment can cause lost time and/or injury; therefore, they can be addressed later. The term "LTI" is "Lost Time Incident" which is often associated with such reporting.

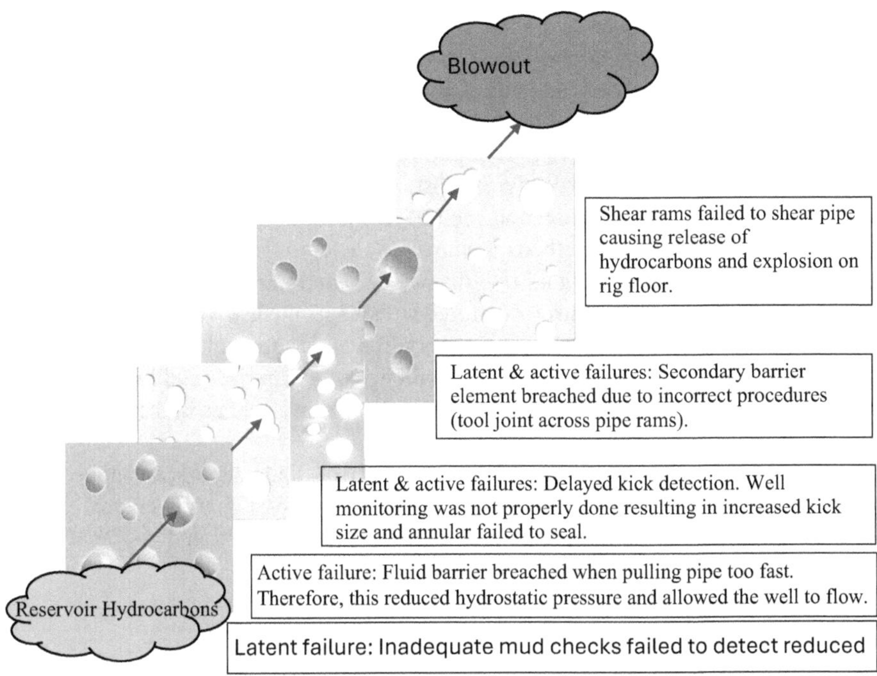

FIGURE 1.8 Swiss Cheese Model for explaining how a blowout occurs in a petroleum or geothermal well.

1.4 WELL CONTROL TRAINING OBJECTIVES

Although the importance of well control has been grasped by the drilling industry for more than half a century, statistics confirm that well control incidents still occur in large numbers. Human error is still the most contributing factor, and this signifies the requirement of more emphasis on proper and more strict training of the drilling crew or well control learners. Therefore, there are many schools in universities and companies worldwide, training drilling personnel to prevent future drilling catastrophes. Recently, there have been several modifications in the instruction and training methodologies, particularly in terms of subsea procedures and utilizing modern well control simulators. Similarly, in this handbook, the principles and procedures are explained for both surface and subsea-BOP stacks. It should be borne in mind that all well control procedures revolve around maintaining the primary well control or restoring it in case of losing it (secondary well control).

1.5 SCOPE AND LEARNING MILESTONES

This book will be aligned with meeting the objectives of primary and secondary well control. Tertiary well control (controlling blowouts) is out of scope of this handbook and is not covered.

Therefore, this book first covers the consequences of loss of well control (LOWC) incidents, risk management and barriers, well control concepts and fundamentals, management of drilling muds and pits, data which should be prerecorded for possible well control events, causes of kicks and LOWC incidents, warning signs and kick indicators, kick detection, barriers required for safe drilling operations, well shut-in and diverting methods and procedures, post shut-in monitoring including pressures and activities, kill sheet completion methodology, and well control/kill methods. Next, modern well control methods are covered including early kick prediction and detection by data-driven artificial intelligence (AI) and industrial internet of things (IoT) and automated well control. Next, recent advancements in flowmeters are discussed, as an essential measure in accurate outflow measurement equipment. In offshore operations where riser is used, the challenge of gas in riser and its modeling is discussed and a simple model is presented. Next, two case studies in the Middle East and Australia are presented.

In brief, learners are expected to learn well control fundamentals and apply them to their well operations work including successfully passing well control exams held by International Well Control Forum (IWCF) or International Association of Drilling Contractors (IADC).

Specifically, learners are supposed to:

- Know the consequences of loss of well control events or blowouts.
- Explain human error and Reason's Swiss Cheese Model.
- Explain risk management procedures in drilling operations.
- Explain well control drills to rehearse well control events.
- Explain barriers required for safe well operations.
- Explain main well control methods during operations in petroleum and geothermal wells.
- Explain how to apply primary and secondary well controls.
- Explain the basic geology of underground formations and how formation properties matter in well control events/situations.
- Explain the hydrostatic pressure and gradient concepts.
- Identify reasons and origins for abnormally high-pressured formations and explain how to predict them in advance.
- Explain the U-Tube principle, how a drilling well is simulated for the principle, and how it applies to account for changes in bottomhole pressures and circulating drill-pipe pressures in situations like pumping pills.
- Explain bottomhole pressure and equivalent circulating density (ECD) and the factors impacting them.
- Explain the appropriate selection of casing setting depths and trip and kick margins.
- Explain surge and swab pressures and the impacting parameters which can cause well control events.
- Explain formation strength and how it is practically measured/estimated, including Leak-Off Test (LOT) and Formation Integrity Test (FIT).
- Explain how to evaluate Maximum Anticipated Surface Pressure (MASP), Maximum Allowable Annular Surface Pressure (MAASP), and Maximum Allowable Mud Weight (MAMW).

- Explain the concept of kick tolerance and its evaluation/calculation method.
- Explain the ballooning concept and its importance in discriminating it from a real kick flow situation.
- Explain gas laws to predict the behavior of gas in the wellbore, both when the well is open (gas expansion) and when the well is shut in (gas migration), and also explain gas influx behavior in case of water-based and oil-based muds.
- Know types of drilling fluids, different methods to measure their densities/weights, and mud pit volume and return volume management.
- Become familiar with the prerecorded data to be collected before a kick occurs, including Slow Circulating Rate pressures.
- Know the causes of kicks and the required measures to prevent them.
- Discriminate warning signs from kick indicators and measures to confirm or reject a well control event using a flow check.
- Explain shut-in procedures based on well conditions and companies' policies and rig types (fixed or floating rigs).
- Complete different types of kill sheets (vertical or horizontal wells and surface or subsea stacks) and use the practical calculations for well kill.
- Explain different well kill methods and identify the right well kill method to take the well under control and know about influx-out and influx-in kill methods.
- Explain different stages of Driller's, Wait and Weight, and Concurrent methods to circulate the kill out of the hole in addition to Volumetric and Bleed method in cases where circulation of the kick influx is not possible.
- Know the situations and procedure of using the influx-in or bullheading well kill method.
- Explain the wireline-related operations and their well control setup.
- Learn about modern well control techniques compared with traditional ones.
- Learn about artificial intelligence (AI) methods for prompt kick prediction and detection.
- Learn about industrial internet of things (IoT) techniques in well control.
- Get familiar with automatic well control systems.
- Learn about new developments and advances in outflow meters for early kick detection.
- Learn about mud gas separators (MGSs) and their appropriate design to prevent loss of well control events.
- Learn about the gas in riser (GIR) challenges and their modeling.
- Learn about two case studies of loss of well control events in the Middle East and Australia.

1.6 RISK MANAGEMENT

1.6.1 Definition

Risk management is the identification, evaluation, and prioritization of *hazards* and *risks* followed by measures required to minimize, monitor, and control the probability or impact of unfortunate events causing harm or to maximize the realization of opportunities. In other words, risk management helps to ensure that drilling crew identify hazards in advance and control associated risks in a systematic way in order to reduce the chance of harm. In detail, risk management first identifies and prioritizes the hazards and risks (i.e., risk assessment); next, it identifies who can be harmed and how; then, it analyzes the risks and decides on good control measures or actions for risk mitigation; finally, it collects further information to review and update the assessment.

Lack of or inadequate risk assessment processes allow some of the hazards and risks to remain unidentified and consequently increase the chance of harm. Hazard is defined as an object which can potentially cause harm (hydrocarbons, chemicals, hoisting or rotating equipment, electrical equipment). Risk is the chance of being harmed, which is delineated by probability and severity of the possible harm. Harm generally covers physical injury or psychological harm to personnel, damage to company's reputation, environmental impact or pollution, and damage to or loss of the equipment. The priorities of protection are as follows:

- The first priority of protection from harm is the people or personnel,
- The second priority is the external environment (air, water, and soil), and
- The third one is the assets (equipment and infrastructure).

Significant incidents or accidents have a large negative impact on companies' reputation and can even bring some companies to approach collapse. Elimination of all risks is not possible, but their mitigation to very low or acceptable levels is possible by using appropriate methods.

1.6.2 Drilling Risk Management

Drilling is a complex activity and subject to several hazards, some of which are location and activity dependent. Thus, the drilling risk management should be commensurate with the site, water depth (in offshore), available information, and the complexity of the situation. As drilling begins, new information becomes available, and some predicted hazards may still pose risk, while others may not. New hazardous situations may be encountered or identified, and the characteristics of those already identified could change. Therefore, the risk management should be carried out periodically at all stages of the project, prior to, during, and post-drilling (Yasseri, 2017). It is important to maintain a positive attitude to risk management and take it seriously rather than considering it just as paperwork to be completed.

1.6.3 Drilling Hazard Identification and Prioritization

Drilling hazards are objects or sources which can cause harm. They can be categorized into several categories; also there may be several paths through which a hazard can threaten the operation. The main hazards are grouped under the distinctive headings as shown next. The categories and subcategories are shown in Table 1.2.

1. Geohazards
2. Equipment and material

TABLE 1.2
Identification of Drilling Hazards and Categorizing Them (Yasseri, 2017)

C1 – Geohazards	C11 – Formation Pressure
	C12 – soil and rock types and strength
	C13 – Shallow faults
	C14 – Gas hydrate
	C15 – Multiple geohazards
	C16 – Top-hole geology
	C17 – Shallow soil (for jack-ups), sediment type, and strength
	C18 – Salt or mud diapirs
C2 – Equipment and Material	C21 – Material suitability and defects, fabrication defects
	C22 – Equipment used (robustness and dependability, maintenance)
	C23 – Effect of ageing, wear and tear, worn or fatigued part
	C24 – Operational and resource limits
	C25 – Operational limit, failure to meet qualification, and code compliance
	C26 – Late changes to well design and procedures
	C27 – Equipment quality (special equipment, delay, damaged)
	C28 – Spare and material availability
	C-29 – Unsuitability and unforeseen site condition, injuries, toxic emission
C3 – Human Elements	C31 – Skill and knowledge-based mix, training, experience
	C32 – Workload work coordination; shift duration
	C33 – Quality of working environment
	Q34 – Communication, language, barrier, openness
	C35 – Performance evaluated and suitability and training; fit for the job
	C36 – Personnel exposure (qualification, experience, required presence, shift)
	C37 – Tiredness, boredom
	C38 – Situational awareness
C4 – Design, Technology, and Operation	C41 – Technology, readiness maturity
	C42 – New technology (e.g., packers & liner hangers)
	C43 – Downhole monitoring
	C44 – Kick tolerance
	C45 – Deviation versus hole size, hole size contingencies
	C46 – Cementing of long casing strings
	C47 – Well access and workover requirements, well design, and job complexity
	C48 – Blowout contingency
C5 – Automation and human–machine interface	C51 – Software error
	C52 – Temporary disabling safety devices to get round annoying alarms
	C53 – Information overload
	C54 – Design of human–machine interface
	C55 – Failure of data processing function; failure of information support function
	C56 – Failure of surveillance function; failure of communication function
	C57 – Expert system which bypasses the operator involvement
C6 – Local conditions	C61 – Water depth
	C62 – Local weather
	C63 – Current, waves, tsunamis, hurricanes, ice, rain, storm surge, tropical cyclones
	C64 – Wind and water-borne debris

(*Continued*)

TABLE 1.2 (Continued)
Identification of Drilling Hazards and Categorizing Them (Yasseri, 2017)

	C65 – Requires special equipment
	C66 – Existing infrastructure, surface, and subsurface
	C67 – Shallow water flow
	C68 – Preservation and sanctuaries
C7 – Organizational elements	C71 – Mix of cultures and compatibility (e.g., working to different procedures)
	C72 – Organizational learning
	C73 – Personnel selection; coordination
	C74 – Training program, process, and formalization
	C75 – Safety commitment, perception, and enforcement
	C76 – Time and cost constraints
	C77 – Bonus systems and benefits upon performance
	C78 – Experience with operators or contractors
	C79 – Operational aspects (language barriers, local marine traffic, shore proximity)
C8 – Well Integrity	C81 – Mechanical wellbore instability (rock type and strength)
	C82 – Wellbore geometry (hole inclination and azimuth)
	C83 – Man-made related stress, poor hole cleaning, excessive drilling vibration
	C84 – Drilling into prestressed rock, excessive wellbore pressure, vibration
	C85 – Shale type and instability, time-dependent swelling, reaction between fluid and shale
	C86 – Shale hydration mechanism, forces holding plates together, pore pressures, stresses
	C87 – Inadequate well planning (wrong drilling fluid, wrong inclination and azimuth)
C9 – Uncertainty	C91 – Phasing and planning
	C92 – Scope change
	C93 – Complex procedures
	C94 – Management Of Change (MOC) (design, operating conditions, equipment substitution, plans, personnel)
	C95 – Safety critical equipment

3. Human elements
4. Design issues, technology, and operation
5. Human–machine interface
6. Local environment
7. Organizational elements
8. Well integrity
9. Uncertainty

1.6.4 Control Measures and Their Hierarchy

Following the identification and prioritization of the drilling hazards, several control measures must be essentially set to mitigate risks. Control measures should not be limited to just mandating the use of personal protective equipment (PPE). But rather,

TABLE 1.3
An Example of Hierarchy of Controls in Drilling

Hazard Example	Control		
	Example	**Type**	**Effectiveness**
Swabbing a kick	Pump out of hole	Hazard elimination	Most effective
	Minimize number of trips	Hazard substitution	
	Reduce (unnecessary) BHA size	Task/equipment redesign	
	Pump a mud cap	Separation of hazard from people	
	Pull by swab calculation	Administrative controls	
	"BOP" equipment	Personal Protective Equipment (PPE)	Least effective

controls should be robust and efficient enough to first focus on the elimination of the hazards. Therefore, hierarchy of controls is an essential part of risk management so that safety is regarded more, and the possibility of harm is mitigated. It is essentially recommended to consider using controls which are higher in the ranking of controls. As an example, a drilling hazard is swabbing a kick which can occur during tripping out of the hole. To deal with these hazards, several controls are designed/set in a hierarchy order as shown in Table 1.3.

1.6.5 Levels of Risk Management

In well operations including like any other activity, risk management is carried out in the following two levels.

- The first level of risk management is comprehensive and covers the design of a new drilling operation (predrilling) in which many hazards are involved. This can be conducted during drilling operations for review, and subsequently modifications can be made using the information obtained in real time. This is a broad and detailed process conducted by drilling safety engineering team. Risk analysis is made quantitatively using several methods such as Hazard and Operability Studies (HAZOPS) and Failure Models and Effects Analysis (FMEA).
- The second level is the assessment of task-specific risks at an operational level. This is mostly a qualitative risk analysis using the assessment sheet, which is managed by supervisor and workers to ensure that hazards are identified, and risk is managed to an acceptably low level.

1.6.6 Management of Change (MOC)

There are uncertainties in a drilling hazard due to changes in plans, which can occur during drilling operations. In fact, a work arising from temporary and permanent changes to procedures, system, equipment, etc., cannot proceed unless a management of change (MOC) is completed. MOC is essentially completed as a control measure of the risk management.

An MOC must recognize the change and cover the reasons for that (accounting for safety or business benefits). It should be signed and authorized by a named authority. If planning or preparation is required prior to the implementation as part of MOC (such as extra training and communication between the crew), it must be conducted beforehand. Finally, during implementation, the change must be monitored to make sure no unforeseen issues are happening.

1.7 WELL CONTROL DRILLS

Practicing different well control events can help the drilling crew master to perform the tasks confidently in such situations. Well control drills allow practicing and understanding the importance and the procedure of well control procedures as well as check the equipment reliability. The purposes of drills and the steps included in the procedures are stated in well control management documents. According to standards, drills must be announced to the company supervisor/man and the tool pusher in advance. When such drills are performed, the time taken to perform the drills must be measured so that a benchmark performance guide for the drill can be established.

Some examples of drills that a drilling crew would be expected to perform are:

- *Pit Drill:*
  1. Company man and tool pusher apply an artificial gradual increase in the apparent pit level by manually raising a float in the active mud pits. It looks like a pit gain (and a kick influx) to the mud engineer. The crew would then need to follow the right well control procedures (which would lead to a flow check).
  2. The driller and the mud engineer are expected to detect the pit gain.
  3. The driller and the floormen must pick up the drill-string until the tool joint clears the BOPs.
  4. Stop the pumps.
  5. Perform a flow check.
  6. Report the result of the flow check to the company man and/or tool pusher.
  7. Record the time for the crew to react and conduct the drill on the Daily Drilling Report (DDR).

- *Trip Drill:* During tripping, the drill may be performed either when the drill-string is in the open-hole section or in cased hole. If the drill-string is yet near the bottom or in the open hole, it may be required that we do *not* shut in the well. If the drill-string is in cased hole, the well must be shut in with the following procedure:

1. Stop tripping operations and install the Kelly (or top drive) and start circulating the drilling mud.
2. Having been asked by the company man to shut-in the well, the driller is expected to do the task as follows:
3. The driller and the floormen should first pick up the drill-string until the tool joint clears the BOPs (this is called "space-out").
4. Stop the pumps.
5. Apply the shut-in procedures according to the company policy (refer to Chapter 8 for more information). For example, for hard shut-in, first, close the annular preventer. Then, open the High Closing Ratio (HCR) valves on the side outlets of the BOP.
6. Record Shut-In Casing Pressure (SICP) at the annulus and Shut-In Drill-Pipe Pressure (SIDPP) as the surface pressure at the drill-pipe.
7. If the rig is of a floating type, double check to ensure the drill-string is spaced out relative to the BOPs, close, and lock hang-off rams and hang-off pipe. Next, check that the Kelly cock or the Drill-Pipe Safety Valve (DPSV) is made up/installed on the drill-pipe in the rotary table. DPSV is also called Full-Opening Safety Valve (FOSV).
8. Notify the company man that the well is shut in and mention the shut-in pressures.
9. Record the time taken for the crew to shut in the well on the Daily Drilling Report (DDR).

- *Choke Drill:* This drill is done after running and cementing the surface casing string or any subsequent casing string (say 9 5/8" casing), prior to drill out of the casing shoe track. Some backpressure is artificially applied on the choke which causes an increase in bottomhole pressure. Since the choke drill is performed before drill out of the shoe track, there is no concern regarding fracturing of the underground formations. The drill establishes equipment performance and allows the crew to gain proficiency with choke operation.
  1. Close the annular preventer.
  2. Start the pump and create some backpressure behind the annular preventer to be stabilized (to simulate the SICP).
  3. Resume mud circulation while operating the choke to maintain the backpressure on the casing pressure (in surface-BOP cases). For subsea-BOP wells, subtract the choke-line friction (CLF) from the casing pressure and maintain that magnitude during the operation. The mud outlet from the choke is discharged into the trip tank to accurately monitor the flow rates.

- *Stripping Drill:* Stripping is defined by IADC as the process of running the drill-string into or pulling it out of the well under "Kick" conditions. Normally, stripping is through a closed annular BOP, but the drill-pipe may be run ram-to-ram by carefully closing, bleeding-off pressure, and opening rams to let tool joints and collars pass.

A Stripping Drill is performed in combination with the volumetric well control once a week per crew. Stripping Drills are not recommended for operations involved with subsea-BOP stacks. The drill should only be conducted after a casing string (e.g., 9 5/8" casing or production string) was already run, cemented and pressure tested, and prior to drilling out the shoe track. The purpose is to establish the equipment reliability and make all the crew members practice and be familiar with the stripping operations.

1. Run in the hole (RIH) to, for example, 1,000 ft above the depth of top of cement (TOC).
2. Close the annular preventer.
3. Install the *inside BOP* on the drill-pipe. Open the fully open safety valve (FOSV) or stabbing valve on the drill-pipe.
4. Apply 500 psi pressure on the annular side through the kill-line.
5. Reduce the closing pressure of the annular preventer to the minimum for avoiding leakage. This is done to consider the effect of the applied pressure.
6. Connect the line from choke manifold to trip tank. The trip tank should be half full. Check if the trip tank drains into the strip tank.
7. Set up the flow line so that any fluid that leaks through the annular preventer goes into the trip tank.
8. Make up/install the next stand.
9. Strip the stand in hole and maintain pressure on the annulus constant by bleeding off through the choke. Run the pipe slowly to avoid pressure surges. Measure the volume of mud bleed off and ensure that the total volume is equal to the closed-end volume of the stand stripped into hole.
10. When a complete stand is run in the hole, close in at the choke manifold by closing the valve behind the choke.
11. Drain the closed-end volume of one stand out of the trip tank into the stripping tank.
12. Make up the next stand.
13. Continue stripping until the crew is well familiar with the operation.

- *Diverter Drill:* The Diverter Drill is practiced for training the crew to deal with shallow gas during top-hole drilling. The criteria for the assessment of the Drill performance include the correct execution of operations and short response time by the crew.

  By manipulating the pit level sensors and/or the flow sensor, a (counterfeit) shallow gas flow is simulated (artificially) when the drill-string is inside the conductor pipe and prior to drilling out its shoe track. The crew should follow the sequence of operations as given next:

  a) During Drilling with Drilling Bit On-Bottom:

  1. Detect the “kick” by noticing the pit gain and alert the crew.
  2. Stop drilling and immediately start pumping drilling mud at highest possible flow rate. If heavy kill mud is available in the mud pits, pump that.
  3. Close the diverter, which automatically initiates the following sequence:

        - The diverter valve is opened.
        - The flow line (mud return) valve is closed.
        - The diverter element (annular preventer) is closed.
    4. Prepare the switchover to kill mud pumping with maximum flow rate.

  b) *During Tripping:*
    1. Detect the "kick" and alert the crew.
    2. Close the diverter.
       Note: In case of shallow gas flow, there are two actions to be essentially done: a) restarting the pump and circulation of the mud and b) closing the diverter. According to well-respected standards, the latter option should be chosen first to guarantee greater safety.
    3. Hang off the drill-pipe in slips and then stab the top drive or Kelly on to the drill-pipe again.
    4. Start pumping kill mud at the greatest possible flow rate.

- *Hang-Off Drill (Subsea-BOPs Only):* Hang-off is defined as an action whereby the weight of that portion of the drill-string below a ram-type BOP is supported by a tool joint resting on the closed pipe rams. Therefore, in this drill, following prescribed procedures, the crew should place the drill-string in position for hang-off. One hang-off should be made before drilling out the surface casing to ensure that all necessary equipment is on hand and in working conditions. Actual hang-off is not normally performed on subsequent drills. This drill can be conveniently performed in conjunction with the pit drill.

REFERENCES

Holand, 2017. *Loss of Well Control Occurrence and Size Estimators, Phase I and II.* Report No. ES201471/2. Office/Division Program, TAP, Project Number. 765. Category, Deepwater.

Noaa, 2017. *Explosion Triggered Economic, Environmental Devastation, and a Legal Battle.* https://www.noaa.gov/explainers/deepwater-horizon-oil-spill-settlements-where-money-went

Yasseri, S., 2017. Drilling Risk Identification, Filtering, Ranking and Management. *International Journal of Coastal and Offshore Engineering, IJCOE*, 1 (1), 17–26.

2 Barriers

Rahman Ashena

2.1 INTRODUCTION

During drilling and completion/workover operations, it is important to prevent the formation fluid flow into the wellbore which can subsequently convert to a surface-flow blowout or an underground blowout. Otherwise, the safety of operations would be severely jeopardized. Barriers form the bedrock of well integrity management, providing layers of defense against a spectrum of operational challenges and external pressures.

Therefore, prior to well spud and commencement of any well operation, well barriers should be considered and set or defined in the well plan for each stage of well operations. In understanding the concept of well barriers, the primary and secondary barriers should be identified, and a barrier envelope must be defined and described. Plans and criteria for testing barrier elements and test documentation must be described.

A well barrier envelope is defined as the combination of one or several well barrier elements defined as any physical component(s) or operational practice that prevents the formation fluids from flowing unintentionally from a formation into another formation or to escape at surface. A well barrier element is defined as a physical element which by itself does not prevent flow but, in combination with other elements, forms a well barrier.

This chapter delves into the critical importance of well barriers in different types of well operations, elucidating their role in upholding operational reliability, environmental stewardship, and regulatory compliance.

2.2 PRIMARY AND SECONDARY BARRIERS FOR SAFE OPERATIONS

Except for drilling shallow or top holes, most regulations or regulatory agencies focus on the two-barrier principle. This principle should be strictly followed in all regulated regions or regions where regulations/standards must be regarded including the United States, the United Kingdom, the Middle East, Europe, etc. The two independent barriers are commonly referred as the primary and the secondary barrier envelopes. A primary barrier is described as the first object preventing flow from a source/formation. The primary barrier is normally the barrier closest to the potential source of the flow, i.e., the hydrocarbon or geothermal reservoir. A secondary well barrier is described as the second object preventing flow from a source.

 DOI: 10.1201/9781003473770-2

In shallow or top-hole drilling of wells, before the BOP can be installed on the wellhead, there is a one-barrier or single-barrier situation which is the hydrostatic drilling fluid pressure. There is no mechanical device to close in the well. This is due to very limited risk of encountering hydrocarbon-bearing or abnormal formations, and there is the possibility of fracturing the formation if flow is prevented. Thus, as a substitute solution, if the well flows in this situation, the fluids would be diverted by a diverter to a distance far from the rig; alternatively, the rig should be relocated quickly.

In deep drilling operations, the hydrostatic pressure from the drilling fluid is regarded as the primary barrier, and the combination of BOP, wellhead, drill-pipe, casing, and cement is considered as the secondary barrier envelope. The well situation is controlled by a hydrostatic pressure (primary barrier). If the hydrostatic pressure becomes lower than the formation pore pressure for some reason, a kick flow (or well control event) may occur, and, consequently, the control of the well must be regained by activating the secondary barrier (the BOP stack) and then circulating the well with a higher density drilling fluid to establish a new competent primary barrier.

In production or injection wells, both the primary and secondary barriers are mechanical barriers. In a flowing well, the barriers closest to the reservoir are usually regarded as the primary barriers. In such wells, the primary barrier envelope would typically be the packer that seals off the annulus, the tubing below the surface-controlled subsurface safety valve (SCSSV), and the SCSSV. The secondary barrier envelope would then be the tubing above the SCSSV, consisting of the Christmas (X-mas) tree main flow side, the casing/wellhead, and the annulus side of the X-mas tree.

One or more of the Secondary Barrier Elements may be unavailable, which may be because the barrier itself failed (e.g., leakage in a wellhead connector) or failed to activate (e.g., failed to close the BOP), or specific operations made the barriers unavailable (e.g., BOP was nippled down to energize the casing seals). If the secondary barriers are not available and a kick flow occurs, the kick may develop into a loss of well control (LOWC) event or blowout.

Barriers can be categorized into two broad phases of a) drilling, coring, and tripping and b) completion and workover operations as described here:

a) Drilling, Coring, and Tripping: Some important well barrier elements during drilling and tripping are listed as follows (see Figure 2.1):

- *Drilling fluid barrier:* The drilling fluid/mud must be of sufficient density and pressure to counterbalance the formation pressure. Since it can act as the first object/barrier to prevent flow from the source (formation), it is called the Primary Barrier (Figure 2.2).
- *Casing/liner and cement:* Properly cemented casing strings act to seal the formations and prevent possible fluid flow from them. Similarly, possible liners should have integrity and be well-cemented particularly at their liner laps to prevent any leakage of fluid flow from the formations (see Figure 2.1).

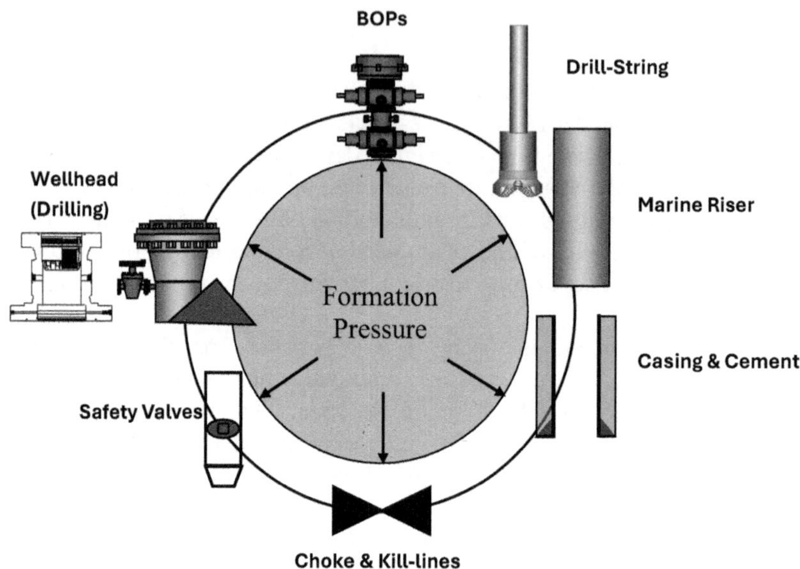

FIGURE 2.1 Well barrier elements compositing the well barrier envelope of a well.

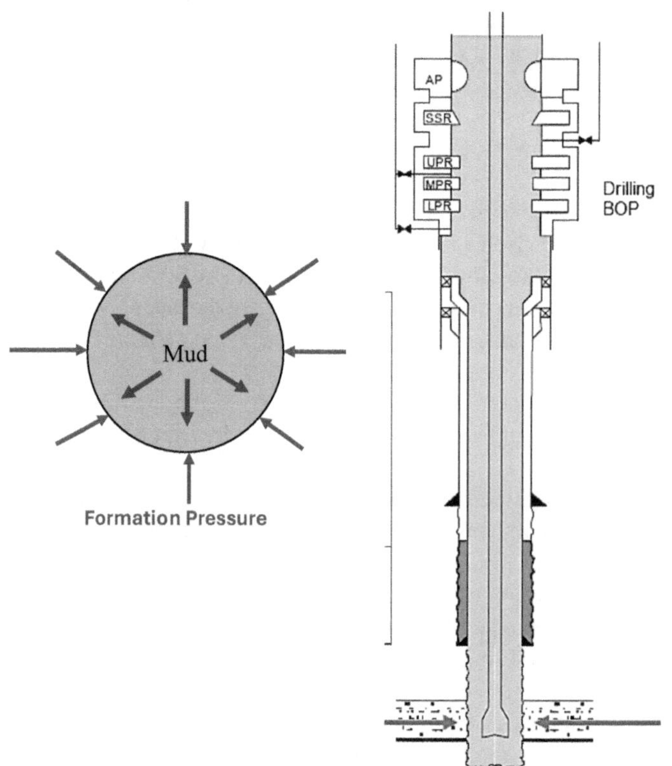

FIGURE 2.2 The drilling mud is the first object preventing kick flow from the formation (primary barrier) as shown in the cross-sectional and plan views.

- *Drill-string:* The drill-string must have integrity (no washout, etc.) to prevent flow from the formations to the surface.
- *Marine riser:* In subsea operations, the marine riser must have integrity and be well connected to the BOPs to prevent the loss of mud column pressure.
- *Stab-in safety valves:* To prevent the possible flow of formation fluids (in case of a kick flow) through the drill-string, a drill-pipe safety valve (DPSV) also called Full-Opening Safety Valve (FOSV), or stab-in valve, is stabbed in on the drill-string to safely seal the drill-string. Usually, several safety valves must be available at the rig floor, and the required number depends on the size and thread type of the drill-pipes and collars in the hole.
- *BOPs:* Blowout Preventers (BOPs) are considered important "Secondary Barrier Elements" at the surface to prevent possible flow through the annulus to the surface (see Figure 2.3). In case of flow during drilling or tripping, BOPs should be closed. Typical drilling BOPs consist of annular preventer(s), pipe rams, and shear/blind rams. In 2013, the blind-shear ram preventer was mandatory for deep-hole offshore operations based on North Sea standards including Norwegian standard, NORSOK D-010. This is regarded as an emergency device and should be activated in very urgent cases since closing this preventer will significantly complicate the well operation required to regain the hydrostatic control of the well.

  In case of a well control event during wireline well logging operations, wireline BOPs (internal BOPs/IBOPs, stuffing box, or lubricators) can be closed around the wireline to shield the well. Alternatively, the wireline should be cut, then the blind rams are closed.
- *Wellhead (i.e., spools and casing hangers):* During drilling operation and prior to well completion, wellhead consists of only the casing head housing, spools, and casing hangers/slips. Later, at the end of well completion and beginning of production, when the Christmas tree is installed, the wellhead may be also called the *lower wellhead* assembly. The integrity of the wellhead is significantly important as it is a barrier against possible flow through the annulus between casings. Following the landing of the casing string, the casing annulus is sealed by setting casing hanger/slips in the spool, which would prevent the possible flow of formation fluids from the annular space between the two casings. Following that, the next spool is jointed, and the spool body together with the casing hanger can be pressure tested to ensure about their integrity.

b) *During Completion/Production and Workover Operations*
   - *Fluid barrier:* During completion and workover operations, a reservoir/productive zone is exposed which can make the well prone to kick or later to loss of well control/blowout. During workovers, the reservoir is exposed nearly all the time (i.e., there is a potential possibility for kick flow). For drilling, a productive zone is exposed only for a short period of the total drilling duration.

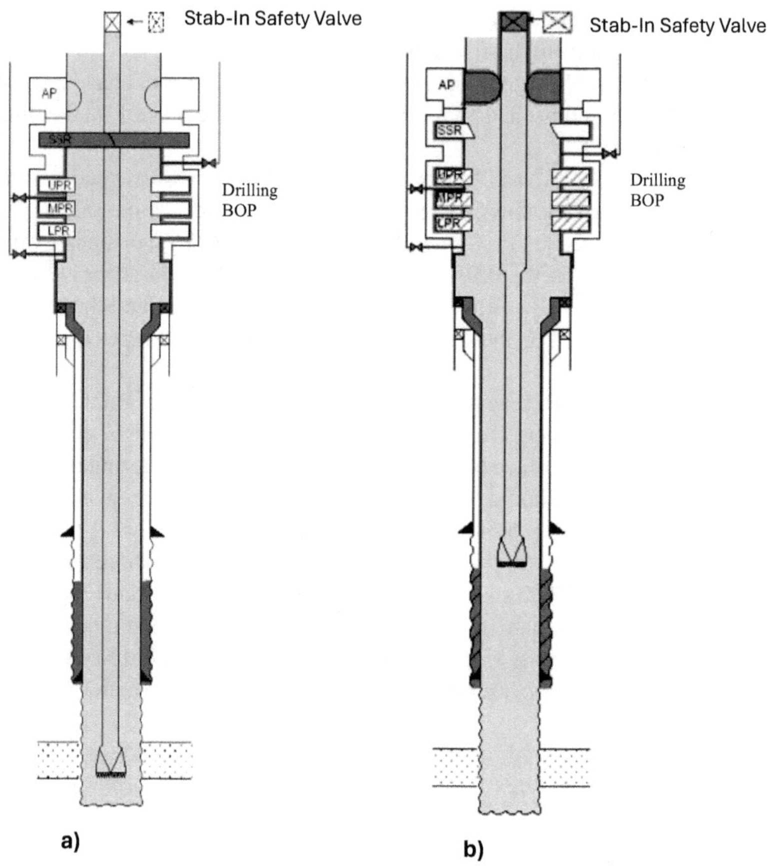

FIGURE 2.3 Well barrier elements for drilling, coring, and tripping: a) With shear rams as the secondary barrier element and b) with annular preventer as the secondary barrier element (Know Energy Solutions, 2019).

During completion or workover phases, the completion/workover fluid must be of sufficient weight to counterbalance the formation pressure, as the Primary Barrier. Therefore, prior to commencing workover operations, the well is killed using a heavy-enough fluid. Next, in the completion/workover phase, a solids-free fluid should be displaced in the hole; thus, there will exist no mud filter cake which acts as a seal against the formation during drilling (in case of existence of an open-hole section) and can be formed around. This means that during workovers, there are normally continuous losses to the formation, so efficient fluid loss controllers are of great importance. In a workover operation, the well can be closed in with higher pressures than during drilling because formation breakdowns on shallow casing shoes are less likely to occur. Thus, bullheading is a recommended kill method for workover kicks with high success probability, compared to drilling kicks. Bullheading is squeezing/forcing the wellbore fluids or influx, i.e., oil or gas, back into the formation. In case bullheading may not be possible (e.g., because of not high-enough surface facilities pressure ratings), there are two methods: a) install a coiled-tubing unit (if available) and displace the wellbore fluid with the heavy fluid and b) lubricate and bleed (add some heavy fluid into the well and bleed off the light fluid out of the well to the production line). As their disadvantage, the processes are timely.

- *Plug in tubing and tubing hanger:* After the well is killed, it is usually required to nipple down the wellhead and possibly install the drilling BOPs if further drilling is required. However, it is not allowed to nipple down the BOPs unless other barriers are placed for making the well safe. To shield the well, a plug (together with a prong) must be placed in the completion string (in the No-Go-Nipple, located near the bottom-end of the string). This job is done by running wireline through the Christmas Cap with a lubricator installed on its top. Next, another plug (called *two-way-check-valve*) is placed in the tubing hanger. With these two additional safety elements, it is possible to safely nipple down the wellhead. At this time, if further drilling is required as part of workover operations, it is possible to nipple up the drilling BOPs.
- *Cement and bridge plugs:* In critical cases, especially gas wells or wells with extremely high pressures or H_2S content, after killing the well, one or two cement plugs (of say 150 ft/45.7 m thickness) are installed at a critical point in the wellbore, usually at a deep point (e.g., near the liner lap). If this is not done and just plugs are placed in the tubing and tubing hanger, gas might accumulate below the plug in the tubing, which would be dangerous when it is intended to retrieve the plug.

  Another application of cement plugs is when it is aimed to repair the bottom pipe rams. In this case, it is necessary to set one or two cement plugs first. Then, the RTTS packer is set at the depth of say 100 ft. Then, the bottom pipe rams can be safely disconnected and repaired.
- *Casing/liner and cement:* During completion and workover, similar to the drilling phase, properly cemented casing strings (particularly the production casing/liner) can seal the covered formations and prevent possible fluid flow from them. Similar to casings, liners should have integrity and be well-cemented including at their liner laps to prevent any possible fluid flows.
- *BOPs:* In workover operations, BOPs (if installed) are important as "Secondary Barrier Elements" at the surface to prevent possible fluid flow. In case the drilling BOPs are installed during workover operations, they can be closed to prevent flow through the annulus. In case of wireline or coiled-tubing jobs, the wireline or coiled-tubing BOPs are used, respectively; alternatively if they do not operate, then wireline can be cut and then the blind rams are closed.

- *Completion string:* A completion string must have integrity, i.e., with no possible holes. Thus, no leakages or pressure drops should be observed during pressure tests of its components so that it can prevent the possible flow of formation fluids to the surface.
- *Wellhead (including the tubing hanger):* In well completion and workover operations, a lower wellhead assembly (spools and casing hangers) is already installed (with the tubing hanger being located on its top); then, at the end of operations, the upper wellhead assembly (usually called Christmas tree) is installed.

  The tubing hanger is a component of the completion string which should properly seal around the tubing string; therefore, no possible fluid can flow through the annulus to the wellhead. The tubing hanger also allows the setting of a two-way-check-valve (which is simply a plug) in its bore; therefore, it acts a barrier against any possible flow through the completion string before we nipple down the BOPs and set up the upper wellhead on the spools.

  Following the installation of the upper wellhead, also called Christmas tree (consisting of the bottom and top master valves, dipping valves, and Christmas Cap), its components must be pressure tested. In case it is required to repair the top master valve, we can close the bottom master valve; therefore, the fluid column pressure and the top master valve act as the two required barriers.

2.3 BARRIER ACCEPTANCE CRITERIA AND TESTING

Well barrier acceptance criteria are technical and operational requirements which must be fulfilled to qualify the well barrier for its intended use. These include functions, required number, and testing of well barrier elements.

The function of each well barrier must be clearly defined. In other words, the purpose and the role of the barrier should be clarified.

As mentioned before, at least two barriers are required during all well operations, i.e., primary and secondary barriers. This is also applicable for abandoned or suspended wells. If this is not the case, operations must stop, and barriers must be reinstated. In shallow or top holes, when drilling out the hole of the surface casing where there are no BOPs, the single barrier is allowed; instead, a diverter is used to divert possible flow.

It is required that barriers be pressure tested and documented periodically. During testing the barriers, enough pressure should be applied in the flow direction. In case this is impractical, there are two possibilities: Either the pressure on the downstream side of the well barrier can be reduced to the lowest practical pressure (i.e., inflow test or negative pressure test), or the pressure can be applied opposite the flow direction (positive pressure test), provided that the well barrier element is constructed to seal in both flow directions (e.g., a plug in the tubing hanger).

All well integrity tests must be documented and accepted by an authorized person. The authorized person can be the drilling supervisor, well intervention supervisor, tool pusher, driller, or the equipment and service provider's representative. The chart and the test documentation should contain type of test, test pressure, test fluid, system or components tested, estimated volume of system pressurized, and volume pumped.

2.4 INFLOW TEST

In case it is impossible to test a barrier by applying enough pressure in the flow direction, inflow test (called by API) is an appropriate test method by lowering the pressure on the downstream side of the well barrier. The "inflow test" is also called "negative pressure test (NPT)". Inflow tests are considered more reliable than (positive) pressure tests (PPTs), particularly for liner laps. There are field cases where PPTs were not able to not detect the leakage but inflow tests were able to. Common barriers that are usually inflow tested are:

- Liner laps – after running liners and cementing.
- Plugs for plug and abandonment (P&A) – both mechanical plugs and cement plugs.
- Cement or other plugs set prior to pulling Christmas tree (prior to workover operations).

An inflow test is successful/negative when there is no change in downhole pressure, and thus fluid flow across the cement barrier can be concluded during the opening of the downhole valve and subsequent shut-in. It is recommended to carry out this test a few days after cementing so that the cement is used in its in-situ conditions; for a liner, such a period would be practically needed due to the required drill-out jobs to clean cement above the liner lap or above the previous casing.

Preparations for the inflow test are as follows:

- A choke manifold is necessary at the rig to connect to the drill-string and monitor from the drill-pipe in a controlled way. The manifold is connected to the main choke manifold for further safety.
- An inflow test bottomhole assembly is required.

A risk assessment is required to be performed before conducting the inflow test.

The procedure of the inflow test, which is similar for onshore and offshore operations, is as follows:

1) Run the bottomhole assembly (i.e., the inflow test assembly) in the hole with drill-pipes to a specified depth about the barrier (e.g., liner lap). In the inflow test, assembly includes a packer and the valve is closed.

Note: There is just a specified height of cushion water above the closed valve.

2) Set the packer to isolate the annulus from the drill-pipe.
3) Open the valve in the inflow test BHA assembly to start the three-hour open period. Monitor pressures and flow through the drill-pipe. It is expected that flow and pressures drop to zero after a while.

4) Close the valve and monitor pressures and possible flow.
5) Repeat steps (3) and (4) to check the well flow conditions and barrier integrity twice.

Note: The BOPs are open; therefore, it is possible to monitor the level of mud level in the annulus. Any drop of the mud level signifies a lack of integrity in the packer or drill-string.

6) Unset the packer and make reverse circulation.
7) Pull the string out of the hole.

Figure 2.4 illustrates a schematic of a successful inflow test in a land rig across the liner lap with the shoe track not drilled out. The test is successful because both the drill-pipe pressure (DPP) and casing pressure (CP) show zero. It is expected that due to thermal expansion, the DPP shows some value in the beginning.

The challenging part of conducting an inflow test is in establishing how to verify no flow across a cement barrier at depth, say 8,000 ft. In such cases, there is a long fluid column between the barrier and the pressure sensor at the surface. Thus, the matter is that the test results are adversely affected by thermal expansion and contraction of the long fluid column in the wellbore or by rig movements in subsea rigs. As for the thermal expansion, when the light and relatively cold fluid is pumped and gets in touch with the underground formations/wellbore, the difference in temperature can result in an unexpected flow. It is thus recommended to allow enough observation time to monitor. If the test period is shot, fluid flow can be observed, and it may appear that the barrier has failed (no integrity). To counteract these effects, it

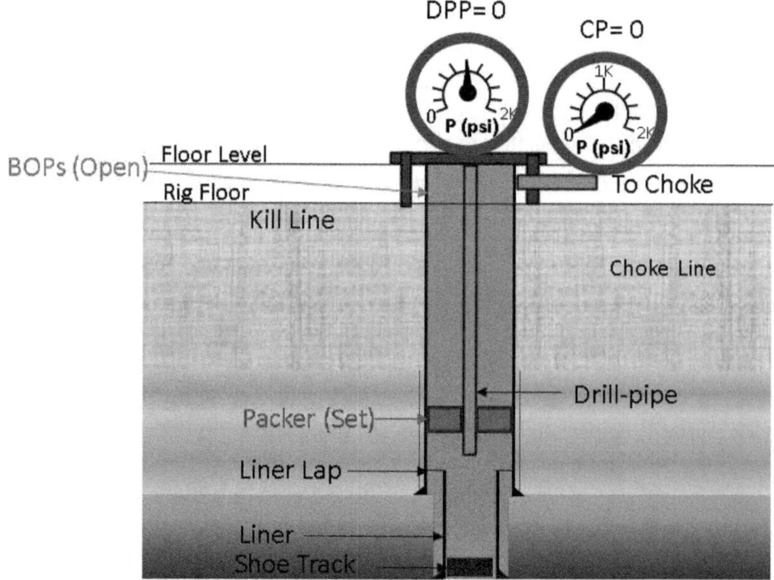

FIGURE 2.4 Schematic of a successful inflow test in a land rig across the liner lap with the shoe track not drilled out.

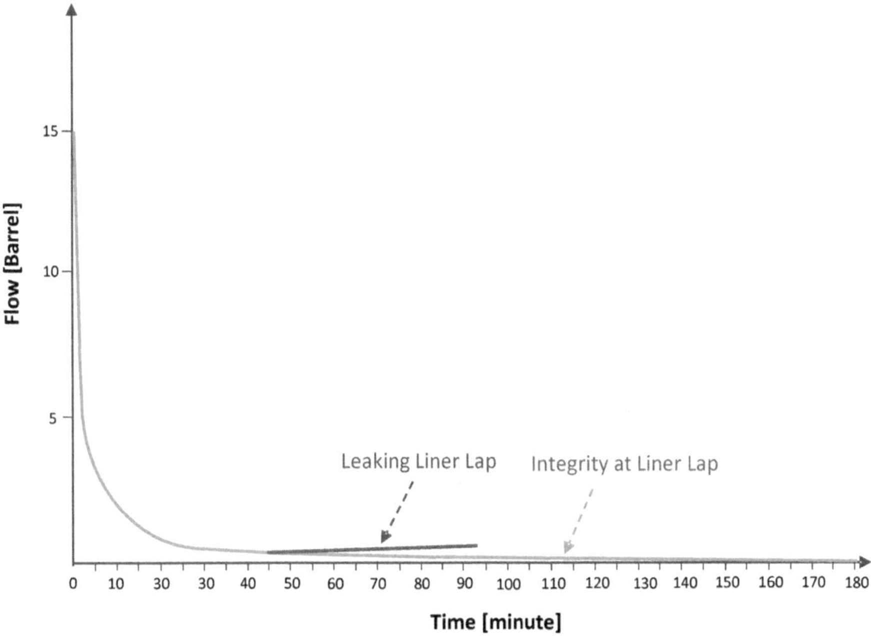

FIGURE 2.5 An example inflow test result of monitoring for three hours of an open period. There are two cases of no integrity and with integrity at the liner lap.

is recommended to allocate enough test observation time to ensure a reliable trend for pressure versus time is obtained, and the test should take at least three hours (180 minutes) for two times/trials, or even longer as per client's requirements. In a successful test showing integrity at the barrier, the flow out of the well should show a decreasing trend toward zero (see Figure 2.5).

2.5 EXERCISES

Exercise 1:

An inflow test is in progress. The drill-string has been displaced with a light fluid, and the pressure has been bled off the drill-pipe. The volume that flowed back from the drill-pipe during bleed-off was three times more than calculated. Pressure returns on the standpipe when the drill-pipe is shut in.

What is your evaluation of the test?

a. Not enough information to make an informed decision.
b. Successful test. Continue with planned operations.
c. Failed test. Shut the well in and notify the management.

Answer: c. There is a lack of integrity.

Exercise 2:

A liner string has been run and cemented and positive pressure tested in 13 pound per gallon (ppg) mud. The liner packer/hanger will later be barriered by a smaller casing. The drilling program calls for a reduction in mud weight before drilling out the shoe to 12 ppg. The packer depth is 8,000 ft. What is the pressure that the packer/hanger should be tested at?

Answer:

The inflow test should be done with the differential pressure of:

$$\Delta P_{\text{diff}} = (13-12) \times 0.052 \times 8000 = 416 \text{ psi}$$

Exercise 3:

There is a 7,000 -ft well with 1,000 -ft open-hole section plugged back into $9-5/8"$ casing with a cement plug. It is intended to perform an inflow test of plug to 700 psi below the formation pressure at 7,000 ft.

Evaluate the height of cushion water required in the drill-pipe run for inflow testing.

Recommend a depth for the string of the inflow test assembly to be run.

- Cement plug extends 300 ft into casing; hence it is 1,200 ft of cement plug in total.
- Casing shoe is at 7,000 ft.
- Mud weight (MW) in well is 11 ppg.
- Cushion water density is 9 ppg.
- Formation pressure is 9.5 ppg EMW at 7,000 ft.

Answer:

The depth of top of cement plug is 7000 – 300 –6700 ft.

The formation pressure at 7,000 ft is:

$$P_f = 0.052\ EMW \times TVD = 0.052 \times 9.5 \times 7000 = 3{,}456\ psi$$

Since 500 psi differential was decided to be across the barrier, for barrier verification, the pressure should be reduced to: $3{,}456 - 700 = 2{,}756$ psi.

To obtain the bottomhole pressure (BHP) of 3,456 psi at 6,700 ft (top of the cement plug), the height of the cushion water is found to be:

$$2{,}756 = 0.052 \times h_{cw} \times 9 \rightarrow h_{cw} = \frac{2{,}756}{0.052 \times 9} = 5{,}889\ ft$$

The depth for the inflow test assembly to be run should exceed this depth. For example, it can be 6,300 ft.

REFERENCES

Holand, 2017. *Loss of Well Control Occurrence and Size Estimators, Phase I and II*. Report No. ES201471/2. Office/Division Program, TAP, Project No. 765. Category, Deepwater. https://www.bsee.gov/research-record/loss-well-control-occurrence-and-size-estimators

Know Energy Solutions, 2019. *Well Barriers Concepts – A Definitive Guide*. www.knowenergysolutions.com/iwcf-well-control/wellbarriers-concepts (Last Accessed in February 4, 2020).

Noaa, 2017. *Explosion Triggered Economic, Environmental Devastation, and a Legal Battle*. www.noaa.gov/explainers/deepwater-horizon-oil-spill-settlements-where-money-went

Nytimes, 2020. *BP Shortcuts Led to Gulf Oil Spill*. www.nytimes.com/2011/09/15/science/earth/15spill.html (Last Accessed in May 16, 2024).

Yasseri, S., 2017. Drilling Risk Identification, Filtering, Ranking and Management. *International Journal of Coastal and Offshore Engineering, IJCOE*, 1 (1), 17–26.

3 Drilling Fluids and Pit Management

Rahman Ashena

3.1 INTRODUCTION

This chapter discusses fundamentals of drilling fluids useful for well control applications. Therefore, functions as well as types of drilling fluids are presented with a practical field example. Techniques used for evaluating mud weight/density and mud contaminants and their effects on mud properties are then discussed. Finally, pit management methods are covered including instrumentation for the detection of well control events.

3.2 FUNCTIONS OF DRILLING FLUIDS

Drilling fluids have the following functions:

- *Cleaning the well from cuttings:* Before rotary drilling became widespread, cable-tool drilling was in use (years 1880 to 1953) by which a hole was drilled by the repeated blows generated by lifting and dropping a heavy chisel bit (called *tool*) on rocks or underground formations.

  One of the disadvantages of cable-tool drilling was that drilling fluids could not be used for cleaning the borehole. Because of this advantage and some other ones, the traditional method was replaced by *rotary drilling*. In rotary drilling, the most primary function of drilling fluids is to remove cuttings from the wellbore (clean the wellbore) so that drilling can progress without any obstruction. To accomplish this, it is important that the hydraulic energy is transmitted to the bit.

  Hole cleaning is challenging when mud pumps are not running/off (e.g., when the pipes are out of the hole or during well logging) as cuttings or solid additives in the mud may settle down. Some consequences of solids settling are pipe getting stuck or even losing the well in severe cases. Therefore, drilling muds should essentially have proper formulation with some gelation characteristics so that solids can suspend in the mud (and not settle down).
- *Preventing well control issues:* The pores in underground formations are contained with pressured fluids which are under pressure (called formation pore pressure). On the other hand, the column of drilling fluid in the well applies hydrostatic pressure on the wellbore.

DOI: 10.1201/9781003473770-3

Under normal drilling conditions, the hydrostatic mud pressure must exceed the formation pore pressure to prevent the entrance of a formation influx into the wellbore (kick). Kick influxes, if not controlled, may lead to catastrophic blowouts causing loss of the well control/blowouts. Therefore, drilling muds must have sufficiently high weight (but not extremely high which can cause formation fracture) to control formation pore pressures and prevent kicks.

- *Maintaining wellbore stability:* When a hole is made into underground formations, their in-situ stress conditions are disturbed making the hole prone to mechanical instability (collapse or fracture). The other type of instability is chemical which occurs during drilling some types of formations, e.g., shale formations which may swell in case of being in contact with water-based drilling muds causing tight holes. Tight holes manifest themselves usually during tripping operations and may cause drill string/pipe to get stuck. To establish mechanical stability, the drilling mud pressure must be sufficiently high to prevent formation collapse, but it should not be too high to cause formation fracture. The composition of the drilling mud should be chosen carefully to prevent chemical instability (e.g., use saltwater muds or oil-based muds/OBMs for drilling formations containing shales).
- *Minimizing reservoir formation damage:* Drilling operations expose the reservoir formation/pay-zone to drilling muds and any solids and chemicals contained, which may bring about damage to formation rocks. Some invasion of mud filtrate and fine solids is inevitable to the formation which may be the hydrocarbon-bearing reservoir formation. This not only disturbs the native fluids contained in the rock (by the formation of emulsion or water blocks) but also may cause damage to the rock, e.g., plugging of pores by fine particles or clay swelling causing reduction of rock permeability. The extent of mud invasion depends on the overbalance pressure, formation rock permeability, and mud properties. Therefore, by carefully designing the drilling mud properties and additives for drilling the reservoir formation (e.g., by using polymers which allow the formation of filter cake around the wellbore), mud invasion and its consequent formation damage can be minimized. Such a drilling fluid is called "Drill-In Mud (DIM)".
- *Cool, lubricate, and control corrosion of the bit and the drilling assembly:* During drilling, the bit and drill-string rotate at relatively high rotary speeds. A considerable weight is also applied on the bit. The circulation of drilling mud through the drill-string, through the bit and up the wellbore annulus, contributes to reducing the friction and cooling the drill-string. In addition, the drilling mud also provides a degree of lubricity to aid the movement of the bottomhole assembly (BHA) and drill-pipe through high angles created intentionally by directional drilling and/or through tight spots resulted from swelling shale. Oil-Based Muds (OBMs) provide a very high degree of lubricity and, thus, are the preferred mud types for highly deviated wells if regulatory agencies allow using them due to their environmental concerns.

The formulation and composition of drilling muds should be appropriately selected so that the corrosion of the bit and the drill-string can be controlled. To do this, the mud should have some alkalinity (maintain pH, e.g., between 9 to 11). In drilling formations containing some H_2S, it is recommended to use H_2S scavengers.

- *Ensuring formation evaluation:* As the drilling mud is in continuous contact with the wellbore, it can provide significant information about the formations being drilled. It is important that the drilling fluid formulation is appropriately selected to preserve the cuttings as they travel up the annulus; otherwise, if the cuttings are not preserved, the quality and quantity of data obtained from cuttings analysis would be adversely limited.

  Next, the drilling mud should be appropriately selected so that much data can be collected downhole by the tools located on the drill-string (mainly Measurement While Drilling, MWD, sensors, and Logging While Drilling, LWD, sensors) and through wireline-logging operations performed when the drill-string is out of the hole. An optimized drilling fluid system that helps produce a stable, in-gauge wellbore can enhance the quality of the data transmitted by downhole measurement and logging tools as well as by wireline tools. Finally, the volume of mud return can help to indicate possible well control event (kick) or mud loss.
- *Minimizing risk to personnel, environment, and equipment:* Drilling should be made with least possible risk to the drilling crew, environment, and the equipment. Regarding drilling fluids, it is required that their testing and continuous monitoring are conducted by specially trained personnel (e.g., mud analyzers). The documentation of the fluid or additives should be carefully done. It is important to be aware of the safety hazards associated with handling of any type of fluid or additives.

  Drilling fluids formulation and its handling must be friendly to both natural and human communities where drilling takes place. To ensure that, it is mandatory to document all the additives and fluids used in the muds. Regulatory agencies should check whether the used muds comply with their established regulations. As some examples, in some European countries, use of oil-based muds is forbidden. In almost everywhere, drilling muds (particularly oil-based muds) are not allowed to be disposed freely to the environment. Use of waste management systems is particularly recommended so that disposed cuttings are not hazardous to the environment. Next, some wastewater is refined which can be used for making new mud (minimizing water used at the rig site). In addition, use of waste management systems enables using smaller sized waste/reserve pits (dug on the ground in land operations) which are more environmentally friendly.

  Some other examples of minimizing risks associated with drilling muds are explained next:

  It is required that mud pumps and their connection lines are checked constantly to observe possible signs of leakages or wear from abrasion or chemical corrosion (prevent mud leakage). Next, the elastomers used in Blowout-Preventer (BOP) equipment must be checked for compatibility with the proposed drilling fluid system. Otherwise, it may cause the BOP elastomers to leak or fail during a well control event which can lead to a very harsh situation/blowout.

3.3 TYPES OF DRILLING FLUIDS

Drilling muds fall into four main categories of Water-Based Muds (WBMs), Oil-Based Muds (OBMs), Synthetic-Based Muds (SBMs), and Pneumatic-Based Muds (PBMs). PBMs are mostly applicable in underbalanced drilling cases. Since overbalanced drilling is the main attention, explanation of the other mud types is focused as given in the following sections.

3.3.1 Water-Based Muds (WBMs)

Most wells worldwide are drilled using Water-based Muds (WBMs). WBMs are drilling fluids with their base fluid being either fresh water or saltwater (brine, saturated brine or seawater). The type of base fluid selected depends on the well conditions (including onshore or offshore) and the formation pressures. Fresh water may be used as the base fluid when the formation does not contain shale (clay minerals) or salt; otherwise, saltwater must be used. In offshore operations, seawater is preferred because it reduces the chance of gas hydration, and it is more economical.

Water (freshwater or saltwater) does not mostly have enough viscosity to efficiently transport the cuttings to the surface and control possible mud loss. This is usually obtained by adding viscosifiers (viscosifying agents/additives) such as bentonite or polymers. Bentonite is an aluminum phyllosilicate clay consisting of montmorillonite clay sheets which hydrate and disperse in fresh water. However, it is rather inefficient in saltwater due to limited swelling in such water. Polymers are high- and low-molecular-weight long-chain molecules which provide viscosity and control the fluid loss, which can be used either in low- and high weight muds.

Water may not have enough weight to control formation pore pressures. The mud should have enough weight (density) to overcome pore pressures of underground formations and enable safe drilling (probably except for top-hole formations). If the pore pressures of the formations are rather low, water (fresh or brine) by itself may be able to overcome the pressures. NaCl-saturated water has the weight of 10 pound per gallon (ppg); therefore, it can overcome, for example, the formation pore pressure of 700 psi at the true vertical depth of 1,500 ft, still, with the overbalance pressure of 80 psi. The $CaCl_2$-saturated water has the weight of 11.5 ppg and can overcome higher pressured formations. The $CaBr_2$-saturated water has the weight of 14.2 ppg, but it is not usually used in drilling due to its high cost; instead, it is used in special cases, e.g., as completion or packer fluids. If saltwater cannot overcome formation pore pressures, weighting agents/materials (barite and frobar) should be added.

WBMs can be classified to top- and deep-hole cases:

- *Top-Hole Drilling:* In petroleum drilling, top-hole sections (near surface intervals) consist of large holes like the first borehole to be drilled which is usually very large (e.g., 26-in) or subsequent large boreholes (e.g., 17 ½-in). Drilling top-hole sections is typically conducted using a very simple WBM made up of the base fluids of water or brine (base fluid in land/onshore) or seawater (base fluid in offshore) with few commercial additives added.

  For drilling the 26-in hole, usually, some bentonite (say 16–20 pound per barrel/Ib/bbl mud) may be added to water for viscosity, enabling an efficient transport of huge amount of cuttings to the surface. Some little soda ash, $Na_2CO_3\left(0.5-1 \text{ Ib/bbl}\right)$, may be added to remove calcium ions (hardness) and prevent disruption of the mud rheology.

  In the 17.5- in hole, for example, in some Middle Eastern land operations, little additives may be added to the mud since pore pressures are still low, and viscosifying additives are not added if natural clays usually may enter the mud and give it enough viscosity to raise the cuttings. However, it may be required to improve fluid loss control and hole cleaning effectiveness by adding some little (e.g., one Ib per bbl of mud) commercial bentonite (to fresh water) or attapulgite (to salt water). In case of poor hole cleaning in top holes, concentrated bentonite pills may be periodically pumped. For example, a batch of 10–20 bbl may be pumped each time you drill a stand (~ 90 ft).

  In top-hole sections, a large amount of cuttings enter the mud which must be separated at the surface before pumping it again. Therefore, a properly designed *solid control* system is required to efficiently remove solids from the mud system and help maintain drilling efficiency. The solid control equipment used in this hole section consists of shale shakers, desanders, and desilters.
- *Deep-Hole WBMs:* A deep hole is a hole drilled after running, cementing, and installing the BOPs. After cementing the surface casing, drilling the next hole section may be continued with a WBM. As greater mud weight may be required for drilling this hole section if the pore pressure gradient is greater, some weighting materials or agents (such as *barite* or *hematite/frobar*) must be essentially added. However, to prevent weighting materials and cuttings from settling (called *barite sag*), greater mud viscosity is required, which is provided by adding polymers (preferentially) or bentonite. Polymers contribute to the encapsulation of low-colloidal clay solids for more efficient removal at the surface, which in turn decreases dilution requirements. Polymer muds can be designed for high weights ranging from 10 to as high as 23 ppg. However, the main limitation of using polymers in deep high-pressure, high-temperature (HPHT) wells was their rather low tolerable temperatures (i.e., polymer molecules decompose at high temperatures). Recently, specially developed high-temperature polymers were made available with the tolerable temperature of up to 350 degrees Fahrenheit. HPHT wells are

defined as wells with BOPs and well control equipment assigned a pressure rating of greater than 15,000 psig or a temperature rating greater than 350 °F. Usually, deep wells have HPHT characteristics. Deep wells are defined as wells with depth exceeding 13,124 ft (4,000 m).

Another classification of WBMs is based on using thinners/dispersants. WBMs fall into two general categories of *non-dispersed* and *dispersed* systems:

- *Non-Dispersed Systems:* A WBM in which *thinners/clay dispersants* are not used is called a *non-dispersed system.* Clay dispersants are additives used to disperse/deflocculate the clays in the mud. The most common non-dispersed mud is a simple water system (mentioned above) which is usually used for top-hole drilling. In such hole sections, natural clays that are incorporated into non-dispersed systems are managed through dilution, flocculation, and solid control.
- *Dispersed Muds:* Though mud viscosity should be enough for good hole cleaning, it should not be extremely high; otherwise, it would cause extremely high frictional pressure losses in the circulation path. There are some consequences, e.g., the pressure loss cannot be overcome by the mud pumps (fluid flow cannot be initiated or continued). For example, the required circulation pressure losses to overcome would become 3,500 psi, whereas the pressure rating of the pump may be only limited to 3,000 psi. As another problem, swabbing may occur during tripping (due to high bottomhole pressure losses during pulling out of the hole), which can cause swabbed-in influx in the hole. In such a case, pump pressure rating may not allow providing enough pressure/energy to circulate the mud. As drilling muds are mostly non-Newtonian (except when water is solely used), viscosity is represented by the two terms of Yield Point (YP) and Plastic Viscosity (PV). The YP represents the initial fluid resistance against flow caused by electrochemical forces between ions present in the mud. The PV represents mechanical friction caused by the interaction between solids and fluids. Specifically, YP of the mud should not exceed a limit; otherwise, an extremely high pump pressure would be required to start mud circulation (by the mud pumps) which either the pumps cannot provide, or it may cause formation breakdown/fracture.

  Though such an extremely high-viscosity situation can occur in either light or heavy muds, it is particularly critical in super-heavy muds (say 14 to 23 ppg) which have a high percentage of solids' concentration (say up to 50%) including the weighting materials and cuttings. In case of a high concentration of weighting materials (which can have some clay impurities) or in case of drilling formations containing shales or anhydrite, resulting in extremely high YP and viscosity.

  In case the YP and viscosity become too high, dispersants/thinners are added to the mud (i.e., dispersed mud) to deflocculate the clay particles. This improves the rheology control by lowering the YP and viscosity, which in turn increases its tolerance for solids. Widely used dispersants include lignosulfonates (most popular), lignins, and tannins. As thinners have acidic nature, dispersed systems typically require the addition of caustic soda ($NaOH$) or lime ($Ca(OH)_2$) to maintain a pH level of 10 to 11.

3.3.2 Oil-Based Muds (OBMs)

The next type of drilling muds is Oil-Based Muds (OBMs) which are used in drilling some problematic formations such as the reservoir and deep (HPHT) holes. An OBM is considered an emulsion mud (or better to say an invert-emulsion mud). It is composed of diesel or mineral oil (a distillate of petroleum) as the continuous phase and water as the dispersed phase. The ratio of the oil percentage to the water percentage in the liquid phase of an oil-based system is called its oil/water ratio. A common range of oil/water is from 70/30 to 90/10.

Water cannot get dispersed in the oil phase naturally. Therefore, to stabilize the mixture, emulsifiers are used. Primary emulsifiers are surfactants with oil-loving and water-loving ends which make them partially soluble in water and oil. Consequently, they are used to reduce the interfacial tension between the liquid phases (water and oil) to establish a conjunction between the two phases (see Figure 3.1). Figure 3.2 shows the arrangement of a primary emulsifier around a water droplet in a stable emulsion of OBM. Due to some downhole issues or improper mud formulations, the stability of the emulsion may be disrupted. The stability issue can be detected by measuring the *electrical stability* by the mud engineer. If electrical

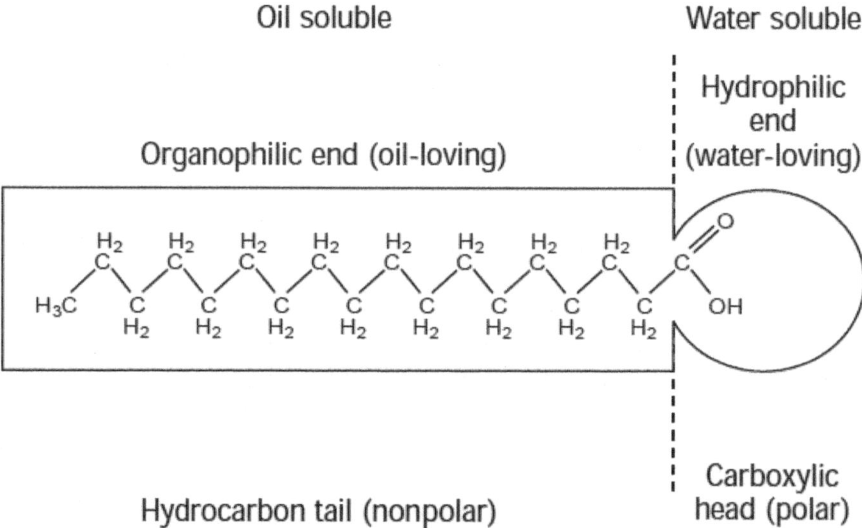

FIGURE 3.1 A primary emulsifier has two organo-philic (oil-loving) and hydro-philic (water-loving) ends which bond the water molecules to those of the oil (MI Swaco, 1998).

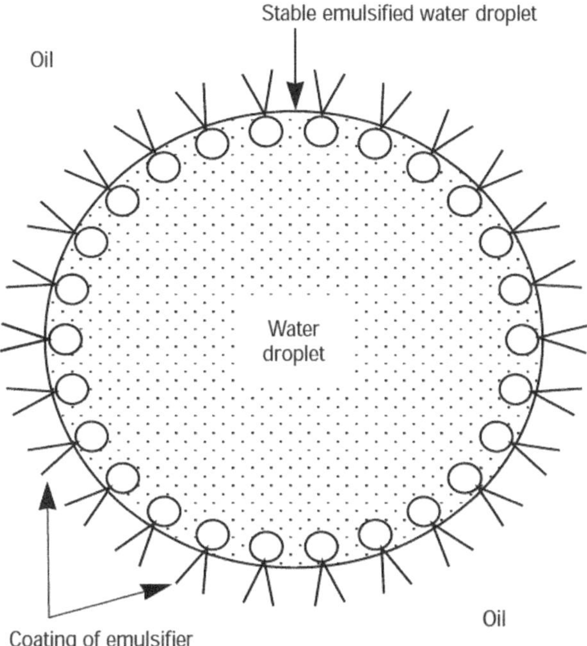

FIGURE 3.2 Arrangement of a primary emulsifier around a water droplet in a stable emulsion (MI Swaco, 1998).

stability decreases, it indicates that, e.g., some formation water has entrained into the emulsion. The stability of the emulsion increases with heat after a few bottoms-up; therefore, the electrical stability would increase from 300 to 550 mV.

Next, as the mud additives and the solid cuttings are mostly water-wet, they have the natural tendency to disperse within the water phase. As water is the dispersed phase and may have a relatively low volume in the emulsion, the solids may stick together, and thus the stability of the emulsion would be disrupted. To prevent this, secondary emulsifiers are added to the mud to make the solid particles oil-wet so that they can tend to disperse within the oil. An industrial trademark of secondary emulsifiers is *Drill-MUL*.

The OBM should have enough viscosity to lift the cuttings up the surface and enough weight/density to overcome formation pressures. To increase the mud viscosity, some viscosifying agents like Drill-Gel (trademark) are added. If greater mud weight is required, weighting agents (such as *calcium carbonate* or *barite*) can be added.

OBMs were developed for their following advantages including:

- *Drilling shaly/salty formations:* Shales would slough and sometimes swell when they are in contact with water. The consequences are tight holes (which can potentially lead to stuck pipes) and excessive mud viscosity. An efficient method of inhibiting shale swelling (preventing shale swelling) is using OBMs. This is probably the main advantage of using OBMs.

In addition, use of OBMs is a good option for drilling salt-bearing formations as salt would not be leached out, and caving does not happen.

- *Drilling HPHT wells:* As deeper wells are drilled, higher pressures and temperatures are encountered. Drilling in high-pressure, high-temperature (HPHT) conditions is a challenging job. The drilling mud should be designed to provide stable rheological properties (along with its high weight). High temperatures may have a more detrimental effect on mud rheology because they can lower the viscosity and gel strength and adversely contribute to greater barite sag/settling (Ahmad, 2017). This is a common issue in drilling HPHT wells, which causes many problems such as mud weight reductions, well-control problems, stuck pipe, and additional drilling fluid losses. Several research works prove that the detrimental effect of temperature on WBMs' rheology and barite is much greater than OBMs. Therefore, OBMS are really recommended for HPHT applications if not violating the environmental regulations (Amani et al., 2012).
- *Directional/horizontal drilling with mitigated torque and drag and stuck pipe:* Torque and drag are the results of friction caused by a moving drill-string in the wellbore. Torque occurs when rotating the pipe along the wellbore, and drag occurs when moving the pipe up and down. In highly deviated and horizontal wells, excessive torque and drag are real issues. Oil-Based Muds (OBMs) give a higher degree of lubricity, contributing to reducing the torque and drag and stuck pipe occurrences. Therefore, such a mud type are preferred for highly deviated/horizontal wells.
- *Lower corrosion* OBMs are effective against all types of corrosion. Therefore, using such muds effectively contributes to drill-string protection against corrosion.
- *Drilling the reservoir with mitigated formation damage:* Generally, mud filtration and its invasion occur much less in OBMs than that in WBMs. Even if some OBM fluid is invaded into the formation, as the continuous phase is oil/diesel, it does not cause a lot of disturbance and damage to the reservoir rock as does the water.
- *Resistance against contaminants:* OBMs are resistant to contaminants such as salt, anhydrite, and CO_2 and H_2S gases. Therefore, if the mentioned contaminants enter the mud, its rheological properties would not be disrupted.

On the other hand, OBMs have some disadvantages, which limit their usage, as follows:

- *Environmental issues:* OBMs have the potential to pollute the soil, sea-water, and groundwater aquifers. The extent of biological effect of OBMs is much greater than that of WBMs. Recently, there have been increasing environmental concerns about using OBMs. The use of OBMs is either prohibited or severely restricted in many places such as European countries and Qatar. In some areas, OBMs should not be disposed on the ground (onshore) or seawater (offshore). But rather, they should be reconditioned and reused in another well nearby. The cuttings should undergo special treatment prior to burial.

- *Cost:* Cost is one of the concerns limiting the use of OBMs. The initial cost per barrel of an OBM is much higher than a conventional WBM. However, as OBMs can be reconditioned and reused, the costs on a multi-well program may be comparable to using WBMs.
- *Well control issues:* Because of considerable solubility of gas (including methane, H_2S, and CO_2) in OBMs which delays gas expansion except near the surface, kick detection in OBMs is more difficult than in WBMs. For more information on this topic, refer to Section 4.14.3.
- *Well logging limitations:* As OBMs have very low electrical conductivity, there are limitations in well logging applications (mainly the resistivity logs). Therefore, data from underground formations with OBMs cannot be obtained as efficiently as with WBMs.

3.3.3 Synthetic-Based Muds (SBMs)

Due to the environmental concerns about the toxicity of diesel or mineral oil leading to strict environmental regulations against OBMs, synthetic fluids were developed which are considered more environmentally friendly. Esters, ethers, olefins, and other synthetic solvents have much lower toxicity with better HSE characteristics than either diesel or mineral oil. Synthetic-Based Muds (SBMs) are invert-emulsion muds in which the continuous phase is a synthetic fluid rather than diesel/oil. This and other more minor changes in formulations have made synthetic fluids in muds more environmentally acceptable. Despite high initial mud costs (even greater than OBMs), SBMs are really popular in offshore areas because of their environmental acceptance and approval by some regulatory agencies to dispose of cuttings into the water.

3.3.4 Practical Example

In this section, a real field well is presented with the formations and lithologies for all hole sections (see Figure 3.3).

Table 3.1 to Table 3.4 list the mud type, components, and properties practically used for each hole section.

In the 26-in hole, the formation lithology is conglomerate and sandstone. Because of very large size of the top-hole, a simple bentonite mud must be prepared. Therefore, first, some little soda ash ($0.5 - 1$ Ib per bbl of mud) is added to precipitate the calcium ions, which will allow a good hydration of bentonite. Next, *caustic soda,*

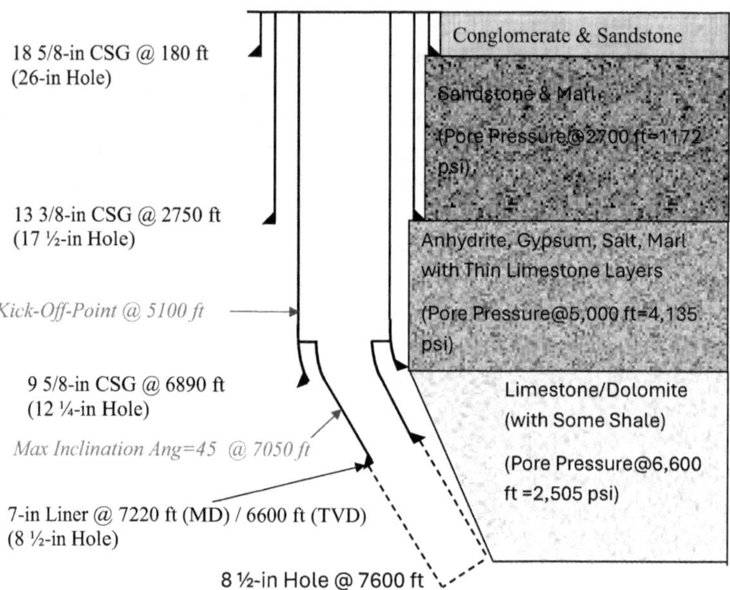

FIGURE 3.3 Well schematic of a real field case with formation encountered in drilling each hole section including their lithologies and pore pressures.

NaOH (0.5 – 1 Ib/bbl), is added to both raise the pH to 10.5–11 (to prevent corrosion in the surface system and downhole tubulars) and allow more better hydration of the bentonite to be added (because of the sodium ion effect). Next, bentonite (20–30 Ib per bbl of mud) is added to make the mud viscous (with the Marsh Funnel Viscosity of 100 sec per quart or 946 cc). Therefore, the mud can transport the voluminous cuttings of this hole successfully. The weight of the selected mud was 8.48 ppg, which is slightly greater than the freshwater weight (8.34 ppg), because of the increased effect of bentonite on the weight. The rheological parameters of the mud in this hole indicate a high yield point (YP) of $30-50$ Ib/100ft^2 (because of the effect of bentonite), but it shows a minor effect on the plastic viscosity (PV) of 5 cp because the mechanical interactions/collisions between the particles are very limited in this very large hole size. Table 3.1 presents all details of this mud.

In the 17.5-in hole, the formation lithology is sandstone and marl which are intermittently repeated in the intervals of this hole. The estimated pore pressure is 1,172 psi at 2700 ft; in other words, the pressure gradient is 0.43 psi/ft (i.e., the formation is normally pressured). Therefore, a simple water mud can overcome the formation pressure. With the large size of this top-hole section, fresh water is combined with some *natural gum* polymer added (say 5 Ib/bbl). To ensure proper hole cleaning, occasionally (e.g., after drilling a stand or so), a bentonite pill of 20 Ib/bbl concentration is pumped. In case of drilling shale or marl in the formation which can cause probable balling of the bit and stabilizers, the rate of drilling progress may become so low (e.g., $<$ 3 ft/hr). As a solution, a saturated saltwater pill may be occasionally

TABLE 3.1
The Mud Type, Additives, and Parameters for Drilling the 26-in Hole Section (Practical Example); the Formation Lithology of This Hole Is Conglomerate and Sandstone

	Additive/Parameter	Concentration/Value
Additives	1. Soda Ash	0.5–1 Ib/bbl
	2. NaOH	0.5–1 Ib/bbl
	3. Bentonite	20–30 Ib/bbl
Weight:		
Rheology	Marsh Funnel Viscosity	100 sec/qt
	Yield Point, YP	50–70 $Ib_f/100\ ft^2$
	Plastic Viscosity, PV	5 cp
	Gel Strength, GS [$Ib_f/100\ ft^2$]	–
API Filtrate Volume:		No control (NC)
pH:		10.5–11
Solid Percent:		Not calculated (NC)

pumped to wash and remove the clay around the bit and stabilizers. The rheological parameters of the mud in this hole indicate an intermediate yield point (YP) of 15 $Ib/100ft^2$ due to the effect of polymer/*natural gum* added. However, the plastic viscosity (PV) is still low with the value of 10 cp because the mechanical interactions/collisions between the particles are still limited in this rather large hole. The gel strength (GS) of the mud is low. The pH is 7.5 which near that of fresh water. The voluminous cuttings from this hole must be screened out using the solid control equipment of shale shakers, desanders, and desilters. Table 3.2 presents all details of this mud.

In the 12 ¼- in hole, in this region, the formation lithology is really complex because the hole is composed of a sequence of anhydrite, gypsum, marl, salt, and thin limestone layers which intermittently appear in between. As salt is contained as reflected by the lithology, salt water should be used instead of fresh water; otherwise, salt formation would be leached out resulting in washout and caving. Next, the pore pressure of this formation is highly abnormal (the estimated pore pressure is 4,135 psi at 5,000 ft; in other words, the pressure gradient is 0.83 psi/ft). Therefore, the mud weight should be essentially as high as 16.7 ppg to provide 240 psi overbalance pressure. This is considered a heavy or super-heavy mud. Heavy mud is defined by some drilling experts as weights greater than 10 ppg. To reach this high weight mud, 530 Ib/bbl barite must be essentially added (refer to Section 4.4.3 for the method of calculation). In this high-weight mud, a high percentage of solids of weighting materials (40% in this case) and cuttings would be present in the mud. Therefore, a high-enough viscosity and gel strength and fluid loss control must be essentially provided by adding *starch* mostly and some natural gum and Carboxyl Methyl Cellulose (CMC). In mud weights lower than 17 ppg, *green-type starch* (from *potatoes*)

TABLE 3.2
The Mud Type, Additives, and Parameters for Drilling the 17.5-in Hole Section (Practical Example); the Lithology of the Formation of This Hole Is Sandstone and Marl

	Additive/Parameter	Concentration/Value
Additives	1. Soda Ash	1–2 Ib/bbl
	2. Bentonite	*Zero* (for mud) 20 Ib/bbl (for Bentonite Pill pumped intermittently for hole cleaning)
	3. Salt:	*Zero* (fresh water is used in this case) But, depending on required mud weight, up to 70,000 ppm may be used in other cases. *Note:* Use 320,000 ppm for salt water which is pumped in case of bit balling.
	4. Natural Gum	5–10 Ib/bbl
	5. NaOH	0.5–1 Ib/bbl
Weight:		8.42 ppg
Overbalance Pressure:		$=(0.052\times8.42\times2700)-1172=10$ psi
Rheology	Marsh Funnel Viscosity	35 sec/qt
	Yield Point, YP	15 $Ib_f/100\ ft^2$
	Plastic Viscosity, PV	10 cp
	Gel Strength, GS	1 (Initial)/2 (Final) $Ib_f/100\ ft^2$
API Filtrate Volume:		No Control (NC) or < 10 cc/30 minutes
pH:		7.5
Solid Percent:		6–13

is added which is more effective in fluid loss control than increasing viscosity, useful for deep holes. If *red starch (corn)* is added in this case, it would cause excessively high viscosity which is unfavorable for deep holes. More very significant points associated with super-heavy/high-solid percent muds are to prevent excessive water/mud filtrate loss and try to limit it below 1 cc/30 minutes in the filter press apparatus. Otherwise, the excessive solids would settle down (barite sag) which can lead to stuck pipes. The pH is maintained between 9 and 9.5, adding preferentially lime, $Ca\,(OH)_2$. Table 3.3 presents all details of this mud.

In the 8.5-in hole, the formation lithology changes to the reservoir being composed of limestone and dolomite with some shales. In such a reservoir with shales, mud is required with least formation damage and is also shale inhibitive. An oil-based mud (OBM) is used which is allowed by the regulatory agency in the area, provided that special waste management system is used with the mud not being disposed to the environment, and special treatment is made on the cuttings prior to disposal. The OBM is made by oil and water emulsified using emulsifiers, and the ratio of water to oil percentages is $70/30$ which is the most commonly used OBM composition.

TABLE 3.3
The Mud Type, Additives, and Parameters for Drilling the 12 ¼-in Hole Section (Practical Example); the Formation Lithology of This Hole Is a Combination of Anhydrite, Gypsum, and Marl with Intermittent Thin Layers of Limestones

	Additive/Parameter	Concentration/Value
Additives	1. Soda Ash	1–2 Ib/bbl
	2. Salt	320,000 ppm (saturated)
	3. Starch (Green Type)	10 Ib/bbl
	4. $Ca(OH)_2$	1–2 Ib/bbl
	5. Natural Gum	2–4 Ib/bbl
	6. CMC High Vis (If Available)	1–2 Ib/bbl
	7. Barite	530 Ib/bbl
Weight:		16.7 ppg
Overbalance Pressure:		$= (0.052 \times 16.7 \times 5000) - 4135 = 207$ psi
Rheology	Marsh Funnel Viscosity	55 sec/qt
	Yield Point, YP	20 $Ib_f/100\ ft^2$
	Plastic Viscosity, PV	50 cp
	Gel Strength, GS	1 (initial)/2 (final) $Ib_f/100\ ft^2$
API Filtrate Volume		< 1 cc/30 minutes
pH		9–9.5
Solid Percent		40%

The formation is subnormal-pressured (lower than normal pressure) due to decades of oil production/depletion, and in this case the estimated pore pressure is 2,505 psi at 6600 ft, or the pressure gradient is 0.38 psi/ft. Therefore, the lower weight of the OBM compared with a possible WBM has this additional advantage that it would not cause too much overbalance. With this 8-ppg OBM (which is considered a light mud), the overbalance pressure is limited to 240 psi. To obtain this mud weight, 26 IB/bbl *calcium carbonate* ($CaCO_3$) powder is added as the weighting material (refer to Section 4.4.3 to see the calculation method for OBM). The observed weighting materials/solids percentage in the mud range between 16 to 24%. Because of this intermediate concentration of solids, a viscosifying agent as well as a fluid loss controller (FLC) is required. The emulsifiers used to stabilize the emulsion contribute to the mud viscosity. The electrical stability of the OBM emulsion measured by the mud engineer is significant in identifying the stability of the emulsion of OBM. It usually ranges from 300 to 500 mV. If the electrical stability drops suddenly, it indicates that the stability of the mud is disturbed/lost. The problem is probably due to the entrance of some formation water into the emulsion. The mud engineer is responsible for solving the problem by adding some emulsifiers to the emulsion. Table 3.4 presents all the details of this mud.

TABLE 3.4
The Mud Type, Additives, and Parameters for Drilling the 8.5-in Hole Section (Practical Example); the Formation Lithology of This Hole Is Limestone and Dolomite with Some Clay

	Additive/Parameter	Concentration/Value
Additives	1. Diesel	–
	2. Primary Emulsifier (Drill-VERT)	9 Drum (each drum = 1.375 bbl/220 litre)
	3. $Ca(OH)_2$ (to Water)	10 Ib/bbl
	4. Fluid Loss Controller (FLC)	8–10 Ib/bbl
	5. $CaCl_2$-Saturated Water	320,000 ppm (in water)
	6. Secondary Emulsifier (Drill-MUL)	4 Drum
	7. Viscosifier (Drill-Gel):	1–3 Ib/bbl
	8. Weighting Agent (Limestone Powder)	26 Ib/bbl
Weight		8 ppg
Overbalance Pressure		$=(0.052\times8\times6600)-2505=240$ psi
Rheology	Marsh Funnel Viscosity	35 sec/qt
	Yield Point, YP	5–10 $Ib_f/100\ ft^2$
	Plastic Viscosity, PV	10–15 cp
	Gel Strength, GS	4.5 (initial)/7 (final) $Ib_f/100\ ft^2$
API Filtrate Volume		< 4 cc/30 minutes
pH		9–9.5
Solid Percent		16–24%
Oil/Water		70/30
Electrical Stability		400–500

3.4 MUD WEIGHT MEASURING TECHNIQUES

It is greatly important to frequently make accurate measurements of mud weight and rheological parameters (yield point and plastic viscosity). The measurements must be made both at the suction tank (inlet to the well) and at the shale shaker (outlet from the well). Comparing the two measurements, they should be identical; otherwise, it signifies that there is a problem happening downhole (e.g., a kick has entered the wellbore) or the mud is not balanced. The frequency of measurements depends on well conditions, e.g., each one hour.

Therefore, it is significant to measure mud weight as accurately as possible. First, it is important to calibrate the mud balance regularly using a freshwater sample of known weight, and the reading should be 8.337 or say 8.34 ppg on the "Ib/gal or ppg" scale. Next, there are two techniques of measuring mud weight of conventional and pressurized mud balances. A *pressurized mud balance* provides more accurate mud weight measurement rather than a conventional mud balance.

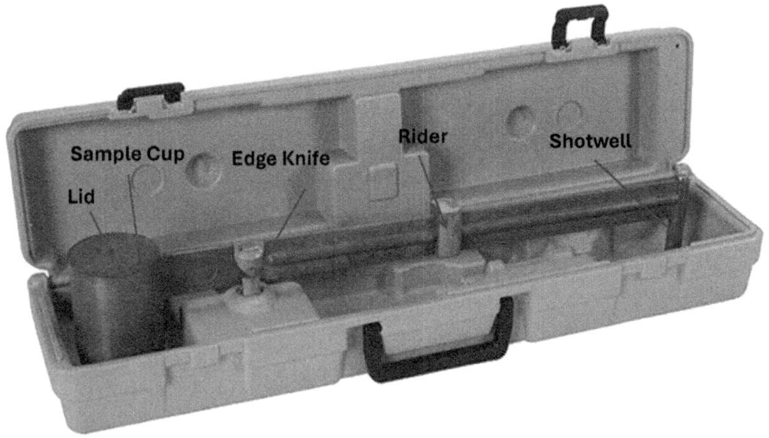

FIGURE 3.4 A conventional mud balance.

- *Conventional Mud Balance:* Figure 3.4 shows the apparatus to measure mud weight conventionally at atmospheric conditions with the following procedure.
  1. First, place the mud balance *base* on a flat surface.
  2. Measure the temperature of the mud and record it on the mud report form.
  3. Fill up the dry, clean *sample cup* with mud.
  4. Place the *lid* on the cup. Ensure that some mud is expelled from the hole of the lid to ensure the cup is full and will free any trapped gas/air.
  5. Cover the hole with a finger and wash all the mud from outside of the cup.
  6. Place the balance on the *knife edge* and move the *rider* along the outside of the *arm* until the cup and the arm are balanced.
  7. Read the mud weight.

- *Pressurized Mud Balance:* A source of error in conventional mud weight measurement is the entrained gases in the mud which cause the measurement to be lower than the real one (see Figure 3.5). The possibility of existence of entrained gases in the mud is high especially in polymer muds or when the mud circulation rate is high. Pressure contributes to eliminating the reducing effect of entrainment gases (Figure 3.6). Therefore, the pressurized mud balance (Figure 3.7) is developed to measure mud weight under sufficient pressure so that the effect of entrained gas bubbles in the mud is eliminated. The difference between the measurement by a pressurized mud balance and a conventional mud balance can be found. In the pressurized mud balance, the plunger is connected to the sample cup in order for the hand pump to pressurize the cup (Figure 3.8). Then, the measurement of mud weight is made similar to the conventional mud balance. Comparison

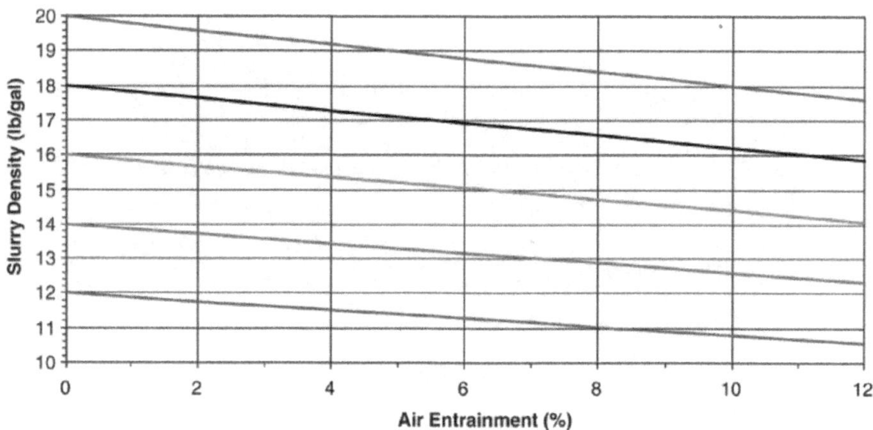

FIGURE 3.5 Effect of gas/air entrainment on mud weight at atmospheric conditions (www.fann.com/content/dam/fann/Manuals/Pressurized%20Fluid%20Balance.pdf, last accessed in March 1, 2020).

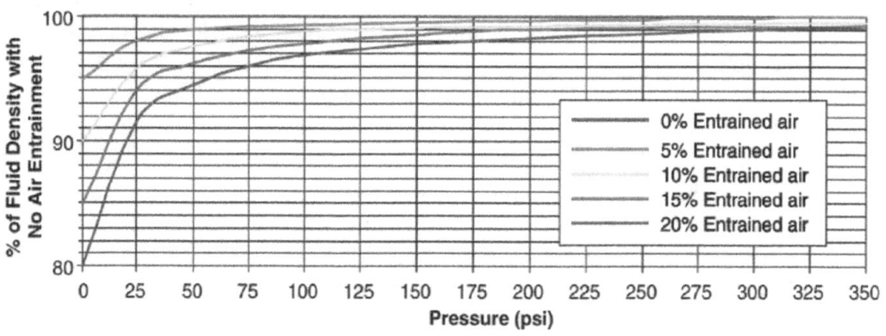

FIGURE 3.6 Mud weight reduction: Air entrainment versus pressure.

of mud weight measurements shows that the accuracy of measurements by pressurized mud balances can be up to 0.2 ppg greater than conventional mud balances.

It should be noted that the mud weight (MW) out and MW into the well should be the same. Usually, the mud engineer reports one of them if they are the same, but it is better if both values are always reported in the DDR.

3.5 POTENTIAL CONTAMINANTS AND THEIR EFFECTS

A contaminant is a foreign material entering the mud system causing undesirable changes in mud properties (such as weight, viscosity, and filtration). These changes cause disturbance to the chemical equilibrium of the mud. Contaminants come

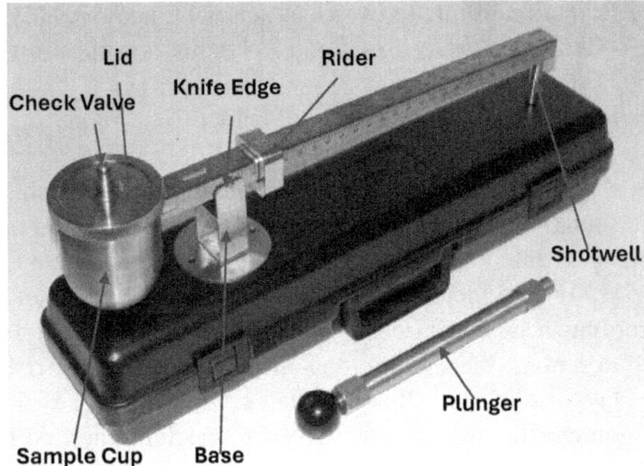

FIGURE 3.7 Components of a pressurized mud balance (www.ofite.com/publications/instructions/281-100-70-instructions/file (last accessed in March 1, 2020).

FIGURE 3.8 Connecting the plunger to the sample cup (in a pressurized mud balance) in order to later pressurize the mud using the hand pump; www.ofite.com/publications/instructions/281-100-70-instructions/file (last accessed in March 1, 2020).

either from the formation or due to overtreatment with additives (by the mud engineer's mistake). Generally, water-based mud systems are the most susceptible to contamination.

The most common contaminants are listed below:

- *Calcium, Ca^{2+}:* Calcium is a major contamination to fresh water-based bentonite muds. The contaminant is either in the makeup water used for mud making or enters the mud upon drilling anhydrite $(CaSO_4)$ / gypsum $(CaSO_4.2H_2O)$ or cement (containing CaO). It can also enter the mud by adding too much lime $(Ca(OH)_2)$ to the mud. In such muds, the calcium ions tend to replace the sodium ions on the surface of dispersed montmorillonite clays in the mud (this is called *base exchange*). This causes the sodium montmorillonite clays to convert to calcium ones. At low concentrations of the contaminant, it causes some flocculation, i.e., bonding of the clay surfaces from their edges, resulting in some increase in the apparent viscosity. At high concentrations, it leads to aggregation (precipitation and elimination of the clay particles) and consequently less resultant viscosity and greater filtration. At such conditions, thinners cannot solve the problem. Treatment of calcium must be significantly designed depending on the source of the calcium ion. If the source is from the makeup water or due to drilling anhydrite or gypsum, soda ash (Na_2CO_3) should be added. If the source is drill-out of cement or addition of lime, sodium bicarbonate ($NaHCO_3$) or SAPP (Sodium Acid Pyro-Phosphate) should be used.
- *Magnesium (Mg^{2+}):* Magnesium ion is encountered when seawater is used as makeup water (offshore wells) or in case magnesium water from the formation flows enters the mud. The adverse effects of magnesium contamination on mud rheology and filtration are similar to those of the calcium.

  As the treatment, little (0.5 Ib/bbl) caustic soda (NaOH) is added to make the ion insoluble as magnesium hydroxide $Mg(OH)_2$. In fact, most of the magnesium ions will be precipitated when pH is increased to 9.5–10.
- *Carbon Dioxide (CO_2):* Many formations drilled contain carbon dioxide (CO_2), which, when mixed with the mud, can produce *carbonate* (CO_3^{-1}) and *bicarbonate* (HCO_3^{-1}) ions. These ions cause disturbed mud rheology and increased filtration.

  Treatment of carbonate or bicarbonate is done by adding lime $Ca(OH)_2$, which removes them from the mud as the precipitate of *calcium carbonate* ($CaCO_3$).
- *Hydrogen Sulfide (H_2S):* When H_2S enters the mud, it is very important to remove it as quickly as possible as it can dangerously harm the drilling crew, as well as can cause severe corrosion damage (including hydrogen embrittlement) to the tubulars. Hydrogen embrittlement is a process which starts by the penetration of the hydrogen ion (H^+), which is a product of H_2S, into the steel body, forming pockets of H_2 gas. The gas would expand and cause the pipe get brittle to fail.

There are two treatment methods for H_2S. The first one is temporary by which caustic soda (NaOH) is added to maintain pH high enough and convert H_2S to two products of NaHS and Na_2S. However, if the pH is not maintained high, these two products can revert to H_2S. The permanent treatment method is by adding *zinc carbonate* ($Zn(CO_3)_2$) for removing the sulfide (S^{-2}) ion as ZnS. In well control events/situations, it is not preferred to displace an influx containing H_2S to the surface; instead, it is squeezed/pushed back into the formation (called *bullheading*) if pressure rating of surface and downhole equipment allow.

- *Oxygen (O_2):* If there is oxygen in the drilling mud, downhole tubulars and surface pumping equipment undergo higher corrosion rates. In severe cases, it may cause fatigue of the drill-string. Oxygen enters the mud mostly at the surface mud tanks. Oxygen entrainment in the mud can be detected using corrosion rings in drill-pipe joints and galvanic probes attached to the standpipe. Rate of oxygen entrainment can be mitigated by proper mud stirring and frequent inspection. In severe cases, an efficient treatment is required, which is done by adding *sodium sulfite* (Na_2SO_3) to precipitate oxygen as *sodium sulfate* (Na_2SO_4).

3.6 PIT MANAGEMENT

The active mud system is a system of mud tanks, lines, pumps, and valves to facilitate the circulation of mud around the well. The active system tanks have some means of allowing mud into the mud tanks as well as have a route out of the tank. The mud tanks in the active system should be monitored continuously to detect any problems occurring and then identify the source of the problem.

During drilling, a higher level of instrumentation of mud pits/tanks is required so that potential well control events (kicks) can be detected quickly so that proper actions can be taken on time. Since equipment failure is more likely in well control events due to additional stresses, a good level of backup instrumentation is required, especially in offshore operations.

Accuracy of measurements of the instruments is of great importance. Pit level/volume measuring devices on onshore and offshore installations must be very accurate. To ensure that, measuring sensors must be properly calibrated and maintained. Ultrasonic sensors are recommended for pit level and outflow rate measurement as compared to traditional flapper-type flowmeters due to their greater accuracy. For more information, refer to Sections 6.4.1 and 6.4.2. Pit volume alarms (for kick and loss) must be switched on. Mud level alarms should be installed in the drillers' doghouse and mud logging unit. They should be switchable so that the active mud pits/system or the trip tank can be monitored separately when required. It is also important to switch on the alarm of the return flow line device to record the rate of mud returns. In most cases, the increase in return flow rate is the first indication of a kick taken during drilling. Particularly in offshore rigs with managed pressure drilling systems (MPDs), it is recommended to use Coriolis sensors for more accurate measurement of the return flow rate (see Section 6.4.1).

Since most kicks occur while tripping, proper design and the use of the trip tank are of great importance. To get as accurate as possible measurements (+/– 0.5 barrel), the tank should have maximum height and minimum cross-sectional area. The level indicators of the trip tank level should be placed where they can be easily read by the driller and recorded by the mud logging unit. In practice, it is found that typical designs of trip tanks do not provide the degree of accuracy required for measuring very small volumes of fluid that have to be bled off from the well when applying stripping or volumetric well control procedures. Under these circumstances, it may be worth considering using a small stripping tank.

3.7 EXERCISES

Exercise 1:

You have successfully shut in on a kick. Which of the following would not be the responsibility of the derrickhand to monitor?

a. Mud weight in all pits.
b. Pressures on choke.
c. Pit levels and measuring devices.
d. Any leaks at pumps or lines.

Answer – b:

The derrickman is not only in charge of working at the derrick during tripping operations, but he also needs to monitor or repair mud pumps and lines, pit levels, and measuring devices. Monitoring choke pressures is not his duty.

Exercise 2:

While drilling ahead, the derrickhand calls the drillers and says that they are going to transfer 20 bbl to the active pit within 5 minutes. After 15 minutes, the driller notices that the pit level has gone up by 25 bbl. What is the safest action to take?

a. Keep drilling, the derrickhand added too much mud and did not know.
b. Call the derrickhand and see if he left a valve open or added too much.
c. Flow check and then call the derrickhand to check the added volume.
d. Shut in well and bullhead 5 bbl of fluid into the formation.

Answer – c:

In case of any suspicion for a kick, time is invaluably important to identify the kick and do the necessary actions. Thus, for safety reasons, the driller and mud engineer should never wait for the derrickhand's explanation and should first do a flow check and then ask the derrickhand.

Exercise 3:

How does wellbore temperature affect mud weight down hole?

a. Increased temperature will increase Mud weight downhole.
b. Increased temperature will decrease Mud weight downhole.
c. Mud weight will not be affected by wellbore temperature.

Answer – b:

Downhole temperature is greater than the surface. This effect causes the expansion of molecules downhole (increased volume), and thus the mud weight bottomhole becomes slightly lower than the surface.

Exercise 4:

How does increasing pressure affect Oil-Based Mud (OBM) density?

a. There is no effect on density
b. Increases density
c. Decreases density

Answer – b:

Generally, a greater pressure means lower volume and thus greater weight/density of fluids, particularly for compressible fluids. Oil-based muds (OBMs) are more compressible than water-based muds (WBMs); thus, the effect of pressure downhole is increasing their density.

REFERENCES

Ahmad, K., 2017. The Effect of High Temperatures and Pressures on the Properties of Water-Based Drilling Mud. Presented at *70th Geological Congress of Turkey Cultural Geology and Geological Heritage*, Hungary.

Amani, M., Al-Jubouri, M., & Shadravan, A., 2012. Comparative Study of Using Oil-Based Mud Versus Water-Based Mud in HPHT Fields. *Advanced in Petroleum Exploration and Development*, 4 (2), 18–27.

MI Swaco, 1998. *Drilling Fluid Manual.* https://www.scirp.org/reference/referencespapers?referenceid=2075847

4 Well Control Concepts

Rahman Ashena

4.1 INTRODUCTION

This chapter is a very comprehensive one covering a broad range of well control topics consisting of primary well control, basic geology, hydrostatic mud pressure and gradient, formation pore pressure, gradient, and equivalent mud weight, abnormally pressured formations and their sources, the U-tube principle, circulating pressure, bottomhole pressure (BHP), equivalent circulating density (ECD), casing setting depths, formation strength estimation methods, estimation of maximum allowable surface pressure (MASP), maximum allowable annular pressure (MAASP), ballooning, gas behavior characteristics, well control considerations in high-angle wells, tapered drill-string considerations, wireline operation types, and their well control setup.

4.2 PRIMARY SECONDARY AND TERTIARY WELL CONTROL

4.2.1 Primary Well Control

Drilling operations have been traditionally carried out in overbalanced conditions, that is, hydrostatic mud pressure exceeding formations pore pressure but in less than the formation fracture pressure. In this type of drilling, primary well control is based on hydrostatic mud pressure as the primary barrier to prevent formation fluids from entering/flowing into the wellbore. In other words, sufficiently high hydrostatic mud pressure acts as the primary barrier against influx entry into the wellbore (kick). Figure 4.1 shows a typical example of primary well control. It is important to maintain primary well control in all well operations from drilling, tripping, logging, etc., by maintaining the drilling mud weight and consequently its hydrostatic pressure high enough to prevent formation influx entrance. For this target to obtain, it is essentially crucial to predict and detect formation pore pressures to design mud weights accordingly.

4.2.2 Secondary Well Control

Primary well control can be unintentionally and undesirably lost in some drilling practices due to failing to maintain the overbalance. Typical situations when this can happen are during drilling into abnormally high-pressured formations, pulling the drill-string out of hole too quickly causing swabbing on trips, gas-cut muds, etc. In such situations of loss of primary control, secondary well control should essentially be implemented by closing *Blowout Preventers* (BOPs) to restrict and control the

DOI: 10.1201/9781003473770-4

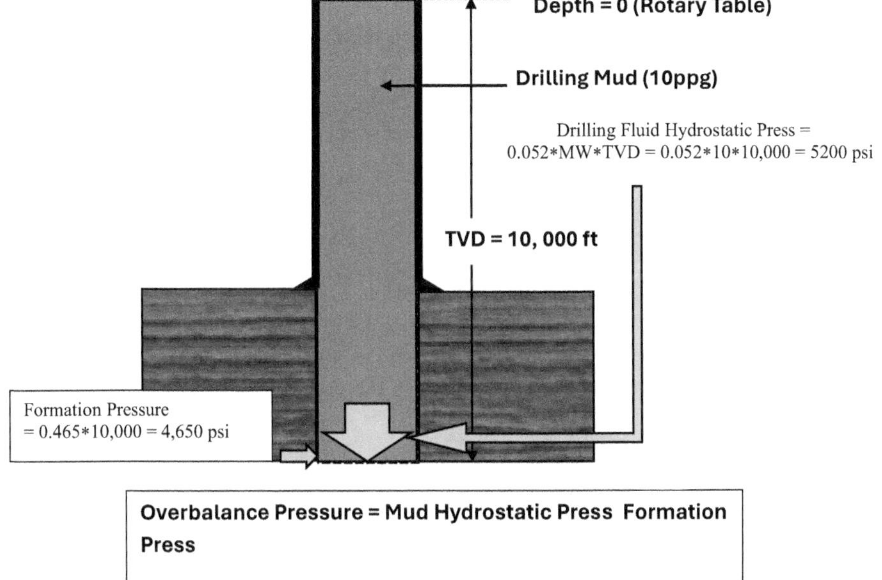

FIGURE 4.1 Primary well control using drilling mud column/hydrostatic pressure to overcome a normal-pressured formation. The overbalance pressure is 550 psi. Therefore, the formation cannot kick or no influx can enter the wellbore.

kick influx. Therefore, after reading/recording the kick size (pit gain) and stabilized well shut-in pressures, well kill procedures must be implemented to extract the kick influx out of the wellbore or alternatively push the influx back to the formation and thus secure the well.

If a gas kick influx in the wellbore is not controlled in time, it rises due to its lower density and expansion at lower pressures. If and when it reaches the surface, it is already converted to a surface-flow blowout. In such cases, if the flow rate of gas and fluid exiting the flow is excessively high, the BOPs may not be able to close and seal the wellbore. There are several examples that BOPs including the blind-shear rams did not operate such as the Macondo's in 2010 or the Walter Oil and Gas blowout on Hercules 265 rig in 2013. BOPs have not been traditionally designed and tested to close against huge rapid flow although some attempts have been made to design such robust BOPs. This point would indicate the beginning of a disastrous blowout.

4.2.3 Tertiary Well Control

In case the formation fluids entering the wellbore (kicks) cannot be controlled by primary and secondary well control means, it is the turn for the third line of defense (called *tertiary well control*) to take the well back to control. In such a situation, when a severe well control event (such as a blowout) is in progress, it is important to handle it using particular techniques, expertise, and/or equipment. Some examples of this means of well control are:

- Drilling some relief wells so that one of them can intersect with the incident well that is flowing and kill the well with heavy weight mud.
- If flow into the well can be possible, pump barite or other weighting materials to plug the wellbore in order to stop flow.
- If flow into the well can be possible, pump cement (i.e., set cement plugs) to plug the wellbore.

4.3 BASIC GEOLOGY

During drilling in pursuit of oil and gas, multiple formation rocks are encountered. A formation consists of a certain number of rock layers that have a comparable lithology or other similar properties. Simultaneously, as a well is drilled through multiple formation types in depth, the pore pressures of the formations usually increase.

4.3.1 Sedimentary Rocks and Formations

Most rocks encountered during drilling to reach the reservoir formation are sedimentary. Sedimentary rocks are formed following sedimentation a) erosion of rocks, b) transportation of rock fragments or debris to marine environments, and c) accumulation and compaction. Figure 4.2 illustrates the mentioned stages of formation of sedimentary rocks. These stages and processes are explained as follows.

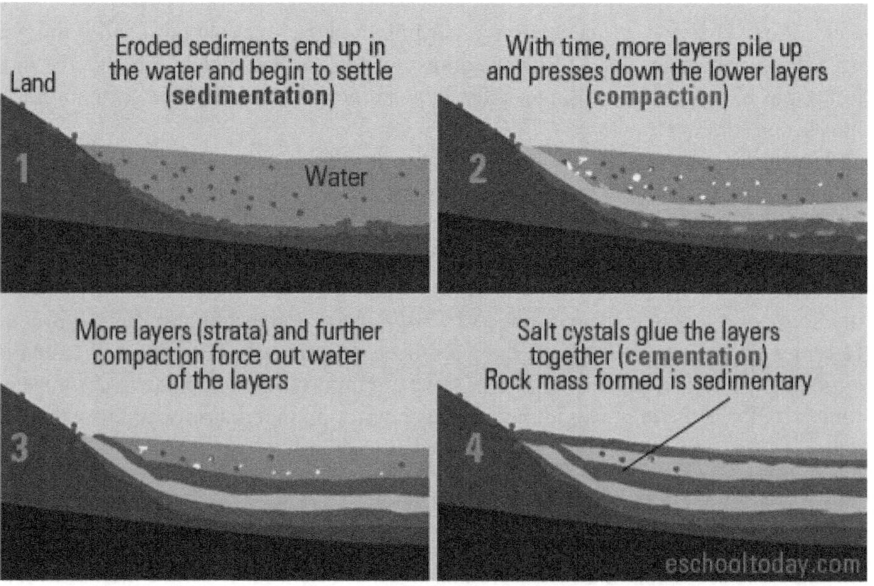

FIGURE 4.2 Sedimentation process consisting of sedimentation, compaction, and cementation phases (www.eschooltoday.com/rocks/what-are-sedimentary-rocks.html (last accessed in Feb 5, 2020),

4.3.1.1 Sedimentation

In general, areas of the Earth's surface, which are above sea level, are affected by the processes of erosion (wearing down of the land masses into particles, fragments, or debris). During geologic age (in the order of millions of years), large and small rock particles (sometimes with organic matter) were transported by rivers, streams, wind, or ice and were deposited on the marine seabed (where they settled or deposited). These sediments are of clastic origin (sands, silts, clays). On the other hand, in the sea, several different chemical ions or substances are also present. When concentrations of these substances are excessively high in the water, they separate from the water and precipitate/sink to the water seabed and form sediments with chemical origin such as carbonates. This sedimentation process continues to create sediments of considerable thickness during millions of years.

4.3.1.2 Accumulation and Compaction

During geologic time, layers of sediments that have settled under the ocean or water bodies accumulate, hardening under the weight of the overlaying deposits and water. This process is known as compaction or compression. Compaction leads to cementation, which is the gluing or cementing of pieces of rock together by some salt compounds. When these harden, they form sedimentary rocks. If organic matter is among the transported particles and sediments, it may settle and be compressed together with the sediments. Therefore, during the formation of sedimentary rocks, this organic matter residue may be converted to crude oil, natural gas, or coal. This is provided that the organic matter is enclosed in impermeable sediments such as shale (called source rock) which does not allow bacterial oxidation, and the matter undergoes adequate heating and compression.

Sedimentary rocks appear in layers called strata or formations. The oldest rocks tend to be at the bottom with newer rocks above them. Some important types of sedimentary rocks include sandstone, shale, limestone, and salts.

4.3.2 Structures and Petroleum Traps

Based on the Principle of Original Horizontality, the layers of sediments are originally deposited horizontally under the action of gravity. During or following rock formation, due to tectonic forces, they are folded, tilted, or faulted.

Formation of some geological structures is favorable for creating petroleum traps. In petroleum geology, a hydrocarbon trap is a geological structure allowing the accumulation of hydrocarbons. Figure 4.3 shows some main structural traps. Following the generation of hydrocarbons in the source rock, they migrate above through porous and permeable formations until they are inhibited by a barrier which is indeed an impermeable rock (called cap rock). In brief, the prerequisites of a petroleum system are the existence of a source rock, the existence of a porous permeable rock formation holding the hydrocarbons within its pore space (called reservoir), and an impermeable formation overlying the reservoir to inhibit hydrocarbon migration (cap rock). The other type of traps is stratigraphic. In a stratigraphic trap, a natural reservoir is formed by confining the hydrocarbons due to changes in porosity and

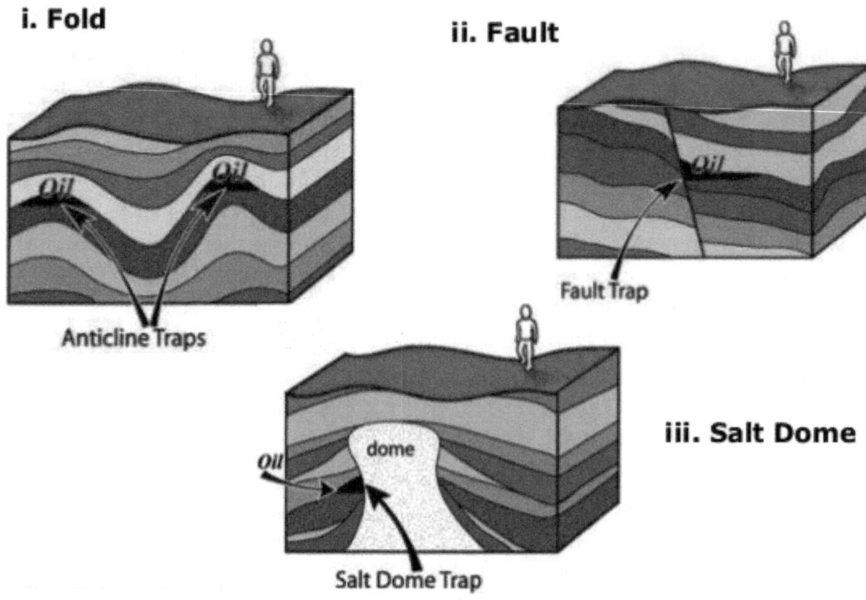

FIGURE 4.3 Structural hydrocarbon traps including folded, faulted, and salt dome traps (www.slideshare.net/MTaherHamdani/hydrocarbon-traps-seals (last accessed in February 5, 2020).

FIGURE 4.4 Some examples of stratigraphic traps. The roof rock or cap rock is here an impermeable shale formation which prevents the upward movement of the hydrocarbons or traps them (https://studfile.net/preview/5342336/page:4/ (last accessed in Feb 5, 2020)).

permeability (and lithology) of the layers rather than as a result of their structural attributes (Figure 4.4).

4.3.3 Formation Properties

Some important characteristics of underground formations are porosity, saturation, permeability, pore pressure, and their structures.

4.3.3.1 Porosity

Porosity is defined as the ratio of pore volume to the total volume of a rock. It is a measure of the empty space available in a rock for storage. This is a significant reservoir property. Porosity is formed in the rock during deposition, called "Primary Porosity", such as the space between the grains, or can develop through the alteration of the rock, called "Secondary Porosity" such as dissolution or dolomitization of carbonates or fractures (see Figure 4.5). Effective porosity is the interconnected pore volume in a rock contributing to fluid flow in a reservoir. Total porosity is the total void space in the rock whether or not it contributes to fluid flow. Obviously, effective porosity is less than the total porosity. Shale formations have relatively high porosity values, but the alignment of the platy grains such as clays makes their permeability very low.

During drilling in porous rock formations, the driller may encounter a sudden increase in drilling rate of penetration (drilling break). High-pressured formations often (but not always) have high porosity values because the formations in these areas may be less compressed/compacted. This important parameter can be evaluated using petrophysical well logs such as neutron, density, or sonic logs.

4.3.3.2 Saturation

As sedimentary rocks are formed under the water (in a marine environment), only seawater is contained initially in their pore spaces. However, in reservoir rocks, following the migration of hydrocarbons, some of the water contained in the pore volume is replaced by oil or gas. Therefore, the proportion water to total pore

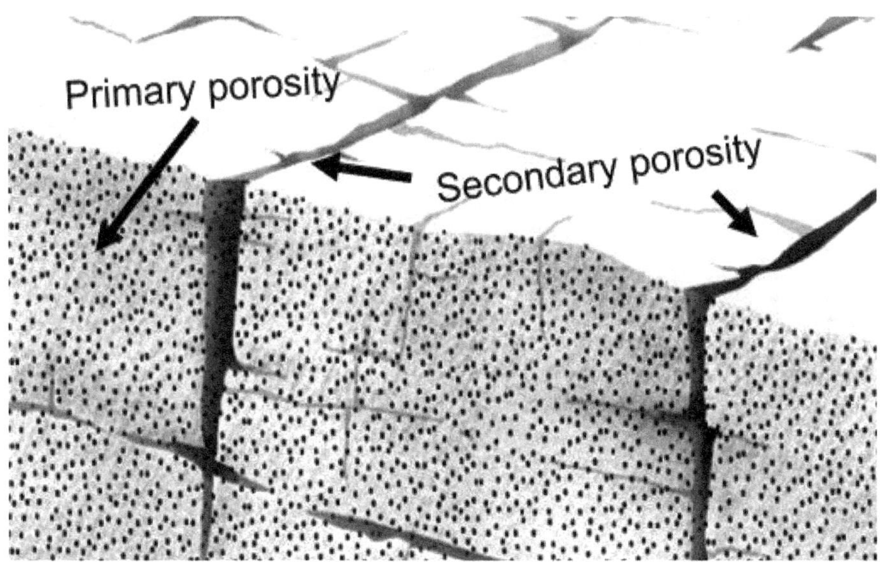

FIGURE 4.5 Primary porosity during deposition and secondary porosity due to fractures, etc. (www.slideserve.com/javen/a-hidden-reserve-groundwater (last accessed in February 5, 2020).

volume is called water saturation; the proportion of oil to total pore volume is called oil saturation; similarly, the proportion of gas to total pore volume is called gas saturation.

4.3.3.3 Permeability

Permeability is another important rock property that describes the rock's ability to transmit fluids through its pores. The larger the pores and the more interconnected they are, the greater the permeability is (Figure 4.6). Formations which transmit fluids readily such as sandstones are described as permeable rocks. However, rocks with small grains such as shales and siltstones have very low permeability values, that is, they are impermeable. Permeability is originally calculated in the unit of Darcy. A porous medium with the permeability of one Darcy permits a flow of 1 cm^3/sec of a fluid with viscosity of 1 cp (water) under a pressure gradient of 1 atm/cm acting across a cross-sectional area of 1 cm^2.

Permeability is considered an important parameter in well control. The greater the rock permeability, the larger the kick influx size, which is more difficult to be controlled. Greater kick size reflects greater the shut-in pressures after closing the well. A sandstone formation with high permeability and porosity is expected to give a much larger kick size than a tight shale formation. Permeability also affects the stabilization time of the well shut-in pressures. The greater the permeability is, the less the stabilization time (Figure 4.7).

4.3.3.4 Pore Pressure

The formation pore pressure is an important property of underground formations which can cause kick flows if it is excessively or abnormally high and not predicted

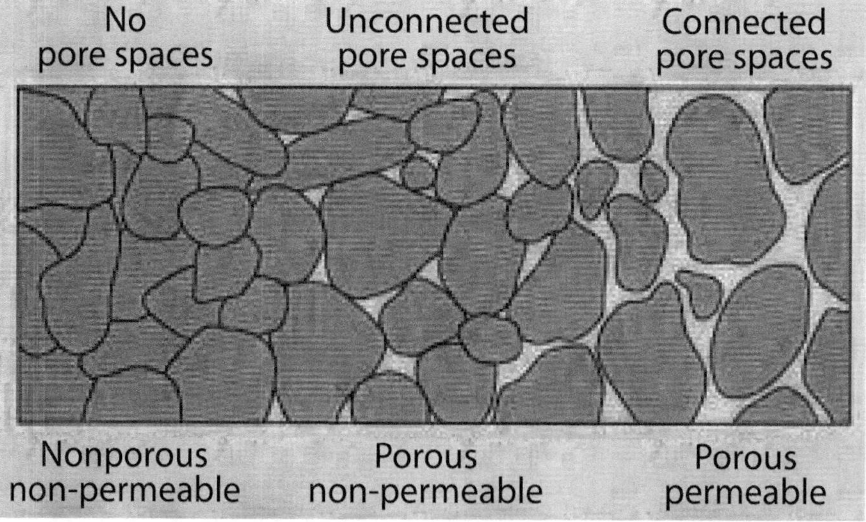

FIGURE 4.6 Connected pores and permeability (http://arnwine.weebly.com/porosity-permeability-percolation.html) (last accessed in February 5, 2020).

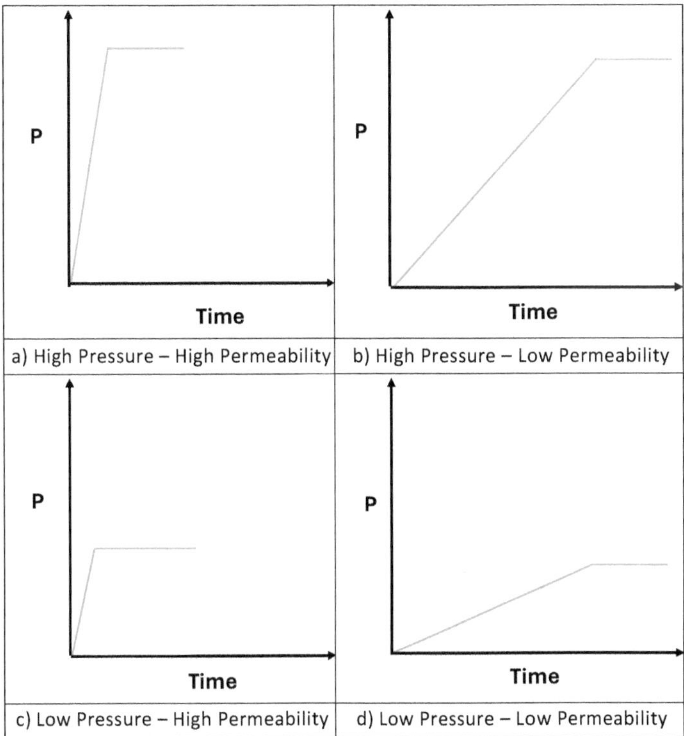

FIGURE 4.7 Effect of formation permeability and formation pore pressure on magnitudes of well shut-in pressures and stabilization time.

in advance to drilling. Pore pressure is related to the depth and geological conditions under which the formation was formed.

Pore pressure is very important for well control operations. First, during drilling operations, the hydrostatic mud pressure should be at least balanced with the formation pore pressure. Next, the greater the pore pressure exceeding the mud pressure (called *underbalance pressure*), the greater the influx size and consequently the magnitudes of stabilized surface shut-in pressure (Figure 4.7).

4.3.3.5 Fluid Type

Depending on the type of fluid contained in the formation (gas, oil, or salt water) entering the wellbore, kick types differ. If gas enters the wellbore, it is called a "gas kick". Because of gas expansion characteristic and its possibility to reach the surface quickly, gas kicks are the most severe.

4.4 HYDROSTATIC MUD PRESSURE AND GRADIENT

4.4.1 Surface-BOP Wells

Land/onshore rigs and bottom-fixed installations in offshore operations typically use surface-BOPs whereas floating rigs usually use subsea-BOPs.

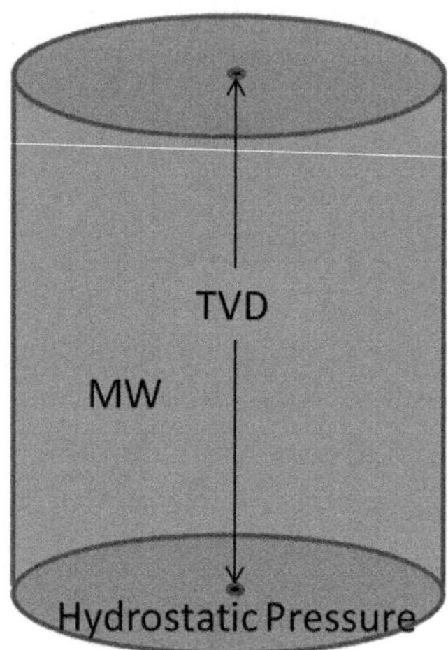

FIGURE 4.8 Hydrostatic mud pressure.

As a basic, pressure is generally defined as the acting force divided by area.

When the BOP is located at the surface (in land operations or fixed offshore rigs such as jack-ups), the hydrostatic pressure exerted by a column of drilling fluid HP_m is proportional to the true vertical height or depth, TVD, and also the mud density/weight, MW (Figure 4.8):

$$HP_m = 0.052 \times MW \times TVD \tag{4.1}$$

It should be noted that in deviated wells (where the measured depth MD is larger than TVD), only TVD is required for calculating the hydrostatic mud pressure (Figure 4.9); therefore, MD is not required for the calculation.

Generally, the pressure gradient *PG* is defined as the pressure divided by TVD. Using Equation (4.1), the mud hydrostatic pressure gradient PG_m is found by:

$$PG_m = \frac{HP_m}{TVD} = 0.052 \times MW \tag{4.2}$$

Depending on the mud type used in drilling a well (gas, oil/diesel, or water), the mud weight changes. Assuming the mud is just water, the greater the salinity, the greater the mud weight (Figure 4.10).

4.4.2 Subsea-BOP Wells and Riser Margin

When the BOP is located at the seabed (i.e., usually when floating rigs are used offshore), the hydrostatic mud pressure is found like the surface-BOP case (based on Equation 4.1). However, as an additional problem for offshore drilling operations, the riser may fail or get accidentally disconnected, which causes loss of part of the hydrostatic mud pressure. When this happens, suddenly, the hydrostatic mud pressure in the well will be reduced by the difference in hydrostatic pressure between the

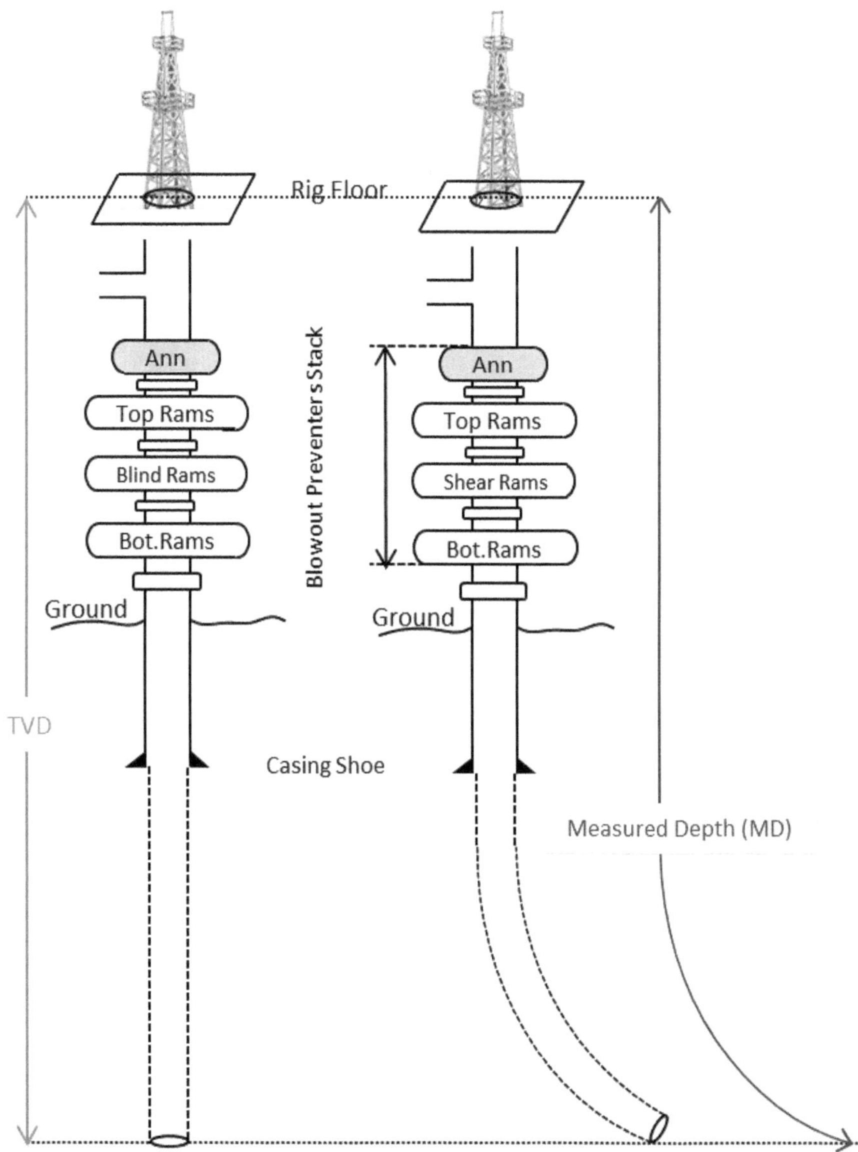

FIGURE 4.9 Well schematics of vertical and inclined wells with True Vertical Depth (TVD) and Measured Depth (MD).

original mud column pressure in the marine riser (up to the return flow line) and the hydrostatic pressure from seawater (see Figure 4.1). This reduction is considerable and may cause underbalance and kick flow into the wellbore.

In case riser disconnection is planned or predicted, in preparation of the removal of the marine riser, the mud weight should be sufficiently increased for a magnitude called *Riser Margin* (while also considering the formation strength). Therefore,

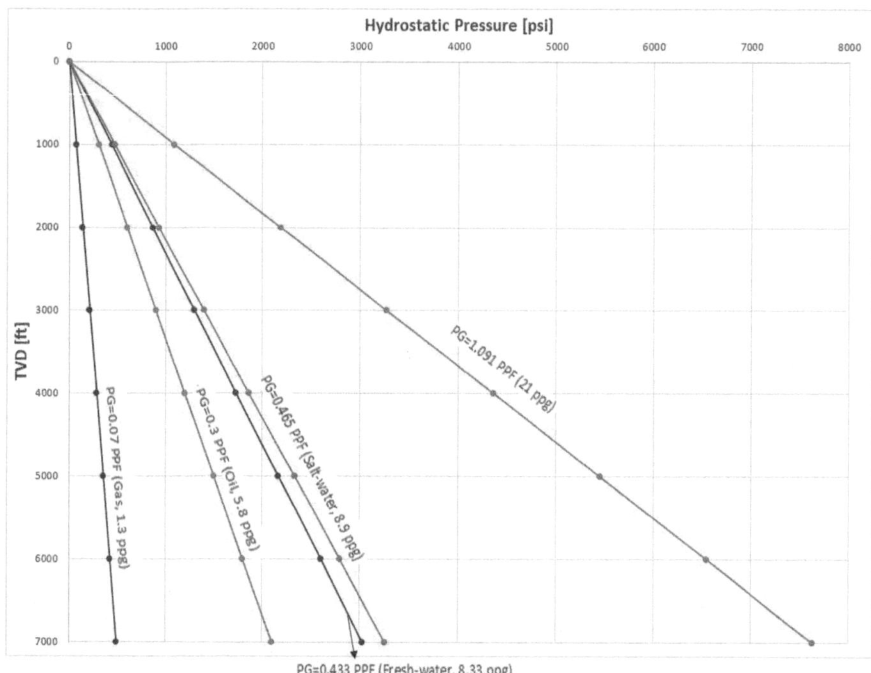

FIGURE 4.10 Hydrostatic mud pressures and gradients for different fluids as the drilling fluid.

following the disconnection of the marine riser, still the hydrostatic mud pressure would exceed the formation pore pressure (primary well control). It is noted that, in shallow depths where the formation strength is low, considering a riser margin could cause formation breakdown and lost circulation. Riser Margin is calculated using the following formula:

$$\text{Riser Margin}\left[\text{ppg}\right] = \frac{\left[\text{Air Gap} + \text{Water Depth}\right] \times \text{MW} - \left[\text{Water Depth} \times \text{W}_{\text{Sea-Water}}\right]}{\text{TVD} - \text{Air Gap} - \text{Water Depth}} \tag{4.3}$$

To understand the topic of riser margin better, an example is given below:

Example:

Using the following well data for a subsea-BOP well case:

- Calculate the drilling mud weight increase (riser margin) required to balance the well in preparation of disconnecting the marine riser.
- Calculate the required mud weight required in the well to ensure the well does not go underbalance if the riser is disconnected.

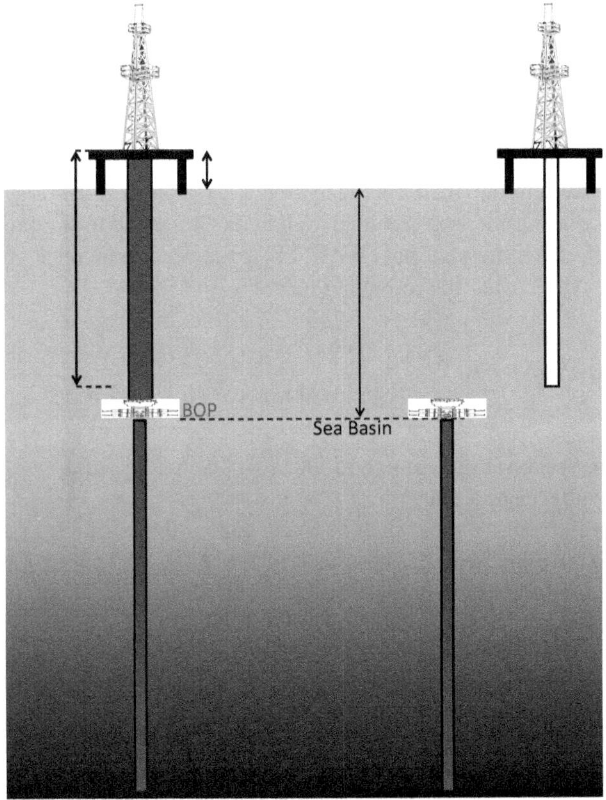

FIGURE 4.11 Mud hydrostatic pressure in two cases: a) Before the riser is disconnected and b) after the riser is disconnected, the hydrostatic mud pressure is less in the second case.

TVD = 10,000 ft
Air Gap = 80 ft
Water Depth = 2200 ft
Mud Weight to Balance Formation Pressure = 13.6 ppg
Seawater Gradient = 0.445 psi/ft

Answer – 15.2 ppg:

Using the formula:

$$\text{Riser Margin}[\text{ppg}] = \frac{[80+2200]\times 13.6-[2200\times 8.6]}{10{,}000-2200-80} = \frac{12088}{7720} = 1.6 \text{ ppg}$$

$$(\text{rounding up of } 1.565\,\text{ppg to } 1.6\,\text{ppg, for further safety})$$

The required mud weight to be used for drilling, prior to disconnecting the riser, is:

$$\text{Required MW} = 13.6 + 1.6 = 15.2\,\text{ppg}$$

Calculations for increasing hydrostatic mud pressure and weight:

Following the entrance of a formation influx in the wellbore, as part of the kill procedure, it is required to increase the hydrostatic pressure of the original mud. This is done by increasing the original mud weight as compared to the kill mud weight. The amount of weighting material (WM), $M_{WM,}$ required to increase the mud weight in a pit is calculated using the following formula:

$$M_{WM}\left[\frac{\text{Ib}}{\text{bbl}}\right] = 42 \times \text{WM Density}\frac{\left(\text{Kill Mud Weight}[\text{ppg}] - \text{Original Mud Weight}[\text{ppg}]\right)}{\left([\text{ppg}]\ \text{WM Density} - \text{Kill Mud Weight}[\text{ppg}]\right)} \quad (4.4)$$

If the weighting material is barite, consider 35 ppg (specific gravity of 4.2) for WM Density in the above equation.

The volume increase of the drilling mud due to the addition of the weighting material (WM) is found from:

$$\Delta V_{\text{m, due to WM}}[\text{gal}] = \frac{\text{Mass of WM}[\text{Ib}]}{\text{Density of WM}[\text{ppg}]} \quad (4.5)$$

Note that the density of barite used can be considered equal to 35 ppg (if it is not given).

4.4.3 Exercises

Exercise 1:

How much Barite is required to increase 100 bbl mud from its original weight of 10 ppg up to a greater mud weight of 12 ppg. Calculate the mud volume increase.

Answer – 8.69 bbl:

$$M_B\left[\frac{\text{Ib}}{\text{bbl}}\right] = 1470\frac{(12-10)}{(35-12)} = 127.826\,\text{Ib/bbl}$$

Therefore:

$$M_B[\text{Ib}] = 127.826 \times 100 = 12{,}783\,\text{Ib}$$

This is equivalent to 127.83→128 Sacks (100-Ib sack) of Barite.

The volume increase is:

$$\Delta V_{\text{m, due to Barite}}[\text{bbl}] = \frac{1}{42}\left(\frac{12{,}783[\text{Ib}]}{35[\text{ppg}]}\right) = 8.69\,\text{bbl}$$

Exercise 2:

The oil-based mud has salt water to oil concentration ratio of 70 / 30. Following the entrance of an influx in the hole, it is required to weigh the mud. How much calcium carbonate powder ($CaCO_3$) is required to reach the greater mud weight of 8 ppg?

Hints: The specific gravity of calcium carbonate used is 2.71 (22.6 ppg). Consider the weight of the salt water to be 9 ppg (specific gravity of water is 1.085 at concentration of 125,000 ppm). Consider the weight of the diesel equal to 7 ppg.

Answer – 26 lb/bbl:

As 30% of the OBM is salt water and the rest is diesel, the original mud weight is:

$$MW_{original} = 0.3 \times 9 + 0.7 \times 7 = 7.6 \text{ppg}$$

Therefore, the weight of the calcium carbonate to have the final mud weight of 8 ppg, is found as:

$$M_{CaCO3} = 42 \times 22.6 \frac{(8 - 7.6)}{(22.6 - 8)} = 26 \text{Ib/bbl}$$

4.5 FORMATION PORE PRESSURE, GRADIENT, AND EQUIVALENT MUD WEIGHT

The formation pore pressure is defined as the pressure applied by the fluids existing within pores of the formation. As a result, the formation pore pressure of a sedimentary rock is expected to be normally pressured, which means the pore pressure can be simulated as a column of salt water from the surface to the formation depth. The underground formation water is much saltier than normal seawater (salt content of normal seawater is about 23,000 ppm chloride); thus, it is slightly heavier than sea-water. This is because, during compaction and exertion of the overlying sediments over-burden pressure, some small water molecules are squeezed out of sediments leaving heavy salts in. Generally, the normal formation pressure gradient differs from region to region. In the US Gulf of Mexico Coast area, for instance, it is 0.465 psi/ft equivalent of pressure gradient produced by a column of water of approximately 100,000 ppm chloride. This value is the criterion of this manual calculation in well control. To meet the primary well control, the mud pressure must be maintained to be greater than the formation pore pressure.

The pore pressure (Ppore or Pformation) can be expressed in terms of the pore pressure gradient, PG_{pore} :

$$PG_{pore} = \frac{P_{pore}}{TVD} \tag{4.6}$$

where TVD is the specified True Vertical Depth at which the formation pore pressure was measured.

Pore pressure can also be expressed in terms of Equivalent Mud Weight (EMW_{pore}):

$$EMW_{pore} = \frac{P_{pore}}{0.052 \times TVD} \tag{4.7}$$

4.6 ABNORMALLY PRESSURED FORMATIONS

If a formation bears normal pressure, it is called normally pressured formations. If the formation pore pressure of a downhole formation is either greater or less than the expected normal pressure, it is called *abnormal formation*. The abnormal formations critical for drilling are *abnormally high-pressured or over-pressured formations*. The origins and reasons for the creation of abnormal pressures in formations are as follows:

- Under-compaction
- Differential fluid pressure
- Traps and hydrocarbon-bearing zones
- Reverse faults
- Diapirism
- Mineralization

4.6.1 Shale Under-Compression

Under-Compaction is defined as the mechanism of producing over-pressures in regions with high rate of shale deposition and thus compaction which does not provide enough time for shale pore water to escape. This factor is considered as the main reason for abnormal formation pressures.

Basically, shales have the unique property of low permeability but relatively high porosity. When shales are initially deposited, their porosity is about 50%. Thus, at the beginning of compaction, water can be easily squeezed out of shale to the surrounding permeable sediments, e.g., sands. This gives rise to a dramatic permeability decrease of the shale. With compaction continued, particularly when rapid deposition has already occurred, there is a time when the water cannot escape fast enough from the shale. At this time, pressure within the shale rises greatly and may even reach the over-burden pressure. The abnormal pressure development depends on the structure, deposition rate, sand to shale percentage, lensing, and faulting. In addition to the over-pressured shale layers, the thin sand layers/zones or even lenses, which are alternatively distributed and enclosed by the shale layers, will be also abnormally high-pressured (Figure 4.12). If a kick flow occurs, it would be from sand layers. During drilling, high-pressured shales do not cause well control events and kicks because of their very low permeability. Instead, the permeable over-pressured sand zones or lenses within massive shales can cause kick occurrences.

4.6.2 Differential Fluid Pressure

Differential fluid pressure refers to the abnormal pressure cases where the hydrostatic mud column pressure is not adequate to control formation pore pressure due to low ground height or formation folding or dipping. A common example of abnormal

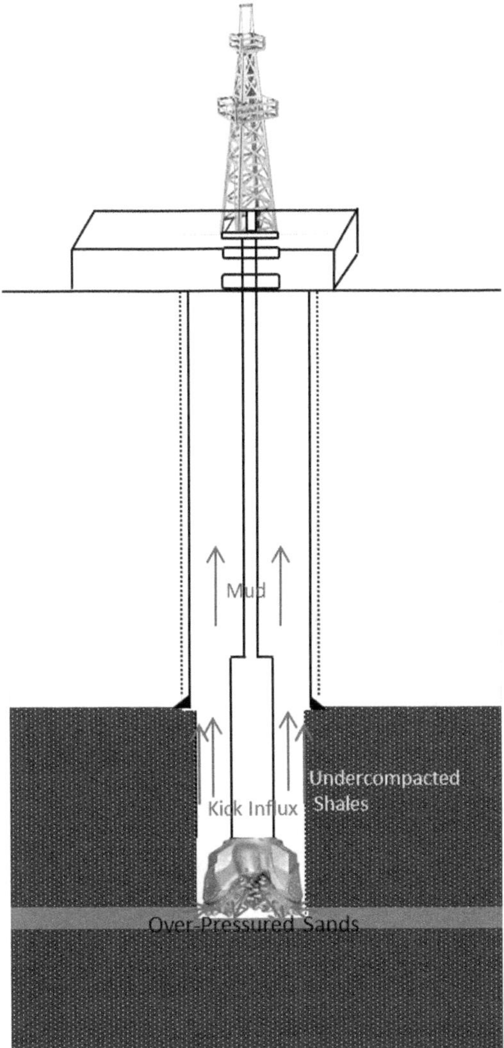

FIGURE 4.12 Effect of shale under-compaction and the enclosed formations which cause abnormal pressures.

pressure due to low ground height is the aquifer formation through which artesian water wells can be drilled (Figure 4.13). As seen in this figure, only a relatively short mud column is available to balance a large column of formation water; thus, a blowout would occur. This low ground height can exist due to surface erosion. In artesian water wells, water blowouts are intentionally allowed to produce saline water.

The normal hydrostatic mud column may not be adequate to control formation pressure in the case of folding or dipping sealed formation. This is a common cause of abnormal pressure in oil and gas fields as formations are dipping or folding. In Figure 4.14, two wells have been drilled to produce oil from the reservoir formation.

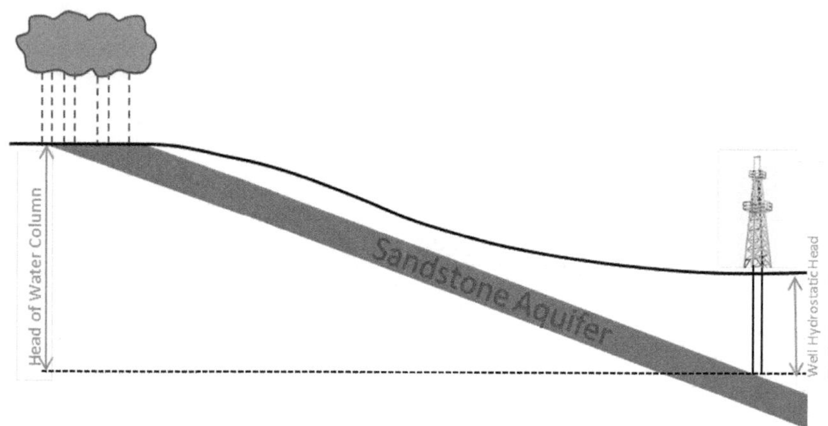

FIGURE 4.13 An abnormal aquifer in an artesian well.

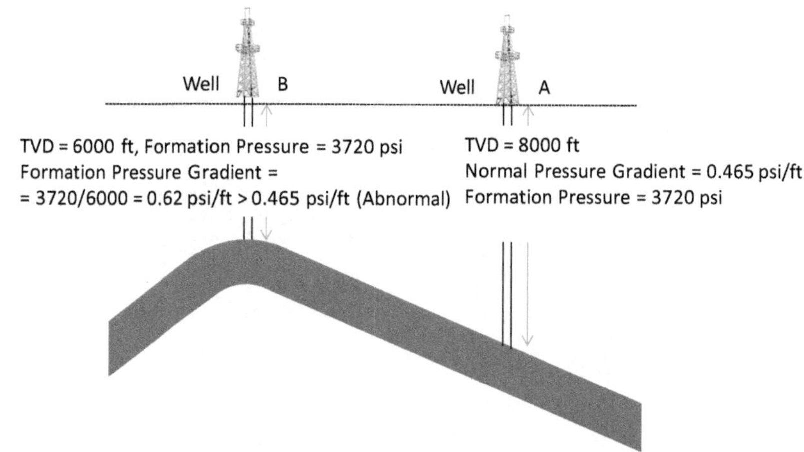

FIGURE 4.14 The differential pressure effect: Effect of folding in differential fluid pressure to cause abnormal pressure for Well B assuming Well A has struck a normally pressured formation.

The reservoir formation is sealed on its top by a cap rock. In this example, Well A is assumed to be normally pressured. If so, Well B is supposed to be over-pressured because the well hydrostatic head is lower than required.

4.6.3 Traps and Hydrocarbon-Bearing Zones

In hydrocarbon zones, part of the water, which is originally existing in a normally pressured formation, is replaced by hydrocarbons (oil and gas) during secondary migration. For hydrocarbons not to escape entirely to the surface, they should be sealed from the top by some impermeable rocks, called a cap rock. Typically, the water zone underlying the hydrocarbon zone is still normally pressured as it is connected to the surrounding water column with normal pressure. It is however different for the hydrocarbon column. It is found by subtracting the hydrostatic column pressure of the hydrocarbon interval from the normal pressure of water zone. Therefore, the formation pore pressure in the hydrocarbon-bearing interval is typically abnormally high in the hydrocarbon interval (see Figure 4.15 and Figure 4.16).

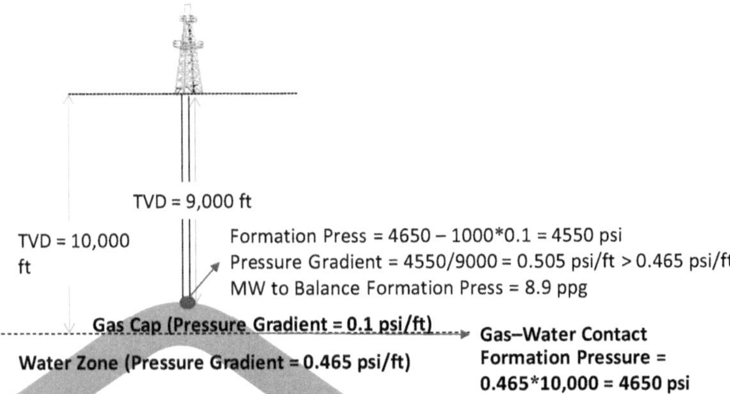

FIGURE 4.15 Hydrocarbon-bearing formation (gas-cap) is abnormally pressured.

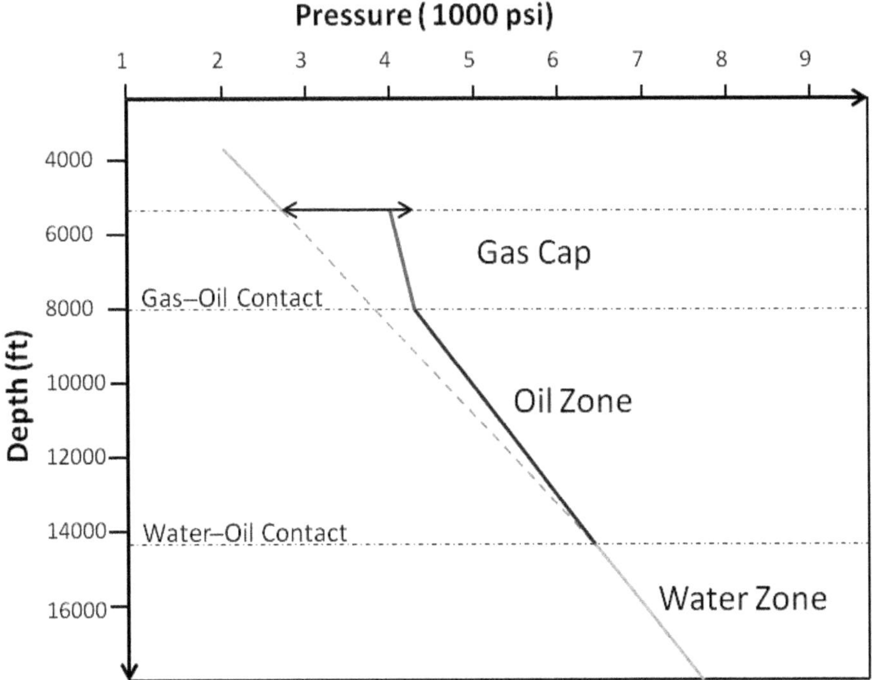

FIGURE 4.16 Trend of pressure versus depth in a reservoir and abnormal pressures in gas-cap and oil zone.

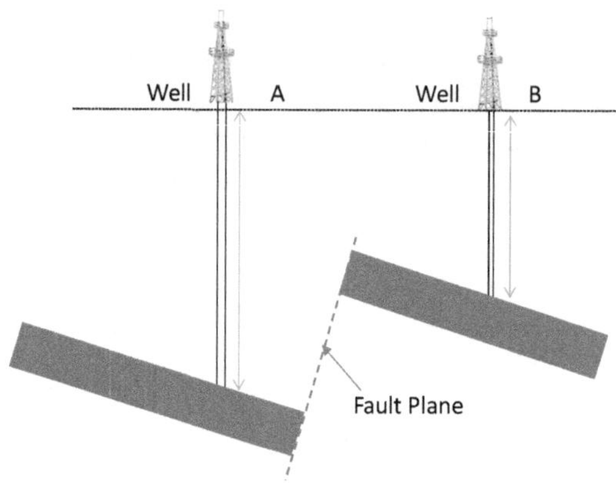

FIGURE 4.17 Effect of reverse faulting on abnormal formation creation for reservoir formation struck by Well B.

4.6.4 Reverse Faults

Due to tectonic forces, faults can occur to cause rock mass breakage and significant displacement. Faults are classified into two general classes of *normal* and *reverse*. In reverse faults, one block of the broken rock mass is moved up against gravity, and the other block may be kept stationary. If the faulted formation contains sufficient hydrocarbons and the surroundings of the raised block have sealing capability, this raised rock block will be over-pressured (Well B in Figure 4.17). This is because the original formation pressure is retained in the raised block but at shallower depth; thus, the pressure gradient would be most likely more than the normal.

4.6.5 Surcharging of Shallow Formations

In case of a naturally occurring fault, the zone of broken rock, at the fault itself, may allow fluid seepage over part of its extent. This can result in shallow formations being charged up (surcharged) with formation fluids at pressures much greater than the normal for their depth (see Figure 4.18). Therefore, this process is a natural phenomenon happening usually in faulted zones.

4.6.6 Diapirism

Diapirism is a geological process by which salt beds move or flow up because of their lighter density than the overlying sediments. This process is more significant particularly through rapid sedimentation. Salts behave differently than rocks due to their lower density and strength as well as their ductile property. It is noted that salt beds contain very little rock frame structure as well. These properties allow them to flow like a viscous plastic under pressure, and thus salt beds can be transformed and developed into salt domes in geological time. Diapirism causes the modification of structures, faulting, and folding as well as compressing of overlying and surrounding

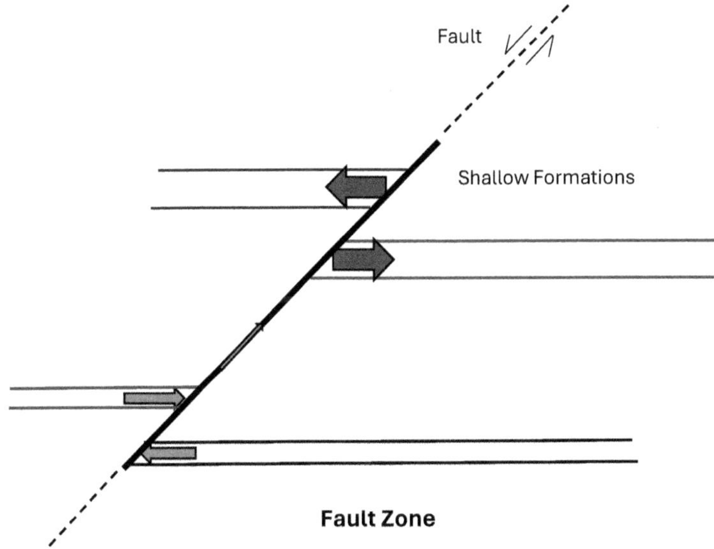

FIGURE 4.18 Surcharging of shallow formations due to the seepage flow from underlying formations through the fault layer.

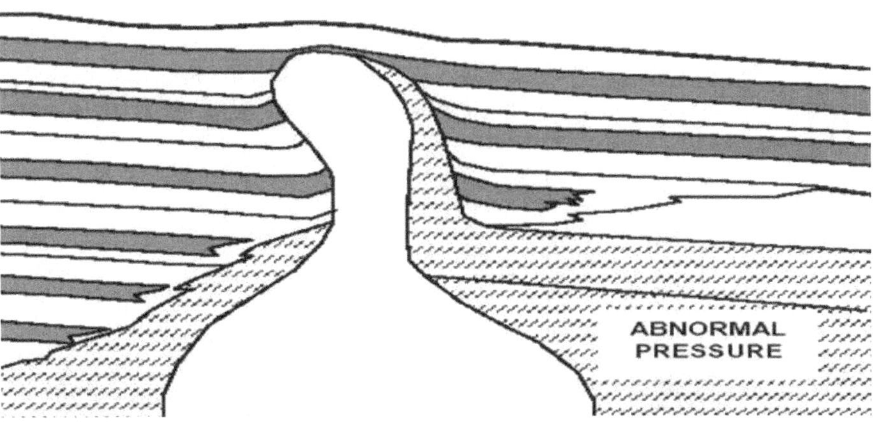

FIGURE 4.19 Effect of diapirism on abnormal pressures in formations above and surrounding the salt dome.

formations which would bring about abnormally high-pressured formations while also sealing the formations (Figure 4.19).

Depending on salt properties such as composition, depth, or temperature, massive salt could bear pressures up to 1 psi/ft. Salt beds can be drilled with lighter mud weights lower than 1 psi/ft because salt flow into the borehole is normally slow, and the rate flow depends on composition and temperature.

The underlying salt bed/layer usually has very little strength. Therefore, the water in the possible shale formations just below the salt bed bears the entire burden of the salt plus the burden of sediments above the salt. The pressure of shale formation below the salt can reach 1 psi/ft. Sand may not occur directly below the salt or below limestone unless there is an overhang. Simultaneously, the formations below the salt bed are older than salt and thus have had more geologic time and have lost much of their water content. Therefore, this effect makes it challenging to choose a proper mud weight to drill formations underlying salts.

4.6.7 Mineralization

Basically, changes in minerals of the sedimentary rocks can change the volume of formation solids and thus yield abnormally high or subnormal formation pressures. Therefore, if the volume of formation solids is increased due to mineralization, the formation pore pressure increases as well. This increase is because an increase in the rock solid volume downhole would impose further compression on the fluid.

For instance, dolomitization causes solid volume to decrease for about 5% – or in other words, pore volume to increase for 5%. Dolomitization is the process whereby some minerals of limestone ($CaCO_3$) are converted to $CaMg(CO_3)_2$ to generate dolomite. It should be reminded that dolomitization occurs prior to hydrocarbon migration; thus, the effect is mostly visible in water-bearing formations.

4.7 PRINCIPLE OF U-TUBE EFFECT

Considering that wells in a U-tube shape are advantageous for analyzing and investigating many well situations; in this analogy, one column of the well represents the drill-string and the other one represents the annulus (Figure 4.20).

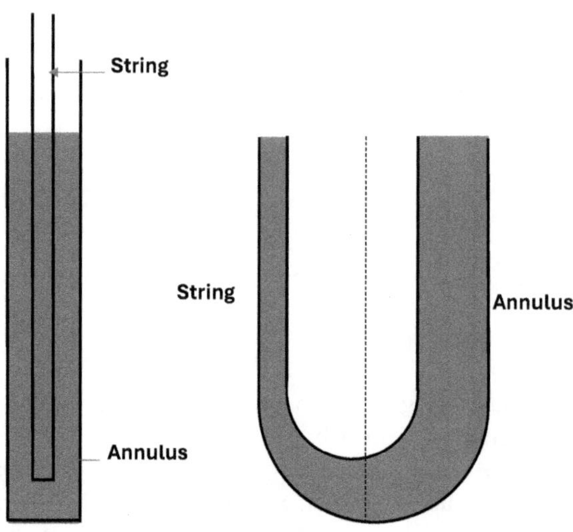

FIGURE 4.20 U-tube analogy.

In static conditions, there are fluids in both the drill-string and the annulus. The atmospheric pressure can be ignored as it works the same on both columns. If there is the same fluid weight in the string and annulus, the hydrostatic pressure would be the same and the fluid would remain static on both sides of the U-tube.

However, if the fluid in one of the U-tube columns is heavier than the fluid in the other one, the static condition would not last but would change. There are two scenarios as explained in the next section.

4.7.1 Heavy Mud/Pill in the Annulus

If there is a heavy pill in the annulus, the annulus hydrostatic pressure downward is greater than that of the drill-string side. Therefore, it will push the mud in the drill-string and displace some of the original mud out of the string, causing an apparent flow at the surface. This flow continues until the fluid level in the annulus drops enough to equalize the hydrostatic pressures on both sides of the U-tube. This phenomenon is called *U-tubing* (shown in Figure 4.21). Through this phenomenon, some of the original mud would be displaced from the annulus to the string, sounding like a mud loss case. This flow at the surface really occurs practically in two common situations:

- Similarly, in case of too fast drilling without enough hole cleaning/cuttings' removal, the mud in the annulus would have too much cuttings and thus would be heavier than that of the drill-string. Therefore, during making connections (static/no circulation in the well), a flow would occur.
- When there is a heavy pill in the annulus and the well conditions become static (e.g., due to pumps failure) and there is no float valve in the string, then flow would happen to the drill-string.

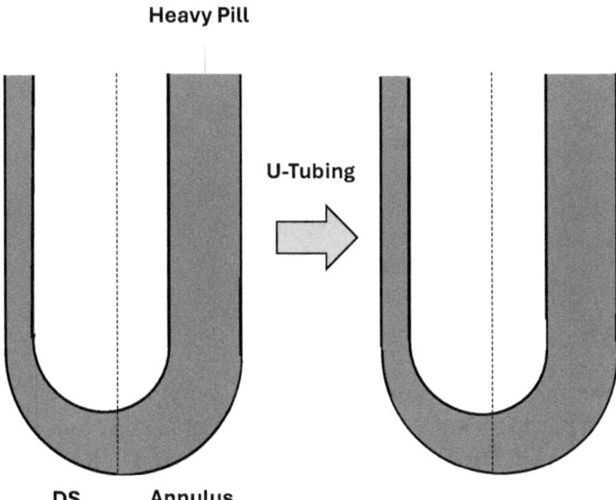

FIGURE 4.21 U-tubing when the heavy weight pill reaches the annulus. There is a tendency for the heavy pill to flow to the drill-string.

4.7.2 Heavy Mud/Pill in the String

If there is a heavy mud/pill in the drill-string, it does exert greater hydrostatic pressure downward than the annular side. U-tubing will push the mud from the drill-string to the annulus (i.e., U-tubing). This displaces some of the original mud from the string to the annulus, showing as a pit gain (in static conditions/pumps are off). If circulation is ongoing (with pumps on), due to the U-tubing, a decrease in pump pressure and an increase in pump strokes are observed. This extra pit gain in the annulus resembles a kick situation.

4.7.3 Heavy Slug in the String

Another example of U-tubing occurs when a heavy slug is pumped in the drill-string which is a common practice prior to pulling out of the hole (POOH) in wet tripping situations (no float valve above the bit). The heavy slug exerts greater hydrostatic pressure at the string side of the U-tube. Therefore, it would continue pushing some mud out of the string during tripping. Placing the slug in the string would extract part of the mud until a specified mud level drop occurs in the drill-string. The specified mud level drop is usually designed to be the average length of one stand pulled. Therefore, the slug is purposefully pumped before starting the tripping operations to make the pipes be pulled dry without any mud spill and disturbance for the roughnecks in handling the mud during the POOH. The calculation related to slug pumping is as follows:

The pit gain due to pumping slug is:

$$\text{Pit Gain}_{(\text{due to slug})} = V_{\text{slug}}\left(\frac{W_{\text{slug}} - MW}{MW}\right) \tag{4.8}$$

The distance of drop in the drill-pipe due to pumping slug, $\Delta L_{\text{mud-DP (due to slug)}}$ is:

$$\Delta L_{\text{mud-DP(due to slug)}} = \frac{\text{Pit Gain}_{(\text{due to slug})}}{\text{Cap}_{\text{DP}}} \tag{4.9}$$

where W_{slug} is the slug weight/density.

4.7.4 Other Applications

- The U-tube principle has applications in accounting for observing the "decrease in pump pressure and increase in pump strokes" as a warning sign of kick occurrence (kick influx entrance into the well). Refer to Section 6.4 for more information.
- The U-tube effect can be used to relate well shut-in pressures to one another (shut-in drill-pipe pressure (SIDPP) and Shut-In Casing Pressure (SICP)). Refer to Section 9.2.2 for more information.

4.7.5 Exercises

Exercise 1:

A well is shut in. What is the casing pressure in this static U-tube?
Well Information:

- Drill-pipe pressure reads 0 psi.
- Well depth = 12,000 ft TVD/12,225 ft MD
- Drill-string is full of 8.3 ppg water.
- Annulus is full of 6.0 ppg gas/water mixture.

a. 1,435 psi
b. 1,462 psi
c. 3,744 psi
d. 5,179 psi

Answer – a:

Since there are two different fluid types and weights/densities in the annulus, due to the U-tube effect, some casing pressure manifests itself at the surface. The casing (annulus) pressure is found by subtracting the annulus hydrostatic pressure from the drill-string hydrostatic pressure:

$$CP = 0.052 \times 12 \times (8.3 - 6) = 1{,}435.2 \sim 1{,}435\,\text{psi}$$

Exercise 2:

Before pulling out of the hole, a 20-bbl heavy slug is pumped and followed by 15 bbl of regular mud. Using the following well data (Table 4.1), how far will the mud level drop in the string when the well has equalized?

Answer – 205 ft:

The pit gain due to pumping slug is:

$$\text{Pit Gain}_{(\text{due to slug})} = V_{\text{slug}}\left(\frac{W_{\text{slug}} - MW}{MW}\right) = 20\,\text{bbl}\left(\frac{13-11}{11}\right) = 3.63\text{ bbl}$$

TABLE 4.1
Given Well Data for Exercise 2

Depth of hole (RKB)	10,400 ft
Drilling mud weight (MW)	11 ppg
Heavy slug weight (W_s)	13 ppg
Drill-pipe capacity (Cap_{DP})	0.01776 bbl/ft
Surface line volume	10 bbl

The distance of drop in the drill-pipe due to pumping slug is:

$$\Delta L_{\text{mud-DP(due to slug)}} = \frac{\text{Pit Gain}_{\text{(due to slug)}}}{\text{Cap}_{\text{DP}}} = \frac{3.63\,\text{bbl}}{0.01776\,\text{bbl/ft}} = 205\,\text{ft}$$

4.8 CIRCULATING PUMP PRESSURE/STANDPIPE PRESSURE

4.8.1 Pressure Losses and SPP

The dynamic or circulating pump pressure is the total pressure required to circulate the mud from the surface through the drill-string to the bit nozzles and then back through the annulus to the surface (Figure 4.22). The pump circulating or dynamic pressure is equal to summation of all frictional pressure losses in the whole circulation path consisting of the surface lines, drill-string, bit nozzles, and the annulus. The Standpipe Pressure (SPP) or Drill-pipe Pressure (DPP) is equal to the circulating pump pressure minus a pressure loss in the pump surface lines. DPP is related to the mud weight and circulation/pumping rate. There are two rules of thumbs for these relationships:

$$\text{DPP}_2 = \left(\frac{\text{MW}_2}{\text{MW}_1}\right) \times \text{DPP}_1 \tag{4.10}$$

$$\text{DPP}_2 = \left(\frac{\text{Q}_2}{\text{Q}_1}\right)^2 \times \text{DPP}_1 \tag{4.11}$$

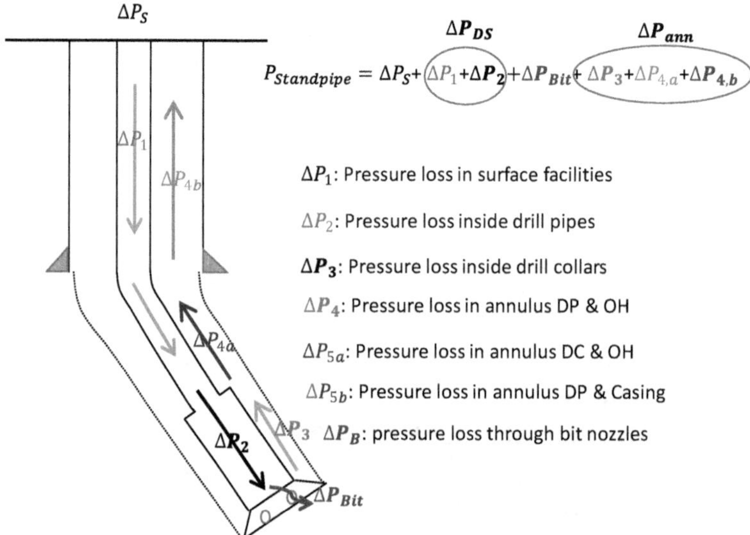

FIGURE 4.22 The hydraulic circulating path and frictional pressure losses summing up to the standpipe or total system pressure loss. The mud weight in both the drill-string and annulus columns is considered uniform/the same.

The greatest frictional pressure losses occur first in the bit nozzles, and then in the drill-string. The pressure loss in the drill-pipe is considerable because of its low inside diameter. However, the pressure loss in the annulus is minimal because of its rather large annular cross-sectional area which makes the flow regime laminar. The annular pressure loss is important for the calculation of the bottomhole pressure (see Section 4.9.1). As a rule of thumb, the annular pressure loss (ΔP_{ann}) is typically considered 10% of the standpipe pressure (provided that the standpipe pressure is 1,700 psi or greater).

4.8.2 Effect of Pills

Therefore, the general SPP relation, which shows possible U-tubing effect, is:

$$SPP = \Delta P_s + \Delta P_{DS} + \Delta P_{Bit} + \Delta P_{Ann} + (P_{H,Ann} - P_{H,DS}) \tag{4.12}$$

During pumping/circulating a heavy mud/pill in the drill-string, the pump pressure or standpipe pressure decreases. This is technically due to the U-Tubing contribution in the same direction of mud circulation (i.e., from the drill-string to the annulus). With this contribution, less pump pressure or standpipe pressure (SPP) is required to circulate the mud (though the ΔP_{DS} term is a bit greater when the mud is uniform in the drill-string and annulus). Next, when the heavy pill starts entering the annulus, the U-tubing effect is in opposite direction to the pumping/circulation direction. Therefore, greater SPP than normal is required to circulate the mud.

Conversely, with light mud/pill in the drill-string, greater SPP is expected (due to the U-Tube effect from the annulus to the string). Next, when the light mud is in the annulus, lower SPP is expected. Table 4.2 summarizes the effects of different weight pills on SPP.

4.8.3 Exercises

Exercise 1:

In drilling with the mud weight of 10 ppg and circulation flow rate of 500 gal/minute, the circulating pressure (standpipe pressure) was measured equal to 2,800 psi. If the mud is displaced with a lighter mud of 9 ppg, but the circulation rate is increased to 550 gal/minute, what is the estimated circulating pressure?

TABLE 4.2
Effects of Heavy and Light Muds/Pill and Locations on SPP

Pill Type	Location	Effect on SPP
Heavy Mud/Pill	Drill-String	Decrease
	Annulus	Increase
Light Mud/Pill	Drill-String	Increase
	Annulus	Decrease

Answer – 3050 psi:

Combining the two Equations (4.10) and (4.11), we have:

$$DPP_2 = (\frac{MW_2}{MW_1}) \times (\frac{Q_2}{Q_1})^2 \times DPP_1$$

$$DPP_2 = \left(\frac{9}{10}\right) \times \left(\frac{550}{500}\right)^2 \times 2{,}800 = 3{,}050\,\text{ps}$$

Therefore, the circulating pressure is expected to increase.

Exercise 2:

Drilling a hole section and running/cementing of the casing were done to the true vertical depth of 10,000 ft. For drilling the next hole section, the current mud with the weight of 14 ppg must be replaced with a saltwater mud of 9 ppg to drill the reservoir formation. Evaluate the increase in the SPP (due to the U-Tube effect) when the light mud reaches the bit nozzles. Interpret this effect in your displacement procedure if the maximum pressure rating of the mud pumps is 3,000 psi.

Ignore the effect of lower mud weight in the drill-string on pressure losses.

Answer:

Since the light mud is in the drill-string, the U-Tube effect (from the annulus to the drill-string) increases. Therefore, the maximum effect of U-Tube on SPP reaches when all the drill-string is filled with the light mud (light mud at the bit nozzles):

$$\text{SPP Increase Due to U} - \text{Tubing} = 0.052 \times 10{,}000 \times (14 - 9) = 2{,}340\,\text{psi}$$

This U-tube pressure is 2,340 psi. There would be some circulation pressure losses as well. Thus, it is very likely that the pressure rating of the pumps (3,000 psi) would be exceeded, and the pump would fail. This signifies that the pumps cannot handle the displacement safely. To solve the problem, displacement should be done in several stages (e.g., three):

1) First, run the drill-string to 3,000 ft and displace the mud with light mud. The SPP increase due to U-Tube would reduce to 7,800 psi for each stage.
2) Run the drill-string to 6,000 ft and displace the mud as the second stage.
3) Run to 9,000 ft and displace the mud.

4.9 BOTTOMHOLE PRESSURE AND EQUIVALENT CIRCULATING DENSITY

4.9.1 Dynamic BHP and ECD

Bottomhole Pressure (BHP) is the pressure exerted on the formation wall by the drilling fluid. If the mud is static, e.g., while tripping or logging, then BHP is equal to the hydrostatic mud pressure. However, while drilling, having circulation, the annular pressure loss acts as a back-pressure to the hydrostatic mud pressure. In other words, the circulating/dynamic bottomhole pressure is equal to the hydrostatic mud pressure (HP_m) plus the frictional pressure loss in the annulus (ΔP_{ann}) :

$$BHP = HP_m + \Delta P_{ann} \tag{4.13}$$

Therefore, the equivalent circulating density (ECD) of the mud is found as follows:

$$ECD = MW + \frac{\Delta P_{ann}}{0.052 \times TVD} \tag{4.14}$$

Obviously, ECD is greater than the static equivalent mud weight (Equation 4.14). The magnitude of difference is typically less than half ppg except for slim-holes or muds with high viscosity where it can be greater. Figure 4.23 presents an example of drilling pressure losses with BHP and ECD calculation.

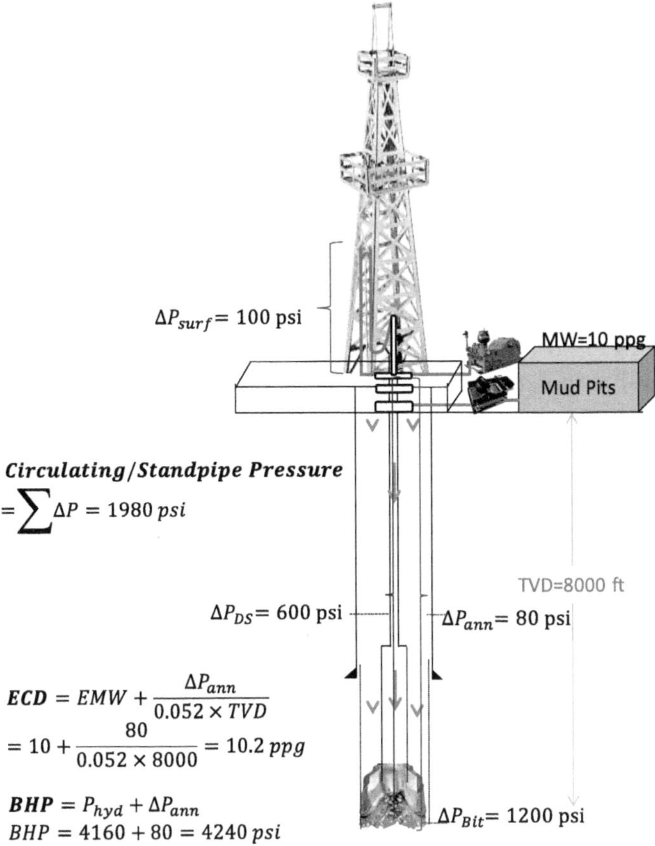

FIGURE 4.23 Bottomhole Pressure (BHP) and Equivalent Circulating Density, ECD, in a circulating mud (dynamic condition).

As the primary well control in overbalanced drilling, the hydrostatic BHP must be always greater than the formation pore pressure in order to prevent any kick. The magnitude of the overbalance is considered between 70 and 200 psi depending on the drilling company's policy. If so, the dynamic pressure will be undoubtedly greater than the formation pressure.

4.9.2 Effect of the Slug

Based on the U-tube principle, the slug creates a mud level drop in the string, which facilitates tripping operations as it eliminates spill and handling of the mud in the string during POOH. The slug is heavier than the mud; however, the added hydrostatic pressure because of its heavier weight of slug is eliminated or counteracted by the mud level drop. Therefore, the static BHP remains unchanged.

4.9.3 Effect of Pills

When a heavy mud or pill is pumped in the drill-string, the circulating bottomhole pressure remains unchanged. However, when the pill starts being circulated into the annulus, the BHP starts increasing. When all the heavy pill is already in the annulus and reaches its maximum height there, the circulating BHP reaches its maximum.

Conversely, if a light mud or pill is pumped in the drill-string (e.g., in order to replace/displace the current mud with a lighter mud), the circulating BHP remains unchanged as long as the mud is in the drill-string. As soon as the light mud starts being circulated into the annulus, the BHP starts dropping. When all the light mud is already in the annulus, the circulating BHP has reached its minimum value. Precautions should be taken to predict enough hydrostatic overbalance; otherwise, an underbalanced situation can occur which allows entrance of a kick influx. Table 4.3 shows the effects of pills on BHP and DPP or SPP. It is seen that when the pill is in the annulus, both SPP and BHP show similar increasing or decreasing trends.

It may look surprising that the circulating BHP remains the same when the pill is in the drill-string (as claimed above), whereas the BHP increases or decreases

TABLE 4.3
Effects of Heavy and Light Muds/Pills on BHP versus DPP Depending on Their Locations in the Wellbore

Mud/Pill Type	Location	Effect on BHP	Effect on SPP
Heavy Mud/Pill	DS	Const	Dec
	Ann	Inc	Inc
Light Mud/Pill	DS	Const	Inc
	Ann	Dec	Dec

Note: Ann, annulus; Dec Ann, annulus decrease; DS, drill-string; Inc Ann, annulus increase.

respectively when the heavy or light pill reaches the annulus. The above phenomena can be accounted for in two ways:

- Based on the U-tube principle, circulating BHP can be found from the annulus column side. When the pill is in the drill-string side, the hydrostatic mud column in the annulus ($P_{h,Ann}$) and the frictional pressure loss in the annulus (ΔP_{Ann}) remain unchanged; therefore, the BHP remains constant. However, when the heavy pill reaches the annulus, both $P_{h,Ann}$ and ΔP_{Ann} increase. Therefore, BHP also shows an increase.
- The circulating BHP can be justified from the drill-string side as well, as:

$$BHP = P_{h,DS} + \Delta P' \tag{4.15}$$

Using Equations (4.13) and (4.15), we have:

$$\Delta P' = \Delta P_{Ann} + \left(P_{h,Ann} - P_{h,DS}\right) \tag{4.16}$$

- When the heavy pill is in the drill-string side, there is an increase in the hydrostatic mud pressure in the drill-string ($P_{h,DS}$) for hydrostatic pill–mud pressure difference. On the other hand, $\Delta P'$ decreases for the same amount. Therefore, BHP remains constant as long as the pill is inside the string.

  However, as the pill reaches the bit nozzles and raises the annulus, the annulus hydrostatic pressure ($P_{h,Ann}$) and its frictional pressure loss (ΔP_{Ann}) increase; $\Delta P'$ has decreased for the same magnitude of the hydrostatic pill–mud pressure difference. Therefore, the BHP remains constant.

  When the pill reaches the annulus, $P_{h,Ann}$ increases. Thus, with increasing $\Delta P'$, BHP increases as well.

4.10 CASING SETTING DEPTHS

A significant part of the well planning process is the selection of casing setting depths as the main section of well construction. During planning, this job is done tentatively, by the geology department, once all the geological formations, lithological columns, and formation pressure profiles have been collected from offset wells already drilled nearby. Selection of setting depths is based on the prevention of kick and formation fracture for each hole section. Figure 4.24 shows a typical field example of casing shoes.

As well control measures in casing design, two safety factors should be considered (see Figure 4.25): a) *trip margin* (to prevent swabbing during pulling the drill-string out of the hole) and b) *kick margin* (to prevent formation fracture/breakdown in case of a kick occurrence). In case the casing setting depths are selected wrongly, it can pose serious well control risk and consequences for drilling the subsequent hole section. Imagine the casing shoe is selected shallower than it should be. This is because the heavy mud weight may cause fracturing of the formation below the casing show (which should have been covered by the casing). Conversely, in case of undesirably deeper selection of the setting depths, there are again well control and economic consequences. The well control consequence is that a kick influx may enter the wellbore before reaching the selected casing shoe depth. The economic

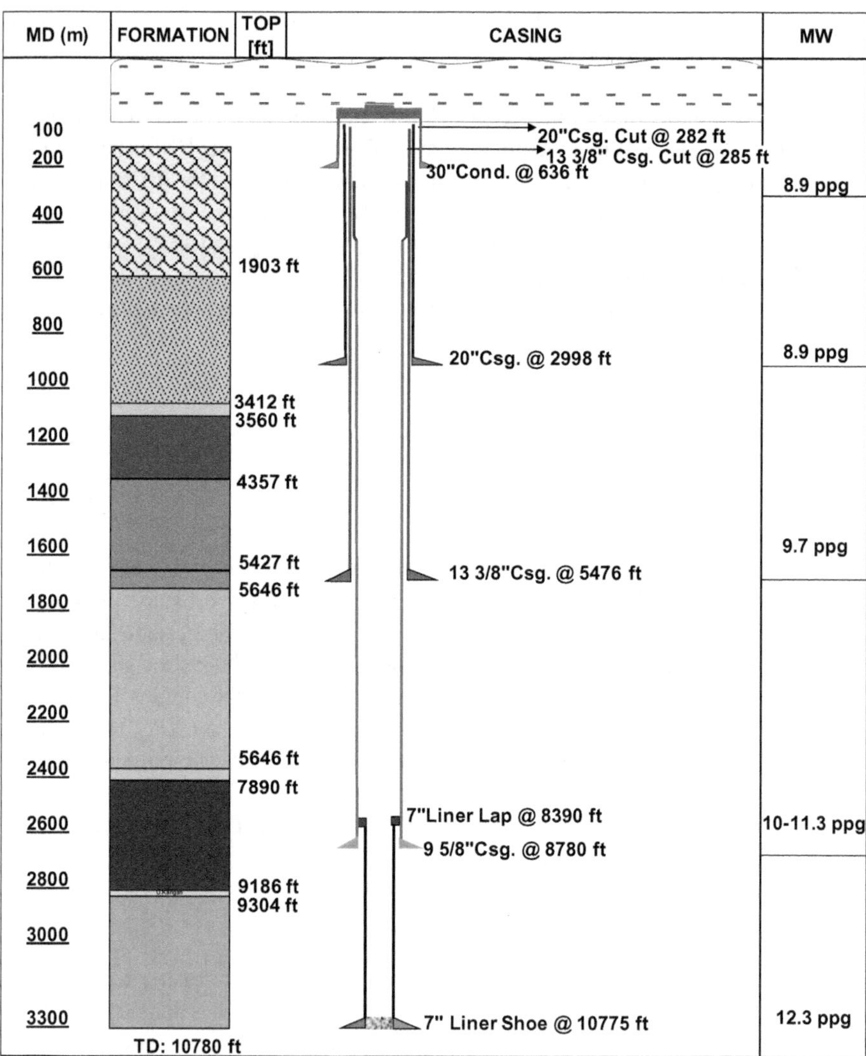

FIGURE 4.24 A typical casing program with selected shoe depths.

consequence is that extra rig time was taken to drill deeper (more than necessary), and a longer than necessary casing string was run.

4.11 SURGE AND SWAB PRESSURES

When the drill-string is pulled up out of the hole (POOH) during a trip, the mud must flow down past it to replace the vacuumed volume left behind. However, the upward movement of the drill-string during POOH will cause partial mud lifting due to the friction between the moving pipe (solid) and the stationary drilling fluid (liquid). Consequently, it causes a reduction of the bottom of the hole. This reduction is called

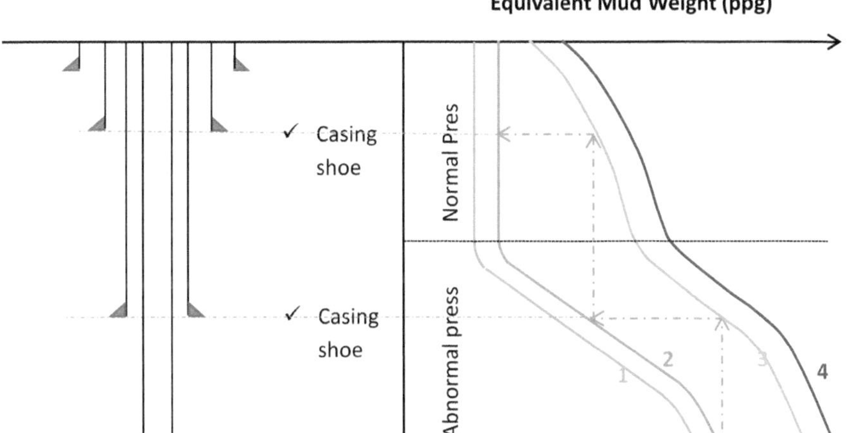

1. Pore pressure gradient, determined as $P_{pore}/(0.052 \times TVD)$
2. Pore pressure gradient + trip margin

1–2: Trip margin (required to compensate pressure loss occurring due to swapping), 0.2–0.5 ppg

3. Fracture pressure gradient minus kick margin

2–3: Safe mud window

4. Fracture pressure gradient, determined as $P_{frac}/(0.052 \times TVD)$

3–4: Kick margin **(required to allow some prevention of fracturing in case kick occurs)**, 0.2–0.5 ppg

FIGURE 4.25 Consideration of well control safety margins in casing design: Trip margin (to prevent swabbing) and kick margin (to prevent formation fracture/breakdown at the shoe in case of a kick).

swab pressure (Figure 4.26a). If the swab pressure is excessive, it can cause the BHP to get less than formation pressure (underbalance).

When the drill-string is run in the hole (RIH), the mud must flow up past the string. However, the downward movement of the drill-string during RIH will cause partial mud pushing down due to the friction between the moving pipe and the stationery drilling fluid. Consequently, it causes an increase in the bottomhole pressure. This pressure increase is called *surge pressure* (Figure 4.26b). Figure 4.27 shows the variations of surge pressure for a typical case of running in the hole.

The surge and swab pressures depend on several impacting factors. Any factors contributing to greater frictional lifting forces on the mud around the pipe as it is lowered (run in the hole, RIH) or raised (pulled out of the hole, POOH), respectively, increase/surge and decrease/swab pressures. In other words, greater energy would be required to replace the mud in the vacuumed space. These factors consist of:

- *Higher pipe tripping rate:* The higher the pipe tripping velocity, the greater the surge pressure (for running in the hole) and the greater the swab pressure (for pulling out of the hole).

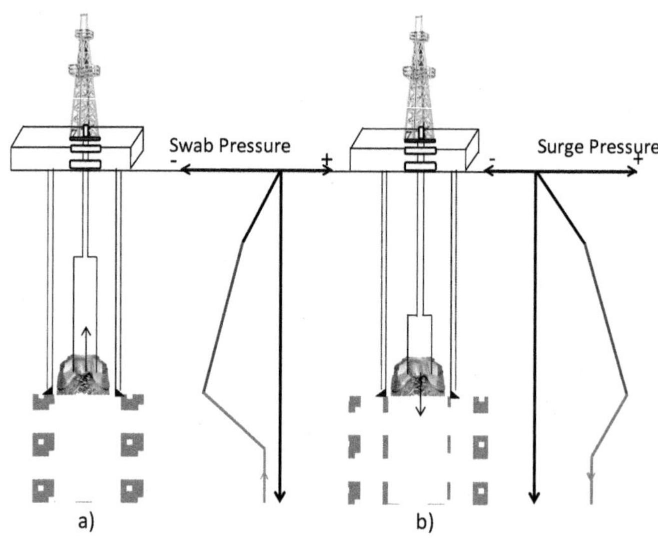

FIGURE 4.26 a) Swab pressure while pulling out of hole and b) surge pressure while running in hole.

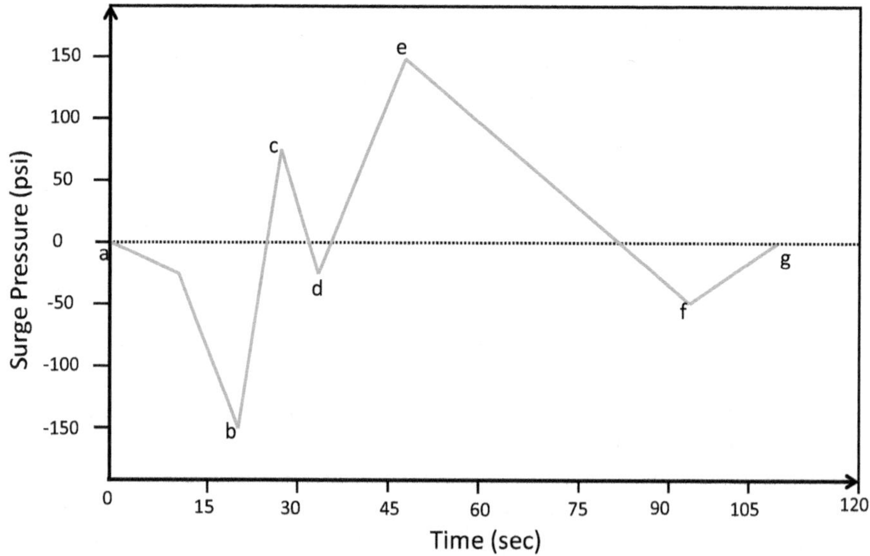

FIGURE 4.27 Variations of surge pressure while running a stand of pipe in the hole.

- *Lower clearance between pipe and wellbore:* The smaller the clearance between the pipe and wellbore (e.g., in slim-holes), the greater the surge and swab pressures.
- *Higher hole inclination angles and dog-legs:* While pulling out through deviated sections with high dog-leg-severities (DLS), the BHA is dragged against the top portion. Therefore, the BHA picks up the debris and causes reduction of clearance between the string and the wellbore.
- *Greater mud viscosity, gel strength, and mud weight:* An increase in either viscosity or gel strength or the density of the mud causes an increase in the surge and swab pressures. It is noted that viscosity is the most critical of all the impacting factors influencing the most.
- *Tight spots:* Tight spots are intervals where the wellbore has become smaller in clearance than expected. This can occur if the formation has swelled due to the presence of shales or unconsolidated nature of the formation or due to bit balling in marl-bearing formations or if proper hole cleaning or mud conditioning is not done during drilling. In these spots, the surge and swabs pressures would be excessively larger than other intervals.
- *Greater number of (or longer) stabilizers:* A packed-hole BHA is a BHA with several stabilizers having lower clearance between the BHA and the wellbore. Thus, such BHA designs cause greater surge and swab pressures than a pendulum BHA with only one stabilizer.
- *Balled-up bit or stabilizers:* Drilling in shale or marl formations can cause the bit and/or stabilizers to become balled up. Balled-up bits cause excessive swab pressures. If the bit is balled up and we pull out of the hole, it can cause swabbing, and a kick influx would enter the wellbore below the bit (swabbed kick).
- *Float valve in the drill-string:* Surge and swab pressures also increase if a float valve is installed in the bit sub just above the bit) or in the string. When running in the hole, the drilling fluid has three routes it can take: It can be displaced up the hole (annulus), move into a formation, or go through the bit nozzles and up the string. When there is a float valve in the bit sub, there would be no path through the bit nozzles and string, and thus surge and swab pressures would inevitably increase.

4.12 FORMATION STRENGTH, MASP, AND MAASP

4.12.1 Fracture Pressures/Gradients

An important information required for proper well planning and casing setting depth selection is the fracture pressures of the formations to be drilled. Fracture pressure is the pressure at which the formation is fractured or broken down and would then accept the whole mud from the wellbore. This pressure constitutes the upper limit of the safe mud weight window. Fracture initiates in the formation when the effective minimum principal stress (i.e., minimum principal stress minus pore pressure) at the wellbore wall (in the direction of maximum horizontal stress) reaches or exceeds formation rock tensile strength. It is noted that the profile of safe mud weight is required for proper selection of casing setting depths and for knowing the maximum allowable pressure tolerable at the surface in case of a kick situation.

In case the wellbore mud pressure becomes equal or greater than the formation fracture pressure, the formation would break down, causing induced fractures and mud losses. It leads to the loss of hydrostatic pressure and potentially loss of primary control. Fracture pressure is related to (but is lower than) the over-burden pressure. The magnitude of over-burden pressure gradient varies depending on the location. In onshore regions, it ranges from 1 to 1.3 psi/ft, with greater values observed in clastic rock formations. In offshore regions, e.g., Gulf of Mexico, the over-burden gradient may be lower than 1 psi/ft due to the effect of sea-water. Lower magnitudes of over-burden pressures are observed particularly at shallow drilling depths and in deepwater drilling. This is because very shallow onshore or offshore formations are usually unconsolidated. In deep offshore drilling (depth > 4000 m/13,124 ft), the effect of seawater interval is decreased, and the gradient approximately reaches 1 psi/ft. Figure 4.28 shows the over-burden pressure gradients in several offshore locations. The over-burden pressure gradient is one 1 psi/ft or equivalently 19.2 ppg.

The magnitude of the fracture pressure (P_{Frac}) is lower than the over-burden pressure, and it can range between, e.g., 0.65 and 0.85 psi/ft. It is usually estimated using magnitudes of the in-situ stresses and formation pore pressure (P_{pore}):

$$P_{Frac} = 3\sigma_h - \sigma_H - \alpha P_{pore} \tag{4.17}$$

where σ_h is the minimum horizontal stress, σ_H is the maximum horizontal stress, and α is the Poisson's ratio.

Thus, the fracture pressure gradient ($PG_{w,t}$) is (in field units):

$$PG_{Frac} = \frac{3\sigma_h - \sigma_H - \alpha P_{pore}}{(0.052 \times TVD)} \tag{4.18}$$

It is noted that the lost-mud pressure or (downhole) leak-off-pressure is lower than the fracture pressure and is equal to the minimum horizontal stress:

$$P_{Loss} = \sigma_h \tag{4.19}$$

The lost-mud pressure gradient is (in field units):

$$PG_{loss} = \frac{\sigma_h}{(0.052 \times TVD)} \tag{4.20}$$

The safe mud window is constructed between a lower limit and an upper limit. The lower limit is either the formation pore pressure gradient or the shear pressure gradient (whichever greater). The upper limit is constituted by the lost gradient. Previously, the fracture gradient was considered as the upper limit, but here we are more exact. An example of a critically narrow safe mud weight window is shown in

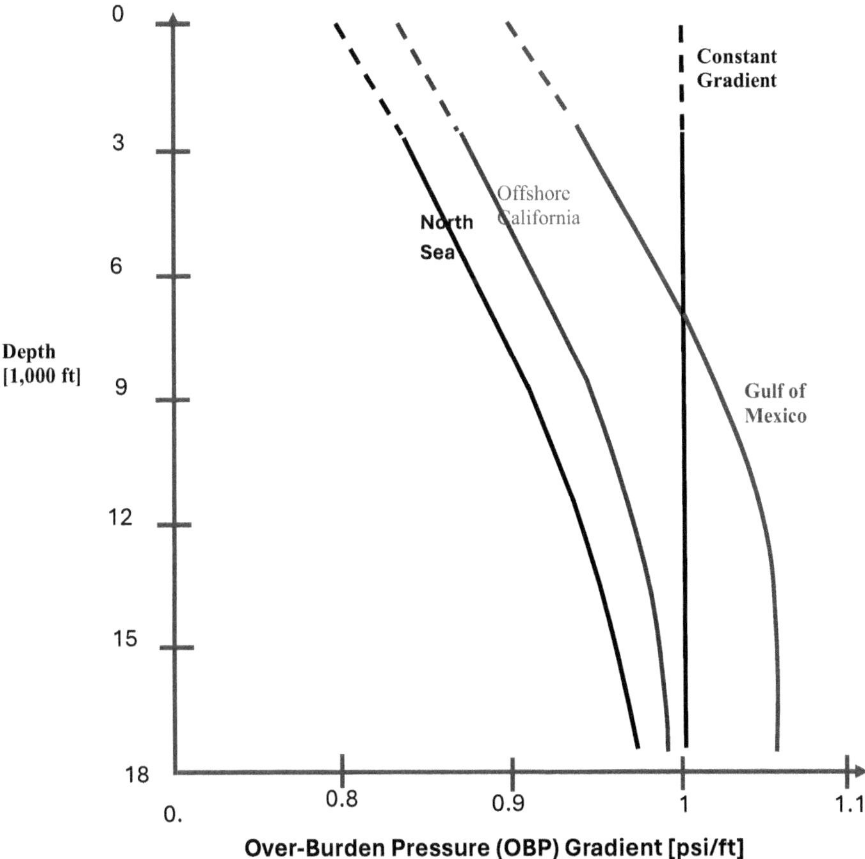

FIGURE 4.28 Over-Burden Pressure (OBP) gradients in several offshore locations. The OBP gradient may be considered constant for onshore locations.

Figure 4.29. Drilling in such a narrow mud window is challenging using a fixed mud weight as the driller would either experience loss or kick.

In offshore drilling operations, the fracture pressure gradient is lower than in onshore drilling, as was the case with over-burden pressure gradient. This is attributed to the effect of water. Figure 4.30 gives a typical example in which an offshore case and an onshore case are compared including the calculations. The 1,500-ft water depth causes the fracture gradient in offshore case to reduce from 0.8 psi/ft (onshore case) to 0.75 psi/ft.

4.12.2 Maximum Anticipated Surface Pressure (MASP)

Maximum Anticipated Surface Pressure (MASP) is an important design parameter for surface equipment in drilling and completion operations (particularly in case of encountering well control events). This parameter is defined as the maximum

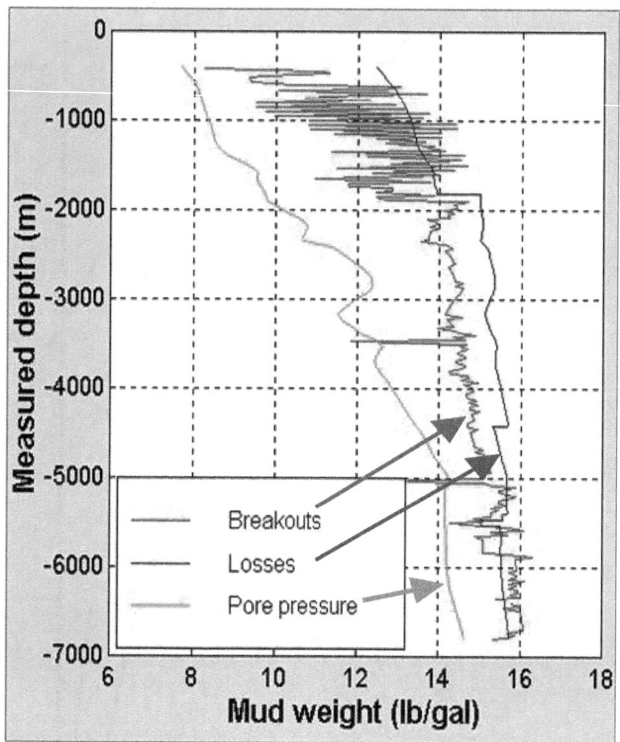

FIGURE 4.29 An example of critically narrow safe mud window in the North Sea. Drilling in such a formation would face the crew with many challenges using a fixed mud weight (https://slideplayer.com/slide/4589429/ (last accessed in January 11, 2020).

internal pressure that the BOPs and related equipment would be reasonably subjected to during well control operations. The pressure rating of the BOPs and equipment should be selected minimum 500 psi greater than MASP, which is also set so as a safety margin for possible bullheading.

MASP at the BOP must be calculated/estimated using a "worst-case" scenario assuming full evacuation of drilling mud to gas from the section total depth (TD) to the BOP/wellhead. The calculation process is explained in an example as follows. Imagine we plan to vertically drill a gas-bearing formation at the TD of 10,000 ft (MD/TVD) with the anticipated formation pore pressure of 5,500 psi. The previous casing shoe is at 7,000 ft, with the formation fracture gradient of 0.8 psi/ft (using Leak-Off Test, LOT results). Follow the following stages to find MASP (also shown in Figure 4.31):

1. As the "worst-case" scenario, assume that a gas influx enters the wellbore and rises until it reaches the surface. At this moment, our surface equipment is exposed to the surface pressure (e.g., we may proceed to closing

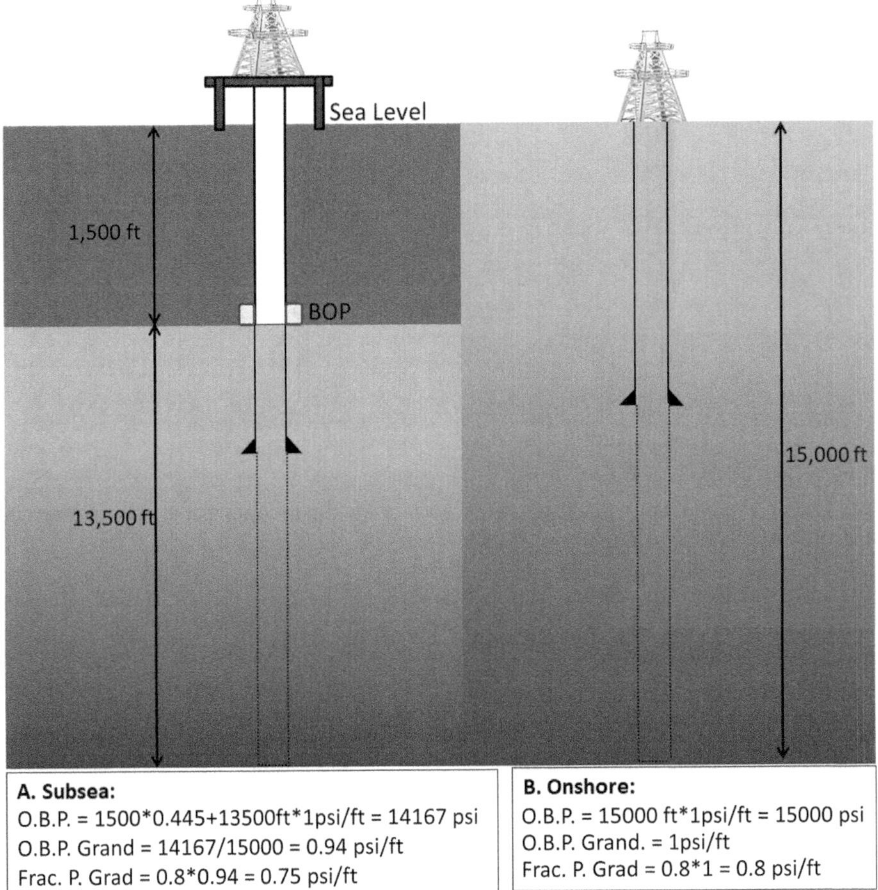

FIGURE 4.30 Comparison of an offshore (subsea) and onshore drilling operation on overburden and fracture pressure gradients. The subsea/offshore over-burden and fracture gradient are lower than in onshore ones.

BOPs). Consider the gas pressure gradient of 0.1 psi/ft. Thus, surface pressure would be:

$$\text{MASP}_1 = 5{,}500 - (10{,}000 \times 0.1) = 4{,}500\,\text{psi}$$

2. As another "worst-case" scenario, assume that a gas influx has already entered the wellbore, risen, and reached the casing shoe and caused formation fracture (at the shoe). After fracturing at the shoe, it has continued rising to the surface. As the formation has fractured, the basis criterion of surface pressure calculation would be the casing shoe (not the formation pressure). Therefore, MASP is found as:

$$\text{MASP}_2 = 7{,}000 \times 0.8 - (7{,}000 \times 0.1) = 4{,}900\,\text{psi}$$

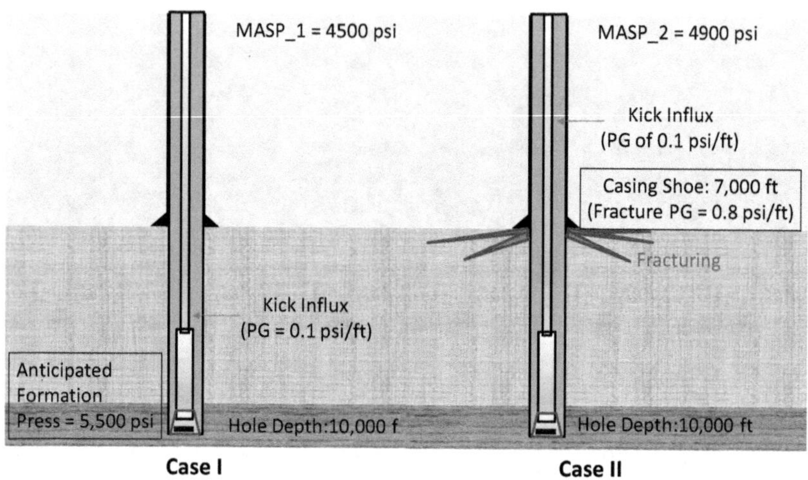

FIGURE 4.31 Stages of calculation/estimation of Maximum Anticipated Surface Pressure (MASP). MASP is the minimum of MASPs found in Cases 1 and 2, that is, 4,500 psi.

3. MASP is considered as the minimum of the above two calculated values. Therefore:

$$\text{MASP} = 4500\,\text{psi}$$

4.12.3 LOT and FIT

4.12.3.1 Leak-Off Test (LOT)

To determine a real prediction for actual fracture pressure of a formation, there are two tests that can be carried out: The Leak-Off Test (LOT) and the Formation Integrity Test (FIT) with the best being the more reliable one. As a prerequisite for meaningful results from these tests, it is important to have a uniform mud weight (in the annulus and drill-string) that is measured accurately, as well as accurate pressure data records. To have these measurements, the drilling mud should be essentially well-conditioned via bottoms-up. For accurate mud weight measurements, representative mud samples should be taken. In addition, pressure gauges should be already well calibrated and should have appropriate scales.

A Leak-Off Test (LOT) is a test conducted to determine the pressure at which a rock formation will take place in fluid or leaks off. "Leak off" means the formation is just about to be fractured (but not really fractured). The determined pressure gives a real lower estimated value for formation fracture pressure and indicates the limit of pressure that can be practically applied to the wellbore over the next section of hole to be drilled. Using this test, the strength of the formation at the casing shoe where it is assumed to be the weakest point of the formation is estimated, and the integrity of cement placed/set behind the casing is also checked (i.e., integrity of the cement/casing). However, in a long open-hole section, the original LOT results may not necessarily determine the weakest point in that section; therefore, in case a weaker formation has been penetrated, a further LOT should be conducted. In treating LOT results, some elements of errors should be considered including the effect of temperature on mud properties such as mud weight and rheological properties, as well as the effect of mud cake.

4.12.3.2 Procedure

A LOT is conducted to measure the surface annulus pressure that causes some minimal leak-off or penetration of the mud into the formation. This surface pressure is called Leak-Off Pressure (LOP). The formation fracture pressure is estimated as the summation of LOP and the hydrostatic mud pressure. To carry out this test, the following procedure is followed in practice (see the setup in Figure 4.32):

1. Drill 10–20 ft below the casing shoe to expose the new formation.
2. Circulate the mud bottoms-up to condition the mud and to make the mud weight uniform in the hole.
3. Shut in the well using the BOPs.
4. Pump into the well at very slow rate increments (e.g., 0.5 bbl/minute).
   Note: Cement pumps (pump trucks) are preferred for this purpose, instead of rig pumps, because of their smaller fluid displacement and their better control in pumping slow rates.
5. Measure/record the stabilized annulus/casing pressure at each pumped volume. The stabilized pressures can be found using the following procedure: The first measured volume is pumped into the well, until an increase in annulus/casing pressure is observed. At this point, stop pumping for say one minute, until the surface annulus pressure gets stabilized. Record this pressure. Resume pumping at the next increased volume and follow the same procedure until all the pressure points are plotted. Figure 4.33 shows the results of a field example LOT.
   Note: Annulus/casing pressures are recorded using very accurate pressure gauges connected to the casing/choke-line.
6. The formation pressure (indeed bottomhole Leak-Off Pressure) is found by adding the LOP to the hydrostatic mud pressure:

$$P_{Frac} = LOP + 0.052 \times MW \times TVD \tag{4.21}$$

4.12.3.3 Formation Integrity Test

In areas with known formation strengths (in brown fields), it may be required to not exceed a given pressure in a specified hole section due to downhole or surface facilities' pressure ratings or restrictions. Therefore, in lieu of a full LOT, a Formation Integrity Test (FIT) may be conducted. In this test, instead of pressuring the wellbore in a test to determine a "leak-off" pressure magnitude, when a predefined pressure is reached, the test is stopped. The selected pressure for FIT is an expected pore pressure in the formations that will be drilled in the next hole section below the casing shoe. Figure 4.34 shows the output pressure curve of a field example FIT for the 8.5-in hole (below 9 5/8-in casing) at the depth of 9,300 ft.

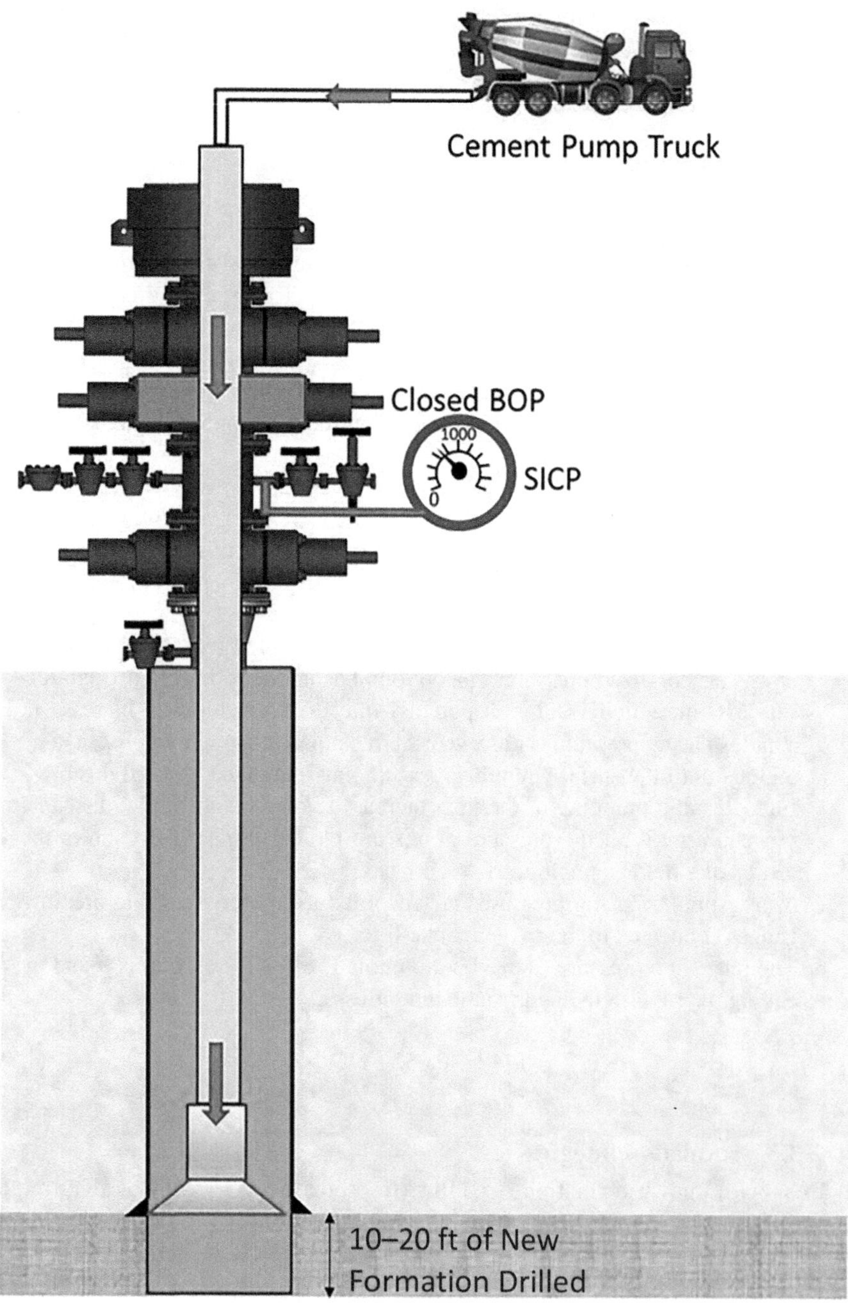

FIGURE 4.32 Schematic of the set up for performing the Leak-Off Test (LOT).

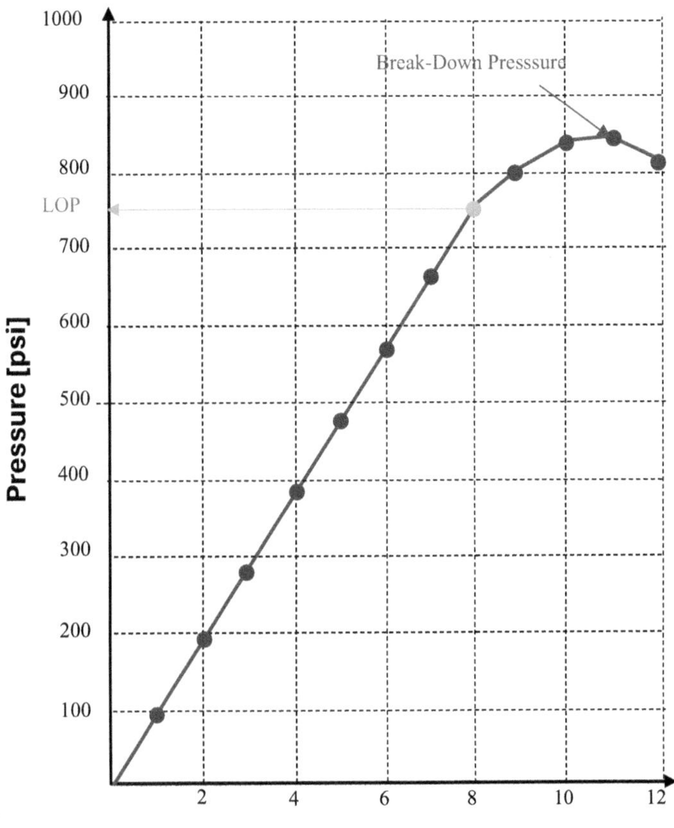

FIGURE 4.33 The results of a field example LOT for the hole section below 13 3/8-in casing at the depth of 3,300 ft. The surface Leak-Off Pressure (LOP) is identified as 750 psi. This is the point after which the curve deflection can be observed.

4.12.4 MAMW

Using the LOP, it is possible to calculate the maximum allowable shoe pressure. When this pressure is found, it is possible to calculate the equivalent drilling mud weight or the Maximum Allowable Mud Weight (MAMW). The MAMW is the maximum weight/density of the drilling mud that can be used for drilling a hole section without fracturing the formation. If a mud has a greater mud weight, the formation would be fractured/broken down. The magnitude of MAMW is based on LOT results and is found as:

$$\text{MAMW} = \text{MW}_{\text{LOT}} + \frac{\text{LOP}}{0.052 \times \text{TVD}_{\text{Casing}}} \tag{4.22}$$

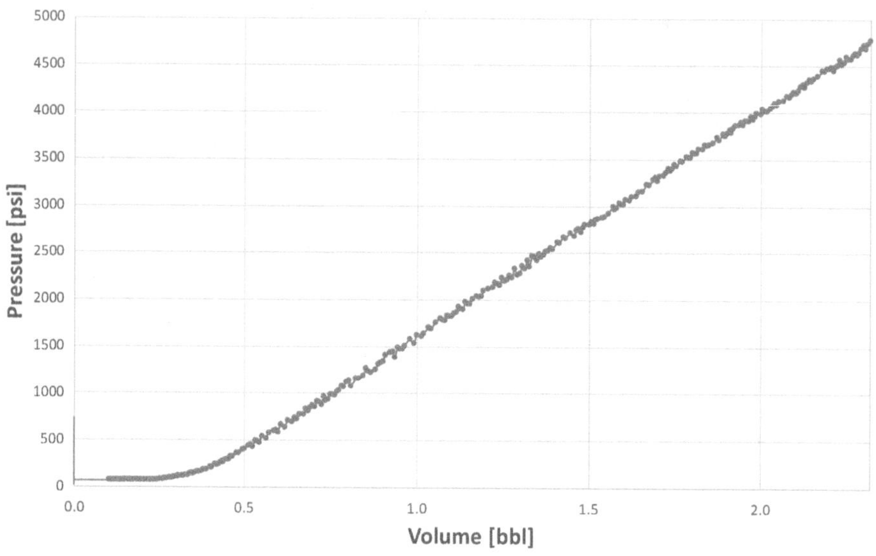

FIGURE 4.34 A field example of results of the Formation Integrity Test (FIT) in the 8.5-in hole (below 9 5/8-in casing) at the depth of 9,300 ft.

As a common practice, MAMW is round down to tenth or one decimal point (e.g., round down 13.991 ppg to 13.9 ppg). This is done for margin of safety to provide some overbalance.

4.12.5 MAASP

Maximum Allowable Annular Surface Pressure (MAASP) is the upper limit allowed for the surface pressure during drilling operation. If this pressure is exceeded, the formation will be fractured at its weakest point (which is the casing shoe in most cases). In other words, the addition of MAASP to the hydrostatic mud pressure at the casing shoe equals the formation fracture pressure (or better to say "Mud Lost Pressure" at which mud loses to the formation). It is noted that formation fracture pressure almost determines the formation strength.

As mentioned, the initial MAASP is calculated using LOT results. MAASP is calculated by subtracting the hydrostatic mud pressure in the casing from the maximum allowable pressure at the shoe. Therefore, the initial MAASP is found by:

$$\text{Initial MAASP} = 0.052 \times \left(\text{MAMW} - \text{MW}_{\text{Current}}\right) \times \text{TVD}_{\text{Casing}} \tag{4.23}$$

For safety, MAASP is round down to an integer (e.g., 874.56 psi to 874 psi).

As confirmed by Equation (4.23), with increasing hydrostatic mud pressure (mud weight) inside the casing, MAASP decreases. Therefore, if the well is displaced by a lighter mud, MAASP would increase.

An important task in well kill operations is to monitor the initial MAASP so that the crew do not allow surface pressures from reaching this pressure in all situations including during circulating out the kick influx. If the initial MAASP is approached before the kick is circulated into the casing, the responsible management must take the safe action. The safe action may be stopping the circulation and instead applying the volumetric method to prevent further increase in casing pressure while evacuating nonessential personnel out of the rig floor. Once the kick influx is inside the casing, there is a risk that the initial MAASP be exceeded as the influx is rising and expanding; to prevent that, the bottomhole pressure must be maintained constant.

During kick influx circulation, MAASP does not remain equal to the Initial MAASP. This occurs when the influx enters the casing and replaces the original mud. Therefore, there would be lower weight/density fluid in the casing, which means greater MAASP based on Equation (4.23). Table 4.4 shows changes in MAASP during influx circulation. It is noted that an increase in MAASP means more room and flexibility for the well control crew or lower possibility of formation fracture.

Example

The mud weight used for leak-off test is 9.5 ppg. The calculated Maximum Allowable Mud Weight (MAMW) is 14.3 ppg. The casing TVD is 3600 ft, and the total depth TVD is 4,300 ft.

If the current mud weight is 11 ppg, evaluate the initial MAASP.

Answer:

Using Equation (4.23), we have:

$$\text{Initial MAASP} = 0.052 \times (14.3 - 11) \times 3600 = 617.76 \rightarrow 617\,\text{psi}$$

4.12.6 Kick Tolerance Relation with MAASP and Kick Intensity

Kick tolerance is the maximum tolerable volume of (gas) kick influx (also called kick size) for which it is safely possible to shut in the well and circulate the influx out of the well, without exceeding the formation strength at the casing shoe.

To find the kick tolerance, the following two cases should be considered and evaluated. Next, compare and select the smaller value as the kick tolerance.

TABLE 4.4
Changes in MAASP during Gas Influx Circulation (Well Kill Operations)

When	MAASP Behavior
1. Gas influx is below the casing shoe.	Initial MAASP
2. Gas influx is entering the casing.	Increase from initial MAASP until gas reaches the choke
3. Gas is displaced out of the choke.	Decrease.
4. Kill fluid replaces the original mud.	Decrease

4.12.6.1 At Initial Shut-in Time

To evaluate the kick tolerance in this case, the criterion is the maximum tolerable height of influx in annulus without fracturing the formation at the shoe, $H_{inf,max}$. It is found by:

$$H_{inf,max}[ft] = \frac{(MAASP - 0.052 \times K.I. \times TVD)[psi]}{(PG_{mud} - PG_{influx})\left[\frac{psi}{ft}\right]} \quad (4.24)$$

where *K.I.* is the kick intensity [ppg], defined as the difference between pore pressure's equivalent mud weight (corresponding to the anticipated pore pressure of a possible kicking formation) and the original mud weight. In other words, the kick intensity expresses the magnitude of the anticipated underbalance in equivalent mud weight. It is defined as:

$$\text{Kick Intensity (K.I)} = \frac{\text{Anticipated}_{pore}}{0.05 \times TVD} - MW \quad (4.25)$$

where P_{pore} is the formation pore pressure.

Knowing the magnitude of $H_{inf,max}$, the kick tolerance is found:

$$\text{Kick Tolerance} = H_{inf,max}\,[ft] \times Cap_{Ann}\left[\frac{bbl}{ft}\right] \quad (4.26)$$

4.12.6.2 When the Gas Influx Is Circulated with Its Top Reaching the Casing Shoe

The same maximum influx height is used, but the kick influx may not be located across the drill-collars. This should be considered in calculation of kick tolerance in Equation (4.26). Next, since gas has risen in this case, and has reached the shoe, it is important to use Boyle's Law to convert gas volume at the shoe to the gas volume at the bottomhole. This conversion is done because the gas volume important for use as the kick tolerance is the initial volume at the bottomhole when kick initially occurs. See a solved example next.

4.12.7 Exercises

Exercise 1:

Using the following LOT data in Figure 4.35, estimate the formation fracture pressure.

Answer – 1,778 psi:

Using the given data in Figure 4.35, the LOP point is the deflection point from the straight line, that is, 900 psi. Therefore, the formation fracture pressure is estimated as:

$$P_{Frac} = LOP + 0.052 \times MW \times TVD = 900 + 0.052 \times 9.5 \times 3600 = 1778.4 \rightarrow 1778\,psi$$

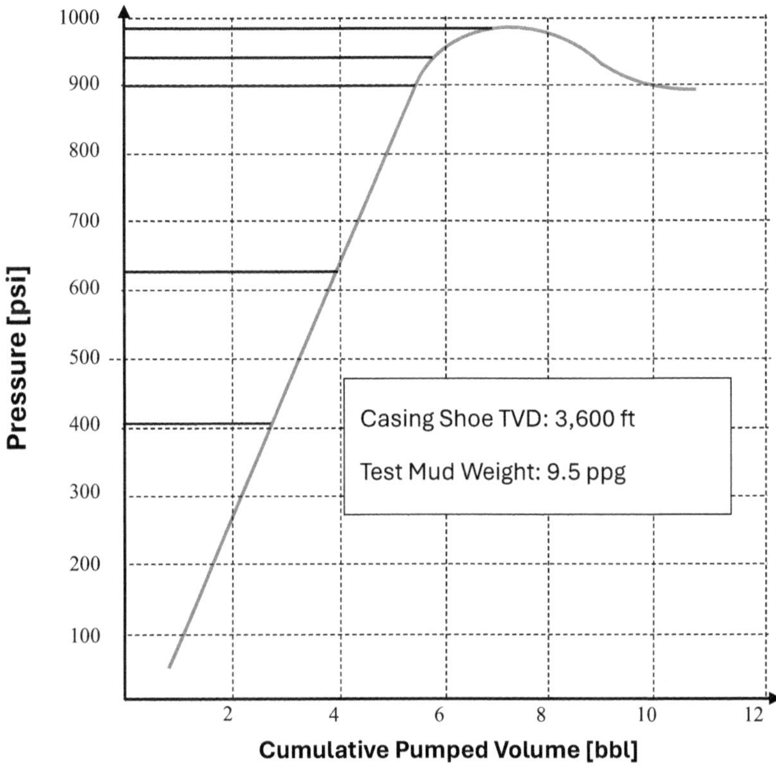

FIGURE 4.35 Leak-Off Test (LOT) data given for the calculation exercise.

The fracture pressure was rounded down to an integer (for safety).

Exercise 2:

At what pressure (at surface) does leak off begin to take place in the graph given in Figure 4.36?

a. 800 psi
b. 1,050 psi
c. 1,200 psi
d. 900 psi

Answer – d:

As shown in Figure 4.36, the deflection/deviation of pump pressure curve from the straight line occurs at about 900 psi. Therefore, this is considered as the surface leak-off pressure at which fluid leaks off to the formation.

Exercise 3:

What is meant by a kick tolerance of 25 barrels?

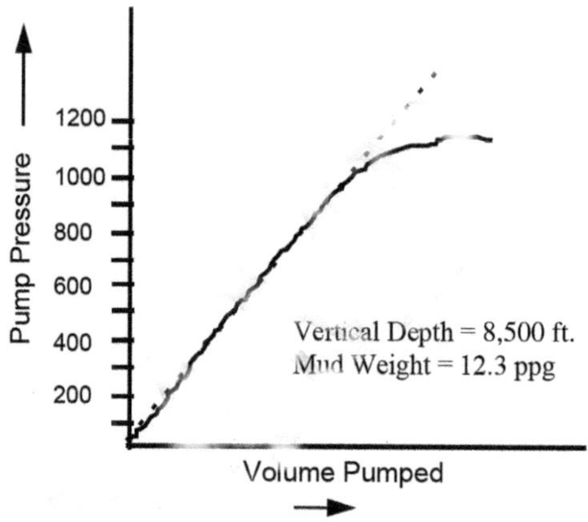

FIGURE 4.36 Leak-Off Test pressure result.

a. A gas kick of 25 bbl is the maximum that can be shut in and circulated out without possible lost circulation.
b. A saltwater kick of 25 bbl is the maximum that can be shut in and bull-headed without possible lost circulation.
c. A gas kick of 25 bbl is the maximum that can be circulated out without bursting the casing at surface.
d. A saltwater kick of 25 bbl is the maximum that can be circulated out without bursting the casing at surface.

Answer – a:

Kick tolerance is the maximum kick influx volume that if it enters the well, surface well pressure reaches MAASP. It can be expressed as either the maximum intensity or, more commonly, the maximum influx volume that can be taken into the wellbore and subsequently circulated out without fracturing/breaking down the formation. Reaching MAASP means formation breakdown at the shoe which means another problem, i.e., lost circulation may be followed.

Exercise 4:

Using the following data, evaluate the kick tolerance:

- Maximum Anticipated formation pressure (Equivalent Mud Weight) = 13 ppg
- Planned total depth (TD) mud weight = 12.5 ppg
- Casing shoe depth = 6,500 ft MD/5,000 ft TVD
- Leak-off test (Maximum Allowable Mud Weight) at casing shoe = 15 ppg
- Hole depth = 10,100 ft MD/10,000 ft TVD
- Bit size = 8.5 inch
- Drill-collar: OD = 6.5 inch, length = 1,000 ft, annular capacity = 0.1047 bbl /ft
- Drill-pipe: OD = 5 inch, annular capacity = 0.1215 bbl /ft
- Influx gradient (gas) = 0.1 psi/ft

Answer – 132.5 bbl:

1) First, MAASP is found:

$$\text{MAASP} = 0.052 \times (15 - 12.5) \times 10{,}000 = 1{,}300\,\text{psi}$$

2) Next, the kick intensity is found:

$$\text{Kick Intensity} = 13 - 12.5 = 0.5\,\text{ppg}$$

3) Using the above results, we can find the maximum allowable influx height in the annulus, $H_{inf,max}$:

$$H_{inf,max}[\text{ft}] = \frac{(\text{MAASP} - 0.052 \times \text{K.I.} \times \text{TVD})[\text{psi}]}{(PG_{mud} - PG_{influx})[\frac{\text{psi}}{\text{ft}}]}$$
$$= \frac{(1{,}300 - 0.052 \times 0.5 \times 10{,}000)}{(0.052 \times 12.5 - 0.1)} = 1{,}891\,\text{ft}$$

4a) *Initial Shut-In:* Knowing $H_{inf,max}$ and the length of the drill-collars (1,000 ft), 1,000 ft of the influx is in the annulus across the drill-collar and the rest (891 ft) would be located across the drill-pipes. Thus, the kick tolerance is:

$$\text{Kick Tolerance}_1 = 1{,}000 \times 0.1047 + 891 \times 0.1215 = 212.9\,\text{bbl}$$

4b) *When Top of Influx Reaches the Shoe:* The kick volume has reached the shoe; thus, all the influx height is across the drill-pipes:

$$\text{Tolerable Kick Volume at Shoe Pressure} = 1{,}891 \times 0.1215 = 229.7\,\text{bbl}$$

This volume should be converted to bottomhole conditions using Boyle's law ($P_1V_1 = P_2V_2$) as follows:

$$\text{Shoe Pressure}(\text{at Leak} - \text{Off} - \text{Test}) = 0.052 \times 15 \times 5{,}000 = 3{,}900\,\text{psi}$$

$$\text{Bottomhole Pressure}(\text{at time of kick}) = P_{pore} = 0.052 \times 13 \times 10{,}000 = 6{,}760\,\text{psi}$$

Therefore,

$$\text{Kick Tolerance}_2 = \frac{229.7 \times 3{,}900}{6{,}760} = 132.5\,\text{bbl}$$

Comparing the kick tolerances, Kick Tolerance$_2$ (132.5 bbl) is the smaller one and is thus selected as the selected kick tolerance. Therefore, if such a high magnitude of kick size (pit gain) is taken, the well would be in severe risk of formation fracture. This example shows the important of early kick detection to avoid uncontrollable well control events.

Exercise 5:

A rig crew shut in on a 2.5 ppg kick intensity with a 20 bbl gain. Using the kick tolerance window in Figure 4.37, can the crew successfully shut in and circulate the kick out of the well without fracturing the well's weak point?

Answer – No.

It is not possible to circulate such a kick safely because when it reaches the shoe, it will cause formation fracture there (see Figure 4.38).

Exercise 6:

The drilling program requires a formation integrity test (FIT) to 15.2 ppg Equivalent Mud Weight (EMW) at the casing shoe.

- Shoe depth = 4,000' (TVD)/ 5,500' (MD)
- Mud weight = 9.5 ppg

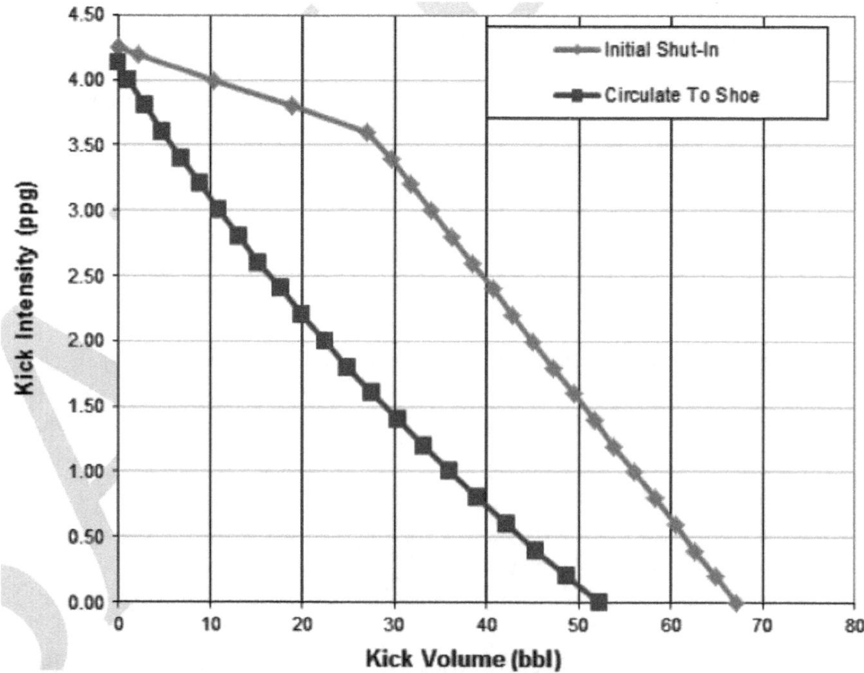

FIGURE 4.37 Kick tolerance curve.

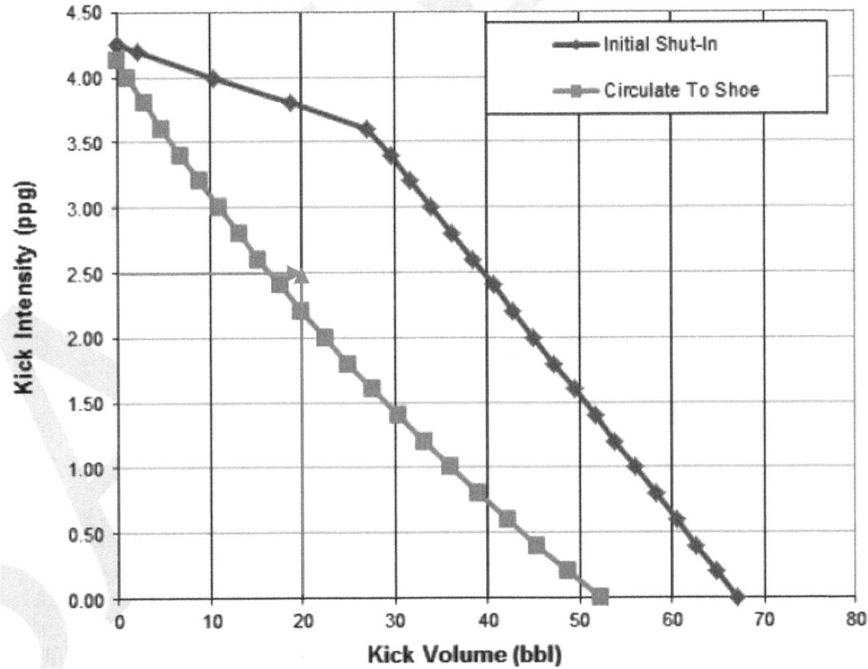

FIGURE 4.38 Kick tolerance curve (analyzed).

What surface pressure is required to test the shoe to 15.2 ppg EMW?

a. 1,976 psi
b. 1,186 psi
c. 1,630 psi
d. 2,382 psi

Answer – b:

The test pressure at the surface is found by subtracting the hydrostatic mud pressure from the formation integrity pressure:

$$\text{Test Pressure} = 0.052 \times 4000 \times (15.2 - 9.5) = 1186\,\text{psi}$$

Exercise 7:

During a trip out, it is calculated that 5 bbl has been swabbed into the well. A flow check is negative. It has been decided to continue with the trip. If the kick is gas what may happen in the well?

a. Gas may migrate and expand causing reduction in hydrostatic pressure.
b. Gas will stay in position and be pushed back into the formation when you run back in the hole.

c. Gas will migrate but not expand so hydrostatic pressure will stay constant.
d. Gas will migrate and expand causing hydrostatic pressure to increase.

Answer – a:

Any gas influx in the wellbore would migrate due to its lower weight/density, and because of lower pressure at shallower depths, it would expand as well.

Exercise 8:

Using the LOT data in Figure 4.35, evaluate MAMW:

Answer:

MAMW is found as:

$$MAMW = 9.5 + \frac{900}{0.052 \times 3600} = 14.307 \rightarrow 14.3\,\text{ppg}$$

4.13 BALLOONING

Wellbore ballooning, also called *wellbore breathing* or *supercharging*, is a term used to describe the small volumetric change in the active drilling fluid system during drilling operations, which shows itself as the reversible process of active drilling fluid volume gain and loss. Ballooning can be considered a drilling problem since it can cause some confusion and waste of rig time as non-productive time (NPT). It is very crucial for drilling crew and decision-makers to understand the major mechanisms and factors contributing to the ballooning phenomenon to avoid confusion with conventional losses and kick flows. Otherwise, NPT cannot be avoided.

Usually in ballooning, mud loss occurs into a fractured formation with pumps on because dynamic/flowing bottomhole pressure (due to equivalent circulating density) exceeds the Fracture Closure Pressure (FCP) causing the fractures to open. Conversely, when the pumps get off (e.g., during making connections), fractures close and thus push the previously entered mud in the formation back to the wellbore. The phenomenon of wellbore ballooning does not occur only in fractured formations, and there are other reasons. Generally, there are five mechanisms for ballooning (shown in Figure 4.39):

1. The opening and closing of natural fractures intersected during drilling,
2. The opening and closing of induced fractures at the near wellbore region,
3. Thermal expansion and contraction of the drilling mud,
4. Compressibility of the drilling mud, and
5. Elastic deformation of the borehole (open hole) and the cased hole.

Following a flow-back, the first action to take (even when ballooning is suspected) is to treat as if a kick has occurred and perform a flow check (explained in Section 5.5.1) and then probably proceed to well shut-in. Next, using the symptoms, the crew

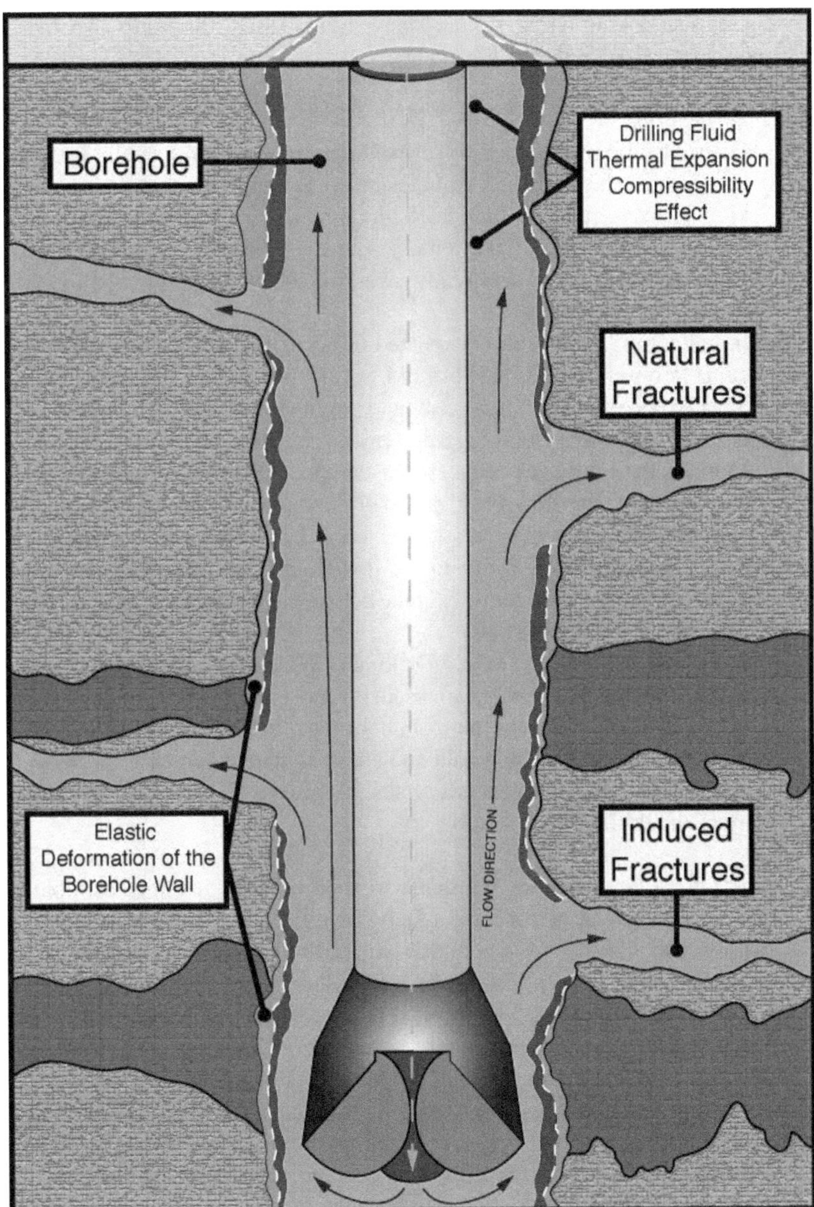

FIGURE 4.39 The main causes of wellbore ballooning (Elmgerbi et al., 2016).

should distinguish or discriminate it from a real kick flow. Finally, following the distinguish, they should handle it properly (which will be discussed in Section 9.3). As already mentioned, ballooning causes some time to be wasted (as NPT) for performing a flow check and well shut-in and distinguishing from a kick and its possible handling using the symptoms.

To distinguish a ballooning from a conventional kick, there are two methods which should be considered simultaneously:

- *Fingerprinting:*
    - Gather ballooning-related offset well data including the geological layers, depths, the observed volumes/rates of mud losses and subsequent kick influxes, and how long the ballooning occurrences lasted until flow stopped (e.g., 2, 5, or 10 minutes). Therefore, during drilling, it is possible for the crew to be already aware of where and how a ballooning occurrence may appear.
    - In ballooning, mud loss is expected to occur first before flow-back (pit gain) is observed. If mud loss did not happen before the flow, this is, most likely, not ballooning. Use mud engineers' observations and measurements as well as mud logging sensors' readings.
    - Monitor the flow-back rates (when pump is off) versus time. In a ballooning well, the flow-back rate will decrease over time. However, in a kicking well, it will increase over time. In a ballooning situation, the more the mud loss is (prior to the flow-back), the faster the flow-back will occur and the longer it will take before the well reverts back to static (i.e., flow completely stops).
    - In the case of Logging While Drilling (LWD) where the resistivity and gamma ray logging sensors are used, it is possible to distinguish between fractured formations (prone to ballooning). When drilling in the fractures, the resistivity log would show a shift. The magnitude of the shift on an electric log plot depends on the fracture width and relative resistivity of the drilling mud.

- *Step-Down Pressure Test:* A reliable method of distinguishing ballooning from a kick flow is performing a step-down pressure test which includes opening and closing of the choke alternatively. With ballooning, it is expected that the Shut-In Casing Pressure will decrease after each shut-down step, whereas with a real kick flow, it is expected that the shut-in pressure increases. The reason for the pressure decline in ballooning is that with each opening of the choke, the fracture width and closure stress decline, when ballooning occurred due to existence of formation fractures. If ballooning has occurred due to other reasons (such as thermal expansion/contraction or compressibility of the drilling mud), again, the shut-in pressure would decline.

4.14 GAS BEHAVIOR

Gas has two significant behavioral characteristics, i.e., gas expansion (following the gas law) and gas migration. In addition to that, this section covers different gas solubilities and behaviors in water-based muds (WBMs) and oil-based muds (OBMs) with some considerations on hydrogen sulfide.

Gas Law: The Ideal-Gas Law is the result of combination of Boyle's, Charles, and Avogadro's Laws:

$$PV = nRT \tag{4.27}$$

where P is the gas pressure [Pa or psi]; V is the volume of gas [m^3 or ft^3]; n is the number of moles of gas; R is the universal gas constant [$8.315\frac{Joule}{K.mole}$ or $10.73\frac{psi.ft^3}{R.lbmole}$]; and T is the absolute temperature [[K or *R*].

For real gases, Equation (4.27) is modified as follows:

$$PV = ZnRT \tag{4.28}$$

where Z is the gas compressibility factor (or simply Z-factor).

Knowing "nR" is constant for a specified mass of gas, using Equation (4.28), it is possible to compare gas properties in two different conditions:

$$\frac{P_1V_1}{Z_1T_1} = \frac{P_2V_2}{Z_2T_2} \tag{4.29}$$

4.14.1 Gas Expansion

When a gas kick influx enters a wellbore, it rises due to its lower weight/density than the mud weight. As it rises and assuming the well (annulus) is still open to atmosphere, its pressure drops. Due to pressure drop, it can freely expand (volume increase).

A typical example is given here for better understanding. Imagine 1 barrel of gas kick influx enters the wellbore with the TVD of 15,000 ft and mud weight of 8.72 ppg. The influx would continue rising. Equation (4.29) is used to find the gas volume at the middle depth and the surface. It can be simply assumed that gas is ideal, and thus its Z-factor is 1. The bottomhole condition is considered as the first point (1), and either the middle depth or the surface is considered as the second point (2). A required parameter for finding the gas volume is the Z-factor (Z_2) which is found using Figure 4.40. When the gas reaches the middle of the well (i.e., 7,500 ft), it has only expanded to 1.5 bbl. However, when it reaches the surface, its volume has dramatically expanded to 320 bbl.

Table 4.5 shows the data of this example and the gas expansion volumes.

Similarly, Figure 4.41 shows the data and gas expansion volumes at the middle depth and the surface schematically. It also shows how the bottomhole pressures (BHPs) change as gas expands. With the 5-in drill-pipe (DP) in the 9 5/8" casing, the annular capacity is 0.0487 bbl /ft. Therefore, at the middle depth, due to the increase of 0.5 bbl gas volume in the wellbore, the head loss of the mud is "$0.5/0.0487 = 10.3$ ft" (replaced by gas). Therefore, the BHP of the mid depth is "$8300 - 10.3\ (0.0528.72 - 0.1) = 8296$ psi". At the surface, due to the added 319 bbl gas volume in the wellbore, the head loss of the mud is "$319/0.0487 = 6550$ ft"

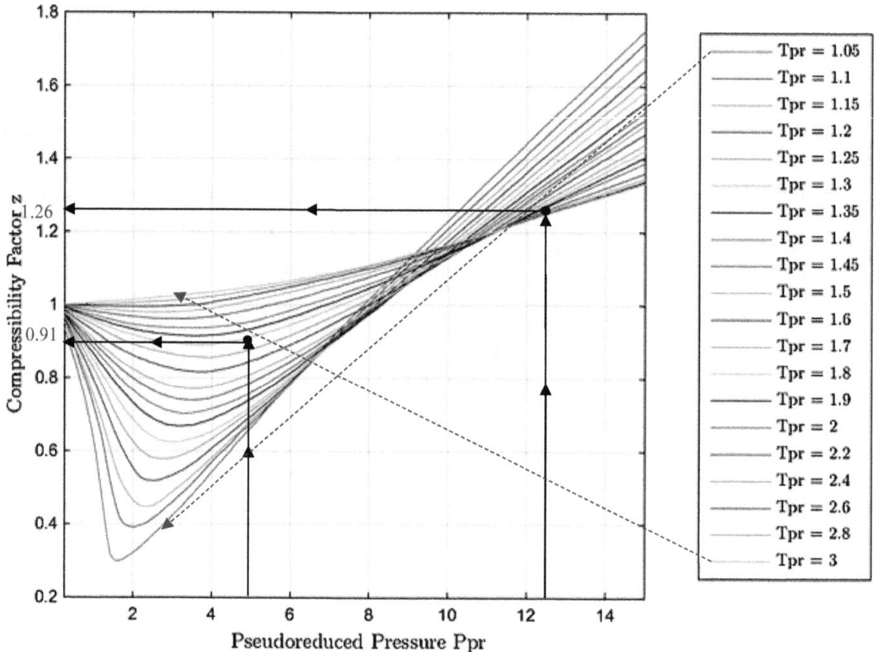

FIGURE 4.40 Z-factor estimation for a gas kick influx at bottomhole conditions and half well depth. Assuming methane as the kick influx, the critical pressure and temperature are respectively 673 psi and −116 °F (574 °R). At the bottomhole conditions, the reduced pressure and temperature are respectively 12.3 and 2.17. Therefore, the Z-Factor at bottomhole conditions is 1.26. At the middle depth conditions, the reduced pressure and temperature are respectively 5 and 1.88. Therefore, the Z-Factor at the middle depth conditions is 0.91.

TABLE 4.5

Required Gas Expansion Parameters (in Equation 4.29) for Finding Gas Expansion Volumes; Considering the Geothermal Gradient of "1°F/70 ft" and the Surface Temperature of 80°F, the Temperatures at the Middle Depth and Bottomhole Are Found

Parameter		Bottomhole	Middle	Surface
TVD [ft]:		15,000	7,500	Zero
(Gas Influx) Pressure [psi]:		8,300	3,400	14.7
Temperature [°F/°R]:		194 °F/754 °R	187 °F/647 °R	80 °F/540 °R
Z-Factor:	$P_r = P/P_{pc}$	12.3	5	–
	$T_r = T/T_{pc}$	2.17	1.88	–
	Z	1.26 (Figure 4.40)	Z = 0.91 (Figure 4.40)	Z = 1 (Ideal Gas)
Volume of Kick [bbl]		1	1.5	320

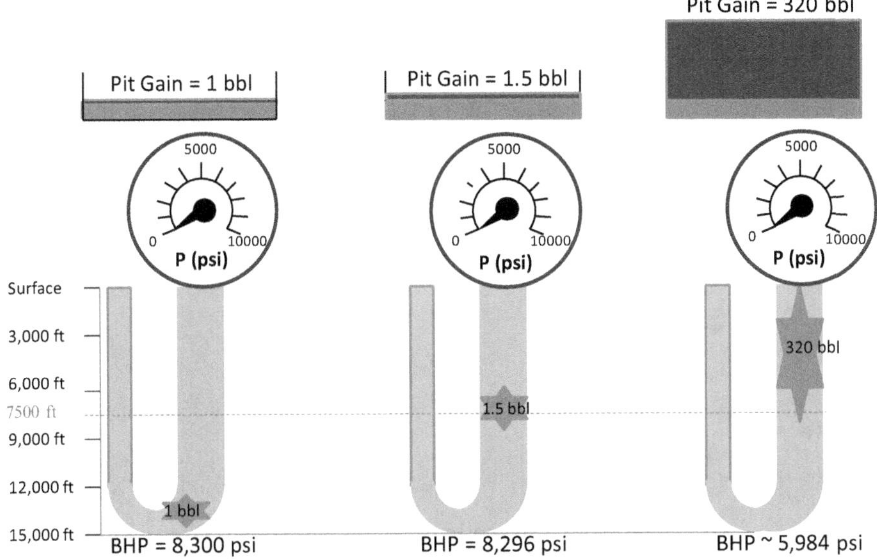

FIGURE 4.41 Gas expansion volumes (pit gains) and Bottomhole Pressures (BHPs) as the kick influx rises freely from the bottomhole to the surface. It is to be noted that the well/annulus is open (i.e., not shut-in by the BOPs).

(replaced by gas). The BHP is "8300 – 6550 (0.0528.72 – 0.1) = 5984 psi". It is noted that this modeling is a simple one and has some limitations for a fully representative simulation of the actual case.

4.14.2 Gas Migration

Following well shut-in, the gas influx continues rising to the surface because of its lower weight/density than the mud weight. The difference is that the gas influx is not allowed to expand as the wellbore volume is restricted. However, the influx would maintain the formation pressure within itself until it reaches the surface (Figure 4.42). Therefore, when the influx reaches the surface, the BHP is equal to the formation pressure plus the hydrostatic mud pressure, which is probably extremely high for the formation to withstand before it can be fractured. Consequently, it is important not to waste time and start killing the well as quickly as possible after well shut-in.

Gas flow also occurs in open wells (well not shut-in) based on gas expansion; however, in that case, the pressure of gas influx drops as it rises and thus it expands. Gas flow in open wells is not usually called "gas migration" though it may be called so.

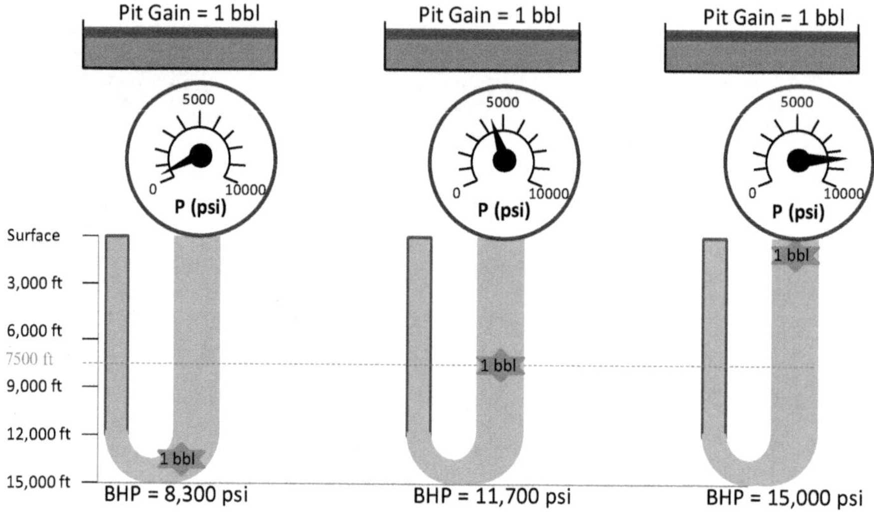

FIGURE 4.42 Casing/annular pressure and pit gain volume with kick influx migrating up from the bottomhole to the surface (with the well already shut-in). When the kick reaches the surface, the BHP would be 15,000 psi (provided that the formation does not already get fractured at the shoe). As the well is shut-in, the influx volume remains unchanged during its migration.

4.14.3 Gas Behavior in Water and Oil-Based Muds

Early detection of gas kicks is significantly important for well control purposes. The behavior of hydrocarbon gases in water-based muds (WBMs) is different from those in oil-based muds (OBMs). It is important to be aware of this difference.

Hydrocarbon gases have much greater solubility in diesel or oil. The amount of methane gas soluble in diesel is approximately 100 times greater than the amount of gas soluble in water at the same temperature (Thomas et al., 1984). Figure 4.43 shows the solubility of different gases in diesel.

Because of the above fact about gas solubility in oil-based muds (OBMs), the observable parameters at the surface (pit volume and return flow rate) are smaller in the beginning and do not change as rapidly as they do in WBMs. In OBMs, gas influx entering the wellbore dissolves in OBMs and limits the initial increase in pit volume (pit gain) and increase in return flow at the surface (initially). Later, assuming the kick is not detected, the "OBM with its dissolved gas" is pumped freely in the wellbore and its pressure drops, and gas expansion cannot occur as much as if it were not dissolved (Figure 4.44). However, in the last 1,000 ft, gas expansion occurs suddenly. On the other hand, in WBMs, the initial increase in pit volume (at the surface) is equal to the real-gas influx volume that entered the wellbore. Later, assuming the kick is not detected, as the mud together with the influx is pumped freely up the wellbore, gas expansion occurs due to pressure drop (Figure 4.44). Summarizing these facts, kick detection is more challenging in OBMs than in WBMs.

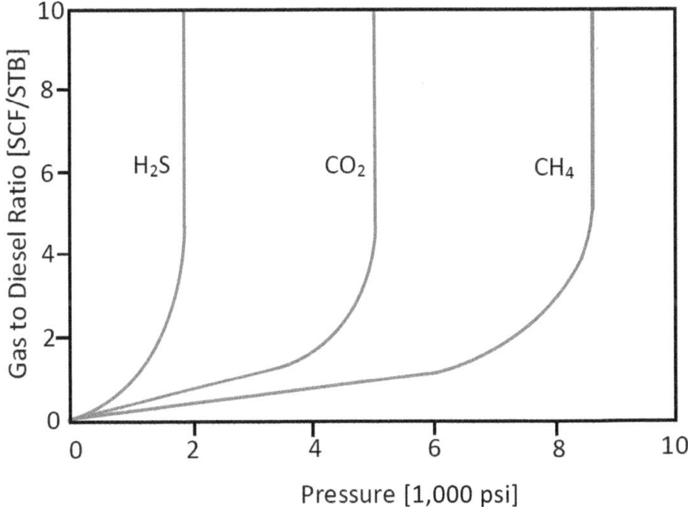

FIGURE 4.43 Comparison of solubility of methane, CO_2 and H_2S in diesel versus pressure (at 250 °F).

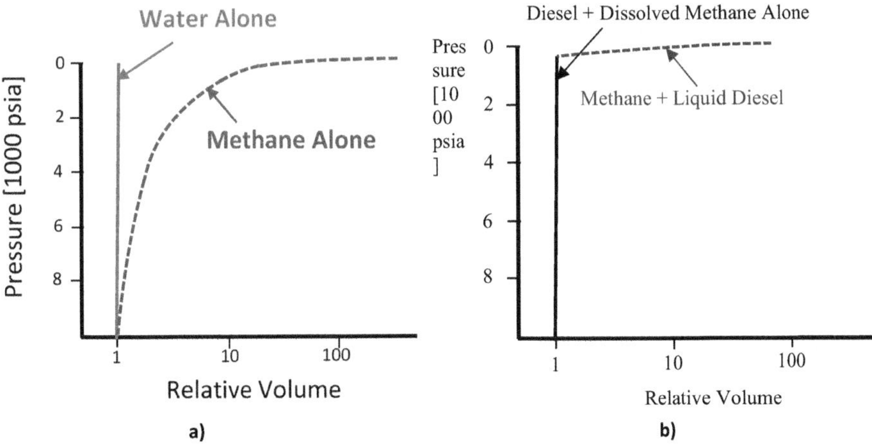

FIGURE 4.44 Expansion of a gas influx in a) Water-Based Muds (WBMs) and b) Oil-Based Mud (OBMs) due to pressure drop (Thomas et al., 1984).

4.14.4 Hydrogen Sulfide

H_2S is a highly poisonous gas that should be avoided or managed very carefully; otherwise, it can be highly fatal. Proper planning can contribute to the accurate prediction of H_2S and proper measures to avoid corresponding incidents. Proper selection of drilling/well control equipment must be essentially performed in advance. Otherwise, improperly selected drill-pipes may become brittle and undergo fatigue (losing barrier), or/and the BOPs may not be able to resist the effects of H_2S. In case of H_2S possibility, special training must be already provided to the personnel, and appropriate breathing equipment must be available at the rig site.

In case of encountering a suspected influx with H_2S, the preferred method of well kill is bullheading the influx fluid back to the formation rather than circulating it to the surface. This is because circulating the H_2S-bearing influx cannot be safely handled by the rig equipment and crew.

4.14.5 Exercises

Exercise 1:

A well is shut in, and a gas kick migrates 1,000 feet with no expansion in the annulus.

Well information:

- Gas volume = 20 bbl
- Total depth = 6,000 MD and TVD
- Casing shoe = 2,200′ MD and TVD
- Open hole = 8.5″
- Mud weight = 8.33 ppg

What is the increase in bottomhole pressure due to gas migration? Assume no losses to the formation.

a. 433 psi
b. 952 psi
c. 2,165 psi
d. 2,598 psi

Answer – a:

When gas migrates with time (with the BOPs closed), bottomhole pressure (BHP) increases, with the increase being equal to the displaced mud hydrostatic pressure:

$$\Delta \text{BHP} = 0.052 \times 1000 \times 8.33 = 433 \text{psi}$$

Exercise 2:

A gas kick is being circulated out using the Driller's method.

What will happen to the bottomhole pressure if the gas bubble is not allowed to expand as it is circulated up the hole?

a. It will increase.
b. Stay the same.
c. It will decrease.

Answer – a:

If gas bubbles do not expand during its rising up, bottomhole pressure would increase. It is, however, noted that such a thing (no expansion) during circulation by Driller's method would never happen. Therefore, it is not a good question.

Exercise 3:

In an open well, what situation could cause the following problems?

- Pit gain
- Reduced hydrostatics
- Riser/casing unloading
- Decrease in bottomhole pressure

a. Ballooning
b. Gas migration
c. High equivalent circulating density
d. Partial lost circulation

Answer – b:

Partial lost circulation would cause mud loss (opposite of pit gain). High equivalent circulating density (ECD) may cause leak off or fracturing of the formation leading to leak-off or loss. Therefore, choices c and d are wrong. Both "ballooning" and "gas migration" cause pit gain at the surface; however, ballooning would not most likely cause reduced hydrostatics or decrease in bottomhole pressure as it occurs maybe due to closure of fractures when pumps get off. It is also noted that gas migration in an open well (no closed BOPs) may cause casing unloading (onshore and offshore) and riser unloading (deepwater offshore). Therefore, choice b is correct.

Exercise 4:

A kick has been taken at TD and will be circulated out using the Drillers Method. You have one active pit (10 feet deep) with 180 bbl capacity.

Kick size = 10 bbl
TD/TVD = 4,800 feet
Mud weight in well = 9.9 ppg
SIDPP = 250 psi
SICP = 350 psi

Maximum predicted surface casing pressure during kill = 500 psi

What is the maximum mud level allowed in the pit before starting the circulation?

a. 5.5 foot deep
b. 6.5 foot deep

c. 7.5 foot deep
d. 8.5 foot deep

Answer – c:

This is a tricky but nice question. It appears that it is related to Driller's method, but it is related to gas law. When the gas influx is downhole, its volume is 10 bbl. However, using ideal-gas law ignoring temperature changes, the expanded volume at the surface can be found.

First, the downhole reservoir pressure and volume are (P_2 and V_2) found as:

$$P_2 = P_{hyd} + SIDPP = 0.052 \times 9.9 \times 4800 + 250 = 2721\,psi$$

$$P_1V_1 = P_2V_2 \rightarrow 2721 \times 10 = 500 \times V_2 \rightarrow V_2 = 54.42\,bbl$$

Therefore, the additional pit gain which will occur when all the gas reaches the surface is:

$$\Delta V = V_2 - V_1 = 54.42 - 10 = 44.42\,bbl$$

Each foot height of the pit tank is equivalent to several barrels. Or the height capacity of the mud pit is:

$$\text{Height Cap}_{tank} = \frac{180\,bbl}{10\,ft} = 18\frac{bbl}{ft}$$

Therefore, the mud pit level would rise as much as:

$$\Delta h = \frac{\Delta V}{\text{Height Cap}_{tank}} = \frac{44.42}{18} = 2.46 \sim 2.5\,ft$$

Thus, the maximum height of the tank before gas circulation should be:

$$L_{max,tank} = 10 - 2.5 = 7.5\,ft$$

4.15 WELL CONTROL IN HIGH-ANGLE WELLS

Well control in high-angle wells does not usually occur due to abnormal pressure situations (as high-angle or horizontal wells are drilled in brown fields (development wells). However, in such wells, kicks can occur due to swabbing or trip gas issues. Therefore, it is important to pump out when spacing out (to place the tool joint of the drill-pipe out of rotary table) during connections or tripping, as well as when performing flow checks wherever required, particularly at the casing shoe and half-way in the horizontal section (Devereux, 1998). Generally, well control in high-angle wells has several differences and complexities compared with vertical wells. These differences are in the shut-in and well killing phases.

Detection of gas kicks in highly deviated or horizontal wells is more challenging than in vertical wells. In horizontal wells, in static conditions (i.e., no mud circulation), an entered gas influx in the wellbore will not expand until it can reach the deviated section of the hole (Figure 4.45). As already mentioned, most gas influxes in high-angle and horizontal wellbores are swabbed gases. Therefore, after pulling out of the hole, the crew would run the drill-string (probably with a new bottomhole assembly, BHA) back in the hole. Then, when they pump the mud (mud circulation), the gas influx is displaced to the deviated section where it can expand due to lower pressure. At this point, the kick influx manifests itself as pit gain and can thus be detected by the rig crew.

There are different observations in shut-in pressures in deviated versus vertical wells. In vertical wells, following the well shut-in, the surface Shut-In Casing Pressure (SICP) is greater than the shut-in drill-pipe pressure (SIDPP), with the magnitude of this difference depending on the volume of the kick influx. In high-angle wells, following the well shut-in, there is a much smaller difference between the shut-in pressures than is in vertical wells. In horizontal wells, as long as the influx is in the horizontal section, shut-in pressures are equal (no difference). However, when the kick influx is circulated into the deviated/build section of the annulus, the gas influx starts expanding which causes the difference between the drill-pipe and casing pressures to increase significantly. Likewise, when the gas reaches the vertical hole section of such wells, a larger gas expansion and thus a more significant difference between shut-in pressures would occur.

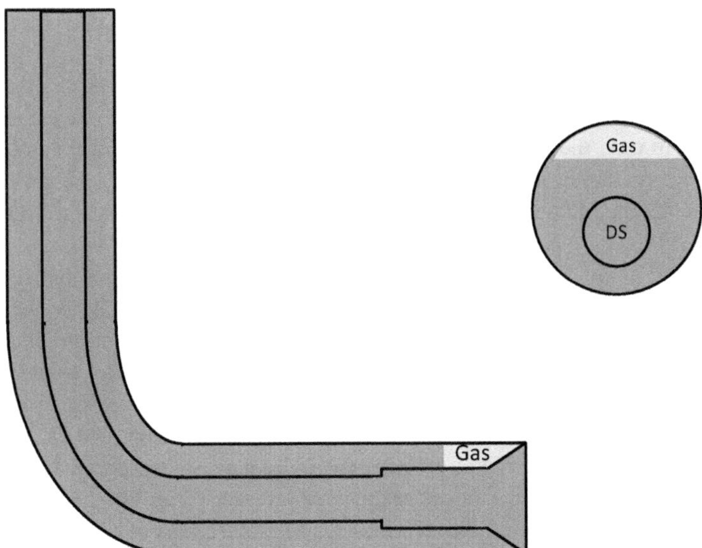

FIGURE 4.45 When a gas influx is taken in a horizontal section, it remains in the high side in static conditions (well schematic view and cross-sectional view). When mud circulation is resumed and the gas influx reaches the deviated section where it can expand, detection would be possible.

In terms of well killing, normal well control principles apply to high-angled deviated wells which are maintaining a constant bottomhole pressure during a balanced kill method. Generally, the calculations associated with well kill, which include completing kill sheets for such wells, are more complex than those of vertical wells and require further time. The complexity pertains to the calculation corresponding to the second circulation of the Driller's method or Wait and Weight method (see Section 11.4). If well conditions are favorable for Driller's method (e.g., the open-hole section is not so long), the Driller's method is the preferred well control method for high-angled wells because of its simplicity; thus, the possibility of making mistakes and becoming underbalanced is lower.

Sometimes, well kill must essentially be conducted using the Wait and Weight method, e.g., due to long open-hole section or low Maximum Allowable Annular Surface Pressure (MAASP) tolerable at the surface. As the heavy kill mud is pumped through the drill-string to reach the bit nozzles, the drill-pipe pressure drops. The rate of drill-pipe pressure drop in deviated wells is not the same as in vertical wells. In the conventional vertical well control calculations, the dropping rate of the Initial Circulating Pressure (ICP) of the drill-pipe (after circulating out the kick and just before pumping the heavy mud) is constant, that is, it drops proportionally to the measured depth over total depth, until it reaches the Final Circulating Pressure (FCP) when kill muds reach the bit nozzles. Therefore, the trend of ICP drop from the surface to the Kick-Off-Point (KOP) differs from that of the KOP to the End-Of-Build (EOB) section and from that of EOB to the total depth (TD). Therefore, these trends must be calculated separately. In the horizontal section of the well, the pressure drop of the drill-pipe is expected to be minimally negative (less than a few psi depending on the horizontal section length, which is due to the pressure loss in the horizontal section). Chapter 10 describes these calculations in detail.

If conventional calculations for vertical wells are used interchangeably for deviated wells, the BHP would become greater than required, which can cause induced fracturing in case the formation strength is not high enough. Figure 4.46 compares the two conventional and deviated-well calculation methods in a well of 2,000 ft measured depth (MD)/1,000 m True Vertical Depth (TVD) with the kick of point (KOP) of 850 ft.

Unwanted angle deviations and undulations in the trajectory and path of the horizontal section may occur, which may allow the accumulation of some gas influx in the pockets. However, if the hole angle exceeds 90 degrees, the possibility for gas accumulation is greater than in the horizontal. Indeed, some gas would definitely remain in the pockets (created due to undulation in the wellbore path) in the high side. It is difficult or impossible to circulate out such gas (Figure 4.47). That is the reason why it is recommended to keep the hole angle 1–2 degrees lower than 90 degrees. In case of a well control event in holes exceeding 90° inclination angle, bullheading is usually preferable (if working pressure ratings of the surface facilities allow the application of high-surface pressures).

It is noted that attempts to circulate gas influx out of the horizontal section with normal Kill Rates (Slow Circulating Rates) might not lead to complete success due to the flow being laminar, and the high weight kill mud tending to flow more along

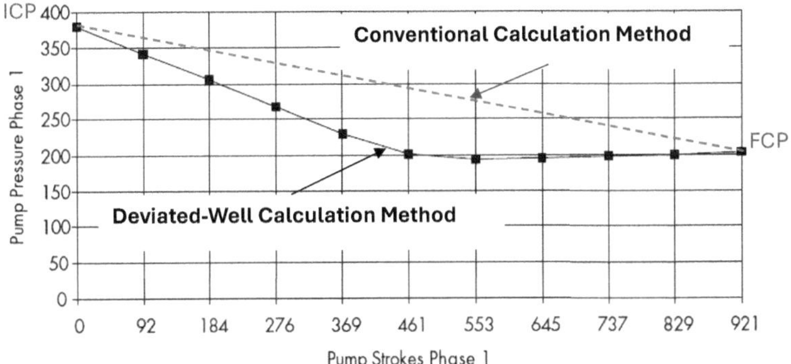

FIGURE 4.46 Comparing the deviated well versus conventional calculation methods for predicting the drill-pipe pressure during pumping the kill mud through the drill-string (modified from Devereux, 1998). The well is 2,000 ft in Measured Depth (MD)/1,000 ft TVD with the KOP of 850 ft. For a deviated well, the deviated-well calculation method must be used for predicting the drill-pipe pressure during pumping the kill mud through the drill-string. In this method, the trend of ICP drop from the surface to the Kick-Off-Point (KOP) differs from that of the KOP to the End-Of-Build (EOB) Section and from EOB to the Total Depth (TD). Thus, they must be calculated separately. In the horizontal section, the pressure drop of the drill-pipe is minimally negative due to pressure losses in the horizontal section.

the low side of the horizontal hole. This has an adverse effect on circulating the gas influx which is mostly located on the high side (Figure 4.48).

4.16 TAPERED DRILL-STRING

A tapered drill-string is a string consisting of two or more different sizes of drill-pipes. In most tapered strings, a larger diameter/size pipe is placed at the top, and the smaller diameter pipe is placed at the bottom. A tapered drill-string affects several parameters consisting of BOP rams' sizes, crossover sizes for drill-pipe safety valves (DPSVs) (in case of a well control event/kick), trip monitoring values, and well control calculations (such as the curve of ICP to FCP values versus the drill-string strokes).

An important item regarding tapered drill-string is the required well control equipment due to the existence of several drill-pipe sizes. Figure 4.49 shows a tapered drill-string in the hole with three drill-pipe sizes. In this example, three fixed pipe rams are required so that they can close around different size pipes in emergency cases. However, as space may be a concern, instead, a set of variable-bore rams are used together with a set of fixed pipe rams. Due to space limitations and more time value in offshore operations, in case a kick flow occurs during tripping of the drill-string, a DPSV must be immediately installed. It is noted that the DPSV must always be available and kept open in a handy place on the drilling rig floor. If the drill-string is tapered, or a casing string is being run, then one or two crossover(s) on the FOSV or stabbing valve with proper threads must be also

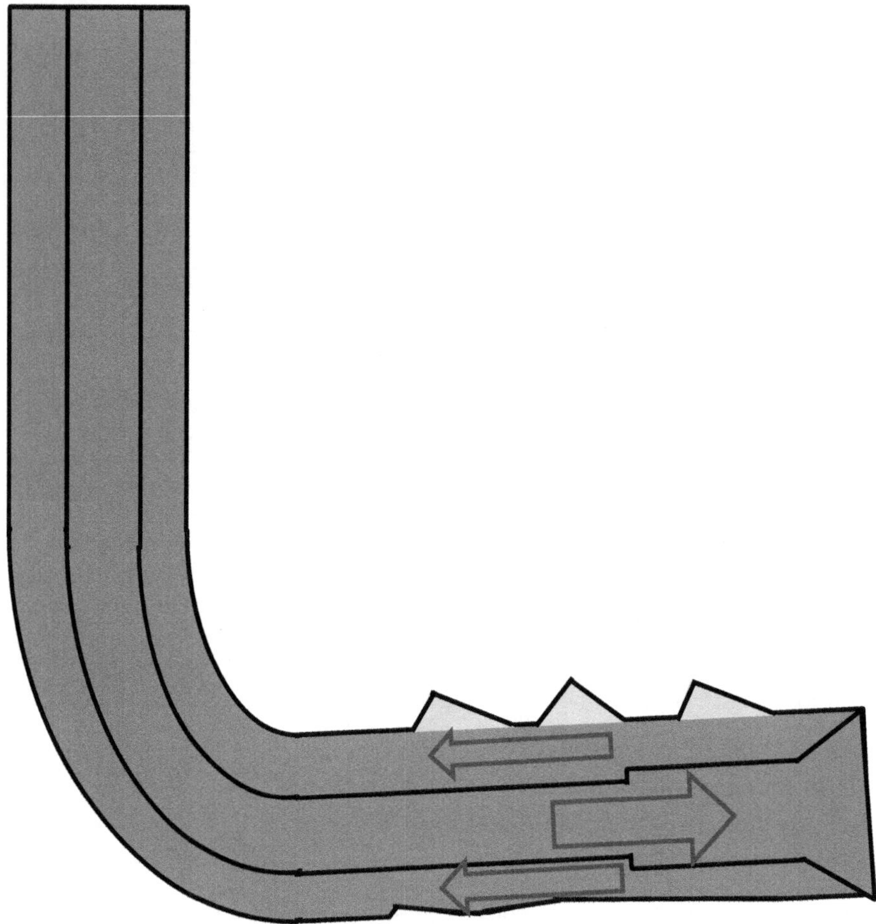

FIGURE 4.47 In highly deviated holes exceeding 90 degrees, some influx would definitely stay in the pockets in high side (created due to undulations in the trajectory) which is difficult or impossible to circulate out.

available and accessible at the rig floor. It is noted that FOSV is also called DPSV or stab-in valve.

Well control calculations for a tapered string have some differences with that for fixed-size drill-pipe. In tapered drill-pipes, the volume calculations cannot be done using the formula on the kill sheet. Instead, the user should calculate separately and transfer the drill-string and annulus volumes to the kill sheet. Next, in well control calculations related to the Wait and Weight method or the second circulation of the Driller's method, for a fixed bore-size drill-pipe, the drill-pipe pressure drops proportionally to the pump strokes. It means, in single-sized drill-pipes, the drill-pipe pressure drops from the Initial Circulating Pressure (ICP) to Final Circulating

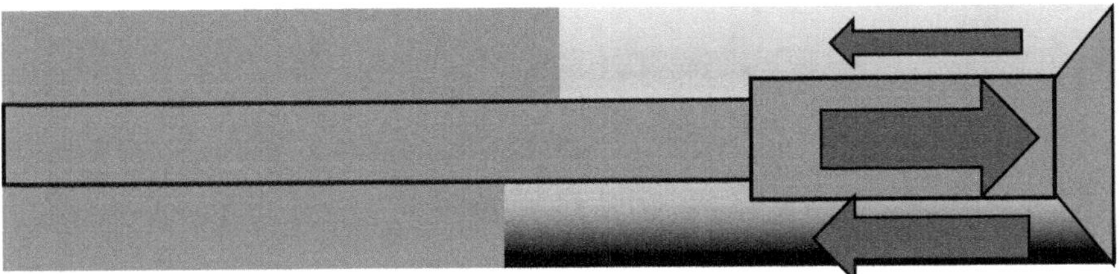

FIGURE 4.48 In the horizontal hole, at the kill rate (which is a Slow Circulating Rate, SCR), the flow regime is laminar; thus, the heavy weight kill mud has more tendency to flow to the low side of the horizontal section. This has some adverse effect on circulating the gas influx out which is mostly located on the high side of the wellbore.

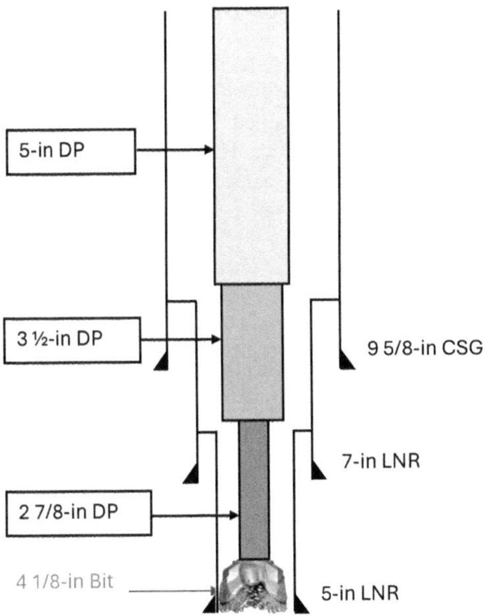

FIGURE 4.49 A tapered drill-string in a workover well. The smallest size drill-pipe is run first to enter the smallest space in the liner (5 in).

Pressure (FCP) in a straight line with the pump strokes. However, if the drill-string is tapered, the different capacity of the pipes affects the pressure drop rate. An example with three components and sizes of the drill-string (consisting of 5-in, 3.5-in, and 2 7/8-in drill-pipes) was already shown in Figure 4.49. However, this effect is not usually considered on commonly used kill sheets. In fact, in kill sheets, the drop rate of the drill-pipe pressure is assumed straight and lower than the required DPP (see Figure 4.50). Hopefully, as some safety factor is considered in ICP and FCP calculation, this drop in BHP may not usually cause underbalance. It is noted that a conventional drill-string can be considered tapered due to the drill-collars; however, the effect is not considerable because of comparatively small length of drill-collars.

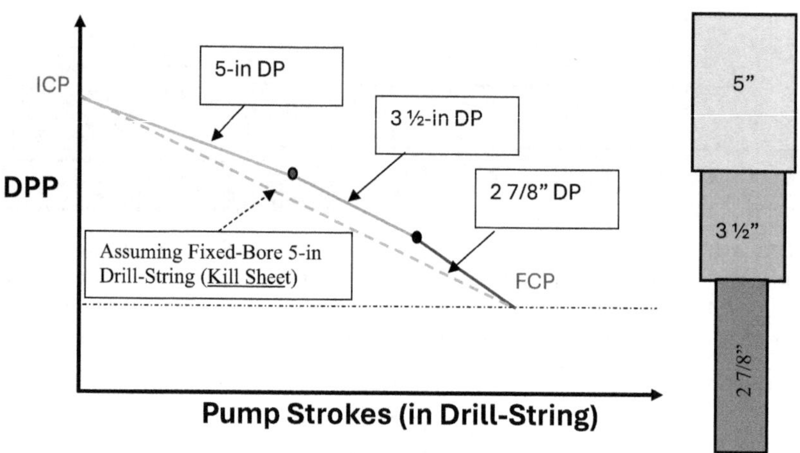

FIGURE 4.50 As the kill mud is pumped from surface to the bit, the drop rate of Drill-Pipe Pressure (DPP) versus pump strokes is assumed to be straight in typical kill sheets (regardless of if the drill-pipe is single-sized or tapered). However, the DPP should drop with different rates in different drill-pipe sizes and that is considered a limitation of typical kill sheets.

REFERENCES

Devereux, S., 1998. *Practical Well Planning and Drilling*. Tulsa: PennWell Corp. ISBN: 13978-0878146963:.

Elmgerbi, A., Thonhauser, G., Prohaska, M., et al., 2016. General Analytical Solution for Estimating the Elastic Deformation of an Open Borehole Wall. *International Journal of Scientific & Engineering Research*, 7 (1), 1056–1068. www.researchgate.net/publication/294581642

Thomas, D. C., Lea, J. F., & Turek, E. A., 1984. Gas Solubility in Oil-Based Drilling Fluids: Effects on Kick Detection. *Society of Petroleum Engineers*. https://doi.org/10.2118/11115-PA

5 Causes of Kicks and LOWCs

Rahman Ashena

5.1 INTRODUCTION

This chapter discusses causes of kicks and loss of well control events (LOWC) events/blowouts. In general, when the primary well control in drilling provided by hydrostatic mud pressure is lost, a kick influx would enter the wellbore (kick occurs). Basically, kicks occur due to geological reasons, wrong drilling practices, and equipment failure. Wrong drilling practices such as improper tripping and swabbing originate from well operations by the drilling crew. In specific, the main kick causes can be classified as:

- Abnormal formation pressures,
- Insufficient mud weight,
- Improper tripping and insufficient hole filling-up,
- Shallow gas formations,
- Fast drilling of gas-bearing formations,
- Surging and swabbing,
- Improper casing/liner running,
- Lost circulation,
- Riser disconnection and riser gas (subsea),
- Equipment failure, and
- Human error.

Causes of LOWC events fall into external and normal groups. The external causes of LOWC events are imposed from outside such as natural disasters like storms or hurricanes, earthquakes, fires, or malicious activities like military or bombing or collisions by boats or ships. The normal causes include too late kick detection, equipment failure, and human error.

5.2 ABNORMAL FORMATION PRESSURES

5.2.1 Effect

In Section 4.6, abnormally pressured formations and their origins were discussed. Abnormally high-pressured formations may cause underbalance in drilling conditions. As a consequence, sudden unexpected kick influxes would enter the wellbore,

DOI: 10.1201/9781003473770-5

particularly in exploration drilling. The abnormal formation can be located anywhere (shallow or deep). The shallow case, which is usually due to shallow gas pockets, will be discussed later.

5.2.2 Measures

Therefore, the prediction of possible abnormal pressures should be performed, both during well planning and operations. Abnormally pressured formations can be under-compacted.

In particular, the measures are as follows:

- Prevention is better than cure. It is advised to predict abnormal formations in advance (using offset well data). Therefore, the mud weight can be increased in advance prior to drilling the formation. Thus, there are several methods to detect abnormal pressures such as sudden decrease in dc exponent trend line, sudden decrease in density log, and decreasing trend in sonic transit time with depth. Some other general signs of abnormal pressures that are sensible while drilling are sloughing shales, tight spots, elliptical hole sections, etc.
- Conversely, if an abnormally pressured formation has not been predicted or detected in advance, and kick influx would enter the wellbore suddenly, one should proceed to well control. Well kill operations should be implemented which include calculations, planning, and operations to kill the well with kill mud weight.
- In top-hole drilling where shallow gas can enter the wellbore, the best option is to divert the gas away; otherwise if the well is closed, the formation would be probably fractured at the shoe.

5.3 INSUFFICIENT MUD WEIGHT

5.3.1 Effect

In overbalanced drilling and completion, hydrostatic mud pressure should be greater than the formation pore pressure. Thus, hydrostatic mud pressure and density/weight are considered as the means of primary well control. Therefore, for each hole section, appropriate determination, maintenance, and adjustment of the mud weight are highly crucial to prevent potential kicks.

If wells can be drilled with a high enough mud overbalance, kicks will be less likely than if wells are drilled with a low overbalance.

Assuming the formation pressure is not abnormal, the hydrostatic mud pressure can become inadequate due to a) accidental dilution in surface tanks, b) gas-cut mud during drilling gas-bearing formations, and c) contamination by formation fluids.

5.3.2 Measures

- To prevent any possible kick, considerable attention should be paid to prevent excessive mud dilution at the surface.

- Routine mud weight checking should be essentially carried out to identify any possible contamination of the mud. Any indications of disturbed mud weight should be immediately announced by the mud engineer to the driller and drilling supervisor. Otherwise, after kick influx, circulating the kick out and increasing the mud weight to kill the mud are required. In case of kicks, proceeding to preparing heavyweight muds should be made quickly.

5.4 NARROW SAFE MUD WINDOW

Hydrocarbon reservoirs in high-pressure, high-temperature (HPHT) wells in the North Sea and other deepwater wells in the US Gulf of Mexico Outer Continental Shelf (GoM-OCS) or elsewhere typically have a low margin between their pore and fracture pressures (narrow safe mud window). In these wells, more frequent kicks are expected than in wells where a large margin between the pore and fracture pressures exist.

5.5 IMPROPER TRIPPING AND INSUFFICIENT HOLE FILL-UP

5.5.1 Effect

Many kicks occur during tripping operations due to improper tripping practices such as moving the pipe too quickly or failing to fill up the hole properly during pulling out of the hole, or running in the pipe too quickly into the hole causing excessively high surge pressures.

5.5.2 Measures

- It is essentially important to follow correct tripping procedures before and while tripping. Before tripping, proceed to weight the mud for the trip margin. It is important to allocate a *trip margin*. Next, the trip sheet should be ready. Estimate safe tripping rates.
- During tripping, the hole must be kept full all the time; otherwise, a kick influx may enter the wellbore. This is particularly important during pulling out of hole – fill up the hole with mud for the lost metal volume. Use a *trip tank* for more accurate monitoring of any possible discrepancies in volumes. *Note:* A trip tank is a long and rather narrow tank with typical 50- to 100-bbl capacity and with a high accuracy of 0.5 bbl. As its accuracy is much greater than that of typical mud pits, its use is recommended for accurately monitoring volume displacement during tripping.

5.5.3 Recommended Tripping Procedures

The recommended tripping procedures of the drill-string fall into "run in hole (RIH)" and "pull out of the hole (POOH)".

5.5.3.1 Run In Hole (RIH)

Safe procedures for running in the hole are as follows.

1. Proper tripping velocity should be determined considering surge calculations.
   Note: The surge pressure during running in the hole causes the bottomhole pressure to increase, particularly at the casing shoe as the weakest point of the formation (Figure 4.26). While running in one stand of drill-pipe, the downhole pressure changes (Figure 4.27). If the RIH rate is too high, a too great surge pressure is developed downhole which may cause fracturing of the formation at the shoe.
2. Prior to tripping, the trip tank must be correctly connected to the *fill-up line*. The return line is also lined up to the trip tank. Do not use the mud pits during tripping.
3. In addition, the trip sheet must be ready by the driller to record the number of pipes pulled out.
4. Just prior to running in the hole, perform a flow check to detect any possible kick flow with the following procedure (for the sake of well safety).
   - Pick up the drill-pipe to space-out,
   - Stop the pumps,
   - Line up the well to the trip tank,
   - Record the volume of mud already in the tank,
   - Check for flow for several minutes depending on the company policy (e.g., 15 minutes).
5. If the flow check is confirmed positive (there is flow out of the hole with pumps off), a kick has definitely occurred. Therefore, it is required to proceed to well shut-in and then kill (refer to Chapters 7, 8, and 11).
   Note: When the drill-string is out of the hole, in case of a confirmed kick, the proposed method of well kill is the volumetric method (Section 11.6).
6. If there is no flow observed (with pumps off), the flow check is confirmed as negative. Record the depth, time, and duration of the flow check. Then, proceed to running in the hole
   Note: During RIH, we can perform a flow check anytime there is a doubt about equality between the volume of pipe run in the hole and the mud volume returned to the trip tank.
7. Run the drill-pipes in hole until the bit reaches the casing shoe.
   - During running in the hole, keep the annulus full using the trip tank. Monitor the volumes to detect any possible kick flow or mud loss.
   - Fill up (inside) the drill-pipes, e.g., every ten stands. If this is ignored, it may pose risk for collapse of drill-pipes.
   - When the bit reaches the casing shoe, it is important to perform a flow check for a possible kick flow. If the flow check is positive (i.e., which indicates a kick flow), first shut in the well preferentially by closing the annular preventer. Then, *strip* the drill-pipes in the hole (i.e., running the drill-pipes in the hole with annular BOP closed) to reach the bottomhole so that it is possible to circulate the kick out of the hole. If the flow check is negative, continue RIH.
8. Continue to run in hole until the bit reaches the bottom. With the bit on-bottom, make another flow check to ensure about well security.

9. Make a bottoms-up (i.e., a circulation of mud from the bottomhole to the surface) to make the mud weight in the drill-string and annulus uniform and clear the well off possible debris.

5.5.3.2 Pulling Out Of the Hole (POOH)

Pulling the drill-string out of hole is essentially done in case of a dull bit, wire-line logging, at the end of drilling a hole section, etc. To prevent any kicks during tripping, it is essential to be familiar with and apply safe POOH procedures as presented next.

Figure 5.1 illustrates an overall picture of the setup for POOH.

1. All the drilling crew must be able (already trained) to apply well control procedures in case of possible kicks or well control events during tripping. Information on maximum POOH rate must be provided to the crew by the drilling supervisor. All other instructions should be already clarified by the tool pusher.
   Note: If mud loss is observed prior to POOH, it is important to cure the mud loss first.
2. Prior to tripping, the mud in the wellbore should be conditioned to remove the mud off any cuttings to make the fluid in the hole uniform (mud conditioning) and reduce possibility of potential swabbing and pipe stuck. Otherwise, as

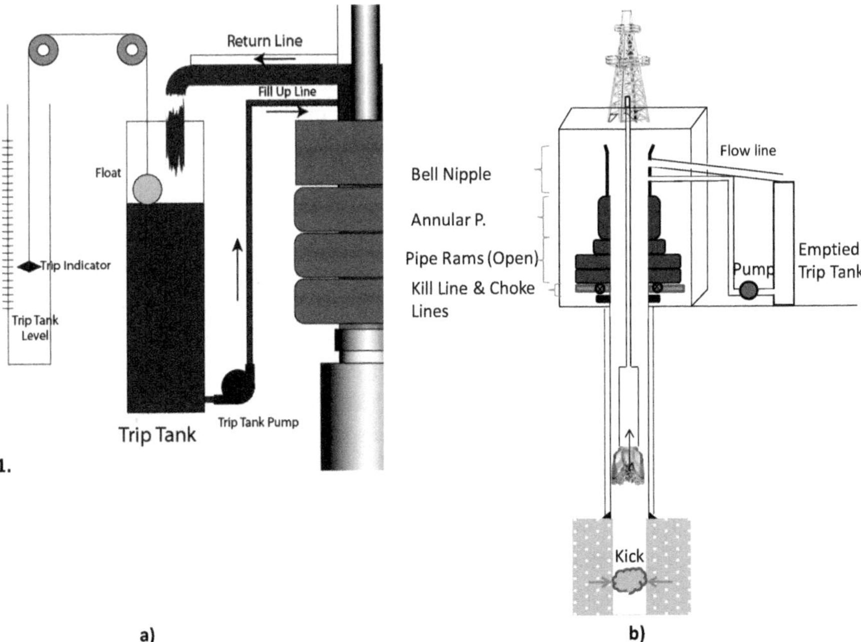

FIGURE 5.1 a) Line-up of return flow line and fill-up line with the trip tank (specifically for RIH) and b) an emptied trip tank can cause kick occurrence during tripping out (kick has occurred).

the drill-string is pulled out of the hole, cuttings would accumulate/pack-off around the bit, causing the balled-up bit and thus a swabbed kick influx below the bit for tripping up. If there is some gas entrained in the mud (i.e., gas-cut mud, due to drilling a gas-bearing formation), the necessity of a bottoms-up is more essential. This goal is reached at least by

- One or several bottoms-up is/are sufficient to condition the mud if casings are not planned to be run after POOH. Following the bottoms-up, obtain and record slow circulation rate pressures (SCRPs).
- If conditioning of mud in the wellbore could not be accomplished by bottoms-up, one or more *short/wiper trip* may be performed in order to clean the solids out of the wellbore. A "short/wiper trip" is the act of raising the drill-string for a specified distance in the open-hole section and then running it back to the bottomhole.
- If it is planned to run a casing string following POOH, a condition trip is essentially required. The difference between a condition trip and a short trip is that the condition trip is done for the whole open-hole, i.e., the whole BHA is raised out of the open-hole with the bit being located at the previous casing shoe.

3. Prior to POOH, recommended tripping rates should be determined considering swab calculations.
   Note: During POOH, swab pressure is created to cause a decrease in the bottomhole pressure. If the POOH rate is very high, too much swab pressure will be developed (called swabbing) which causes an underbalance leading to a kick flow (Figure 4.26). It is noted that the most critical point where kick flow may occur is at the beginning of tripping, at the bottom of the open-hole section.
4. Prior to tripping, the line-up should be performed. To do that, the trip tank must be correctly connected to the *fill-up line*. The return flow line is also lined-up to the trip tank (Figure 5.1).
5. The driller must prepare the trip sheet to record the number of pipes pulled out.
6. Just prior to POOH, to ensure well safety, perform a flow check.
   If the flow check was positive (kick flow was observed with pumps off), it is essential to proceed to well shut-in and subsequently well kill. If the flow check is confirmed negative, get ready for POOH.
   Note: During POOH, a flow check should essentially be made anytime there is a doubt about equality between volumes pulled out and the mud volume displaced from the trip tank during POOH.
7. Pull the drill-pipes out of the hole while continuously keeping the hole (annulus) full using the trip tank. Make sure the trip tank is not empty; otherwise, a kick flow may occur.
   In case of encountering *tight spots* during POOH (i.e., where hole diameter has become less than the expected drilled bit diameter), *wash and ream* the tight interval.
   If the mud is very viscous which can cause swabbing or is expensive (containing precious polymers or additives), a heavy slug (much heavier

than mud, with the volume of say 1–20 bbl) is pumped and placed in the drill-string so that typically 150–200 ft of dry pipe (from the surface) can be resulted. Displacing a slug into the drill-string creates a mud level drop in the drill-pipe, which eliminates handling the mud at the surface and facilitates pulling the string out of the hole. Otherwise, there would be no need to use the *mud box* to handle the mud out of the drill-pipes. Using the mud box is a timely process (in addition to some mud which is wasted as it cannot return to mud tanks). After pumping the heavy slug into the pipe, another flow check is performed. For slug volume calculations, refer to Section 4.7.3.

If swabbing occurred due to too much tripping rate or at tight spots, run back to the bottom by stripping with the BOP closed. Next, kill the well (by circulating the kick influx out and then replacing the mud in the wellbore with heavy weight mud to create enough overbalance pressure).

8. POOH from the bottomhole until the bit reaches the casing shoe.
When the bit is at the casing shoe, it is always necessary to make a flow check to ensure whether the well is safe (any kick influx is ongoing or not). If the flow check is positive, shut in the well and strip in the hole (with preferentially the annular preventer closed) until you can reach the bottomhole to kill the well (to circulate the kick out). If the flow check is negative when the bit is at the casing shoe, the well is confirmed safe, and it is possible to continue POOH.
10. Continue POOH until the top of drill-collars (DCs) or bottomhole assembly (BHA) reaches the drill floor.
11. Pull out the BHA (DCs, etc.) out of the hole with reduced tripping rates.
Note: As the clearance between the BHA and the hole is lower, the tripping rate should be reduced to prevent swabbing.
12. Finally, make the last flow check to ensure about well safety.

5.5.4 Trip Margin

The term *trip margin* was first introduced in casing design to counteract the swab pressure during POOH. Thus, to prevent formation fluids from being swabbed into the well during POOH, a *trip margin* is allocated (see Figure 4.25). This margin is provided by weighing the drilling mud before tripping. By pore pressure prediction, it is possible to determine how large the trip margin should be. However, it typically ranges from 0.2 to 0.5 ppg. Trip margin is evaluated by:

$$\text{Trip Margin} = \frac{\text{Required Overbalance Before Tripping}}{0.052 \times \text{TVD}} \tag{5.1}$$

The weighed mud weight required before tripping (MW$'$) is:

$$\text{MW}' = \text{MW} + \text{Trip Margin} \tag{5.2}$$

5.5.5 Filling-Up the Hole in Dry and Wet Tripping

During tripping, the reduced volume from the hole due to pipe removal must be essentially replenished with the mud to prevent any fall in mud level in the hole. This is done by connecting the return line to the *trip tank*. Prior to tripping, the trip tank should be essentially lined up. Filling up the hole by trip tank can be performed continuously by its corresponding centrifugal pump to ensure the hole remains full all the time. It is also possible to measure fill volumes continuously and accurately. However, a frequent checking of the trip tank level is required to refill the tank some time before it becomes empty. Any attention distraction, particularly during pulling out drill-collars (which have large volume metal), can lead to an empty trip tank and thus reduction of mud level in the hole. It is noted that the drill-collar steel volume per unit length is about five to ten times that of drill-pipe.

The reduction of mud level in the hole is equivalent to a drop in bottomhole hydrostatic pressure. If the crew particularly the driller is not careful, the fluid level in the hole may fall while tripping. When this pressure drop causes BHP to fall below the formation pressure, consequently a kick influx enters the open-hole section provided that the formation is permeable.

Therefore, the number of pipes pulled out must be continuously and accurately recorded in the *trip sheet*. In addition, the volume of "steel" or "steel plus mud removed" (depending on cases of POOH) must be accurately calculated as the fill-up mud volume required for one stand of pipes. Then, this volume is compared with the fill mud volume. Fill mud volume or trip tank volume change can be measured accurately by level meters or electric sensors. If the trip tank volume change is less than the required fill mud volume, a kick influx has definitely entered the wellbore. Conversely, if the trip tank volume change is greater than the required fill mud volume, mud loss has occurred.

Basically, there are two cases of POOH:

5.5.5.1 Dry Pipe POOH

In tripping, dry pipe means that the bottom-end of the drill-string is open end (i.e., there is no float valve or check valve installed in the string, usually above the bit). Thus, during POOH, the mud inside the drill-string can freely exit the drill-string (via the bit nozzles) back to the wellbore. This means that only the steel part of the drill-pipe or drill-collar is extracted from the hole when picking up the pipe above the rig floor. Therefore, only the steel volume (i.e., pipe displacement) must be replenished by the mud in order to keep the hole full of mud. There are some important relations related to dry pipe POOH as explained next.

The required mud volume to fill up the hole during dry pipe POOH, $V_{fill\_dry}\left[bbl\right]$, is:

$$V_{fill\_dry} = L_{DP} \times Dis_{DP} \tag{5.3}$$

where L_{DP} is the length of extracted drill-pipe out of the hole and Dis_{DP} is the drill-pipe metal displacement [bbl/ft].

When pulling drill-pipes out, the mud level drop without filling-up the hole is $\Delta L_{m,drop\_DPdry}\left[ft\right]$:

$$\Delta L_{m,drop\_DP\_dry} = \frac{Dis_{DP}}{Cap_{Riser/CSG} - Dis_{DP}} \times L_{DP} \tag{5.4}$$

where Dis_{DP} is the drill-pipe metal displacement [bbl/ft], L_{DP} is the length of drill-pipe, and$Cap_{Riser/CSG}$ is the capacity of the riser (for subsea) and casing (for operations with fixed rigs) whichever applies.

When pulling remaining drill-collars, the mud level drop (without filling-up the hole) out of the hole dry is $\Delta L_{m,\ drop\_DC\_dry}\ [ft]$:

$$\Delta L_{m,drop\_DC\_dry} = \frac{Dis_{DC}}{Cap_{Riser/CSG}} \times L_{DC} \tag{5.5}$$

where L_{DC} is the length of drill-collars and Dis_{DC} is the displacement of drill-collars.

The mud pressure drop (per foot of dry pulled DP) without filling-up the hole is $\Delta P_{m,drop\_DPdry}\ [psi/ft]$:

$$\Delta P_{m,drop\_DP\_dry} = \frac{0.052 \times MW \times Dis_{DP}}{Cap_{Riser/CSG} - Dis_{DP}} \tag{5.6}$$

The mud pressure drop (per foot of dry pulled DP) without filling-up the hole is $\Delta P_{m,drop\_DCdry}\ [psi/ft]$:

$$\Delta P_{m,drop\_DC\_dry} = \frac{0.052 \times MW \times Dis_{DP}}{Cap_{Riser/CSG}} \tag{5.7}$$

where MW is the mud weight [ppg].

5.5.5.2 Wet Pipe POOH

In tripping, "wet pipe" means that the bottom-end of the drill-string is closed-end (i.e., float or check valve is installed in the drill-string, e.g., in the bit sub). Thus, the mud inside the drill-string cannot exit the drill-string via the bit nozzles while POOH. This means that the total volume of "the steel part and the mud inside" is extracted from the hole when picking up the pipe above the rig floor. Therefore, this volume must be replenished by the trip tank mud in order to keep the hole full. It is noted that in wet POOH or even making connections, the mud inside the pulled pipe must be controlled by the *mud box*.

There are some important relations related to Wet POOH as explained next.

The required mud volume to fill up the hole is $V_{fill\_wet}\ [bbl]$:

$$V_{fill\_wet} = L_{DP} \times (Dis_{DP} + Cap_{DP}) \tag{5.8}$$

where L_{DP} is the length of extracted drill-pipe out of the hole; Dis_{DP} is the drill-pipe metal displacement [bbl/ft]; and Cap_{DP} is the drill-pipe capacity [bbl/ft]. The term "(Dis_{DP}+Cap_{DP})" is also called the closed-end displacement of the pipe.

For drill-pipes, the mud level drop without filling-up the hole is $\Delta L_{m,drop\_DPwet}\ [ft]$:

$$\Delta L_{m,drop\_DP\_wet} = \frac{(Dis_{DP} + Cap_{DP})}{Cap_{Riser/CSG} - (Dis_{DP} + Cap_{DP})} \times L_{DP} \tag{5.9}$$

For pulling the remaining drill-collars, the mud level drop (without filling-up the hole) is $\Delta L_{m,drop\_DCwet}\ [ft]$:

$$\Delta L_{m,drop\_DC\_wet}\ [ft] = \frac{(Dis_{DC} + Cap_{DC})}{Cap_{Riser/CSG}} \times L_{DC} \tag{5.10}$$

The mud pressure drop (per foot of wet pulled DP) without filling up the hole $\Delta P_{m,drop\_DP\_wet}\ [psi/ft]$ is:

$$\Delta P_{m,drop\_DP\_wet}\ [psi/ft] = \frac{0.052 \times MW \times (Dis_{DP} + Cap_{DP})}{Cap_{Riser/CSG} - (Dis_{DP} + Cap_{DP})} \tag{5.11}$$

The mud pressure drop (per foot of wet pulled DC) without filling-up the hole is $\Delta P_{m,drop\_DC\_wet}$ [psi/ft]:

$$\Delta P_{m,drop\_DC\_wet}\ [psi/ft] = \frac{0.052 \times MW \times (Dis_{DP} + Cap_{DP})}{Cap_{Riser/CSG}} \tag{5.12}$$

5.5.6 Exercises

Exercise 1:

A wellbore was drilled with the mud weight of 9 ppg to the true vertical depth of 9,000 ft. Prior to tripping, an overbalance of 200 psi was found essential to counteract the swabbing pressure. Evaluate the *trip margin* and the required mud weight.

Answer:

The required trip margin is:

$$\text{Trip Margin} = \frac{200}{0.052 \times 9{,}000} = 0.427 \rightarrow 0.5 \text{ ppg}$$

For safety purposes (i.e., further overbalance), the trip margin should be rounded up to the tenth.

The mud weight prior to tripping is:

$$MW' = 9 + 0.5 = 9.5 \text{ ppg}$$

Exercise 2:

During pulling out of the hole, the hole fill pump was stopped, and the complete BHA was pulled dry. Calculate the reduction in BHP using the following data:

Hole size	8.5 inches
Length of BHA	600 ft
Internal capacity of BHA	0.006 bbl/ft
Steel displacement of BHA	0.03 bbl/ft
Internal capacity of casing	0.072 bbl/ft
Capacity between BHA and casing	0.035 bbl.ft
Mud weight	12 ppg

Answer:

$$\Delta P_{m,drop\_DCdry}\left[\frac{psi}{ft}\right] = \frac{0.052 \times MW \times Dis_{DP}}{Cap_{Riser/Csg}} = \frac{0.052 \times 12 \times 0.03}{0.072} = 0.258 \text{ psi/ft}$$

$$\Delta P_{m,drop\_DCdry}[psi] = 0.258\frac{psi}{ft} \times 600ft = 155 \text{ psi}$$

Exercise 3:

A vertical well has been drilled to the depth of 7,430 ft.

Casing shoe depth	3,800 ft
Mud weight	12 ppg
Pore pressure gradient (at 7,480 ft)	0.6 psi/ft
Open-hole capacity	0.1458 bbl/ft
Casing capacity	0.1571 bbl/ft
Drill-pipe metal displacement	0.008 bbl/ft
Drill-pipe capacity	0.0177 bbl/ft

Task: How many complete stands can be pulled wet before the well flows? (Assume one stand equals 93 ft).

Answer:

The well starts to flow during POOH when the mud hydrostatic pressure equals the formation pore pressure. At 7,430 ft, the mud hydrostatic pressure is:

$$HP_m = 0.052 \times MW \times TVD = 0.052 \times 12 \times 7{,}430 = 4{,}636 \text{ psi}$$

At this depth, the formation pore pressure is:

$$P_{pore} = PG_{pore} \times TVD = 0.6 \times 7{,}430 = 4{,}458 \text{ psi}$$

Therefore, the pressure drop (due to wet POOH) required just at the borderline of kick flow is:

$$\Delta P_{m,drop\_DPwet} = HP_m - P_{pore} = 4636 - 4458 = 172 \text{ psi}$$

Using Equation (4.11) for wet POOH, the length of DP pulled is found:

$$\Delta L_{DP} = \Delta P_{m,drop\_DPwet}\,[\text{psi}] \times \frac{Cap_{CSG} - (Dis_{DP} + Cap_{DP})}{0.052 \times MW \times (Dis_{DP} + Cap_{DP})}$$

$$= 172 \times \frac{0.1571 - (0.008 + 0.0177)}{0.052 \times 12 \times (0.008 + 0.0177)} = 1{,}409.3 \text{ ft}$$

The number of stands is:

$$No_{st} = \frac{1{,}409.3 \text{ ft}}{93 \text{ ft / st}} = 15.2 \text{ stands}$$

Therefore, if 15 DP stands and a part of a joint are pulled, the well is at the borderline to get underbalanced and kick.

5.6 LOST CIRCULATION

5.6.1 Explanation

5.6.1.1 Effect

While drilling overbalanced in fractured or vugular formations, particularly if they are depleted (highly low-pressured), severe mud loss can occur. Fractures can be naturally occurring or artificially induced. Artificial or induced fractures can be induced due to a number of reasons such as high mud weight or improperly selected casing shoe depth, excessive surge pressure due to too quick run in the hole, tight, or swelled or tight hole intervals, while circulation initiation due to too high gel strength. The severe mud loss in such formations can lead to the loss of all mud returns from the well (i.e., lost circulation) and mud level drop. If the mud level drop is high enough, the BHP can become lower than formation pressure, and thus kick would occur (Figure 5.2b). Such a kick type may soon become severe if the well is not filled, and the mud level drop goes down quickly.

5.6.1.2 Measures

In case of severe mud loss, fill up the hole with water (while measuring the volume pumped) in order to reduce the rate of mud level drop and reduce the underbalance as much as possible. It is recommended to use lower mud weights and mix the mud with lost circulation materials (LCMs).

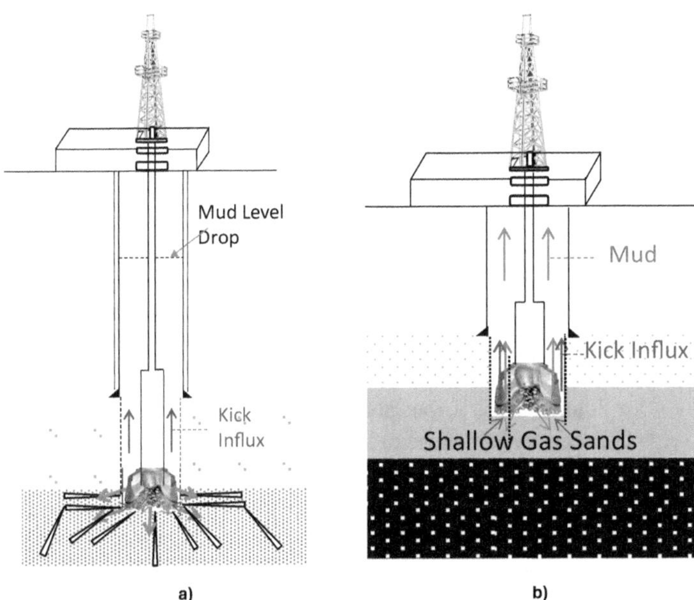

FIGURE 5.2 Kicks occurred a) following lost circulation and b) due to encountering a shallow gas pocket.

5.6.2 Exercise

After drilling with the mud weight of 13.7 ppg in a new formation at 14,750 ft TVD on a surface-BOP stack rig, the driller experienced a total loss of returns. He immediately stopped drilling and started to fill up the annulus with water. The hole took 30.25 bbl of water (with pressure gradient of 0.455 psi/ft) to fill and stop the loss. The annular capacity between the drill-pipe and casing at the surface is 0.0458 bbl/ft.

What is the bottomhole pressure after the mud loss has stopped?

Answer:

First, the head of water (in lieu of mud) in the annulus is evaluated:

$$\text{Head of Water} = \frac{30.25\text{bbl}}{0.0458\,\text{bbl/ft}} = 660.5\text{ft}$$

Therefore, the bottomhole pressure, BHP, is:

$$\text{BHP} = 660.5 \times 0.455 + 0.052 \times (14{,}750 - 660.5) \times 13.7 = 10{,}337\text{ psi}$$

5.7 SHALLOW GAS AND TOP-HOLE DRILLING

5.7.1 Effect

Shallow boreholes or holes drilled before running the surface casing and installation of BOPs are called top-hole sections. They can be, for example, 24-inch and 36-inch holes. As shallow formations are weak in strength and are susceptible to failure, usually lower overbalance pressure is recommended for their drilling. At these shallow depths, particularly in exploration wells, there may be shallow gas pockets or lenses, which are highly porous, permeable and maybe over-pressured, and completely enveloped with mudstone. While drilling in such top-hole sections, penetration into shallow gas pockets is one of the most hazardous situations that can be encountered. In such wells, gas can travel to the surface rapidly, with little warning. This gas, if not diverted, can get ignited to cause explosion of the rig.

In deepwater offshore (seawater > 600 m/1,969 ft), floating vessels normally drill shallow sections of wells in a riser-less manner (without using a riser pipe). The "riser-less" also means that the diverter systems are normally not installed when drilling the shallow section of the well from a floating vessel. In offshore riser-less operations with floating rigs, if the gas reaches the seabed, it rises until it reaches under the rig. This causes gas-cut seawater below the rig which lowers the buoyancy required for keeping the rig above the water surface. Consequently, the floating rig would become tilted and finally sink to the bottom. When drilling without a riser, the drilling fluid is usually seawater with a density of 8.6 lb/gallon (1,030 kg/m^3). Due to the lower density of seawater and consequently lower hydrostatic pressure in the well than is the case with a normal mud, there is a greater probability for loss of well control (LOWC) in case of shallow gas pockets.

Shallow gas was a larger problem in the 1980s and 1990s than to date. In a study from 2000–2015, three of the shallow gas LOWC events ignited two in the US GoM-OCS and one in Indonesia. All three occurred on a jack-up. One ignited immediately, another one ignited after half an hour, and the last one after 26 hours. None caused fatalities (Holand, 2017).

In top-hole drilling, shallow kicks may occur due to one or a combination of the following reasons:

- Existence of a charged pressured layer/formation at shallow depths can lead to entrance of a kick influx into the wellbore. The kick influx can be gas or salt water, but the gas is more severe or dangerous. The charged nature of shallow gas pockets may be due to geological abnormality or artesian effect which is the case in wildcat wells. However, shallow gas-charged sands may also be found in development wells. In development wells, a shallow sand formation may be charged with high-pressure gas from deeper zones of a neighboring well which has integrity issues. Thus, gas has migrated through channels in cement or formation created due to lack of integrity. The causes of lack of integrity in neighboring wells consist of poor cement jobs, casing failures, inadequate/inefficient abandonment procedures or bad abandonment, micro-annulus on the boundaries of a cement sheath, underground blowouts, and injection well operations.
- Very high rate of drilling/penetration in a possible shallow gas pocket leads to unloading of gas (in the cuttings) in the annulus if the mud is excessively gas-cut mud. Thus, the mud pressure may lie below the formation pore pressure (underbalance), leading to a kick influx.

- Overloading the annulus with cuttings (due to too high penetration rate) in narrow safe mud window cases may cause the bottomhole pressure (BHP) to lie above the fracture pressure, leading to mud losses. The mud loss makes the mud level in the wellbore drop. If the hole is not filled up quickly to replenish the mud loss in the hole, a kick influx may enter the wellbore. Depending on the fluid contained in the rock at the loss depth, a saltwater or gas kick may occur.
- Since lower overbalance pressure is taken in top holes, particularly in riser-less offshore operations where salt water is used as mud, swabbing due to fast POOH can quickly make a kick influx enter the wellbore. As a matter of fact, improper hole fill-up during POOH can quickly lead to a gas kick.

5.7.2 Measures

The measures fall into prevention, detection, and diverting shallow gas as explained next.

a) *Prevention:* In top-hole drilling, the following precautions should be taken:
    - To avoid encountering shallow gas pockets during drilling, the exact well spud location is usually based on experience from previous offset wells drilled and seismic surveys. This measure helps, but field experience shows that the probability of failing to predict shallow gas pockets is considerable.
    - It is recommended to prepare a reserve of weighed mud of twice the hole volume prior to spudding well.
    - Adjust the mud weight such that the overbalance pressure is kept low enough to prevent undeliberate fracturing of top-hole formations.
    - Install a float valve in the drill-string (usually above the bit) to prevent any unprecedented flow through the drill-string. If the well kicks through the drill-pipe when the drill-pipe is connected to the mud system (i.e., not when the pipe is disconnected for tripping or adding an extra stand or joint), the pressure may be closed in by a valve located in the drill-string flow path.
    - Keep the rate of penetration, ROP, low enough to a) prevent too much gas-cut mud causing overbalance and b) avoid overloading the annulus with too much cuttings and annular pressure loss; otherwise, dynamic BHP would become too high for the formation to withstand – that is, the formation would be fractured.
    - Drill a pilot hole (a narrower hole, and then use a hole-opener to enlarge the hole size) at a controlled rate of penetration.
    - Pump out of the hole on trips: During trips, keep circulating the mud through the drill-string. The extra annular pressure loss enables the BHP to be greater, and thus it prevents swabbing.
    - In offshore operations, drill with returns to the seabed. Avoid using risers for drilling shallow holes in offshore operations.

b) *Detection:*

- As the bottoms-up time involved is short in top-hole drilling, alertness of the drilling crew is highly crucial in the detection of the kick as quickly as possible. The return flow sensor is the only item of the monitoring system equipment that can provide an early enough warning of a shallow gas kick in progress. This sensor should be maintained and be working continuously, particularly in this hole section.
- Gas detection sensors are highly important in top-hole drilling and should be checked to be working properly.
- If in doubt of any kick, quickly shut down the pumps and perform a flow check. Although pit level gains are valuable indications and probably the most reliable indicator, they are not early enough for noticing the shallow gas influx.
  Note: Typically, ROP in top holes changes excessively, and thus it cannot provide any reliable kick indication.
- In offshore operations, if possible, a camera should be installed to observe for any possible sign of gas. A crew member should be allocated to observe the surface of the sea.

c) *Action (Pumping and Diverting):*

- In deepwater offshore (with drillships or semi-submersible rigs) where shallow formations are drilled riser-less, in case of any gas pocket flow, the vessel will be moved away from the wellsite until all the gas is discharged; then the rig is reverted back to the position. The risk for the rig installation will depend on the water depth as well as on the gas flow rate. Generally, in deep water, the gas will pose limited danger for the installation. Some little gas would dissolve in the water, and the gas which comes to the surface (if any) will be released in a large area, thus no explosive mixture of gas and air will be formed. However, in shallow water, gas released on the seafloor may pose a danger.

  In case of a shallow gas kick in onshore operations or bottom-supported offshore (like jack-ups), the diverter is allowed to be put into use while keeping mud circulation. As long as the gas is diverted properly, there is a minimal risk of possible explosion. However, field experience shows that in some cases of large gas flow, which is usually mixed with sand, the diverter lines were eroded and the gas leaked and exploded.

  Note: Shutting in the well is not an option either in land or subsea operations. This is because first the kick influx reaches the surface rapidly (limited time for operating BOPs). More importantly, in case BOPs were available and closed, there would be a high risk of fracturing the formation at the casing shoe (because of low strength of formations in the top-hole section). This, in turn, creates the possibility of gas coming up around the casing outside the well to the surface. Similarly, in subsea

operations, if the shallow gas reaches the seabed and the crew proceeds to shut in the well using BOPs, the unconsolidated formation can fracture at the shoe, and the gas can penetrate through the fractures causing them to extend further up to the seabed, causing cratering, which, in a worst case, may cause a bottom-supported platform to tilt and capsize. In addition, the gas reaching the seabed would rise to reach under the rig floor. The mixture of seawater and gas (gas-cut sea-water) has lower buoyancy for the drillship to be held on the water. Consequently, the rig floor would become tilted, and it would sink to the bottom. Section 8.4 gives further information on diverters.
- Shallow gas releases from LOWC events occurring when drilling with drillships and semisubmersibles normally happen on the sea floor.

5.8 FAST GAS FORMATIONS' DRILLING

5.8.1 Effect

While drilling a gas-bearing formation, the gas from the cuttings enters the mud and makes it *gas-cut* or gasified. The extent of gas-cut depends on:

- *Gas content in rock:* The more the initial gas content is in the rock, the greater the gas-cut of the mud would be.
- *Rock permeability:* The greater the permeability of a gas-bearing formation is, the greater the penetration of gas into the mud and thus the greater gas-cut.
- *Rate of penetration (ROP):* The more the ROP in a gas-bearing formation is, the greater the gas-cut in the mud.
- *Duration of making a connection:* Generally, the longer the time for a connection and the mud in the tank is exposed to atmosphere, the more air would be mixed with the mud in the mud tanks and thus the greater the air/gas-cut in the mud would be. This general factor is irrelevant to the formation fluid content (gas, water, etc.).
- *Lag time:* The longer the lag time (e.g., time taken for the cuttings to be pumped from the bottom to surface), the more gas can enter the mud from the cuttings and thus the more the gas-cut mud and lightening of mud.
- *Hole depth:* Based on Boyle's law, the deeper the well, the more the gas expansion can occur while liberated gas is rising to the surface. Thus, greater gas-cut would occur to the mud, and greater drop in the hydrostatic pressure would occur.

As the entrained gas in the mud is under the hydrostatic bottomhole pressure, the gas volume is initially minimal. As the gas is circulated up by the mud, the pressure decreases, and thus it expands and pushes large mud volume at the surface. This causes considerable hydrostatic pressure reduction.

5.8.2 Measures

- The ROP in gas-bearing formation should be controlled and maintained lower than typical ROPs in drilling water-bearing formations of the same depth and geological lithology.
- Vacuum degassers should be utilized to remove the gas in the mud before the mud is pumped back to the well. This is done to prevent an increase of gas in the mud and reduction of hydrostatic mud pressure.
- The connection time should not be prolonged to let excessive gas in the hole.
- In case of very fast drilling (too high ROP) in a gas-bearing formation, the concentration of gas in the mud is increased and well control problems may result. The rapid expansion of gas near the surface may cause belching at the bell nipple with a mud loss from the well and a consequent hydrostatic pressure drop. This may induce further *belching* and lowering of the hydrostatic pressure to get lower than that of the formation (underbalance), so a kick may occur. This sequence, once started, rapidly gets out of control.

5.9 SURGING AND SWABBING

5.9.1 Effect

Surging happens when the bottomhole pressure is increased considerably (exceeding the formation fracture pressure) because of running the drill-string too fast in the hole such that downhole mud losses occur. In case the mud loss is converted to a lost circulation situation and the well cannot be filled up properly, the mud level drops in the hole and a kick influx may enter the well.

Swabbing is when the bottomhole pressure drops considerably due to the effect of too quickly pulling out of the hole which causes underbalance and a kick influx to enter the hole (Figure 5.3). Initially, the kick influx volume due to swabbing is very small. If the crew fail to observe this small discrepancy immediately, the gas kick influx expands considerably due to pressure drop after it has risen enough, showing itself as a significant volume discrepancy between the volume extracted and the mud volume filling the hole from the trip tank (large kick size). However, this warning is sensible when the well has approached a serious well control event.

5.9.2 Measures

Basically, to prevent surging and swabbing, the following measures should be taken before and during the tripping operation:

Pre-Tripping:

- Swab and surge pressures should be accurately calculated prior to tripping. To prevent surging (during RIH) and swabbing (during POOH), tripping rate should be calculated low enough. To do this, the effects of possible high mud viscosities, low clearances between wellbore and pipe, balled-up bit/stabilizers, etc., should be considered.

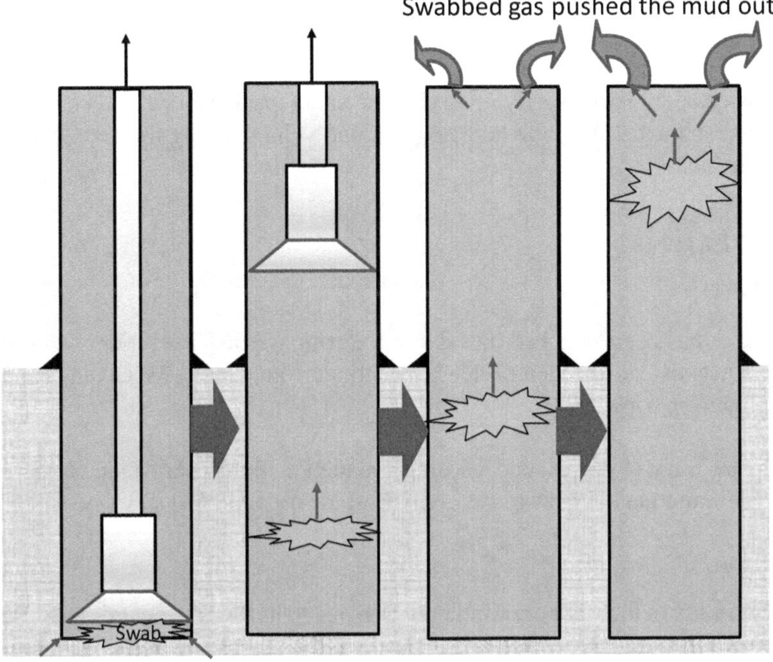

FIGURE 5.3 Swabbing during POOH which causes entrance of a kick influx in the hole expanding with time.

- It must be ensured that sufficient overbalance or trip margin exists prior to tripping to counteract the swab pressure.
- Mud conditioning and bottoms-up prior to POOH should be performed.
- Zone intervals prone to tight spots must be identified using previous trips and offset wells in the region. In case of encountering any indication of tight holes or swelling spots, first wash and ream the hole slowly or do condition trip.

While Tripping:

- The pipe tripping rate must be carefully controlled and maintained below the maximum permissible rate determined by the calculation. In tight spots, reduce the tripping rate. In wet tripping case, the tripping rate should be lower than in dry tripping.
- In top-hole sections, pump out during POOH so that the increased BHP can counteract the swab pressure.
- Proper monitoring of the displacement volume with the trip tank is required at all times.

  In case of possible surging causing induced fracturing and consequent mud loss, the volume of mud displaced out of the hole is lower than the drill-string volume, which can be noticed by a careful monitoring of the trip tank volume changes. In case of swabbing and entrance of a kick influx, the volume of mud required to fill up the hole is lower than the drill-string volume extracted, which can be noticed by careful monitoring of trip tank volumes while implementing appropriate hole fill procedures. This can be detected by immediately observing the small discrepancy between the volume extracted from the hole and the mud volume filling the hole from the trip tank.

5.9.3 Exercises

Exercise 1:

A gas kick has been swabbed into the well during tripping operations. As the flow check is negative, it has been decided to continue with the trip. What may happen in the well during tripping?

a. Gas may migrate and expand, causing a reduction in the hydrostatic pressure.
b. Gas will migrate but not expand, so hydrostatic pressure will stay constant.

Answer a:

Any gas influx in the wellbore would migrate due to its lower weight/density than the mud. Since the annulus is open during tripping, the hydrostatic pressure would drop. It is noted that the crew have taken a wrong decision to trip if they have confirmed swabbing.

Exercise 2:

The driller has swabbed 5 bbl into the well. The bit is returned to the bottom, and bottoms-up is being circulated. The pit level is slowly increasing, as bottoms-up is getting closer to surface. What is a safe action to take?

a. Shut in the well and continue circulation through the choke using the first circulation of the Driller's method.
b. Continue circulating with well open but switch on vacuum degasser.
c. Shut in well, record SIDPP, and use Wait and Weight method to kill the well.
d. Shut in the well and use the volumetric method to remove gas from the well.

Answer – a:

In case of swabbing, it is recommended to use driller's first circulation.

5.10 RUNNING CASINGS OR LINERS

5.10.1 Explanation

The procedures used for running or pulling out the drill-string were already explained in Section 5.5.3. However, it is important to be conscious in running casings/liners because these tubulars are expected to run only once to reach their setting depths and be cemented. In addition, the casing strings and their threads must be in the best shape; otherwise, their integrity and thus the well integrity will be jeopardized. Next, the annular spacing between the casing/liner and the wellbore is much lower than that of the drill-pipe and the wellbore. Therefore, there is a greater possibility for surge (potentially causing formation fracture and mud loss) or pipe stuck. Therefore, the following considerations should be taken care of as follows:

- Receive and prepare the casing joints in advance as drilling is carried on (before the time for running casing). The casing joints must be carefully measured and drifted. Check and clean the threads of the casing strings. Any pipe which fails to drift or has a damaged pipe body or thread must be marked on the body (as OUT) and not used in the operation, to ensure well integrity.
- Check casing accessories such as the casing shoe and float collar or the Latch-Down-Collar (LDC) for liners. A main item to ensure is the type of threads of these two parts of the casing string to match with the casing joints. The floats should not be tested to prevent any back flow.
- Prior to running the casings, the BOP pipe rams should be changed to suit the casing size (for closing around them). Usually, the top pipe rams' type and size should be changed in this case, whereas the bottom pipe rams are unsafe to change because there would be no barriers below them. Next, the pipe rams must be pressure tested. Closing pressures of the annular preventer and the rams should be checked and adjusted against collapse pressure ratings of casing strings (to prevent the collapse of the strings).
- A circulating swage (with a high-pressure low-torque valve on top) should be available on the rig floor to fit the casing to be run. The circulating swage is a short crossover joint used between two sizes or specifications of casing. It acts as an adapter enabling a temporary circulating line to be rigged to the top of the casing string to:
  - allow filling up the casing,
  - allow mud circulation through the casing up the annulus for well control reasons. This is done particularly when the casing reaches the previously run casing shoe, or in case a well control event (kick flow) occurs.
- *Kick Cause 1 (Quick Tripping and Surging):* Casing/liner strings should be run smoothly, avoiding high acceleration and deceleration which can cause unnecessary surge pressures. Therefore, casing/liner running rates should be limited to maximum 45 seconds per joint or to the optimum speed from surge/swab calculations.

  Note: During casing running, the main focus is on the running operation and not on kick detection. This should not be the case; instead, enough care should be simultaneously taken in monitoring flow returns as well as pit volumes by using the trip tank so that it is possible to inform the driller of any observed potential kick/mud loss.
- *Kick Cause 2 (Improper Annulus Fill-up or Float Failure):* During running in the hole, the casing string should be filled up periodically (e.g., every ten stands). If filling up the casing with mud is ignored by the crew and the

float fails, the mud would flow from the annulus into the casing, causing a significant drop in the mud level in the hole, causing an underbalance leading to a serious kick flow situation. In another case, the empty casing may collapse due to the huge pressure difference from inside and outside of the pipe; in such a case, not only an underbalance occurs, but also casing collapse has occurred. Sometimes, to save time to periodically fill up the casing pipes, auto-fill floats (also called self-fill or differential float) in the shoe and float collar may be used to allow automatic filling of the casing string from the annulus during running in the hole. However, it must be checked by the driller and supervisor that the hole is really filled using such floats. Otherwise, the float may not work, and the casing string can be left empty, potentially causing a kick flow and/or casing collapse. Once the casing string reaches the bottomhole, a ball is dropped to convert the auto-fill float into a check valve (i.e., preventing any upward flow from the flapper of the float). Because of the mentioned disadvantages of using auto-fill floats, there is a controversy over using such floats.

- *Handling Kicks During Casing/Liner Running:* When a kick flow is detected, the well should be shut in using the pipe rams (usually the top pipe rams) or the annular preventer (refer to Section 8.3.7 for more information). Next, by connecting the circulating swage on top of the casing/liner and rigging up a circulating line, we can allow mud circulation through the casing bore (and then through the annulus up) to both prevent kick flow through the casing bore and circulate the kick influx out (kill the well).

5.10.2 Exercise

Ten joints of the 13 3/8-inch casing (each joint 40-ft long) are run in the hole with a conventional float valve (in the float collar). The casing capacity is 0.1521 bbl/ft. There was a problem with the fill-up line, and the casing was not filled. If the float valve were to suddenly fail, how would this affect the bottomhole pressure? The mud weight is 12 ppg, and the annular capacity is 0.124 bbl/ft.

Answer:

Due to the float valve failure, the empty volume is:

$$\Delta V_{Empty} = 10 \times 40 \times 0.1521 = 60.84 \text{ bbl}$$

The mud level drop is:

$$\Delta h_{mud} = \frac{\Delta V_{Empty}}{Cap_{Casing} + Cap_{Ann}} = \frac{60.84}{0.1521 + 0.124} = 220.35 \text{ ft}$$

The pressure loss due to float failure is:

$$\Delta P_{float-fail} = 0.052 \times MW \times \Delta h_{mud} = 0.052 \times 12 \times 220.35 = 138 \text{ psi}$$

5.11 POOR CEMENT AND CEMENT SETTING

A too low cement weight/density can bring about kick during and after cementing. Bad cement placement can also cause annular loss and kick. Poor cement quality due to improper design of the type and additives of cement failures can cause lack of proper coverage around the casing and thus kicks during the production phase.

Sometimes kicks may occur after the casing has been run and cemented. The most vulnerable period for the cement is immediately after its placement and prior to its setting. During this period, as cement slurry is developing gel strength, it gets self-supporting and would lose part of its hydrostatic pressure. This loss of hydrostatic pressure causes some underbalance, leading to gas invasion.

In the investigation of causes for "deep" drilling loss of well control (LOWC) events in the US GoM-OCS and the regulated areas (2000–2015), Holand (2017) mentioned that four LOWC events occurred while the cement was setting after running casing. The blowout database shows that this has been a problem in the 1980s and 1990s as well. This type of primary barrier failure is common for shallow LOWC events as well.

5.12 RISER DISCONNECT (OFFSHORE)

If riser disconnection is planned, the mud weight should be sufficiently increased in advance for a magnitude called *Riser Margin* (refer to Section 4.4.2). Therefore, following the disconnection of the marine riser, still the hydrostatic mud pressure would exceed the formation pore pressure. If this is not regarded, following the riser disconnection, a kick flow would most likely occur.

Another point regarding riser disconnection is the necessity of using advanced equipment to maintain control of the BOPs after riser disconnect. Holand (2017) mentions a loss of well control (LOWC)/blowout of a well drilling in 678 meters (2,223 ft) of water which had not considered the riser margin. The well kicked because the riser was disconnected. When disconnecting the riser, the main BOP control was also lost (loss of the secondary barrier). With the actual BOP setup, the primary and secondary barriers were not independent. Today, all US subsea-BOPs have emergency functions and/or alternative control systems; thus, the occurrence of such an event is far less likely. The emergency systems include remotely operated vehicle (ROV) controls or the acoustic backup system.

5.13 RISER GAS

In deepwater and ultra-deepwater wells while drilling or after killing the well in case of a well control event, a small gas influx may pass through the Blowout Preventer (BOP) to enter the marine riser prior to its detection. The gas in the riser would expand during moving up the riser due to pressure drops at lower heights. Gas migration and expansion through the riser may lead to the gas unloading event causing blowout and potential explosions. This occurrence is the most severe in nonaqueous/oil-based muds than in water-based muds because in such muds, gas expansion and its getting out of solution would occur suddenly at shallower depths near the surface. This would reduce the hydrostatic pressure, causing a kick influx. For more information on this incident, refer to Chapter 18.

5.14 EQUIPMENT FAILURE

Equipment failure can also occur during drilling or completion operations as a cause of kick occurrence. It may become a particular problem mainly in exploratory high-pressure, high temperature (HPHT) operations. If some rig equipment such as mud pumps, or trip tank pump fail, it can lead to a kick due to imperfection in primary well control.

5.15 HUMAN ERROR

Generally, human error is inherent in all human activities. Human error is a major cause of many kicks. This error can originate from planning and/or operation phases. Particularly, if the drilling crew are not skilled and well trained enough, they can make multiple mistakes, sequentially leading to occurrence of kicks. For instance, a string of mistakes was made by the crew that resulted in the fatal Macondo blowout and explosion in 2010. However, proper training to increase the rig crew's awareness is required for the prevention of well control events to minimize human errors.

5.16 CAUSES OF LOWC EVENTS

When kicks go out of control, loss of well control (LOWC) events or blowouts occur. Causes of loss of well control (LOWC) events can be external or non-external.

5.16.1 External Causes

The external causes of LOWC events are imposed from outside such as natural disasters like storms or hurricanes, earthquakes, and fires or malicious activities like military or bombing or collisions by boats or ships which historically have occurred mostly in shallow waters of offshore operations.

In a drilling well, an external cause can destroy the secondary barrier or the BOPs, but the primary barrier (mud hydrostatic pressure) should still remain there, unless the damage is very severe, damaging mud pumps and the well is already in an unstable situation (like mud loss).

As for a production well, an external cause only damages the wellhead/Christmas tree barrier or the topside barrier, which is considered as the secondary barrier. For a blowout (particularly surface flow) to occur, the downhole barrier must also fail. Therefore, an external cause will not be the single cause for a blowout, except for wells that are not equipped with a subsurface or downhole safety valve. In case the downhole barrier is not installed (e.g., there is no SSSV in the well), or it fails to activate or leaks, a blowout would occur.

The greatest risk contribution from producing wells in offshore operations stems from LOWC incidents caused by hurricanes (as an external cause). When a hurricane damages the topside barriers, the quality of the downhole barriers (such as tubing, packer, and SCSSV) is important (Holand, 2017).

5.16.2 Normal Causes

Normal causes of LOWC consist of all causes of loss of secondary barriers in addition to the primary barriers (which were explained in Section 1.8 and in this chapter). During drilling, the main causes of loss of secondary barrier are late kick detection, equipment failure, and human failure, as was the cause in the Macondo's blowout.

5.16.2.1 Too Late Kick Detection

If kick is detected too late, the influx fluid flow (if gas) would expand and rush out of the well quickly and may get ignited before the crew can even think of applying the secondary barrier (BOPs). In addition, in case of too late kick detection with a rushing surface flow, most types of shear ram BOPs would not be able to shear off the pipe. Thus, late kick detection is related to equipment failure in most loss of well control (LOWC) events.

5.16.2.2 Equipment Failure

Equipment failure, particularly BOP failure or defects, causes loss of secondary well control and prevents proper control and management of kicks.

BOPs may not operate to seal the well in case of a rushing surface flow. Specifically, the blind-shear rams may not shear off the pipe. For some LOWC incidents like Macondo's, when attempting to close the BOP, the flow rate through the BOP was too high, such that the blind-shear rams could not perform shearing of the pipe well. API-16A had no requirements related to BOP closure under dynamic flowing conditions. Most BOPs were therefore not designed or tested to close and seal under high-rate flowing conditions. Most BOPs are not designed to shear the pipe when it is not centered or buckled, or they may not operate well to shear the drill-collars. Therefore, Bureau of Safety and Environmental Enforcement (BSEE, 2019) has proposed some revisions to clarify BOP system requirements and to modify certain specific BOP equipment capability requirements. Thus, regulatory agencies are stricter to necessitate better functioning BOPs and other secondary barriers.

For offshore LOWCs with normal causes, most of the primary barrier failures involve the surface-controlled subsurface safety valve (SCSSV). For the majority of LOWCs, first there is a surface leak in the X-mas tree (secondary barrier failure), and then the SCSSV fails to close. For others, tubing leaks, casing leaks, multiple casing leaks, and formation breakdown were observed. The failure of SCSSV to close or too slow closure may be related to the controls, the emergency shutdown (ESD), the valve itself, sand in the well, or scale. The closed late incidents are typically incidents where the SCSSV valve is closed after the release on surface is observed.

5.16.2.3 Human Error

Human error plays a very important role in causing LOWC events as was the case with kicks. Bad management and a communications breakdown by BP and its Macondo well partners caused the oil disaster in the Gulf of Mexico. A report by expert staff mentions three companies including BP and Transocean, guilty of poor management.

"Most of the mistakes and oversights that led to the blowout were the result of management failures by these companies", the staff concluded (The Guardian, 2010).

REFERENCES

BSEE, 2019. *Oil and Gas and Sulfur Operations in the Outer Continental Shelf-Blowout Preventer Systems and Well Control Revisions.* www.bsee.gov/sites/bsee.gov/files/2022-proposed-well-control-rule.pdf (Last Accessed in May 16, 2024).

The Guardian, 2010. BP Oil Spill Blamed on Management and Communication Failures. *The Guardian.* www.theguardian.com/business/2010/dec/02/bp-oil-spill-failures (Last Accessed in May 16, 2024).

Holand, 2017. *Loss of Well Control Occurrence and Size Estimators, Phase I and II.* Report No. ES201471/2. Office/Division Program, TAP, Project No. 765. Category, Deepwater. https://www.bsee.gov/research-record/loss-well-control-occurrence-and-size-estimators

6 Kick Warning Signs, Indicators, and Detection

Rahman Ashena

6.1 INTRODUCTION

Quick and in-time detection of well control events or kicks is significant so that proper actions can be taken as quickly as possible to manage and control well control situations. Kicks cannot be detected quickly enough if warning signs/indicators are not known or if they are ignored during drilling operations. In case of lack of (gas) kick detection, the kick influx size will exponentially increase as it rises in the hole. This phenomenon is because of the gas influx expansion as it moves further up to the surface. The larger the kick size is, the more challenging and complex the subsequent well control operations will be. An early kick detection means smaller kick size and its simpler handling or easier well control operations.

In short, there are several possible kick detection scenarios in the order of severity, as follows:

Early kick detection and small kick size: With early kick detection, the kick is detected quickly just after a small size influx (size < 5 barrels) has entered the wellbore. When a small size of influx enters the wellbore, there will be low possibility of its expansion prior to well shut-in. According to procedures, following the detection, the well should be shut in immediately. Then, well kill operations must begin quickly before there is any chance for the gas kick to migrate up and expand significantly. Therefore, it is easier to deal with the kick as lower pressures will be involved during the well kill operations. Thus, it is the driller's responsibility to detect kicks when the influx is small and then to shut in the well early enough.

Not early kick detection and medium kick size: When the kick detection is not too early, the size of the kick influx is medium (5 barrels < size <20 barrels), and its top has not considerably moved up through the annulus. When a medium-size kick influx enters the wellbore, there has been just some possibility for its expansion prior to shut-in. Therefore, its top has not considerably risen or moved up through the annulus. After detection, the well must be instantly shut in. Then, well kill operations must begin quickly before there is any chance for the kick to migrate up much. As some

DOI: 10.1201/9781003473770-6

expansion has occurred, well and surface pressures are greater, and greater shut-in pressures are involved during kill operations. Therefore, in this case, kill operations and kick influx circulation out are not as straightforward as for small kicks.

Late kick detection and large kick size: With late kick detection, a large-size kick influx (> 20 bbl) has entered the wellbore, and thus its top has considerably moved up through the annulus. When a large-size kick influx enters the wellbore, there has been considerable possibility for its expansion. Therefore, its top has considerably risen or moved up through the annulus. In this case, conventional BOPs can still shut in the well successfully, but excessive pressure loads are imposed to the bottomhole, casing shoe, casings, and surface facilities (wellhead BOPs). As the top of the kick influx migrates up with time (prior to shut-in), the imposed pressures are increasing. It is extremely important to shield the well by the quickest and as most careful well kill operation as possible.

Note: Mentioned magnitudes for the kick sizes here are rough approximation. They would differ for each well case. A large kick size is determined on the basis of how the downhole pressures compare to fracture gradient and equipment limits, both initially and during migration.

Too late kick detection and very large kick size: When kick detection occurs too late, a very large size kick influx has entered the wellbore, and its top has reached or about to reach the surface. As the top of the kick influx migrates up with time (prior to well shut-in), mud with extremely high flow rate or velocity is flowing out of the well/annulus. Therefore, at this time, it is likely that most BOPs cannot shut in the well at all because BOPs are not designed to stand rushing flows. On the other hand, if the BOP can shut in the well, it is possible that the imposed pressures will soon exceed the maximum surface tolerable pressures to cause formation fractures at the casing shoe or failure of the surface-BOPs. In such extreme cases, the kick has entirely gone out of control and been already transformed to a blowout.

Therefore, it is critically necessary to be aware of possible warning signs and kick indicators using surface or downhole measurements and to monitor the well drilling parameters vigilantly to observe them. In addition, it is the responsibility of the drilling crew, particularly the driller, to continuously pay careful attention to any possible signs and indicators and take the proper action. It is also noted that not all kick warning signs can be observed for a particular instance. It must be also noted that the observation of primary kick indicators definitely confirms a kick. However, warning signs cannot confirm kick occurrence, but rather just provide an indication, unless primary indicators are visible as well. Therefore, the only definite way to confirm kick occurrence is to shut down the pumps and observe the well (called a flow check).

Therefore, this chapter discusses warning signs and kick indicators consisting of decrease in pump pressure/increase in pump rate, sudden change in rate of penetration (ROP), change in dc exponent, increase in flow line temperature, increase in torque and drag and hole fill, hole fill or sloughing/heaving shales, larger and more angular slender cuttings, lower shale cuttings density, gas-cut mud, increase in total gas levels, increase in drill-string weight, change in mud properties, mud salinity increase, and viable ECD drilling. The primary kick indicators discussed are increase in return flow rate, increase in pit volume, flow when pumps are off (flow check), improper hole fill-up during trips, and flow between stands while running the strings in the hole.

6.2 CLASSIFICATION OF KICK INDICATORS

Basically, kick indicators fall into two categories of primary and secondary, based on their importance and the extent of direct definite indication of a real kick:

- *Warning signs:* Warning signs (also called secondary kick indicators) indicate possible abnormal pressures. They are less definite and reliable indications of a possible kick compared with kick indicators. This is because some of these signs might be observed for reasons other than kicks, such as lithology change or reduction of overbalance pressure or an approaching underbalance. To verify if a kick has occurred, a flow check should be carried out.
- *(Primary) kick indicators:* Primary kick indicators provide rather direct and certain indication that a kick has probably occurred and is in progress. Therefore, a kick is ensured if any of these indicators are observed clearly. However, there is still doubt about vivid observation of primary indicators, and the only way to confirm or reject them is by making a flow check.

6.3 WARNING SIGNS

Warning signs, also called secondary indicators, provide some indications of abnormal over-pressured zones, but they do not provide definite indications about a possible kick. This is because some of these indicators might appear to be observed for reasons other than kicks. Sometimes, some of these indicators appear when mud overbalance pressure is reducing or when an abnormal or over-pressured formation is to be encountered (no kick yet).

The warning signs are explained in the next section.

6.3.1 Decrease in Pump Pressure/Increase in Pump Rate

When a kick influx, particularly gas, enters the wellbore, the mud hydrostatic pressure is reduced in the annulus. If float or check valve is installed in the drill-string (e.g., in the bit sub above the bit), the influx cannot enter the drill-string, and thus the mud weight and hydrostatic pressure in the drill-string would remain unaffected. Even if there is no float valve in the string, the possibility of entrance of part of the influx through the bit nozzle into the string is low.

With the kick influx in the annulus, due to the difference between the hydrostatic pressures in the pipe and annulus, the mud in the drill-string tends to move (*U-tube*) to the annulus (Figure 6.1). This facilitates pumping as the pump does not need to provide as much energy/pressure as before for mud circulation. Thus, a gradual reduction in pump pressure and an increase in pump rate or strokes (i.e., stroke per minute) are observed (see Figure 6.2). This effect is also contributed by the expansion of the gas influx as it rises because the expansion causes additional U-tubing from the drill-string to the annulus; this results in further reduction in mud circulation pressure (pump pressure) and additional increase in the circulation rate.

It must be noted that these effects may not be necessarily well noticeable in case of a kick. Next, the mentioned observation "*decrease in pump pressure and increase in pump rate/strokes*" may also happen in case of a pipe washout (i.e., existence of a hole in drill-string). Therefore, these observations do not necessarily mean a kick

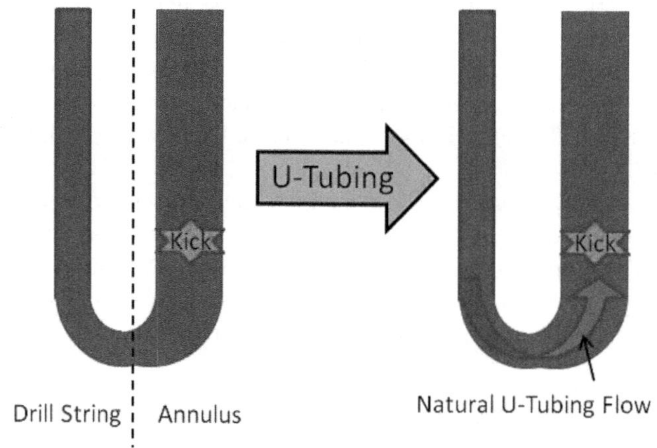

FIGURE 6.1 U-tubing effect due to kick influx into the annulus.

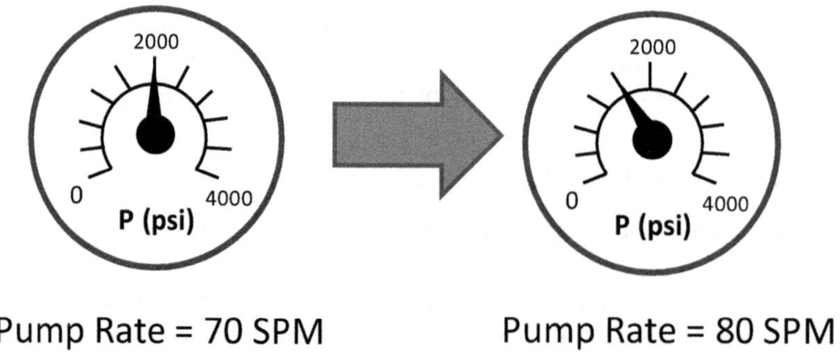

FIGURE 6.2 Pump pressure decrease (from 2,000 to 1,600 psi) and pump stroke increase (from 70 to 80 SPM) due to kick influx entrance into the wellbore.

has occurred; that is why they are considered just as warning signs or secondary kick indicators. Therefore, it is recommended to cross-check this observation with other possible kick warning signs or indicators. Again, a flow check is the most definite indicator to accept or reject a kick.

6.3.2 Sudden Change in ROP

While drilling in a fixed geological formation, ROP is kept almost constant with depth, or it decreases slightly with depth. Basically, ROP only depends on differential pressure (difference between mud pressure and formation pressure) and bit wear, provided that drilling parameters such as weight on bit, rotary speed, hydraulics, and mud properties remain unchanged. In a fixed formation and at a specified depth, if the mud rate is kept unchanged, the pump pressure should not vary, and it is expected that differential pressure is not changed either. However, since the bit wears slightly more with depth, it is expected to observe a slightly decreasing trend of ROP with depth.

In fact, a sudden significant change in ROP, i.e., two or more times the normal, is called a *drilling break*, which is indicative of a change in formation geological/lithology type or abnormal formation pressure or both. Typically, a sudden increase in ROP indicates a reduction in the differential pressure or, in other words, abnormal formation pressure. It should be noted that not only an increase in ROP but also a sudden decrease in ROP, a *negative drilling break*, can be an indicator of a coming abnormal formation. Negative drilling breaks can indicate drilling into a stiff *cap rock* which could be confining gas or oil underneath.

A sudden change in ROP, as a secondary indicator, does not necessarily signal kick occurrence but signifies that an abnormally pressured formation has been encountered. The driller watches out when a drilling break is observed. After stopping drilling, the driller should instantly proceed to making a flow check to confirm or reject the occurrence of a possible kick. It is recommended to cross-check with possible observation of other warning signs or kick indicators.

6.3.3 Change in dc Exponent

The corrected d-exponent, dc exponent, used in mud logging and formation pore pressure analysis in the oil industry, is an extrapolation of certain drilling parameters to estimate a rough pressure gradient for pore pressure evaluation during drilling. Usually, mud logging companies are required to estimate this parameter during drilling at all times. The equation for this parameter is:

$$dc = \frac{\log\left(\frac{ROP}{60RPM}\right)}{\log\left(\frac{12WOB}{1000d_{bit}}\right)} \times \frac{MW_{previous}}{MW_{Current}} \tag{6.1}$$

where ROP is the rate of penetration [ft/hr], RPM is revolutions per minute [one per minute], WOB [weight on the bit] is weight on the bit, d_{bit} [inch] is the bit diameter, and MW is mud weight [pound per gallon/ppg].

Basically, in normally pressured formations, it is expected to observe a steady increasing trend of the dc exponent with depth. However, in transition zones to abnormally over-pressured formations, the dc exponent tends to decrease to lower than expected values, that is, a decreasing trend can be observed. Therefore, a sudden decrease in dc exponent signals penetration into a transition zone or an abnormally pressured formation. This can be recognized using Equation (5.1): Since $\frac{ROP}{60RPM} < 1$, if ROP increases, then $\log\left(\frac{ROP}{60RPM}\right)$ would decrease. Therefore, the dc exponent varies inversely with ROP. In case of a drilling break, *dc* would decrease.

To better recognize the change in the exponent, the calculated values of dc exponent values should be plotted versus depth on a cartesian or semi-log paper. The normal trend line must be determined in the normally pressured zone where it should increase steadily with depth in a normally pressured formation. As mentioned, in case of abnormally pressured formation, the trend line becomes suddenly decreasing (Figure 6.3). In abnormal pressure intervals, the difference between the extrapolated values of dc exponent and those calculated from actual data can be roughly observed. Thus, it is possible to roughly estimate the extent of over-pressure of the abnormal formation.

6.3.4 Increase in Flow Line Temperature

In abnormally pressured formations or even in transition zones, the temperature gradient suddenly increases to about twice the rate of the normal temperature gradient.

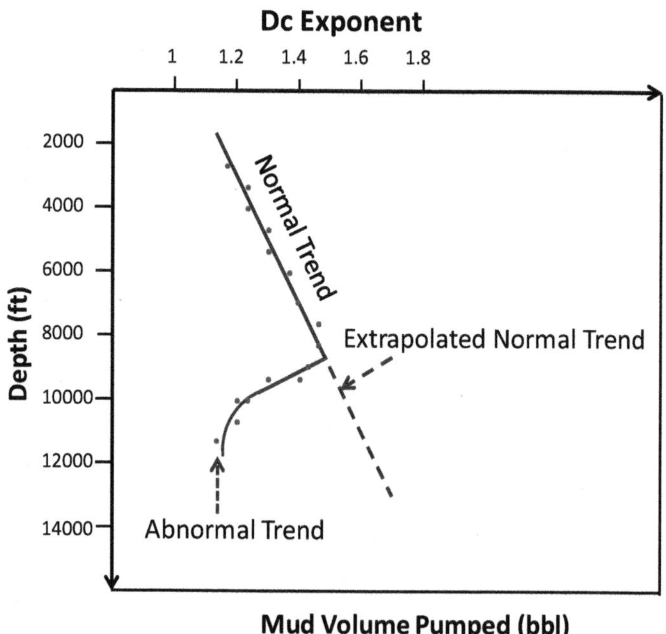

FIGURE 6.3 Utilization of dc exponent for the identification of abnormal pressures.

In such formations, the mud temperature reaching the surface shows an increase at the surface for about 2–10 °F. Thus, an increase in the flow line temperature can be used as an indicator of abnormal pressure and thus to detect such formations. It should be borne in mind that an increase in temperature does not necessarily imply abnormal pressure, because it can be due to other reasons like change in lithology, porosity increase,[1] after run in the hole, or existence of salt domes underneath (Figure 6.4). Thus, this indicator is merely secondary. It should be noted that some consideration should be given to circulation times, trip times, connection times, stabilized temperature after tripping, temperature of the mud at suction pit, etc. In addition, it is crucial to use observation of other indicators (if any) to obtain a cross-checking with this sign.

In offshore operations, due to the cooling effect of the riser and air gaps, the temperature increase may be completely counteracted, and thus this kick indicator may not be efficient.

6.3.5 Increase in Torque and Drag and Hole Fill

Generally, increases in torque and drag during drilling and tripping with the drill-string in the hole are considered as good indicators of abnormal pressure; however, as a secondary indicator, they must be cross-checked with other indicators.

Torque and drag are two drilling parameters which are usually measured at the surface. If logging while drilling (LWD) sensors can be used, torque and drag can be measured downhole which would give a better real representation of these two parameters. It is noted that *torque* is the moment required to overcome the rotational frictional force exerted by the wellbore wall on the rotating drill-string.

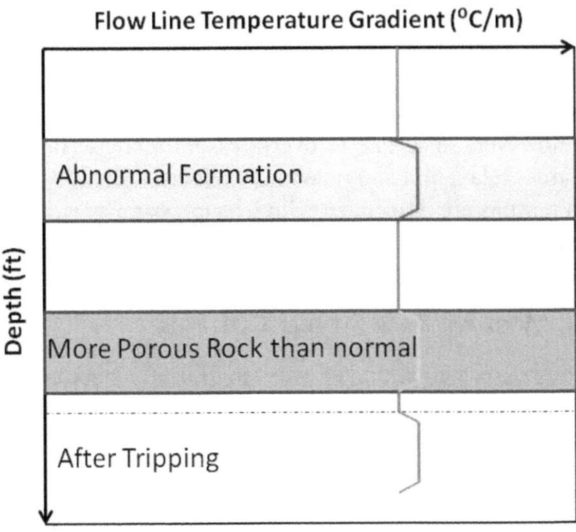

FIGURE 6.4 Log of flow line temperature gradient versus depth to detect abnormal pressures.

The "Moment" is the product of the rotational force and the perpendicular distance from the point of action of the force (related to bit and drill-string diameter). *Drag* is the additional force required to be applied on the drill-string for pulling or lowering the drill-string due to the friction between the drill-string and the wellbore wall.

While drilling in depth, torque may increase as a result of greater contact in depth between the formation wall and the drill-string (and specifically bit and BHA).

There are some sources/reasons for high torque and drag:

- Too quick drilling (too high ROP) generates too much (in amount) and large cuttings which may get into a *hole-fill,* stuck particularly to the BHA and bit, making its rotation and up/down movement more difficult. It is noted that the greater the ROP, the larger the cuttings bitten by bit cutters. Therefore, at the surface, an increase in the amount of cuttings can be observed at the shale shaker, which is considered a shaker evidence that there is huge torque and drag downhole.
- Drilling in formations prone to instability or collapse can cause accumulation of solids in the annulus and hole-fill. Formation which are chemically unstable can lose their stability and slough when drilled. In particular, formations containing shale or marl tend to slough or heave further into the wellbore when the mud overbalance is lower. Mechanical wellbore instability can occur in many formations due to excessively low mud pressures used due to excessively high stress release and weakening of the walls of the hole. The sloughed solids (from the formation) in the annulus would accumulate around and stick to the BHA as well as the bit (again hole-fill). This is called *bit balling*. For the bit is balled up, covered with cuttings. The hole fill not only increases torque but also causes extra drag. The increase in drag can be observed while drilling but mainly during pulling out or lowering the pipe for making connections or trips.
- In case of encountering an abnormally pressured formation and occurrence of an underbalance pressure condition, torque and drag increase. The main reason for this observation is the effect of lower overbalance or differential pressure. However, since due to over-pressure increase in ROP and some wellbore stress relaxation and potential wellbore instability occurrence, the above two reasons are associated with over-pressure as well and contribute to its magnification.

6.3.6 Larger, More Angular Slender Cuttings

In case of an underbalance condition or even small overbalance condition, bit cutters would take larger bites into the formation, that is, the size and shape of cuttings will be different from normal overbalanced drilling. In this case, cuttings are larger in size (as mentioned in Section 6.3.5) and volume. This can be observed/detected at the shale shaker (as the shaker evidence for a possible kick). With small overbalance and particularly underbalance, the shape of cuttings, particularly of shales, will be longer,

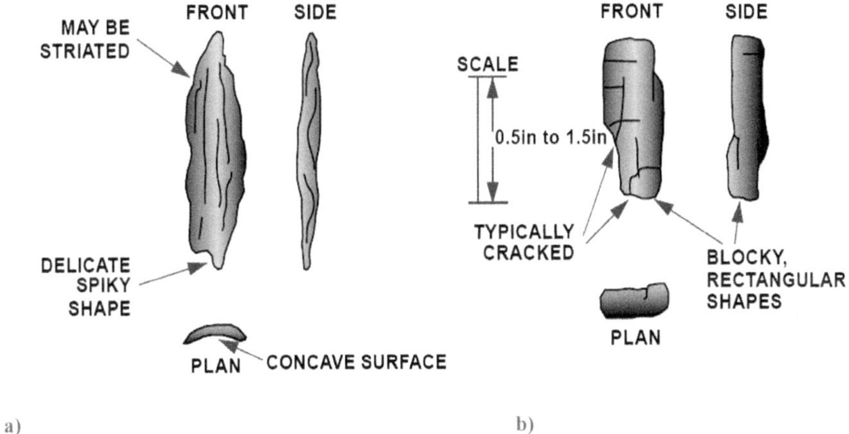

FIGURE 6.5 Shape and size of cuttings particularly shale, due to a) balanced/underbalanced condition and b) overbalanced condition (Well Control School, 2002. *Well Control Manual*, USA).

more slender, angular, and concaved shaped (Figure 6.5). In general, the volume of rock fragments will be greater in near-balanced or underbalanced condition than in the overbalanced condition. Careful observation of cuttings obtained at the shale shaker must be regularly performed by the geologist.

The mentioned change in cuttings' shape and volume occurs due to the fact that fluid trapped at a high pressure within pores of the formation, particularly the shale or low permeable rocks, finds the chance to expand when exposed to the lower mud hydrostatic pressure. Therefore, the formation is cracked by the fluid and sloughs into the wellbore, thereby creating larger and different shaped cuttings or cavings. Increase in the volume of cuttings also occurs due to a) faster penetration rates and b) increased hole volume occurring due to sloughing and caving and the underbalance.

It is noted that if hole-fill occurs due to other reasons, e.g., swelling or low stability of shales, the shale cuttings' shape differs from that when the well is underbalanced.

6.3.7 Lower Shale Cuttings' Density

Drilled cuttings must be carefully collected from the shale shaker, washed, and their densities be measured. Then, the cuttings' density is plotted versus depth to find the normal trend line of increasing density versus depth. Several techniques, such as the graduated density column method or the mud balance method, are available to measure the density of shale cuttings recovered at the shaker.

In case of lower overbalance, or underbalance, the bulk density of shales shows a decreasing trend (this is another shaker evidence for a possible kick). This occurs because abnormally pressured shales are usually under-compacted and thus their volumes are higher than normally pressured compacted shales. It is noted that under-compacted shales are those whose containing fluid has not been dissipated, see Section 4.6.1 for more information.

6.3.8 Gas-Cut Mud

A *gas-cut mud* is defined as a mud that contains gas bubbles, indicating that the returning mud weight is less than the inlet mud weight. Here, the gas-cut mud observation is minded qualitatively, not quantitatively. Since outlet mud samples are taken from near the shaker tank, it can be somehow considered another shaker evidence of possible kick. When the gas-cut mud appears at the return flow, an instant overreaction may be sometimes made to increase the mud weight. However, if no or minimal pit gain has been recorded, this reaction may not be wise. Gas-cut mud may be resourced from cuttings of gas-bearing formations and thus can show itself as *background gas level* (to be discussed later). It is important in such a case to switch on the vacuum degasser to extract gas from the returning mud.

It is interesting to note that, due to the isothermal compressibility coefficient of gas, if the mud is 50% gas-cut at the surface, there is very minimal gas volume bottomhole and the bottomhole pressure will be dropped for only 50 psi compared to when it is located at 10,000 feet depth. If the gas-cut is increased from this value, then it can become dangerous. However, if no pit gain has been observed and the gas-cut has remained constant with time, it is probably just a gas zone being drilled, and thus it is not a hazardous situation. Therefore, gas-cut mud should be paid attention and regarded as a secondary kick indicator along with other possible indicators.

6.3.9 Increase in Total Gas Levels

Another shaker evidence of a possible kick is a (rather quantitative) increase in the gas content/cut of the returning mud, considered as an indicator of bit penetration into an abnormal formation. Thus, gas detectors are essential instruments for the measurement. Gas is typically obtained from a suction line above the gas trap located immediately upstream the shale shaker. The gas detectors can measure any change in the relative content of gas in the returning mud as well as the cuttings, though they do not normally measure absolute gas contents in the mud, but they measure the relative contents.

Generally, an increase in the gas content of mud at return flow line does not necessarily mean an underbalanced condition. It can just mean an increase in the gas content of the formation being drilled (due to possible lithology, porosity, or permeability changes) and thus higher gas released to the mud from the generated cuttings or caving.

As part of the mud logging output, Figure 6.6 illustrates the curve of total gas level versus depth. As shown in Figure 6.7, if a gas chromatography is also utilized at the surface, a single curve may be provided for each gas component versus depth (methane, butane, propane, etc.). In this curve, a rather straight background gas level can be characterized for each depth interval. If drilling is performed near balanced, there is a connection gas detectable too. Otherwise, it cannot be detected.

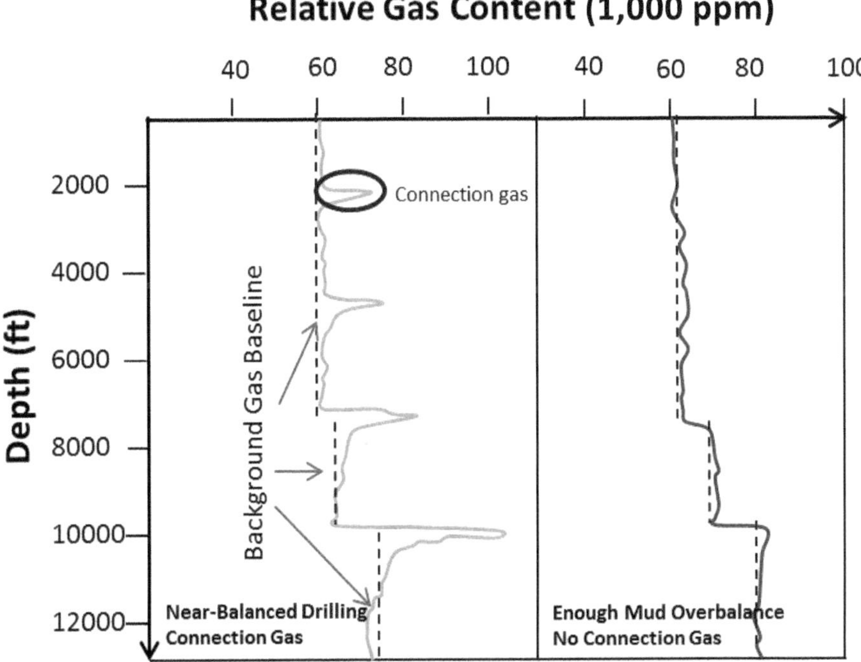

FIGURE 6.6 Log of relative gas content versus depth: a) Near-balanced condition with connection gas and b) high enough overbalanced condition and thus no connection gas.

Different types of gases observed in returning mud are:

- *Background gas level:* Background gas is characterized as an average or baseline measure of gas entrained in the circulating mud. This baseline trend is related to the gas that is released from the cuttings to enter the mud downhole while drilling through a uniform formation interval at a constant ROP. The gas entered from cuttings to the mud while cuttings are removed to the surface is called *drill gas*. In conditions of normal pressure and normal overbalance, background gas should not vary significantly as the hole is drilled. Therefore, a sudden increase in the normal background gas content from one depth interval to the next interval indicates that the flow of formation gas into the mud has occurred due to either underbalance/kick or just change of formation properties and thus higher gas content and consequently more gas entrance to the mud from drilled cuttings.
- *Connection gas*: If drilling is performed with a minimum required mud weight (very small overbalance), swabbing may occur due to quick upward pipe movement during connections, bringing about underbalance allowing the formation gas to enter the hole. There is another possibility for connection gas: The drilling might be conducted in (equivalent circulating

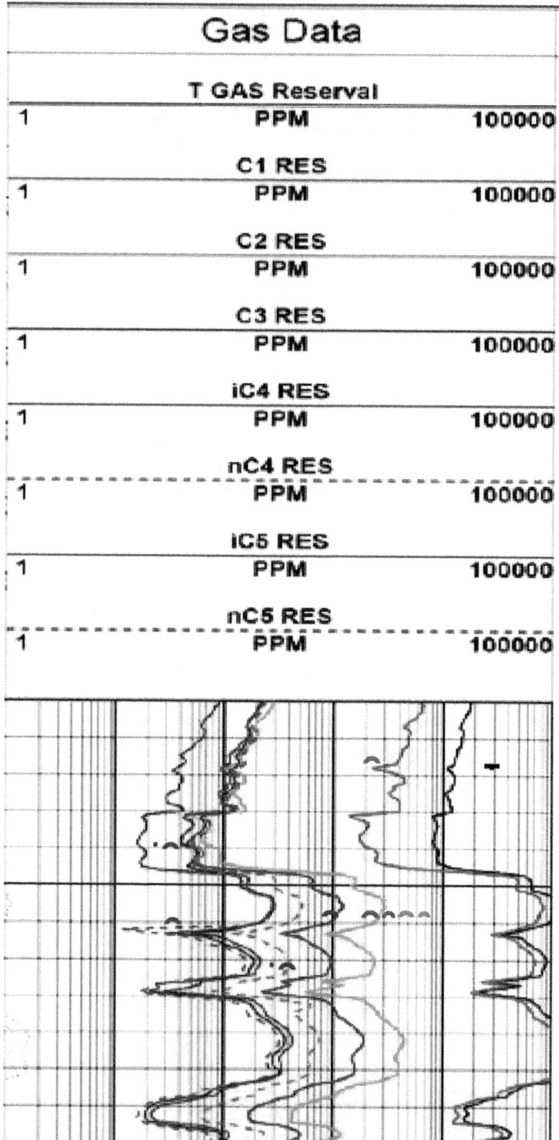

FIGURE 6.7 Log of relative total gas content and its constituent gases (methane, ethane, etc.) versus depth.

density) mode meaning that the static mud pressure is underbalanced, but the dynamic mud pressure is overbalanced; thus, during connection when the well is static, underbalance allows connection gas. Connection gas is defined as a measure of gas swabbed into the hole while pulling up for a connection. The time to observe this gas at the gas connector after the connection can be identified by estimating the period taken to pump the mud from the bottomhole. In the total gas curve, after an increase in gas content due to connection gas, the units should again return to the background levels. Otherwise, an underbalanced condition could exist.

It is possible to eliminate connection gas by establishing a high enough mud weight (and thus the differential mud pressure) or reducing the pulling rate and/or adjusting mud properties. However, connection gas, if and when it happens, can be considered as a good indicator of formation pressure in near-balanced drilling. In other words, an increase in the level of the connection gas is a warning for a possible underbalanced condition.

- *Trip Gas*: Trip gas is very similar to connection gas except that it is a measure of swabbed gas over an entire tripping operation. It is sometimes a normal practice in tripping to first make a short trip of 10–20 stands and circulate bottoms-up and detect possible swabbed gas (if any). Therefore, if very high trip gas is observed during this short trip, the mud should be weighted to increase the mud overbalance/trip margin and/or counteract possible swab pressure later.

6.3.10 Increase in Drill-String Weight

Since the drilling mud provides buoyancy for the drill-string, it reduces its actual weight supported by the derrick. When a kick influx enters the wellbore to reduce the mud weight, the buoyant force provided by the mud is reduced, and thus an increase in drill-string weight may be observed at the surface.

6.3.11 Change in Mud Properties

Mud samples should be continuously taken at the suction pit and from the return line at the shaker box. Next, some mud properties (density, viscosity and rheological properties, and pH) should be measured both at the inlet and outlet. Next, their measurements should be compared to recognize what could have occurred after its circulation taken. When the kick fluid enters the mud, it can change the mud properties. Depending on the fluid type from the formation, there would be different effects on the mud properties. If the kick fluid is gas (without impurities), its entrance into the mud may make little or no change on the mud chemical and flow properties though it would just lower the mud density and additionally make the mud foamy with higher viscosity. However, if the gas contains H_2S or CO_2, it will disturb the mud properties, see Section 3.5.

Entrance of formation brine or oil alters mud properties slightly as it would reduce the mud weight and change the composition and chemistry of the mud. Chemically, formation brine entrance into the mud changes viscosity and rheological properties, increases fluid loss (filtration), and reduces the pH value. It is noted that the extent of any detrimental effects of contaminants is largely dependent upon the amount of undesirable drilled solids in the mud. In freshwater muds or slightly treated muds, the chemical effect of a possible kick is considerable.

In brief, a change (even small) in mud chemical and rheological properties can signify the previous entrance of a possible kick influx into the hole. If inhibitive muds are used, the effect of kick fluid on mud chemistry and flow properties may not be negligible or invisible.

6.3.12 Mud Salinity Increase

Invasion of the formation brine into the mud can sometimes be detected by a change in the average salinity of the outlet mud returning from the annulus. Usually, formation water is three to five times saltier than drilling muds, and thus an influx can be detected by a significant increase in the chloride content of the mud filtrate.

6.3.13 Viable ECD Drilling

The dynamic mud pressure and bottomhole pressure are greater than those in static condition, which are due to the effect of the annular pressure losses in the dynamic condition. It is possible in some cases that the bottomhole pressure exceeds the formation pore pressure when pumps are running/in dynamic condition (overbalance), whereas it is lower than pore pressure (underbalance) when pumps are off/in static condition.

Thus, there is no extra outflow rate observed when mud pumps are running (with mud circulation). However, when the pumps are shut down (and thus mud pressure becomes static), flow is observed at the surface. This usually occurs in narrow safe mud window cases, and it is not a safe drilling mode. Therefore, it is recommended to use managed pressure drilling system (MPDs).

This kick indicator resembles flow check and may have some definite indication. However, it may be confused with ballooning, see Section 4.13 to know how discriminate it from ballooning. Therefore, it is recommended to consider it secondary and cross-check it with other warning signs to verify the kick.

6.4 PRIMARY KICK INDICATORS AND KICK DETECTION

Primary kick indicators, sometimes called kick indicators, provide direct, definite, and reliable indications of a kick. When one of the primary indicators is truly observed, the possibility of an already occurred kick is definite. With the chance of kick occurrence confirmed, there is even no need for making a flow check. Therefore, it is recommended to proceed to well shut-in.

6.4.1 Increase in Return Flow Rate

In case a kick influx gets into the well during mud circulation, there would be an increase in the mud flow rate exiting the well. This increase in flow rate is a definite sign of a probable kick in progress, and thus it is considered as a primary indicator of kick. Conversely, a decrease in return flow rate is a definite indication of mud loss to the formation.

This observation is considered as the first reliable definite indicator of a kick-in progress as there are few other possible causes for an increase in flow rate. The possible causes are ballooning (see Section 4.13 for more information), low pump efficiency, and standpipe drainage. If the possibilities of these occurrences are rejected, any increase in flow rate can be directly related to kick.

Thus, it is necessary to measure the mud outflow rate to obtain an indication of any variations in the return rate. Basically, there is a wide range of mud types, weights, viscosity, solid contents, and other factors. Therefore, it is required that mud flowmeters have the ability to handle these wide ranges of different parameters.

Conventionally, a mechanical *flow paddle* is used as the flowmeter (Figure 6.8). It is simply installed directly on the mud return line to measure the mud flow rate as a percentage of maximum possible. Conventional flow paddle employs a single string plate or paddle extending down into the flow and is deflected upward by fluid impinging on it. The extent of deflection is a function of the impact force of the fluid on the paddle. The impact force is in turn a function of mud height in the line and velocity and thus the mud flow rate. As defects of such flowmeters, the flow paddle is not indeed accurate enough to detect a little volume increase due to small kick influxes. For example, if a kick occurs from rather low-permeability formations, the possibility of observing the small variation in flow rate is low. An important issue with paddle-type flowmeters is that they do not provide absolute flow rate values but

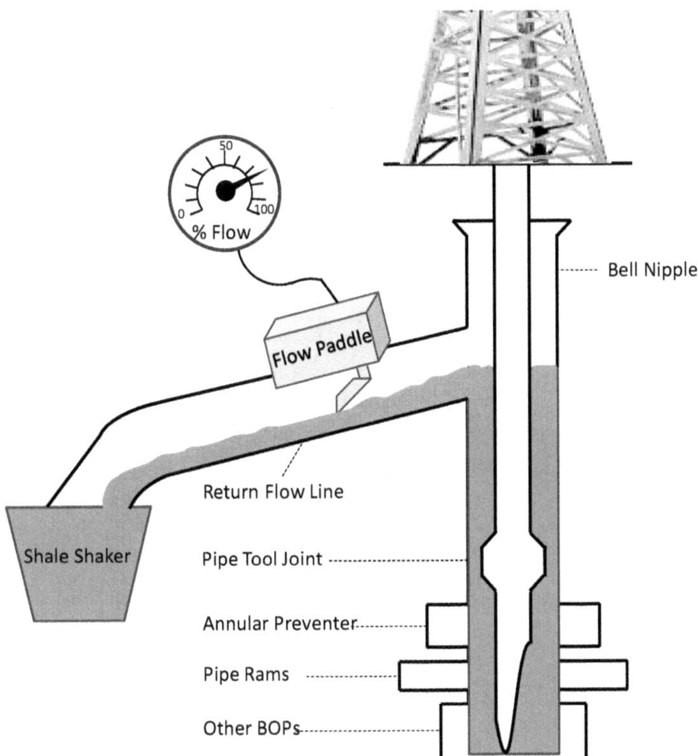

FIGURE 6.8 Conventional flow paddle on the return line to measure return outflow rate.

rather just qualitatively percentage of the output as the fraction of full paddle deflection. This flowmeter is also prone to wear due to contact with the drilling mud and cuttings. However, this flowmeter is still the most used on the drilling rigs because of its low cost and is the requirement of most operators.

Due to the high importance of accurate flow rate measurement for kick detection, particularly in offshore operations, the drilling industry's flowmeter manufacturers have recently improved measurement mechanisms and flowmeter technologies for measuring the rate of mud return flow. In addition, more accurate flowmeters than flow paddles can be utilized such as *ultrasonic flowmeters* or *Coriolis meters.*

An ultrasonic sensor is installed on top over the mud return line to measure the mud level in the return line using ultrasonic waves (Figure 6.9). This sensor can measure outflow rate up to 1,200 gallon per minute (gpm) with the accuracy of 2% (+/– 24 gpm in 1,200 gpm). Unlike the flow paddle, this flowmeter has the advantage of having no moving parts. As the travel speed of ultrasonic waves to the mud surface and back to the receiver is affected by temperature, the temperature is measured via three thermometers. Sometimes, this flowmeter may be blocked by impurities and solids. Therefore, filter screens can be used to protect against blockage of ultrasonic flowmeters.

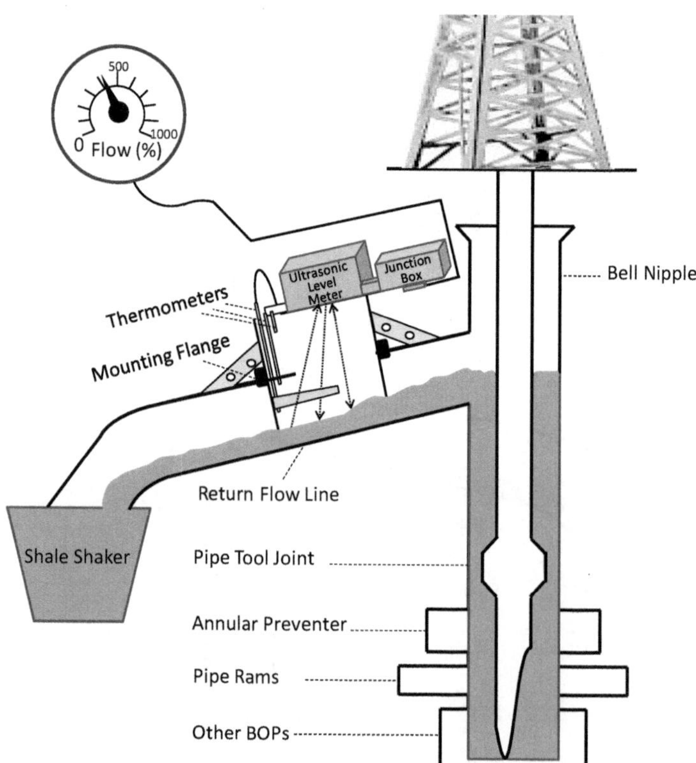

FIGURE 6.9 Ultrasonic return flowmeter equipped with level meter and thermometers on the return line.

Coriolis sensors consist of a manifold which splits and directs the fluid flow into two tubes to measure mass flow rate and back out the outlet side of the return line (Figure 6.10). Coriolis sensors which can be easily installed on a bypass on a conventional bell nipple are able to measure volumetric flow rate with an accuracy of 0.05–0.1%. It is noted that Coriolis sensors measure not only the absolute flow rate but also mass, density, and temperature.

In subsea rigs, the measurement of flow rate and its change is typically more challenging than in land operations because of the effect of pitch (rotation along axis running side to side), roll (rotation along axis running aft to stern), and particularly heave (i.e., vertical) motions. In fact, mud return flow rate continuously changes due to continuous heave, particularly in deepwater rigs (seawater > 600 m/1,969 ft). Therefore, to correct these effects, particularly the heave, two heave position sensors are installed on the *slip joint* to track any displacement or movement of the rig. In specific, to damp out the effect of heave on outflow rate, the measured outflow values are averaged over the rig heave movement. In addition, an angle sensor is employed to contribute to better outflow determination. However, using averaging systems, it is hard to spot a true flow rate change.

For more information on flowmeters, refer to Chapter 16.

6.4.2 Increase in Pit Volume

In case of a kick influx entry into the hole, after an increase in flow rate out of the hole, the volume of mud in the active mud tanks at the surface would inevitably increase. This increase is usually called *pit gain*, which simply occurs because the return mud input from the well to the active mud pits (i.e., outflow rate of return mud) is greater than the output from pits into the drill-string. Conversely, a decrease in mud pits' volume usually indicates a mud loss to the formation.

Conventionally, the volumes of pits are measured by measuring the mud levels in the pits. Therefore, if kick occurs, the mud pit level would rise consequently. The

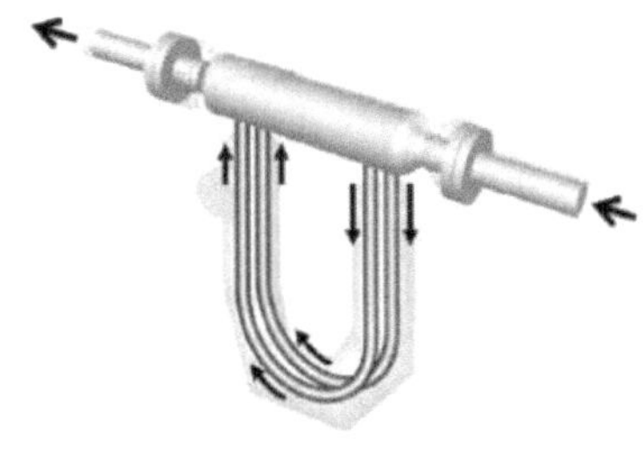

FIGURE 6.10 Coriolis flowmeter installed on a bypass from return line for accurate return flow rate measurement (http://www2.emersonprocess.com/siteadmincenter/PM%20Micro%20Motion%20Documents/Drilling-Fluid-Monitoring-WP-001243.pdf, last accessed in April 3, 2016).

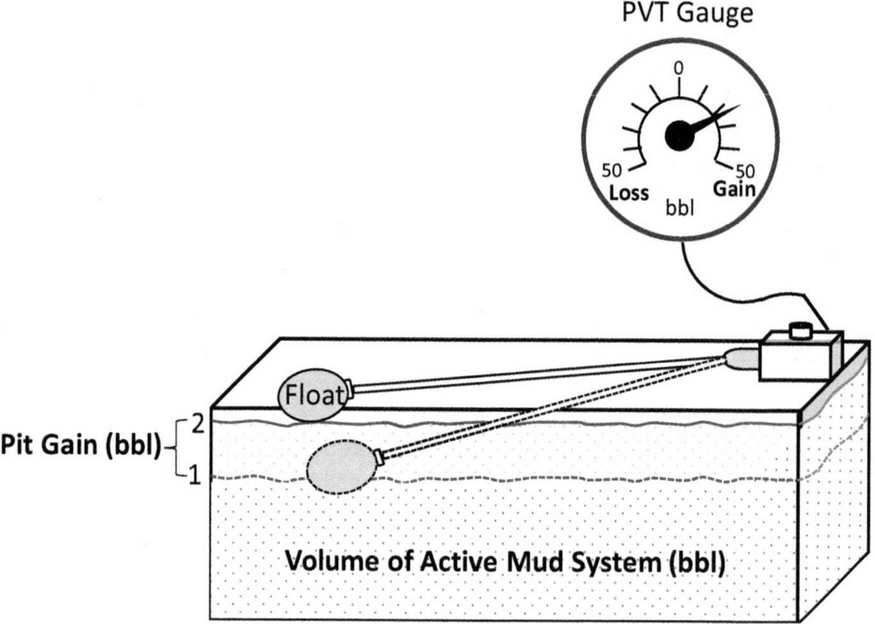

FIGURE 6.11 Conventional Pit Volume Totalizer (PVT) gauge to measure pit gain to and loss from the *Active Mud System*.

mud level is determined using *float(s)* (Figure 6.11) as well as by the mud engineer. The gain volume is calculated by multiplying the measured level increase by the length and width of the rectangular mud tank. Although this method is commonly practiced, it is not indeed accurate because the pit gain is dependent on the measured pit level change, and pit level change is detectable only when considerable gain occurs. Therefore, if the volume of a kick influx (kick size) is small (< five barrels), the detection of pit level increase is almost impossible. To increase the accuracy of measuring mud level of a tank, an ultrasonic level meter can be used in lieu of float system (Figure 6.12). Detecting small kick sizes of less than five bbl is only possible using early kick detection systems including AI-based methods (Chapter 13), accurate outflow meters (see Chapter 16), and downhole sensors.

In floating rigs, the measurement of pit volume is not accurate and reliable enough because the movements of the rig cause undesirable changes of the mud pit level. To mitigate the negative effects of such movements on measurements to acceptable levels, several sensors (e.g., floats) must be utilized for each pit.

It is noted that there are several other factors (other than kick influx and mud loss) which can cause a change in pit level. One of the important factors is mud transfers causing confusion of rig crew to detect kick and loss problems. However, there are only very few factors (except for kick and loss) that can cause changes in rates of return flow. Pit level change is also possible for considerable kick sizes of larger than five bbl. Therefore, the practical use of return flow rate for the purpose of kick

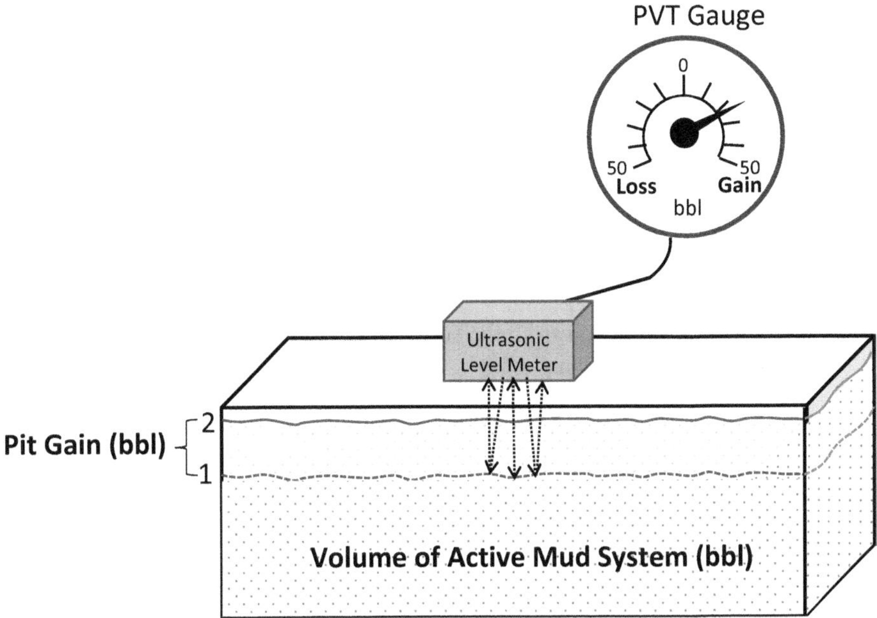

FIGURE 6.12 Ultrasonic level meter for accurate pit gain or loss evaluation.

indication can be greater than the pit volume. The problem with this indicator is lack of proper and reliable flowmeters and cost, whereas mud level can be practically measured by the mud engineer as well as by the sensors. The other factors causing change in active mud system volume are as described next:

Transfers: Sometimes, the mud engineer may proceed to such transfers without noticing the driller or drilling supervisor (wrong). In such cases, it is crucial that the rig is equipped with a software that can consider the mud transfers, evaluate the expected active mud tank volume, and raise alarm in case of unexpected volumes.

Surface additions to the active mud system: When a continuous addition is being made to the mud of a tank, e.g., dilution with water, the additional rate should be determined and monitored so that any possible further increase can be attributed to kick occurrence. Adding significant amounts of additives, particularly weighting material, also changes the total volume of active mud system. These volume changes should be calculated using the corresponding relations and considered in the active pit volume (refer to Equation 3.5).

Withdrawals from the mud system: The continuous utilization of solid control and mud screening equipment on the active mud system while drilling makes a slow continuous reduction in the volume of mud. The rate of loss can be estimated if the mud engineer has enough experience with the equipment. Such a continuous loss easily masks small continuous gains.

Solid control system: It is critically important that the driller, tool pusher, and drilling supervisor notify any actions of switching on and off of the solid control equipment. Consider a condition that solid control equipment is running, a small kick is entering the wellbore, and the pit level is kept constant. If the driller already knows that solid control equipment is in operation, he/she would immediately suspect a coming kick and would take action for well control. Otherwise, if the driller has not been informed by the mud engineer that the solid control equipment is not working, he would misinterpret the condition as normal and would not take any well control action.

Dumping: A portion of waste mud may be dumped as a withdrawal from the mud pit. This reduction from the active mud system must be considered in active pit volume system.

Therefore, to consider the effects of other factors on pit level and for cross-checking purposes with floats and ultrasonic measurements, it is required that the actual mud pit level is visually observed and measured by the mud engineer at regular intervals with notes of any additions and changes made. This provides a real direct reference to the measured values and complements observations for the driller and crew. Therefore, the crew can make the proper action in case of any suspicious observations.

6.4.3 Flow When Pumps Off (Flow Check)

If one or some of the mentioned warning signs is/are observed, a flow check should be carried out to confirm or reject a kick. The correct procedure for a flow check is as follows:

1. Pick up the drill-string and space-out (such that you see a tool joint above the rotary table).
2. Shut down the mud pumps.
3. Line up the well on the trip tank with a known volume already in the tank.
4. Check for any possible return flow.
5. If any flow is observed, it is reported as a positive flow check. If no flow is observed, it is reported as a negative flow check, and drilling can be continued. The depth, time, and duration of the flow check should be recorded. In case the flow check is confirmed positive, it confirms that a kick is in progress.

6.4.4 Improper Hole Fill-Up on Trips

Figure 5.1 shows a schematic of the hole, BOPs, shale shaker, trip tank, and its indicator. It is important that the hole is filled up with mud properly during POOH so that the correct volume of mud is added to the hole to replace the drill-string volume extracted. Similarly, during the process of run in hole, it must be made sure that the correct mud volume is returned to the surface (from the well) as the drill-string is run in hole.

Therefore, while tripping, it is required to compare the change in the trip tank volume with the hole fill-up volume. It is possible that there is a discrepancy between the two volumes, that is, the mud return volume exceeds the volume of the drill-string displaced while run in hole; or the hole fill-up volume is lower than the removed drill-string volume while POOH. In this case, perhaps a kick has occurred. If such a discrepancy is observed, in such cases, the most secure way is to shut in the well and monitor pressures.

In case there is high possibility of swabbing (e.g., due to tight hole or slim-hole geometry), it is recommended to first make a short trip of a few stands with bottoms-up. If the trip was made successfully, then a full trip can be safely performed.

6.4.5 Flow between Stands while Running in Hole

Specifically, when a stand of pipe is run in the hole, some mud returns to the surface, which is expected to be equal to the displaced/inserted pipe volume. After the stand is run in the hole and the slips are set around it, it is expected that the flow diminishes in a few seconds. If the flow does not diminish, a kick may have occurred. It is also possible to double check with other indicators. Subsequently, the well can be shut in.

NOTE

1 Water heat conductivity is one-sixth to one-third of any matrix material. The greater the porosity, the smaller the bulk heat conductivity, and thus the more heat can accumulate in the formation causing temperature rise.

7 Prerecorded Data for Possible Well Control

Rahman Ashena

7.1 INTRODUCTION

Just before tagging the formation by the bit and starting the actual drilling, some data must be essentially recorded which can be later useful for well control purposes. Therefore, such data are measured to be kept in advance for subsequent kill sheet calculations in case a well control event/kick situation arises. The most significant data to be recorded prior to drilling are a) the dynamic drill-pipe pressure loss values at Slow Circulating Rates and b) volumes and strokes. Such data is measured for surface-BOP and subsea-BOP wells with some differences, which are explained in this chapter.

This chapter covers Slow Circulating Rates and dynamic pressure losses for surface and subsea-BOPs, kill and choke-line fluid densities, and calculation of volumes and strokes used for completing the pre-kick kill sheet.

7.2 SLOW CIRCULATING RATES AND DYNAMIC PRESSURE LOSSES

Slow Circulating Rates (SCRs), also called Reduced Circulating Rates, are pump rates giving circulation flow rates ranging from one to five5) barrels per minute. Circulation rate in "gallon per minute (gpm)" is found by multiplying the pump rate (stroke/minute, or SPM) by the pump displacement or pump factor (barrel per stroke or bbl/stroke). Typical values of SCRs range from 30 to 50 SPM.

Well control operations must be conducted at SCRs. There are several reasons for using slow rates:

- Minimize friction pressure losses and annular pressures (to prevent formation fracturing at casing shoe or reaching Maximum Allowable Annular Surface Pressure, MAASP),
- Allow the choke operator for more controlled choke adjustments,
- Provide further time for weighing up the mud (required for the second circulation of Driller's method),
- Allow for degassing of the mud (to suit the working pressure or capacity of the mud gas separator) for disposal of the influx,
- Reduce the chance of choke erosion/washout, and

 DOI: 10.1201/9781003473770-7

- Reduce risk of over-pressuring the system if plugging occurs (by preventing pressure shocks in case of plugging of bit nozzles or choke).

As well control operations are carried out at SCRs, we must be, in advance, aware of their corresponding dynamic pressure losses, called Slow Circulating Rate Pressures (SCRPs). SCRPs must be already taken/measured for all the mud pumps at the rig. They are obtained in the following cases:

- At the beginning of every tour (if possible),
- Any time the mud properties are changed,
- Any time the bit nozzles or bottomhole assembly (BHA) is changed,
- After drilling of at least 1,000 ft of new hole in non-reservoir zones/500 ft of new hole in reservoir formations,
- After significant change in mud pump or surface equipment or after repairs, and
- After tripping back to the bottom and circulations bottoms-up.

SCRPs and their SCRs should be recorded regularly on the updated pre-kick kill sheet. This update is made traditionally manually or by software.

7.3 DYNAMIC PRESSURE LOSSES FOR SURFACE-BOP WELLS

For each pump, minimum two dynamic pressures losses are taken at two Slow Circulating Rates (SCRs). These pressures are also called Slow Circulating Rate Pressures (SCRPs). In a surface-BOP well, the procedure to obtain SCRP(s) is as follows (refer to schematic in Figure 7.1):

1. Bring pump#1 to an SCR value (e.g., 30 SPM).
2. Read the Drill-pipe Pressure (DPP) Gauge at the standpipe. This represents the first SCRP ($SCRP_1$).
3. Raise SCR to, e.g., 30 or 40 SPM.
4. Read the Drill-pipe Pressure Gauge. This represents the second SCRP ($SCRP_2$).
5. Switch off pump#1 and repeat steps "1" to "4" for Pump#2.

7.4 DYNAMIC PRESSURE LOSSES FOR SUBSEA-BOP WELLS

Subsea pressure losses consist of dynamic pressure losses through the riser, through the choke-line, and the Choke-Line Friction Pressure (CLFP).

7.4.1 Dynamic Pressures and Choke-Line Friction Pressure (CLFP)

In subsea-BOP wells, dynamic pressure losses may be taken through the riser and choke-line or kill-line. The dynamic pressure loss through the riser is obtained as the drill-pipe pressure when pumping mud through the drill-string, nozzles, back through annulus and riser (with the BOPs open) to the shale shaker. The

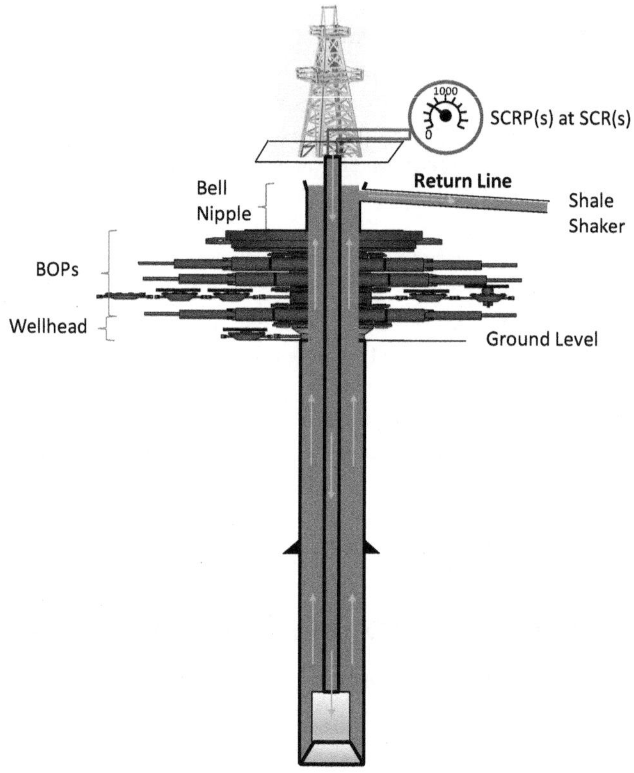

FIGURE 7.1 Slow Circulating Rate Pressures (SCRPs) or dynamic pressure losses at slow rates should be taken at minimum two Slow Circulating Rates. The gauge reading of the drill-pipe pressure represents the SCRPs.

dynamic pressure loss through the choke-line is important during well kill operations, which is the drill-pipe pressure when pumping mud through the drill-string, nozzles, back through annulus and choke-line (with the BOPs closed) to the choke manifold at the surface. The Choke-line Friction Pressure (CLFP) is the frictional pressure loss through the choke-line. This pressure loss is a constituent of the total dynamic pressure loss (or drill-pipe pressure) when fluid is pumped through the choke-line.

The CLFP should be taken at Slow Circulating Rates a) prior to drilling the first casing string following BOP installation and b) after any significant change in drilling mud weight or other drilling fluid properties.

It is important that the CLFP is known for a wide range of circulating rates. This signifies that the extra pressure on the well would be known at a range of circulating rates, and thus the most suitable circulating rate can be selected.

a)
Dynamic Pressure Loss Through
Choke-line
= 900 psi

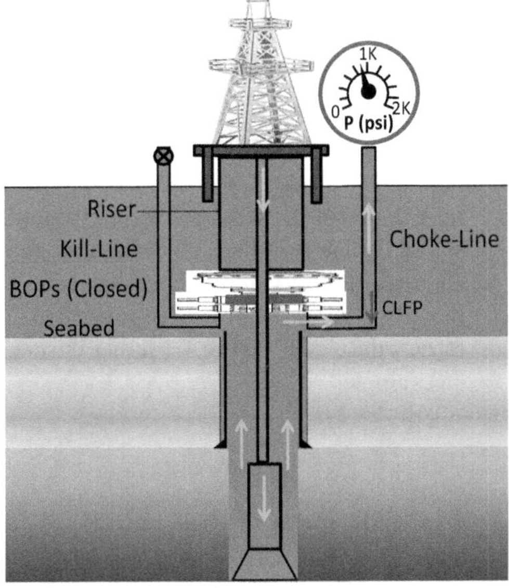

b)
Dynamic Pressure Loss
Through Riser
= 750 psi

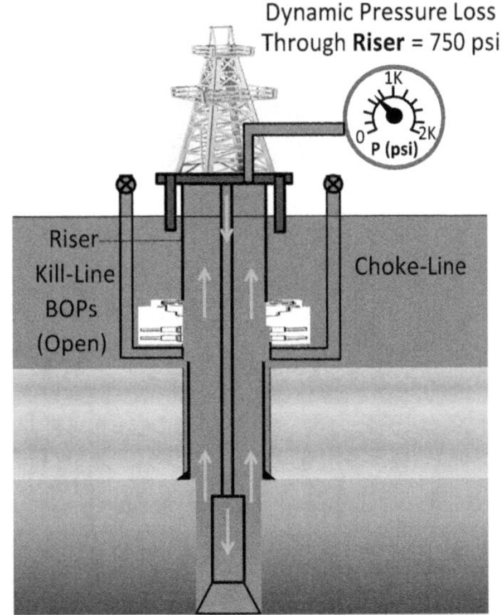

c)
Choke-Line Friction Pressure (CLFP)
= Dynamic Pressure Loss Through Choke-line – Dynamic Pressure Loss Through Riser
→ CLFP = 900 – 750 = 150 psi

FIGURE 7.2 The first method of evaluating the Choke-line Friction Pressure (CLFP) at Slow Circulating Rate (SCR): a) Find the drill-pipe pressure required to circulate the well through a full-open choke with the BOP closed. b) Next, find the drill-pipe pressure required to circulate the well through the marine riser (with the BOP Open). c) Subtraction of the second pressure from the first one which gives CLFP.

7.4.1.1 Methods for Recording CLFP

There are four recognized methods for evaluating the CLFP. The first two cases are the most commonly used prior to drill-out. The third method is used mostly in deviated wells.

1. Read/find the drill-pipe pressure required to circulate the well through a full-open choke with the BOP closed. Next, read the drill-pipe pressure required to circulate the well through the marine riser with the BOP open. Subtraction of the second pressure from the first one gives CLFP. Figure 7.2 shows the methodology schematically and with an example.
2. Circulate the well through a full-open choke with the BOP closed while recording the pressure on the (closed) kill-line. The kill-line pressure represents the CLFP. Figure 7.3 shows the methodology schematically and with an example.

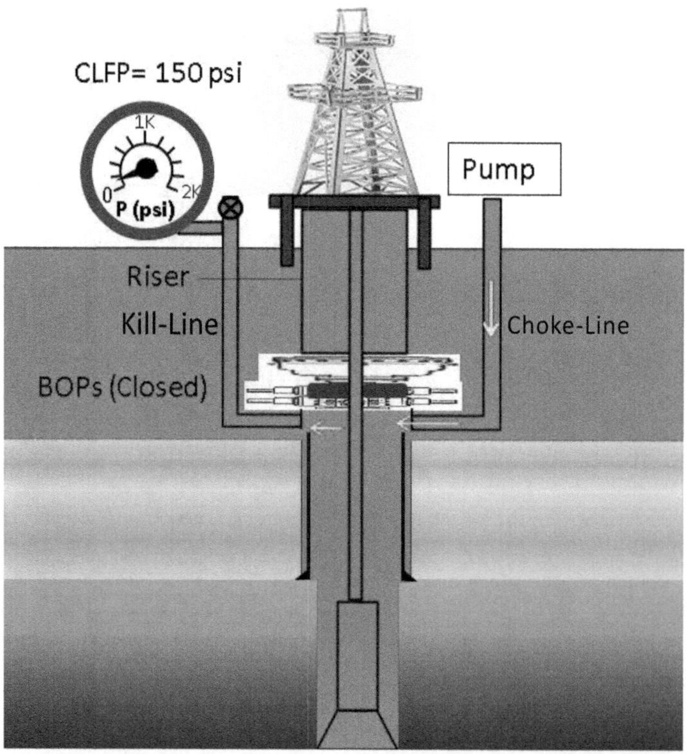

FIGURE 7.3 The second method of evaluating the Choke-line Friction Pressure (CLFP) at Slow Circulating Rate (SCR): Circulate the well through a full-open choke with the BOP closed while recording the pressure on the (static) kill-line. The measured kill-line pressure represents the CLFP.

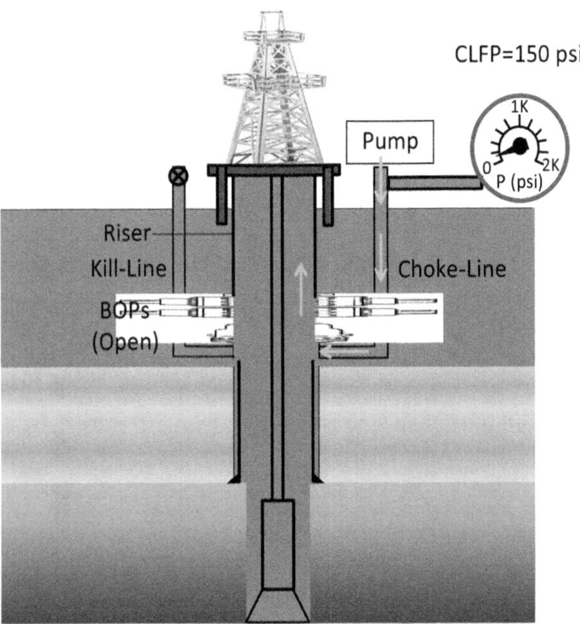

FIGURE 7.4 The third method of evaluating Choke-Line Friction Pressure (CLFP) at Slow Circulating Rate (SCR): Circulate down the choke-line and up the marine riser with the BOP open. The measured circulation pressure represents the CLFP.

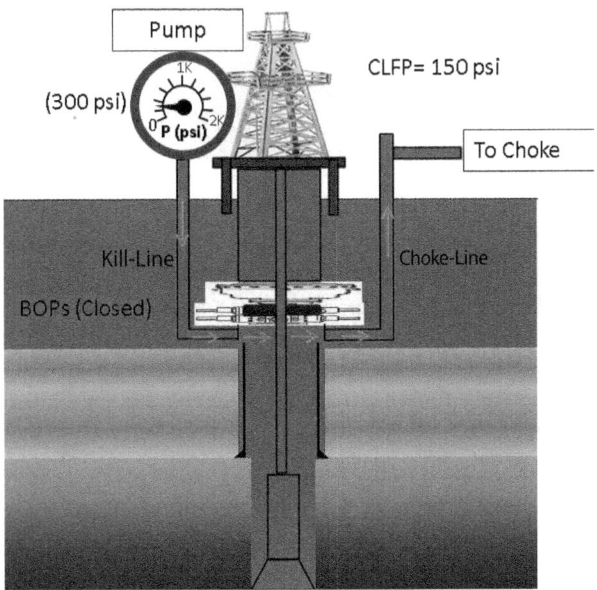

FIGURE 7.5 The fourth method of evaluating Choke-Line Friction Pressure (CLFP) at Slow Circulating Rate (SCR). CLFP is half the measured pump pressure.

3. Circulate down the choke-line and up the marine riser with the BOP open. The pressure required for circulation represents the CLFP. Figure 7.4 shows the methodology schematically and with an example.
4. Circulate down the kill-line taking returns through the choke-line (with-full-open choke) while the riser and wellbore isolated by closing the BOPs. The recorded pressure is double the CLFP. Figure 7.5 shows the methodology schematically and with an example. This method is very practical as well.

It is noted that the pressure readings should be taken at the choke manifold (using choke manifold pressure gauge sensor), rather than the pump pressure gauge. This is recommended to eliminate the pressure losses between the pump and the choke manifold. It is also noted that CLFP should be found at different Slow Circulating Rates.

In case the fluid in the choke-line is replaced by a different fluid weight, the new CLFP is found as:

$$\text{New CLFP} = \text{Old CLFP} \times \frac{MW_{New}}{MW_{Old}} \tag{7.1}$$

7.4.1.2 Reasons for Consideration of CLFP

The magnitude of CLFP depends on its length and inside diameter as well as the properties of the fluid flowing through it. For deepwater operations (seawater > 600 m/1,969 ft), particularly with a heavy viscous mud, the magnitude of CLFP is significant when circulating up the choke-line and should not be ignored. Generally, there are some consequences if CLFP is ignored:

- In case CLFP is ignored, it will lead to a large increase in the pressure applied to the casing shoe, which may cause formation fracturing. CLFP applies a backpressure which acts downward. It signifies that it causes an increase in bottomhole pressure. Therefore, in subsea cases, the MAASP will have to be reduced by the amount of the CLFP whenever circulating through the choke-line.
- There is another consequence if CLFP is not considered. During well kill operations, in the case the SICP is less than the CLFP, the latter pressure cannot be fully compensated by further opening of the choke since this will already be wide open before the desired pump rate is reached. At the point when the kill mud enters the choke-line, it will be obviously impossible to compensate for the additional CLFP if the same pump rate is used. To avoid such a situation, the Slow Circulating Rate must be reduced to a lower rate so that CLFP is reduced accordingly.

7.4.2 Dynamic Pressure Loss through Riser and ICP

Although in well kill situations, CLFP is part of the total Slow Circulating Rate pressure, it is not included in the calculation of ICP. This is the dynamic pressure loss through the riser which is considered for the calculation of ICP. The backpressure effect of CLFP would be applied on the downhole, particularly at the casing shoe. Therefore, by deliberately not including CLFP in calculation of ICP, this choke effect should be compensated by reducing the backpressure applied by the adjustable choke.

In brief, ICP in subsea drilling can be found by adding the dynamic pressure loss through the riser ($\Delta P_{\text{Thru Riser}}$) to the shut-in drill-pipe pressure (SIDPP): $\text{ICP} = \text{SIDPP} + \Delta P_{\text{Thru Riser}}$

7.5 CHOKE AND KILL-LINE FLUID DENSITIES

In deepwater drilling with high drilling mud weights, it is a common practice to prevent barite settling issues in the choke and kill-lines by using water to fill them. In this case, the hydrostatic effect of water should be considered when measuring the Shut-In Casing Pressure (SICP). In fact, the measured SICP would be greater with water in the choke-line than with heavy mud in the choke-line.

7.6 VOLUMES AND STROKES

In any drilling activity, the driller and supervisors should be essentially aware of the values of capacity and metal displacement of different pipes in the hole. These values are required for: a) Calculation of the removed/extracted pipes out of the hole or in the hole during drilling and tripping and b) for possible well control events/situations which would require circulating original or heavy (kill) muds into the drill-string and the annulus. The capacity of a tubular/pipe can be calculated in field units [bbl/ft] as follows:

$$\text{Cap}_{\text{pipe}} = \frac{\text{ID}^2}{1029.4} \tag{7.2}$$

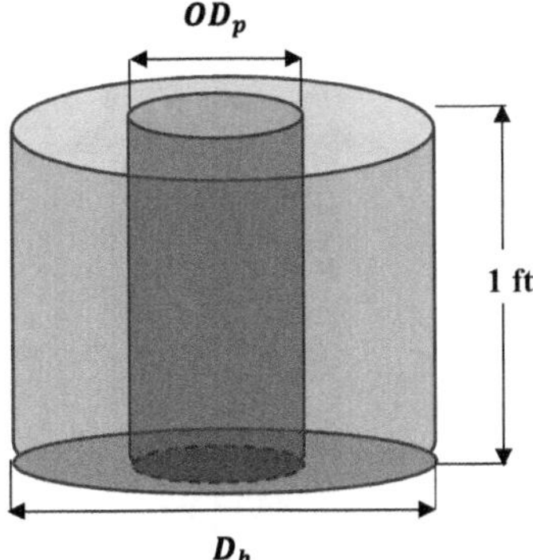

FIGURE 7.6 Annular capacity is the volume enclosed between the hole (with diameter D_h) and the pipe (with the diameter OD_p).

where ID is the inside diameter of the pipe.

The capacity of the annulus (shown schematically in Figure 7.6) can be calculated in field units, as follows:

$$Cap_{Ann} = \frac{D_h^2 - OD_p^2}{1029.4} \quad (7.3)$$

where D_h is the hole diameter and OD_p is the outside diameter of the pipe in the hole.

For mud circulation, the drill-string and the annular volumes must be known.

The drill-string volume, V_{DS}:

$$V_{DS} = \sum_{i=1}^{n} L_i \times Cap_i \quad (7.4)$$

where the subscript "i" denotes each of the drill-string components with a different ID.

Therefore, using Figure 7.7, we have:

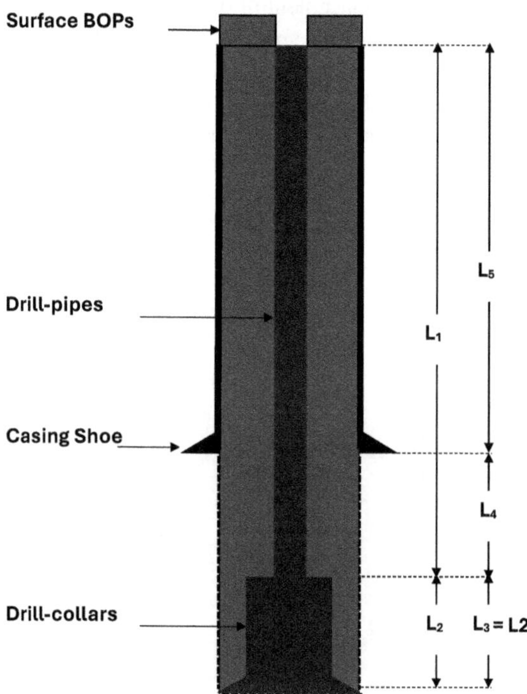

FIGURE 7.7 Schematic of a typical well with surface-BOP well shown for the calculation of drill-string and annulus volumes.

$$V_{DS} = L_1 Cap_1 + L_2 Cap_2 = L_{Dp} Cap_{Dp} + L_{DC} Cap_{DC} \tag{7.5}$$

The annular volume V_{Ann} depends on surface-BOP or subsea-BOP cases. In surface-BOP wells, the annulus volume is found by:

$$V_{Ann} = \sum_{i=1}^{n} (L_i \times Cap_{Ann,OH}) + \sum_{j=1}^{m} (L_j \times Cap_{Ann,CSG}) \tag{7.6}$$

The above equation consists of two terms – the first term is for the annular volume in the open-hole section and the second one is the annular volume in the cased-hole section. Using the schematic in Figure 7.7, we have:

$$V_{Ann} = L_3 Cap_{Ann,3} + L_4 Cap_{Ann,4} + L_5 Cap_{Ann,5} \tag{7.7}$$

In subsea-BOP wells (such as for floating rigs), the choke-line volume should be also considered. Therefore, the annulus volume is found by (see Figure 7.8):

$$V_{Ann} = L_3 Cap_{Ann,3} + L_4 Cap_{Ann,4} + L_5 Cap_{Ann,5} + L_6 Cap_{CL} \tag{7.8}$$

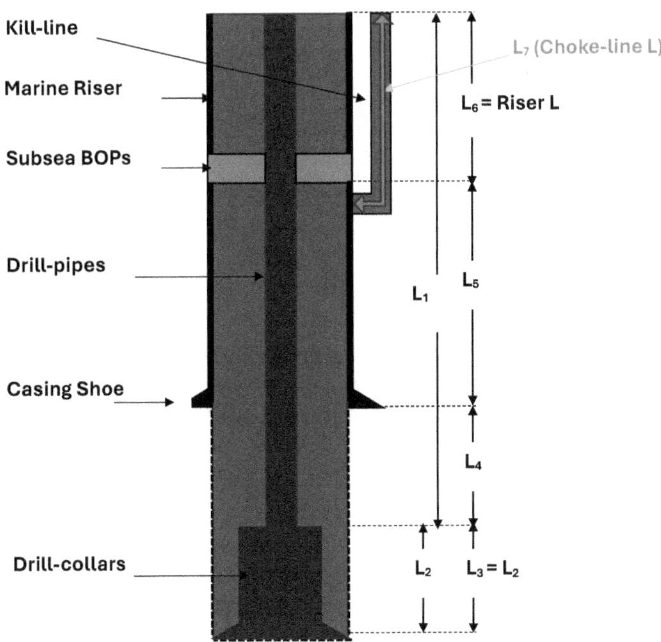

FIGURE 7.8 Schematic of a typical subsea-BOP well shown for the calculation of drill-string and annulus volumes.

where Cap_{CL} is the choke-line capacity.

The total hole volume V_{Hole} is found by:

$$V_{Hole} = V_{DS} + V_{Ann} \tag{7.9}$$

The number of strokes ($No_{strokes}$) required to pump a specified volume of mud (V) is found by dividing the volume by the pump displacement or factor [bbl/stroke]:

$$No_{strokes} = \frac{Volume}{Pump\ Displacement} \tag{7.10}$$

To pump mud to reach the bit nozzles, the required number of strokes is:

$$No_{strokes,DS} = \frac{Volume_{DS}}{Pump\ Displacement} \tag{7.11}$$

To pump mud into the drill-string to the bit and back to the surface, the required number of strokes is found by:

$$No_{strokes,Hole} = \frac{Volume_{Hole}}{Pump\ Displacement} \tag{7.12}$$

The above volumetric calculations must be already made in the pre-kick kill sheet prior to any kick situation.

7.7 PRE-KICK KILL SHEET

A kill sheet consists of two pages (or three pages for deviated wells) to be filled out and completed. The first page is called Pre-Kick Kill Sheet or Prerecorded Kill Sheet which is filled out in drilling before any kick occurs. This is done to save time later in completing the kill sheet in case a kick occurs later. Whereas the second page is filled out after detecting the kick and well shut-in.

During drilling, the pre-kick sheet is filled out using the mentioned prerecorded data consisting of Slow Circulating Rate pressures, formation strength data, drill-string, and annulus volumes and strokes, which are essentially measured, collected, or calculated. An important point about the pre-kick kill sheet is the necessity of updating the mentioned first page frequently as we drill. The pre-kick kill sheet may traditionally be completed at the beginning, in the middle, and near the end of drilling a hole section. Several drilling crew members are responsible to be aware of this simultaneously, i.e., the driller, tool pusher, and the drilling supervisor. This is for double-checking purpose. In modern systems, however, this update is digitized, i.e., handled by software with the data being fed continuously by mud logging sensors to keep the kill sheet always ready at any depth. Chapter 9 will cover steps and examples in kill sheet preparation in reasonable detail.

7.8 EXERCISES

Exercise 1:

In a surface-BOP well, the drill-string is in the cased hole. Using the following data, calculate the capacities, drill-string volume, the annulus volume, and the total well volume.

Casing/Hole: OD = 9 5/8", ID = 8.681, Weight = 47 Ib/ft, MD = 10,000 ft, TVD =9,800 ft
Drill-pipe: OD = 5", ID = 4.25", Length = 9,500 ft
Drill-collar: OD = 7", ID = 2 13/16", Length = 500 ft

Answer:

It is noted that the Measured Depth MD (*not* True Vertical Depth, TVD) should be considered for calculation of volumes. Table 7.1 gives calculations of the drill-string and the annular capacities and volumes.

Exercise 2:

Using the following data for an onshore well, calculate the drill-string and annulus volumes:

Well Total Depth: 12,000 ft
Open Hole: 8 ½"

TABLE 7.1
Calculation of the Capacities and Volumes of the Drill-String and Annulus for Exercise 1

		DP	DC	Total
Drill-string	Cap [bbl/ft]	$Cap_{DP} = \frac{4.25^2}{1029.4}$ $= 0.01754$	$Cap_{DP} = \frac{2.813^2}{1029.4}$ $= 0.00768$	–
	L [ft]	9,500	500	
	Vol [bb]	166.63	3.84	170.47
Annulus	Cap_Ann [bbl/ft]	$Cap_{Ann,DP-CSG}$ $= \frac{8.681^2 - 5^2}{1029.4} = 0.0489$	$Cap_{Ann,DC-CSG}$ $= \frac{8.681^2 - 7^2}{1029.4} = 0.0256$	–
	L [ft]	9,500	500	
	Vol [bbl]	464.75	12.8	477.55
Total Well Volume [bbl]:				648

Casing: OD = 9 5/8", ID = 8.681", Capacity = 0.00768 bbl/ft, Depth = 8,000 ft
Drill-collar: OD = 7", ID = 2 13/16", Capacity = 0.00768 bbl/ft, Length = 500 ft
Drill-pipe: OD = 5", ID = 4.25", Capacity = 0.01754 bbl/ft

Answer:

With the input data shown schematically in Figure 7.9, the calculations for the drill-string and annulus volumes are summarized in Table 7.2.

Exercise 3:

Using the following data for a subsea well, calculate the drill-string and annulus volumes:

Well Total Depth: 12,000 ft
Open Hole: 8 ½"
Casing: OD = 9 5/8", ID = 8.681", Capacity = 0.00768 bbl/ft, Depth = 8,000 ft
Choke-line: ID = 4", Capacity = 0.01554 bbl/ft, Length = 1,550 ft
Riser: Length = 1,500 ft
Drill-collar: OD = 7", ID = 2 13/16", Capacity = 0.00768 bbl/ft, Length = 500 ft
Drill-pipe: OD = 5", ID = 4.25", Capacity = 0.01754 bbl/ft

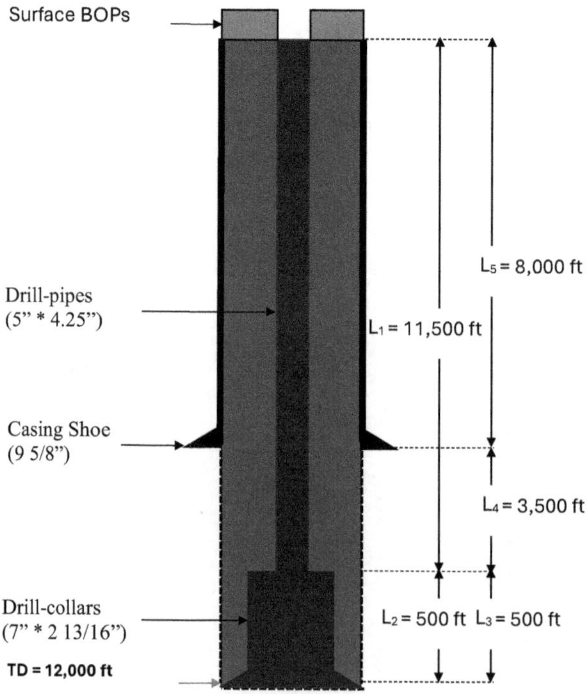

FIGURE 7.9 Well schematic for the calculation of the drill-string and annulus volumes for Exercise 2.

Answer:

With the input data shown schematically in Figure 7.10, the calculation of the drill-string and annulus volumes is summarized in Table 7.3.

TABLE 7.2
Calculation of the Drill-String and Annular Capacities and Volumes for Exercise 2

		DP	DC	Total
Drill-string	Cap [bbl/ft]	$Cap_{DP} = \frac{4.25^2}{1{,}029.4}$ $= 0.01754$	$Cap_{DP} = \frac{2.813^2}{1{,}029.4}$ $= 0.00768$	
	L [ft]	11,500	500	
	Vol [bb]	201.71	3.84	205.58
Annulus (OH)	Cap_Ann [bbl/ft]	$Cap_{Ann,DP-CSG}$ $= \frac{8.5^2 - 5^2}{1{,}029.4} = 0.0459$	$Cap_{Ann,DC-CSG}$ $= \frac{8.5^2 - 7^2}{1029.4} = 0.0225$	
	L [ft]	3,500	500	
	Vol [bbl]	160.65	11.29	171.94
Annulus (CH)	Cap_Ann [bbl/ft]	$Cap_{Ann,DP-CSG}$ $= \frac{8.681^2 - 5^2}{1{,}029.4} = 0.0489$	–	
	L [ft]	8,000	–	
	Vol [bbl]	391.37	–	391.37
		Total Well Volume [bbl]:		768.9

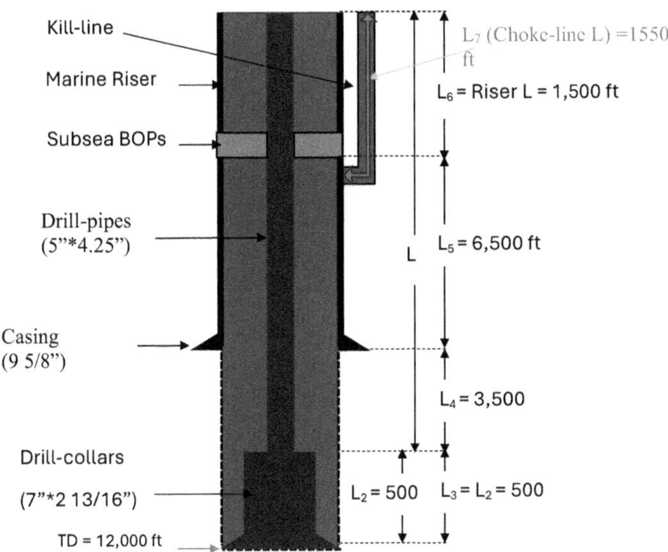

FIGURE 7.10 Well schematic for calculation of the drilling string and annulus volumes – Exercise 3.

TABLE 7.3
Calculation of the Drill-String and Annular Capacities and Volumes for Exercise 3

		DP	DC	CL	Total
Drill-string	Cap [bbl/ft]	$Cap_{DP} = \frac{4.25^2}{1029.4}$ $= 0.01754$	$Cap_{DP} = \frac{2.813^2}{1029.4}$ $= 0.00768$		
	L [ft]	11,500	500		
	Vol [bb]	201.71	3.84		205.58
Annulus (OH)	Cap_Ann [bbl/ft]	$Cap_{Ann,DP-CSG}$ $= \frac{8.5^2 - 5^2}{1029.4}$ $= 0.0459$	$Cap_{Ann,DC-CSG}$ $= \frac{8.5^2 - 7^2}{1029.4}$ $= 0.0225$		
	L [ft]	3500	500		
	Vol [bbl]	160.65	11.29		171.94
Annulus (CH)	Cap_Ann [bbl/ft]	$Cap_{Ann,DP-CSG}$ $= \frac{8.681^2 - 5^2}{1029.4}$ $= 0.0489$	–		
	L [ft]	6500	–		
	Vol [bbl]	317.85	–		317.85
Choke-line	Cap [bbl/ft]			Cap_{CL} $= \frac{4^2}{1029.4}$ $= 0.01554$	
	L [ft]			1550	
	Vol			24.1	24.08
Total Well Volume [bbl]:					719.54

8 Well Shut-In and Diverting Procedures

Rahman Ashena

8.1 INTRODUCTION

If hydrostatic mud pressure in drilling is less than formation pressure, primary well control or barrier is lost, and thus a kick influx enters the wellbore. Except for the shallow/top-hole drilling, when primary barrier is loss, well shut-in is recommended to guarantee well safety as the secondary well control/barrier. Therefore, in case of a kick, the well is shut in by closing the Blowout Preventers (BOPs). The BOP stack (composing all the BOPs) is installed below the bell nipple. They usually consist of an annular preventer on top, several pipe rams (usually two for onshore drilling), blind rams, and blind-shear rams. When the drill-string is in the hole, pipe rams are used to seal around the drill-string and shield the well. In emergency conditions which can lead to catastrophic conditions/blowout, where standard shut-in is not possible, *shear rams* are activated to seal the well by cutting and dropping the drill-string. If there is no drill-string in the hole (e.g., after pulling out of the hole), the blind rams are closed to shut-in and seal the well. Nowadays, particularly in offshore, blind-shear rams are required. Blind-shear rams are essentially used for more safety because the shear rams cut the drill-string, and simultaneously the blind rams would then close/seal the well safely.

Early kick detection is the key to success for successful well shut-in and generally well control, guaranteeing secondary well control. The earlier the well is shut-in, the less the duration of underbalance and thus the smaller the kick size. Smaller kick size contributes to mitigating the severity of the well control event as it would be easier to handle the kick situation. A vigilant careful driller would observe the well continuously to detect a possible well control event quickly. In case of noticing possible kick indicators, e.g., by a subsequent positive flow check, he/she would instantly proceed to well shut-in. If the well is not shut in immediately after the kick, the kick influx (particularly gas) rises and expands. As gas influx expands, it pushes considerable volume of mud out of the hole at the surface and thus causes further loss of mud height and drop in the bottomhole pressure. The excessively greater bottomhole pressure drop means greater underbalance, which would allow further fluid/gas influx to enter the wellbore without any restrictions in a more rapid pace than before. Figure 4.41 shows an example open well case with the mud pump not running (off). The situation would be much worse if the mud pump were running/on because the mud circulation would accelerate rising of the kick influx to the surface.

DOI: 10.1201/9781003473770-8

Well shut-in simply aids in controlling the situation because any further increase in the influx volume is prevented in the wellbore. In other words, well shut-in prevents the expansion of the current kick influx and entry of another influx (Figure 4.42). Following well shut-in, it is extremely important to apply an appropriate well kill method to circulate the kick out. Otherwise, gas migration (while the well is shut in) can cause extremely high pressures at the wellhead and at the casing shoe, giving rise to formation fracture (see the mentioned figure).

In brief, well shut-in provides the following advantages:

- It restricts further kick influx entrance in the well and keeps its volume almost constant. Thus, it keeps the surface pit gain constant.
- It keeps surface annular pressure lower both initially when the well is shut in, and later during influx circulation (when the kick influx is circulated up the annulus through the choke).
- No further decrease in the bottomhole pressure can happen.

It is important to select and apply an appropriate line-up and shut-in procedure considering the following:

- Company policy (i.e., soft or hard shut-in)
- Type of drilling rig (fixed rigs with surface-BOPs or floating rigs with subsea-BOPs)
- Bottomhole operations (drilling or tripping operations)

It is reminded that well shut-in is not always possible in all conditions. In shallow/top-hole drilling or when the surface casing has not yet been run, it is not possible to shut in the well because there is no spool or wellhead on which the BOP stack can be installed. As a substitute, the only possibility would be to divert a possible kick flow to the surface but far away from the rig site. In top-hole drilling, even if the BOP could be installed and closed in case of a kick flow, it would most likely fracture or break down the underground formation, leading to channels/fractures created in the formation by the fluid (most likely gas) to the surface. Therefore, diverting the influx has the advantage of preventing fracturing of the formation and its consequences. This will be more discussed in the last section of this chapter.

8.2 BOPS USED FOR SHUT-IN

8.2.1 Surface-BOPs

Surface-BOPs are used in onshore and shallow water depths. Such BOPs are called "dry BOPs" since they are not submerged in seawater in offshore operations.

The type of BOPs selected by the crew to shut in the well depends on the company policy and whether the drill-pipes are out of the hole or are in the hole. When the drill-pipes are out of the hole, blind rams are used to close the well. In subsea offshore or risky onshore drilling, blind-shear rams are used.

With the drill-pipes in the hole, different drilling companies have different policies and procedures for well shut-in. For well shut-in with surface-BOPs, some companies prefer to shut in the well using the annular preventers rather than pipe rams. They have the following reasons for this policy:

- If spacing-out is not performed and the annular is closed on the tool joint, the well can still be shut in. However, pipe rams do not have such a flexibility to close and seal the well.
- The possibility that in well control events, the crew make a mistake in closing shear or blind rams instead of the annular is really low. However, if pipe rams are to be operated by the crew based on policy, in emergency situations they may make a mistake to confuse pipe rams with shear or blind rams and make well control operations difficult.

On the other hand, some other companies prefer to shut in the well using pipe rams rather than the annular preventer. They have the following reasons for this policy:

- As less hydraulic fluid behind the rams is required than the annular preventer, pipe rams can close the well more quickly. The period of 10–15 seconds is required to shut in the well depending on the size of rams; however, an annular preventer requires 20–30 seconds for well shut-in.
- Pipe rams are more efficient in terms of sealing than annular preventers with less possibility of leakage.
- As the pressure rating of pipe rams is greater than annular preventers, pipe rams provide greater safety for the well.

It is noted that in case the annular preventer is closed but a leakage of fluid occurs causing complete sealing not established, then the pipe rams must be essentially closed. Some companies may always require closing pipe rams after closing the annular.

8.2.2 Subsea-BOP

Wells in deepwater offshore (seawater > 600 m/1,969 ft) are usually drilled with subsea-BOPs installed on the subsea wellhead at the seabed.

For well shut-in with subsea-BOPs, first, the well is always closed using the upper annular preventer. Then, the drill-pipe is positioned such that its corresponding tool joint is just above the pipe rams. Afterwards, the pipe rams are closed as well so that the drill-pipe can hang off on the rams.

8.3 LINE-UP ARRANGEMENTS FOR SOFT, FAST, OR HARD SHUT-IN

8.3.1 Selection

Companies have different shut-in policies, such as soft or hard shut-in. Hard shut-in is recognized as the most acceptable and preferred method of shut-in by most companies, such as OMV, Shell, and BP. However, some companies have adopted soft shut-in procedures as their policy. Hard and soft shut-in line-up methods and procedures are compared as follows:

- Fast shut-in's line-up is similar to that of soft shut-in – just the remote choke is closed in the choke manifold in the hard shut-in line-up whereas it is open in the soft shut-in's line-up.
- Hard shut-in is easier than soft shut-in because the hard shut-in procedure has two steps rather than three, that is, it does not require closing the remote choke valve.
- Hard shut-in is less complicated than soft shut-in and can thus be easily performed only by one drilling man/woman.
- Using hard shut-in, the influx volume/size would be smaller because the influx is restricted mechanically more quickly. Less kick size signifies lower shut-in pressures. Therefore, kick control and management are easier in hard shut-in than soft shut-in.
- In case of soft shut-in line-up, the mud located between HCR (High Closing Ratio) valve and the remote choke valve should be flushed two times per day to ensure it is not plugged with any debris.
- In non-deep wells (< 4,000 m/1,124 ft), hard shut-in may cause excessive *water hammer effect* bringing about formation fracture at the casing shoe. This effect is more severe in hard shut-in because the mud flow as well as expansion of the kick influx are suddenly stopped by the BOPs. As this effect is much lower in soft shut-in, it is recommended to use soft shut-in for shallow hole sections which have formations of low fracture gradients.
- There is slightly greater chance with hard shut-in than with soft shot-in, that the initial Shut-In Casing Pressure may be greater than the Maximum Allowable Annular Surface Pressure (MAASP), i.e., formation fracture can occur at casing shoe.
- In top-hole sections, e.g., say at depths about 200–500 m (656–1,640 ft), well shut-in is not recommended as it may cause fracturing of the formation. Instead, diverting a possible kick flow is the recommended option.

8.3.2 Line-Ups and Valve Positions

Depending on the shut-in policies of companies, the line-up and position of valves in the choke manifold differ. This means that initially during drilling, it is important that a specified line-up and position for the valves are set (including at the choke manifold); therefore, in case of kick, the crew can easily apply the desired type of shut-in. Some companies adjust the line-up and position of the choke manifold valves based on soft shut-in. However, most drilling companies prefer hard shut-in, and the appropriate line-up should be adjusted based on that. The specified

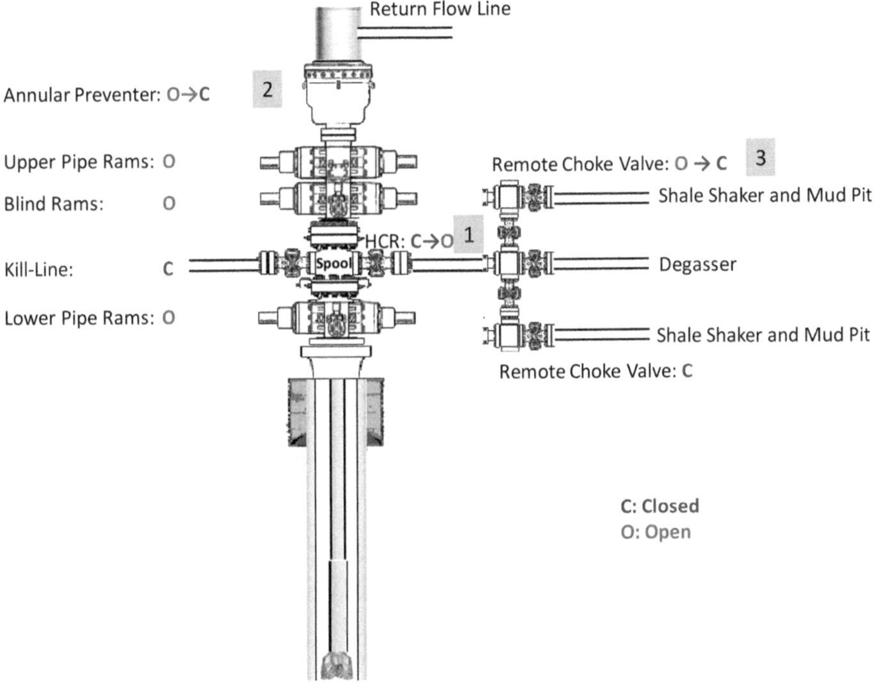

FIGURE 8.1 Soft well shut-in line-up (initially) and procedure in a typical surface-BOP well (with changes shown after the arrows).

line-up and position of valves should remain the same both for drilling and tripping operations.

8.3.2.1 Soft Shut-In Line-Up

If the company policy is soft shut-in, the required choke manifold system is as follows (see Figure 8.1):

- The High Closing Ratio (HCR) valve must be closed.
  Note: HCR is the valve of the choke manifold nearest to the BOP stack. It is controlled hydraulically from the remote.
- The remote choke valve and burn pit line must be open.
  Note 1: The remote choke is the valve on the choke manifold which can usually be controlled manually. Recently, remote valves have been transformed to remotely controlled using pneumatic means.
  Note 2: This state (keeping the remote choke valve and burn pit open) is considered the only difference between the soft and hard shut-in line-up. The flow path from remote choke reaches to mud gas separator, mud pit, or diverter line as dictated in the procedures.

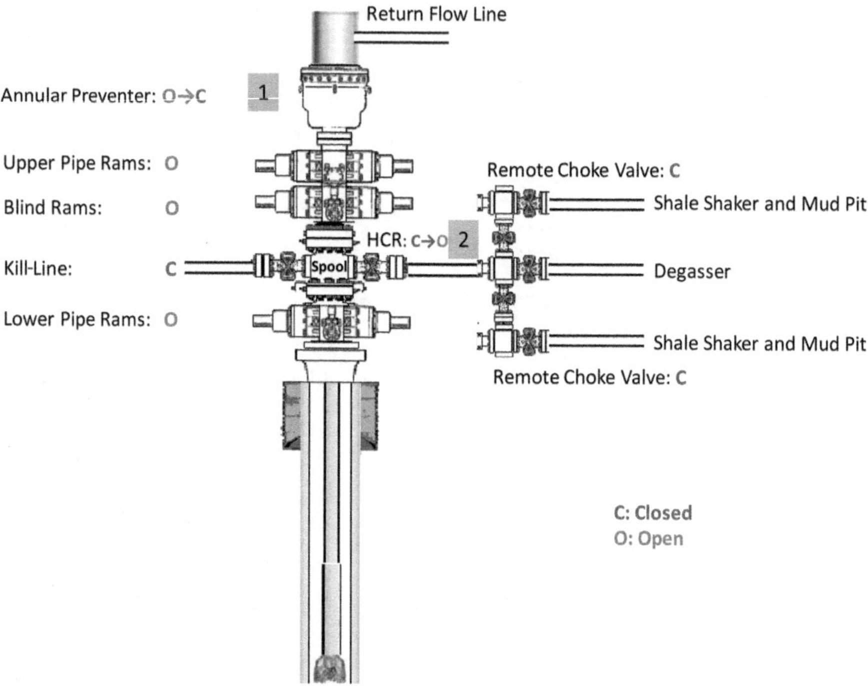

FIGURE 8.2 Hard well shut-in line-up (initially) and procedure in a typical surface-BOP well (with changes shown after the arrows).

8.3.2.2 Hard and Fast Shut-In Line-Up

If the company policy is hard or fast shut-in, the required choke manifold system is as follows (see Figure 8.2):

- The Hydraulic Choke Remote Valve (HCR) must be closed.
  Note: HCR is the valve of the choke manifold nearest to the BOP stack. It is controlled hydraulically from remote.
- The remote choke valve must be closed.
  Note: As another alternative, it is also possible to keep the valve downstream of the remote choke valve closed.
- The line from the remote choke valve reaches to the mud gas separator, mud pit, or diverter line as dictated in the procedures.

8.4 SHUT-IN PROCEDURES

Shut-in procedures are classified depending on drilling rig types and line-up.

8.4.1 Drilling (Surface-BOP)

Shut-in of surface-BOP wells is normally simpler and less complex than that of subsea-BOP wells. During drilling of such wells, the following methods of shut-in are followed:

8.4.1.1 Soft Shut-In

Figure 8.1 clearly illustrates the soft shut-in procedure, with the steps as below:

1. Alert the crew about kick flow and necessity of well shut-in.
2. Stop pipe rotation.
3. Pick up the drill-string to properly space out the drill-pipe's tool joint so that the bottom-end tool joint does not face the BOPs.
4. Close the *full-opening safety valve* on the drill-string.
5. Shut down the mud pumps (no mud circulation).
6. Open the HCR valve.
7. Close the annular preventer or pipe rams.
8. Close the remote choke valve.
9. Ensure whether the well is secure by checking stability of shut-in pressures and integrity of BOPs and surface.
10. Read and record Shut-In Drill-pipe Pressure (SIDPP), Shut-In Casing Pressure (SICP), and pit gain.

8.4.1.2 Hard Shut-In

The hard shut-in procedure is shown in Figure 8.2. The steps in the procedure are as follows:

1. Alert the crew about kick and the necessity of well shut-in.
2. Stop pipe rotation.
3. Pick up the drill-string to properly space out the drill-pipe tool joint from the BOPs.
4. Shut down the pumps.
5. Close the annular preventer or pipe rams.
6. Open the HCR valve.
7. Ensure whether the well is secure.
8. Read and record Shut-In Drill-pipe Pressure (SIDPP), Shut-In Casing Pressure (SICP), and pit gain.

8.4.1.3 Fast Shut-In

Figure 8.3 illustrates the fast shut-in procedure, with steps given below:

1. Alert the crew about kick flow and the necessity of well shut-in.
2. Stop pipe rotation.
3. Pick up the drill-string to properly space out the tool joint.
4. Shut down the pumps.

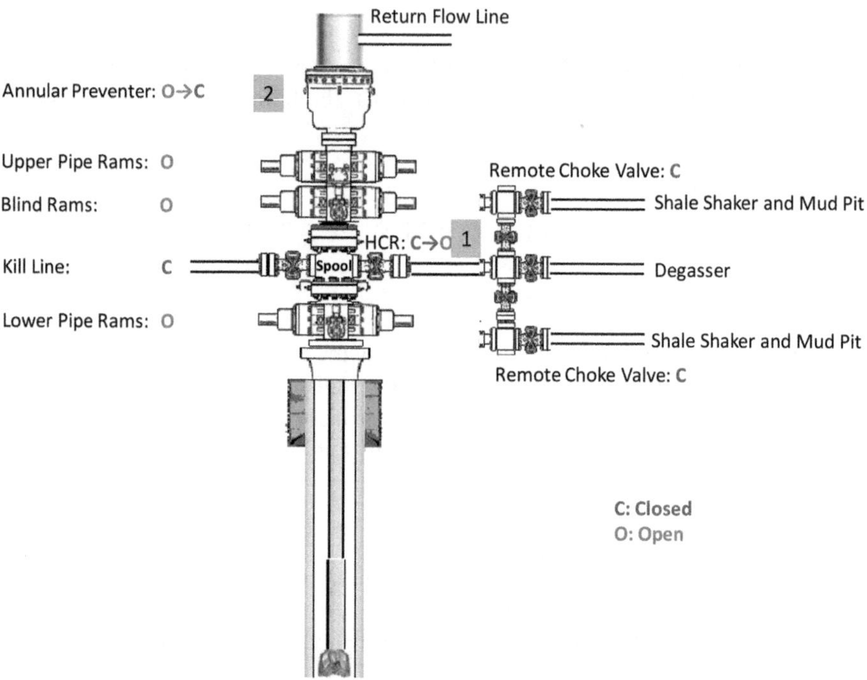

FIGURE 8.3 Fast well shut-in line-up (initially) and procedure in a typical surface-BOP well (with changes shown after the arrows).

5. Open the HCR valve.
6. Close the annular preventer or pipe rams.
7. Ensure whether the well is secure.
8. Read and record Shut-In Drill-pipe Pressure (SIDPP), Shut-In Casing Pressure (SICP), and pit gain.

For all shut-in types, there are some notes to pay attention for drilling surface-BOP wells' shut-in, as follows:

- Do not set the slips after picking up the pipe.
- The type of BOP valve to close depends on company policy dictating whether annular preventer or pipe rams should be closed.
- Upon shutting down the pumps, the valves of the pumps are closed by the seats, effectively encapsulating the fluid within the pump and standpipe, no that fluid can go back to the suction pits. Consequently, the standpipe pressure would indicate the pressure at closure or SIDPP.
- If there is a non-return float valve (above the bit), pump through it for a short time just to open it slightly for hydraulic communication and possibility of reading SIDPP.

8.4.2 Drilling (Subsea-BOP)

Well shut-in with subsea-BOPs is more complex than with surface-BOPs. Generally, it starts with closure of the upper annular preventer. After the annular closure, the drill-pipe tool joint is spaced out such that its tool joint is just above the pipe rams. Then, the pipe rams are closed such that the tool joint can hang off on the rams. This common subsea procedure means it is not possible to lower/strip the drill-pipe into rams; however, the drill-pipe may be raised by the unpredicted floating rig movement to get more kick influx.

The following methods of shut-in are explained stepwise in the next sections.

8.4.2.1 Soft Shut-In

1. Alert the crew about kick flow and necessity of well shut-in.
2. Stop pipe rotation.
3. Pick up the drill-string to properly space out the tool joint in BOP.
4. Shut down the mud pumps.
5. Open the failsafe valve to allow flow to the remote choke panel.
6. Close the upper annular preventer.
7. Close the remote choke valve.
8. Ensure whether the well is secure.
9. Read and record Shut-In Drill-pipe Pressure (SIDPP) on the standpipe gauge, Shut-In Casing Pressure (SICP), and pit gain.
10. Close the "hang-off pipe rams".
11. Slack off on drill-string and land the tool joint on the pipe rams to support some drill-string weight on rams (i.e., hang off the drill-string on the upper pipe rams).
12. Open the upper annular preventer.

8.4.2.2 Hard Shut-In

1. Alert the crew about kick flow and the necessity of well shut-in.
2. Stop pipe rotation.
3. Pick up the drill-string to properly space out the tool joint from BOPs.
4. Shut down the pumps.
5. Close the upper annular preventer.
6. Open the failsafe valve to allow flow to the remote choke panel.
7. Ensure whether the well is secure by checking the stability of shut-in pressures and integrity of BOPs and surface.
8. Read and record Shut-In Drill-pipe Pressure (SIDPP), Shut-In Casing Pressure (SICP), and pit gain.
9. Close the "hang-off pipe rams".
10. Slack off on drill-string and land the tool joint on the pipe rams to support some drill-string weight on rams (i.e., hang off the drill-string on the upper pipe rams).
11. Open the annular preventer.

8.4.2.3 Fast Shut-In

1. Alert the crew about kick flow and necessity of well shut-in.
2. Stop pipe rotation.

3. Pick up the drill-string to properly space out the tool joint from BOP.
4. Shut down the pumps.
5. Open the failsafe valve to allow flow to the remote choke panel.
6. Close the upper annular preventer.
7. Ensure whether the well is secure.
8. Read and record Shut-In Drill-pipe Pressure (SIDPP), Shut-In Casing Pressure (SICP), and pit gain.
9. Close the "hang-off pipe rams".
10. Slack off on drill-string and land the tool joint on pipe rams to support some drill-string weight on rams.
11. Open the annular preventer.

For all shut-in types, there are some notes to pay attention for drilling subsea wells' shut-in, as follows:

Do not set the slips after picking up the pipe.

In subsea wells, failsafe valves act like High Closing Ratio (HCR) valves in surface-BOP wells.

Upon shutting down the pumps, the valves of the pumps are closed by the seats, effectively encapsulating the fluid within the pump and standpipe, so that no fluid can go back to the suction pits. Consequently, the standpipe pressure would indicate the pressure at closure or SIDPP.

The "hang-off" pipe rams are the rams to support drill-string weight. These rams are either the upper pipe rams or middle rams. Hang-off rams should not be the lower pipe rams; otherwise, if a problem or leakage occurs to the kill or choke-line, it would not be possible to secure the well.

If there is a non-return float valve (above the bit), pump through it for a short time just to open it slightly for hydraulic communication and possibility of reading SIDPP.

8.4.3 Tripping (Surface-BOP)

8.4.3.1 Soft Shut-In

1. Alert the crew about kick flow and the necessity of well shut-in.
2. Pick up the drill-string to properly space out the tool joint in BOP.
3. Install and close *drill-pipe safety valve (DPSV)* on the drill-string.
4. Open the HCR valve.
5. Close the BOPs (annular preventer or pipe rams).
6. Close the remote choke valve.
7. Ensure whether the well is secure.
8. Install a check valve or *inside BOP (IBOP) valve* on the top of the DPSV.
9. Install Kelly or top drive on the Inside BOP valve.
10. Open the DPSV. It is safe to open the DPSV because it is now topped by the IBOP valve.
11. If there is a (non-return) float valve (above the bit), pump through just to make it open (required for letting pressure to be transmitted to the top of drill-pipe).
12. Read and record Shut-In Drill-pipe Pressure (SIDPP), Shut-In Casing Pressure (SICP), and pit gain.

8.4.3.2 Hard Shut-In

1. Alert the crew about kick flow and the necessity of well shut-in.
2. Install and close the *DPSV* on the drill-string.
3. Close the BOPs (the annular preventer or pipe rams).
4. Open the HCR valve.
5. Ensure whether the well is secure.
6. Install a check valve or *inside BOP (IBOP) valve* on the DPSV.
7. Install the Kelly or top drive on the IBOP valve.
8. Open the DPSV.
9. Read and record Shut-In Drill-pipe Pressure (SIDPP), Shut-In Casing Pressure (SICP), and pit gain.

8.4.3.3 Fast Shut-In

1. Alert the crew about kick flow and the necessity of well shut-in.
2. Install and close DPSV on the drill-string.
3. Open the HCR valve.
4. Close the BOPs (annular preventer or pipe rams).
5. Ensure whether the well is secure.
6. Install a check valve or *inside BOP (IBOP) valve* (or gray valve) on the DPSV.
7. Install the Kelly or top drive on the IBOP valve.
8. Open the DPSV.
9. Read and record Shut-In Drill-Pipe Pressure (SIDPP), Shut-In Casing Pressure (SICP), and pit gain.

There are some notes to pay attention in tripping as follows:

DPSV is also called full-opening safety valve (FOSV), or TIW valve (from Iron Works company), or stab-in valve. In the Kelly system, the safety valve is called *Kelly cock*.

Company policy dictates which BOP to close (annular preventer or pipe rams).

Inside BOP (IBOP) valve, also called Gray valve (from manufacturer company "Gray"), is simply a check valve not allowing flow from both up and down directions.

If there is a non-return float valve (above the bit), pump through it for a short time just to open it slightly for hydraulic communication and the possibility of reading SIDPP.

If the drill-string is off-bottom during tripping, to better control the well, it may be necessary to strip the drill-string to the bottomhole using snubbing unit or by keeping the annular preventer closed around the pipe. Then, at the bottomhole, it is possible to pump kill mud through the string and circulate the kick influx out.

8.4.4 Tripping (Subsea-BOP)

8.4.4.1 Soft Shut-In

1. Alert the crew about kick flow and the necessity of well shut-in.
2. Install and close DPSV on the drill-string.
3. Open the failsafe valve to allow flow to the remote choke panel.

4. Close the upper annular preventer or pipe rams.
5. Close the remote choke valve.
6. Ensure whether the well is secure.
7. Install a check valve or *inside BOP (IBOP) valve* on top of the safety valve.
8. Install Kelly or top drive on the check valve.
9. Open the DPSV.
10. Read and record Shut-In Drill-pipe Pressure (SIDPP), Shut-In Casing Pressure (SICP), and pit gain.

8.4.4.2 Hard Shut-In

1. Alert the crew about kick flow and the necessity of well shut-in.
2. Install and close DPSV on the drill-string.
3. Close the upper pipe rams.
4. Open the failsafe valve to allow flow to the remote choke panel.
5. Ensure whether the well is secure.
6. Install a check valve or *inside BOP (IBOP) valve* on top of the safety valve.
7. Install Kelly or top drive on the check valve.
8. Open the DPSV.
9. Pump through the non-return valve just to make it open too.
10. Read and record Shut-In Drill-pipe Pressure (SIDPP), Shut-In Casing Pressure (SICP), and pit gain.

8.4.4.3 Fast Shut-In

1. Alert the crew about kick flow and the necessity of well shut-in.
2. Install and close DPSV on the drill-string.
3. Open the failsafe valve to allow flow to the remote choke panel.
4. Close the upper annular preventer or pipe rams.
5. Ensure whether the well is secure.
6. Install a check valve or *inside BOP (IBOP) valve* on top of the safety valve.
7. Install Kelly or top drive on the check valve.
8. Open the DPSV.
9. Read and record Shut-In Drill-pipe Pressure (SIDPP), Shut-In Casing Pressure (SICP), and pit gain.

There are some notes to pay attention in tripping as follows:

DPSV is also called full-opening safety valve (FOSV) or TIW valve (from Iron Works company) or stab-in valve. In the Kelly system, the safety valve is called *Kelly cock.*

In subsea wells, failsafe valves act similar to High Closing Ratio (HCR) valves in surface-BOP wells.

Company policy dictates which BOP to close (annular preventer or pipe rams).

Inside BOP (IBOP) valve, also called Gray valve (from manufacturer company "Gray"), is simply a check valve not allowing flow from both up and down directions.

If there is a non-return float valve (above the bit), pump through it for a short time just to open it slightly for hydraulic communication and possibility of reading SIDPP.

If the drill-string is off-bottom during tripping, to better control the well, it may be necessary to strip the drill-string to the bottomhole using snubbing unit or by keeping the annular preventer closed around the pipe. Then, at the bottomhole, it is possible to pump kill mud through the string and circulate the kick influx out.

8.4.5 Tripping with BHA across BOPs

If a kick flow is encountered when the bottomhole assembly (BHA) is located across the BOP stack during tripping, depending on the kick severity, the following is done:

- *Controllable kick flow:* When it is time to trip the BHA (either during run in the hole/RIH or pull out of the hole/POOH), a flow check should be essentially made to carefully observe/monitor if there is any kick flow or not. Due to ignoring this procedure or for other reasons, a kick flow can stay undetected. In case of any controllable (small size) kick flow:
  1. We should close the annular preventer around the drill-collar (DC).
  2. Make up some drill-pipe (DP) stands to the BHA and run in the hole.
  3. Strip the drill-pipes to the bottom using the snubbing unit or by keeping the annular preventer closed around the pipes (so that we can reach the bottom and kill the well).

- *Severe kick flow:* In case of a very severe kick flow or a surface flow/blowout, we should quickly overcome the situation by dropping BHA into the hole by:
  1. Hanging off the drill-collars by the annular preventer.
  2. Disconnecting the elevator from the BHA/DCs.
  3. Opening the pipe rams to allow the BHA to drop into the hole.
  4. Closing the blind rams.
  5. Bullheading the well using a kill-line (if surface and downhole equipment pressure rating allow).

8.4.6 String Out-of-Hole (Surface and Subsea-BOP Wells)

If a kick flow occurs when the drill-string is out of the hole, then:

1. Open HCR (for surface-BOP) or failsafe valves (for subsea-BOP).
2. Close the blind rams or blind-shear rams.
3. After waiting for pressure stabilization, read and record the Shut-In Casing Pressure and pit gain.

8.4.7 Casing Running and Cementing

8.4.7.1 Casing Running

To shut in the well on a casing string, the procedure given next is followed:

1. Install a cross-over (adaptor) sub on top of the casing string at the rig floor.
2. A DPSV is installed on top of the crossover sub.
3. The drill-string is connected on the DPSV/casing assembly.
4. Space out such that casing couplings and centralizers are not across the ram-type BOPs. It is noted that BOPs cannot properly shut in and seal the well when casing couplings or centralizers are across the BOPs.
5. Shut in the well using the casing pipe rams.

8.4.7.2 Cementing

During cementing and cement slurry displacement, it is important to monitor/compare the volume of cement slurry pumped/displaced inside the hole and the volume of mud expelled out of the hole. If the volume of mud expelled out of the hole is greater than the volume of slurry displaced in the hole, it signifies a kick flow.

When a kick flow is detected during cementing, we cannot shut in the well. Therefore:

1. Continue the displacement of the cement slurry (despite the kick flow) until bumping is observed (i.e., indicating the *top wiper plug* is seated on the *float collar*).
2. Release the pressure inside the casing.
3. Flush the BOPs (from the annulus) to clean them from cement slurry, which also creates some back-pressure. Otherwise, cement would set in the BOP stack, and we would lose the BOPs.
4. With the back-pressure from flushing, shut in the well using pipe rams (the top pipe rams that suit the casing size).
   Note: The back-pressure contributes to minimizing the kick size/volume before the well shut-in.
5. Wait On Cement (WOC): This WOC time may range from two to four days depending on hole depth, temperature, and slurry weight. As the cement is hardened, its hydrostatic pressure reduces which may expose the well to a possible kick flow. Therefore, monitor the well pressures and flow returns in the trip tank during the WOC.
6. If automatic casing hanger/slips were not already run, set the non-automatic casing hanger/slips to seat around the casing in the spool.
   Note: In special cases with possible well control events/situations, it is recommended to use automatic casing hanger/slips because we can ensure that the top of the casing is sealed. Using automatic casing slips has the disadvantage of causing a back-pressure on the expelled mud during the displacement (as it has to pass through the spool outlets), which may cause fracturing of

the formation. Therefore, in some cases, some companies may not allow using automatic casing slips when the formations are unconsolidated (susceptible to fracture).

7. For liners, we will proceed to applying a negative pressure test or an inflow test to verify possible kick flow or ensure the integrity of the cement behind the liner (especially at the liner lap). To know more about this test, refer to Section 2.4.

8.4.8 Wireline Operations

Prior to commencing the wireline operations, the method of shut-in and killing the well should be agreed and determined by involved parties. For each of the aforementioned types of wireline types and setups (see Chapter 12), we follow their proprietary procedures as follows:

a) **No pipes in the hole:**

A-1: No Pressure Control:

- In case of detection of a kick flow:
  1. First, stop logging/wireline job.
  2. Next, we shut in the *rig annular BOP* around the wireline.
  3. Open the HCR valve.
  4. With the fully closed choke manifold valve, read the shut-in pressures and measure the pit gain as required data for well kill.
  5. Inform drilling supervisors that the well is shut in.

- In case of a severe kick flow or a blowout (surface flow):
  1. Cut the wireline under tension. For possible cutting purpose, in advance make sure mechanical or hydraulic cutters are available on the rig floor.
  2. Immediately shut in the *blind rams* to shield the well.

A-2 and A-3: Pressure Control (Using "shooting nipple and IBOP" and "Lubricator"):

- In case of detection of a kick flow:
  1. First, stop logging/wireline job.
  2. Next, shut in the *wireline BOPs* (IBOP or the stuffing box) to seal around the wireline.
  3. Open the HCR valve.
  4. With the fully closed choke manifold valve, read the shut-in pressures and measure the pit gain, which are the required data for well kill.
  5. Inform the drilling supervisor and tool pusher that the well is shut in.

- In case of a severe kick flow or a blowout (surface flow at the surface):
  1. Cut the wireline under tension. For possible cutting purpose, in advance make sure mechanical or hydraulic cutters are available on the rig floor.

2. Immediately shut in the *rig blind rams* to safe the well. In case of A-2 (using shooting nipple and IBOP), instead of closing the blind rams, it is possible to stab in a *safety valve* on the IBOP.

b) Pipes in the hole:

- In case of detecting a kick flow:
  1. First, stop logging/wireline job.
  2. Next, shut in the *wireline BOPs* (stuffing box of lubricator) to seal around the wireline.
  3. Open the HCR valve.
  4. With the fully closed choke manifold valve, read the shut-in pressures and measure the pit gain, which are the required data for subsequent well kill.
  5. Inform drilling supervisors that the well is shut in.

- In case of a severe kick flow approaching a blowout (well flow):
  1. Cut the wireline under tension. For possible cutting purpose, in advance make sure mechanical or hydraulic cutters are available on the rig floor.
  2. To shield inside the pipes, immediately shut -in the already-installed *gate valve* which plays the role of the safety valve.
  3. To shield the annulus (between the pipes and hole), close the rig pipes rams.

c) With wireline BOPs on already-installed wellhead:

- In case of detecting a kick flow:
  1. First, stop logging/wireline job.
  2. Next, we shut in the *wireline BOPs* (stuffing box of lubricator) to seal around the wireline.
  3. Open the two-inch gate valve at the outlet of the swept bent (which is connected to the choke manifold).
  4. With the fully closed choke manifold valve, read the shut-in pressures and measure the pit gain, which are required data for well kill.
  5. Inform drilling supervisors that the well is shut in.

- In case of a severe kick flow or a blowout (surface flow):
  1. Cut the wireline under tension. For possible cutting purpose, in advance make sure mechanical or hydraulic cutters are available on the rig floor.
  2. To shield the well, immediately shut in the *wireline blind rams*.

Note: Instead of following the above two steps, use the *wireline shear rams* to cut the wireline and seal the well.

8.5 DIVERTING PROCEDURES

Possible shallow gas formations (usually sands) encountered in top-hole drilling, particularly for wildcat and exploration drilling, can greatly endanger the well.

The danger is considerable because the surface casing has not been set and thus no BOP stack is already installed on the well, and formation fracture would happen if well shut in were practically possible (which is not possible). Therefore, the proper well control strategy to control shallow gas is based on diverting the flow. as the most appropriate well control method in such holes.

8.5.1 Fixed Rigs

In case of a shallow gas kick in fixed rigs (land/onshore or fixed offshore rigs), the gas can reach below the rig floor and disrupt the buoyancy required to hold the rig above the water. This gas flow would cause tilting and sinking of the rig to the bottom. To divert the gas as the only possible solution, it is first required to leave the diverter installed until the 13 3/8" casing has been run.

An automatic diverter system should first:

- Open an alternative flow path to overboard lines (i.e., the diverter outlets).
- Close shaker valve and trip tank valve.
- Close diverter annular (packing) around drill-pipe.
- If there are two overboard lines, then the upwind (i.e., opposite wind direction) valve should be manually closed.

The recommended diverting procedure in the event of a shallow gas kick is:

1. Pump with maximum rate and start pumping kill mud if available (while circulating through the diverter).
2. Space out so that the DPSV is above the drill floor.
3. With the diverter line open, close shaker valve and diverter packer.
4. Shut down all non-essential equipment.
5. On jack-up and platform rigs, monitor the sea for any evidence of gas breaking out around the conductor.
6. If the mud reserves run out, then continue pumping with water (onshore) and seawater (offshore).
7. While drilling the top-hole section, it is recommended to use a float valve in the drill-string. This will prevent gas entering the drill-string if a kick flow is taken.

8.5.2 Floating Rigs

There are two scenarios for floating rigs with different procedures:

a) *Riser-less (if a shallow gas occurs while taking returns to the seabed):*
   1. Try to control the well by pumping seawater and diverting the gas away from the rig.
   2. If the gas kick is causing danger to the rig crew, drop the drill-string.
   3. Move the rig to a safe position, upwind of gas plume.

b) *If there is a riser pipe and connector, and a shallow gas kick has been taken:* Diverting is not recommended as any solids in the gas influx would quickly erode the overboard lines. Instead:

1. Slack off the drill-string on the bottom pipe rams, and back off the drill-string from the connection above the rams. The drill-string weight will be kept by the BOPs.
2. Unlatch the connector (at Lower Marine Riser Package, LMRP) with an overpull on riser tensioner lines.
3. Move the rig to a safe position away, upwind of gas plume while slacking off on guide lines. The dynamic positioning system (DPS) and particularly tug boats are used for the move.

9 Post Well Shut-In Monitoring Pressures and Activities

Rahman Ashena

9.1 INTRODUCTION

After the well has been shut in, it is important to confirm that procedures have been followed correctly and that the BOP is working with integrity as it should do. Following well shut-in, it is essential to ensure that the BOPs are sealing and not leaking, or, in other words, they are holding pressure, which can be ensured by lining up the "trip tank pump" to circulate above the closed BOPs and measure volumes.

Next, immediately after well shut-in, it is essential to monitor pit gain and shut-in pressures. Using the recorded stabilized data, the kick influx height, formation pore pressure, kick influx gradient, and type can be found, which are covered in this chapter. Using the recorded data, the kill sheet can be completed (which will be covered in Chapter 10). Next, the influx migration rate is discussed.

Next, the effect of trapped pressure on causing error in future kill sheet calculations are discussed, and then the method of eliminating them is mentioned. Next, in case we suspect that the flow (at the surface) was due to wellbore ballooning and it is a kick flow, it is important to know how to discriminate them from each other. Another important item to be discussed is the effect of (non-parted) float valve on shut-in drill-pipe pressure (SIDPP) and the procedure to recover this measurement. Finally, the required line-up to be prepared for well kill is discussed.

9.2 KICK LOG

Immediately after well shut-in, the driller must carefully monitor pit gain and shut-in pressures, called *kick log*. By monitoring the pressures, it is possible to gauge the rate at which pressures build up in the well. Depending on the permeability of the kicking formation, the shut-in pressures stabilize. When the pressures stabilize, they are recorded. Using the recorded stabilized pressures, the drilling supervisor and the tool pusher can complete the kill sheet. Following pressure stabilization, the pressures may rise again due to gas migration which will be discussed later in this chapter. Briefly, the list of items sought from the stabilized readings are:

- Determination of formation pore pressure and Initial Circulating Pressures, ICPs (useful for kill sheet completion as will be discussed in Chapter 10).

DOI: 10.1201/9781003473770-9

- Determination of kick influx gradient and type of kick.
- Observation of any pressure changes and necessity of releasing/bleeding-off pressures.
- Monitoring shut-in pressures not to exceed Maximum Allowable Annular Surface Pressure (MAASP)
- Determination of kill mud weight (useful for kill sheet completion, as in Chapter 10).

9.2.1 Pit Gain

The driller should record the stabilized pit gain which represents the influx volume. Using this volume, it is possible to calculate the influx height:

$$H_{influx} = \frac{\text{Pit Gain}}{\text{Cap}_{Ann}} \tag{9.1}$$

Comparing the pit gain with the annular volume between the drill-collars (DCs) and the hole, it would be clear how long kick influx is located between the DCs and the hole (with Cap_{Ann} being the annular capacity/volume per foot between DCs and the hole) and how long kick influx is located between the drill-pipes (DPs) and the hole (with Cap_{Ann} being the annular capacity/volume per length between DPs and the hole).

9.2.2 Shut-In Pressures and Relationships

Two important shut-in pressures to be recorded post well shut-in are:

- Shut-In Drill-pipe Pressure (SIDPP), and
- Shut-In Casing Pressure (SICP).

The SIDPP represents the drill-pipe pressure after well shut-in which is measured/recorded by the standpipe pressure gauge. Since the fluid in the drill-string is considered uniform, consisting of the original mud, the bottomhole pressure, which equals the formation pore pressure following shut-in, is found by adding SIDPP to the hydrostatic pressure of the original mud column:

$$P_{pore} = \text{SIDPP} + 0.052 \times \text{MW} \times \text{TVD} \tag{9.2}$$

The SICP represents the surface annular pressure after well shut-in which is measured/recorded by the choke manifold pressure gauge. This pressure gives an indication of the imbalance of pressures in the annulus. It should be monitored to not exceed the MAASP. The combination of SICP, SIDPP, and pit gain can be used to find the influx gradient and type which are very important to know for the well kill operation. The magnitudes of SICP and SIDPP are compared as follows:

- In vertical wells, if the drill-string is on-bottom (at the bottomhole), SICP is greater than SIDPP because of the existence of the light kick influx in the annulus. The greater the influx height and the lower the influx density, the greater the difference between SICP and SIDPP. For this situation, the following relationship holds between SICP and SIDPP:

$$SICP = SIDPP + H_{influx}\left(PG_m - PG_{influx}\right) \tag{9.3}$$

- If the drill-string is off-bottom and the influx is below the bit, both SICP and SIDPP are likely to be the same (e.g., $SICP = SIDPP = 400$ psi). Similarly, in horizontal wells, both SICP and SIDPP are equal.
- If a kick has occurred due to swabbing (not due to underbalance), after we run the new drill-string back to the bottom, a pit gain may be observed due to gas influx expansion. In this case, after well shut-in, there is some SICP reading (say 100 psi), but the SIDPP reading is zero. The other reason for zero SIDPP is using a float valve (e.g., at the bit sub) in the drill-string. However, when there is a float valve above the bit, a procedure should be applied to get pressure reading, as mentioned in Section 8.4 and more explained in Section 9.6).
- It is noted that following well shut-in, in case of delay in well kill operations, gas would migrate up the wellbore due to its lower density than the mud. Consequently, SIDPP and SICP both increase for the same amounts.

9.2.3 Influx Gradient and Kick Type

Following well shut-in, it is important to identify the type of the kick influx. If it is a gas kick, the case would be more severe, and well kill operation would be complex. For this identification, the influx gradient must be calculated using the recorded shut-in data.

First, using the recorded pit gain and annular capacities, the influx height H_{influx} can be found.

Next, the influx gradient PG_{influx} is found using the shut-in data H_{influx} and mud pressure gradient PG_m :

$$PG_{influx} = PG_m - \frac{\left(SICP - SIDPP\right)}{H_{influx}} \tag{9.4}$$

Comparing the evaluated PG_{influx} with the ranges listed in Table 9.1, the type of kick influx (gas, oil, or water) can be identified.

9.2.4 Exercises

Exercise 1:

After detecting a kick flow, the well was shut in, and then shut-in data were recorded. Using the following data, evaluate the influx height.

TABLE 9.1
Classification of Kick Influx Types Based on Their Pressure Gradients

Influx Pressure Gradient [psi/ft]	Kick Type
< 0.15	Gas
$0.15-0.4$	Mixture of Gas, Oil, and Water
> 0.4	Water

Pit gain = 40 bbl

Drill-collar (DC) length = 400 ft

Annular capacity between DC and hole = 0.084 bbl/ft

Annular capacity between DP and hole = 0.122 bbl/ft

Answer:

First, the influx height H_{influx} must be evaluated:

1. The fluid volume which can be placed between DCs and the hole is found to be equal to 33.6 bbl (by multiplying the drill-collar length of 400 ft by the annular capacity between DC and the borehole of 0.084 bbl/ft).
2. The pit gain (kick size) is 40 bbl which is 6.4 bbl greater than the volume between the DCs and hole. This 6.4 bbl volume is between the DPs and the hole. Dividing 6.4 bbl by the annular capacity between DPs and hole (0.122 bbl/ft) gives the height of 52.46 ft.
3. Therefore, the total kick influx height is:
   $H_{influx} = 400 + 52.46 = 452.46$ ft

Exercise 2:

After kick detection during drilling with the mud weight of 11 ppg at the depth of 8,000 psi, the well was shut in and the recorded data were collected. Evaluate the formation pressure and the influx height. The influx gradient is assumed to be equal to 0.1 psi/ft.

SIDPP = 400 psi

SICP = 700 psi

Answer:

The formation (pore) pressure is:

$$P_{pore} = 400 + 0.052 \times 11 \times 8000 = 4976 \text{ psi}$$

Substituting the known values, the influx height is the only unknown:

$$SICP = SIDPP + H_{influx}(PG_m - PG_{influx})$$

$$700 = 400 + H_{influx}(0.052 \times 11 - 0.1)$$

$$\rightarrow H_{influx} = 636 \text{ ft}$$

Exercise 3:

After a round trip at 8,960 ft with 10.9 ppg mud, we start the pump and start circulating. An increase in flow is detected, and the well is shut in with zero (0) psi on the drill-pipe gauge and 200 psi on the casing. What is the required mud weight to kill the mud? It is noted that there is no float valve in the drill-string.

Answer:

Since there is no float valve in the drill-string, zero (0) psi pressure on the drill-pipe pressure gauge indicates that the kick has happened due to swabbing (which was noticed after the round trip). Therefore, the mud weight is enough to overcome the formation pore pressure, i.e., 10.9 ppg mud is already sufficient.

Exercise 4:

Following detection of a kick flow, the well was shut in, and shut-in data were recorded. Using the following data, evaluate the influx gradient and type.

SIDPP = 600 psi

SICP = 800 psi

Pit gain = 40 bbl

Drill-collar (DC) length = 400 ft

Annular capacity between DC and hole = 0.084 bbl/ft

Annular capacity between DP and hole = 0.122 bbl/ft

Mud weight = 10 ppg

Answer:

First, the influx height H_{influx} must be evaluated, which was already done in Exercise 1 in this section. Referring to its answer:

$$H_{influx} = 452.46 \text{ ft}$$

Next, the influx gradient PG_{influx} is found using the shut-in data H_{influx} and mud pressure gradient PG_m :

$$PG_{influx} = PG_m - \frac{(SICP - SIDPP)}{H_{influx}} = (0.052 \times 10) - \frac{(800 - 600)}{452.46} = 0.077 \text{ psi / ft}$$

Since $PG_{influx} = 0.077 \frac{psi}{ft}$ is lower than 0.15 psi/ft, the type of kick influx is identified as "gas" (refer to Table 9.1).

9.3 INFLUX MIGRATION RATE

9.3.1 Explanation

Once the well is shut in, it takes a while for shut-in pressures to stabilize before they can be recorded. The time required for pressure stabilization depends on formation permeability. Careful observation or logging of shut-in pressures is essential for the determination of stable shut-in pressures.

Following pressure stabilization and determination of the shut-in pressures, if there is a delay before well kill operations can start, both the SIDPP and SICP again continue to rise slowly and continuously. The cause for this effect is the gas influx migration or percolation up the wellbore (this was introduced in Section 4.14.2). The greater the shut-in pressures, the lower the MAASP and the more challenging the kill operations would be. This is one of the reasons for the necessity of expediting to begin the well kill operations.

Calculation of gas migration or percolation of rate is based on the following relation:

$$\text{Gas Migration Rate}[\frac{ft}{hr}] = \frac{(P_2 - P_1)}{0.052 \times MW} \times \frac{1}{(T_2 - T_1)} \quad (9.5)$$

9.3.2 Exercises

Exercise 1:

While pulling out of the hole with the bit at 2,000 ft off-bottom, the well started flowing and it was shut in. The following data was recorded.

SIDPP = 350 psi

SICP = 350 psi

How will the drill-pipe pressure change as the gas migrates up toward the bit?

Answer:

Due to gas migration, SIDPP and SICP both increase by the same amounts.

Exercise 2:

Following well shut-in and pressures stabilization, the initial shut-in pressures were recorded (SIDPP = 400 psi and SICP = 700 psi). After half an hour, both shut-in pressures rose with the same amount and were recorded (SIDPP = 600 psi, SICP = 900 psi). The mud weight is 12 ppg. Evaluate the gas migration rate. Figure 9.1 shows this example.

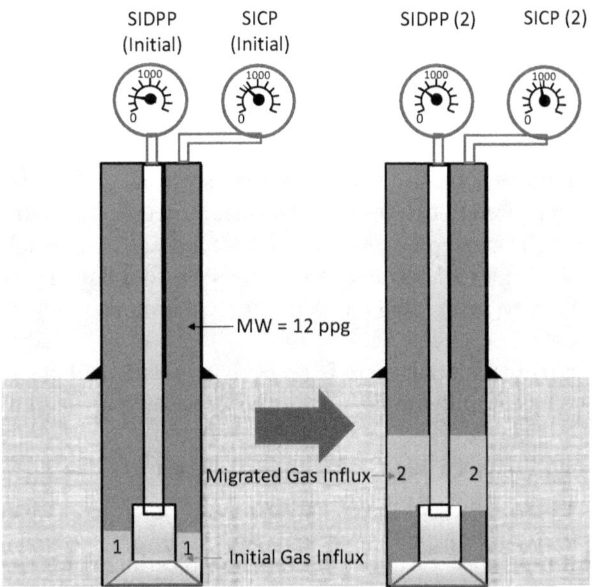

FIGURE 9.1 Shut-in pressures at initial stabilized conditions and after gas migration.

Answer:

The increase in either of the shut-in pressures is:

$$P_2 - P_1 = 600 - 400 = 200\,\text{psi}$$

$$\text{Gas Migration Rate} = \frac{200}{0.052 \times 12} \times \frac{1}{0.5} = 641\,\text{ft / hr}$$

Gas migration in the wellbore and the consequent increase in shut-in pressures must be identified and then properly handled by the Volumetric and Bleed Method. This method includes bleed-off of some mud volume and excess pressure at the choke so that bottomhole pressure remains constant. For more information on the volumetric method, refer to Section 11.7.

9.4 TRAPPED PRESSURES

It is possible that in some circumstances, some pressure in excess of the pressures caused by the kick zone be trapped in the well. This occurrence is observed in the following cases:

- The pumps were left running after the well was shut in.
- The influx is migrating/percolating up.
- The pipe has been stripped into the hole without bleeding-off of the correct/enough mud volume.

As the result of the trapped pressure, surface shut-in pressures do not reflect the actual kick zone pressure. If there is a trapped pressure in the well, the surface shut-in pressures appear constant after well shut-in, and no pressure build-up can be observed. Conversely, if the build-up of shut-in pressures is observed after the well was shut-in, it confirms that there is no trapped pressure in the well.

Existence of trapped pressure has some consequences. The shut-in drill-pipe pressure (SIDPP) is used to determine the formation pressure and thus the kill mud weight. In case of a trapped pressure, this pressure reading is artificially greater than the real one which leads to determination of the kill mud weight greater than what is required. This leads to overkilling the well and even unwanted formation fracturing/breakdown.

If there is an indication that there is probably some trapped pressure, verify the trapped pressure by carefully implementing the following procedure:

1. Using a manual choke, bleed-off a small volume of mud (e.g., 0.5 bbl) from the annulus to a suitable measuring tank.
2. Shut in the well and allow shut-in pressures to stabilize.
   If the drill-pipe pressure does not drop after bleeding mud from the annulus, no pressure is trapped. Beware of the fact that if there is no trapped pressure in the well, each increment of mud bled from the well will cause a further influx into the well. Thus, if no reduction in drill-pipe pressure is detected after bleeding 2 bbl of mud, no more mud should be bled-off.
   An increase in SICP is a definite indication that not only there is no trapped pressure, but also some gas influx has entered the wellbore.
3. If both the drill-pipe pressure and casing pressure have decreased, continue bleeding-off the next increment of mud from the well.
4. When the drill-pipe pressure no longer drops as mud increment is bled-off, record the pressure as the SIDPP and stop bleeding.

9.5 HANDLING BALLOONING

Imagine a well that is shut in due to flow-back from the wellbore, and shut-in pressures were recorded. If wellbore ballooning is suspected as the cause of the flow, this issue must be detected first. Refer to Section 4.13 to understand the ballooning phenomenon.

Following the detection process, if it was detected that the flow-back has occurred due to ballooning, it is important to handle it accordingly. If ballooning has occurred probably due to existing or induced fractures in the formation, the ballooning zone should be cured which is usually done by setting a balance cement plug and then pulling the drill-pipes out of the cement slurry by stripping (using snubbing unit or while keeping the annular preventer closed). It is important not to pull the drill-pipes conventionally (without stripping) out of the cement slurry as the formation may breathe the penetrated cement slurry out, and there were cases when the drill-pipes were stuck in the cement. Following curing the ballooning zone, it is safe to open the well and continue drilling. If the formation is a reservoir, the cement used should be of magnesium type (Magneset) so that it can be later acidized to be removed, and not cause reservoir formation damage.

9.6 OPENING (BUMPING) THE FLOAT VALVE

If there is a float valve in the drill-string, the SIDPP may be zero. However, some float valves have a small hole (3/16") drilled through the float which will allow the pressure to build up slowly on the drill-pipe side (record at the standpipe pressure gauge).

If a non-ported float valve is installed in the string and a kick is taken, the valve should close against the differential pressure, and no shut-in drill-pipe pressure (SIDPP) will be recorded at the standpipe. To open this valve and allow transmission of pressure to the surface, the following procedure is taken:

1. Line up the pump to the drill-pipe.
2. Pump into the hole at a controlled flow rate (very slowly at three to five SPM) while carefully monitoring both the pump and casing pressures.
   Note: Low circulation rates are selected to prevent possible formation fracturing while pumping in a closed well which is under pressure.
3. Record the increase in pump pressure and the volume of mud pumped. The relationship between these two parameters will be linear as the mud in the drill-pipe is compressed. If pumping is continued with the pressure equalized across the float valve, the valve will open. When the valve opens, the pump pressure will increase more slowly than before; this change should easily be recognizable at slow pump rates. Stop the pump when this change is noticed. Another indication that the float valve has opened is when the casing pressure is also likely to show an indication of the valve opening.
4. Isolate the pump at the standpipe. Record the shut-in drill-pipe pressure as the pump pressure recorded immediately just before the float valve opened.
5. If the casing pressure increases at any stage, immediately stop the pump. Isolate the pump. Bleed off the excess pressure from the casing/annulus. As an example, if the casing pressure increased for 100 psi, bleed off the 100 psi from the casing and record the shut-in drill-pipe pressure (if any changes).

9.7 POST SHUT-IN LINE-UP

9.7.1 To Check if BOPs Are Sealing Properly

Immediately after well shut-in, it is essential to ensure that the BOPs are sealing properly (or holding pressure) or not leaking. Therefore, immediately after well shut-in, the Trip Tank Pump should be lined up to circulate over the well above the BOP (called "establishing a cross-wellhead flow") by drilling crew. Therefore, by observing the Trip Tank levels, flowmeter, and shut-in pressures, we can notice if the BOPs are leaking or not. To review the method of line-up of the trip tank to the well, refer to Figure 5.1.

9.7.2 Preparation for Well Kill

To be prepared for killing the well, the appropriate line-up should be made/checked by the drilling supervisor or tool pusher as follows:

Standpipe manifold:

- Only one pump is used as a low circulation rate is required during well kill. The second pump is taken as stand-by.
- The valves in the standpipe manifold (connecting the pumps to the well) should be open. If there is concern about solids settling, we can close a valve connecting the second pump to the standpipe.

Choke manifold:

- The valves across the remotely operated hydraulic/remote choke valve number 1 (including the manual choke valve) must be open, whereas the remote choke valve itself must be closed.
- Ensure the remote choke valve number 2 is closed.
- There are three outlets out of the choke manifold (which are the mud pit line, bleed line, and burn pit line). Only the mud pit line (also called *shaker line*) must be open. The rest two outlets must be closed.

Mud gas separator (MGS) line: The valve(s) from mud pit line to the MGS (or poor-boy degasser) must be open. Therefore, during well kill operations, the gas influx would subsequently reach and exit through the MGS.

10 Kill Sheets' Completion Methodology

Rahman Ashena

10.1 INTRODUCTION

Kill sheets are essential documents for well drilling operations, especially after well shut-in following well control events. These sheets provide crucial calculations and guidelines for regaining well control and restoring safe operating conditions. By detailing specific parameters such as mud weight, pump rates, and volumes, kill sheets enable well engineers and operators to formulate precise plans for kick prevention as well as well interventions for well control. Through meticulous analysis and application of kill sheet data, operators can effectively counteract kicks or potential wellbore influxes, safeguarding personnel, equipment, and the environment. Therefore, following well shut-in and recording the stabilized shut-in pressures, the kill sheet must be completed using the recorded data as soon as possible. This completion must be conducted by the drilling supervisor, the tool pusher, and mud engineer simultaneously so that they can cross-check their calculations. Completing kill sheets is a very important tact for drilling crew and generally well control learners to learn. Nowadays, to prevent errors, kill sheets are getting digitized/computer based by which the software would do all calculations automatically.

A typical kill sheet to be completed is composed of two pages (or three pages in case of deviated wells):

- The first page (which is the pre-kick kill sheet) should be already filled with available rig data before a kick occurred. Following the kick, first, the pre-kick page of the kill sheet must be checked again.
- Next, using the recorded shut-in information, the next pages of the kill sheet must be completed before well kill operations can start.

Kill sheets are classified based on the following two criteria:

- Rig or BOP type (fixed rigs with surface BOPs and floating rigs with subsea BOPs). In fixed rigs (onshore/land, jack-ups or barges offshore), the BOPs are located either on the ground or at the surface of the water. However,

DOI: 10.1201/9781003473770-10

for floating rigs used in deepwater drilling (seawater > 600 m/1,969 ft), the BOPs are located at the seabed, "subsea-BOPs".
- Vertical or deviated wells – wells are either vertical or deviated.

Therefore, there are the following four types of kill sheets:

1. *Surface-BOP – Vertical Well,*
2. *Surface-BOP – Deviated Well,*
3. *Subsea-BOP – Vertical Well, and*
4. *Subsea-BOP – Deviated Well.*

This chapter provides practical examples of kill sheets with stepwise answers for different scenarios of fixed and floating rigs with surface and subsea Blowout Preventers (BOPs) for vertical and deviated wells.

Since completing paper-based kill sheets for early well control learners is accompanied with possible errors in calculations, it is recommended to use a pencil which enables erasing possible errors.

10.2 COMPONENTS OF KILL SHEETS

There are some slight differences between kill sheet components for surface and subsea-BOP wells, which are explained in this section.

10.2.1 First Page (Pre-Kick Kill Sheet)

10.2.1.1 Surface Well

The pre-kick sheet consists of the following components:

- *Current Well Data:*
  - Current mud weight,
  - Casing and casing shoe data (size/diameter and depths of casings, i.e., measured depth, MD, and TVD),
  - Hole data (MDs and TVDs of the borehole drilled),
  - Deviation data (only for deviated holes which includes MDs and TVDs for the following)
    - *Kick-Off Point (KOP),*
    - *End Of Build (EOB), and*
    - *Total Depth (TD).*
- *Formation Strength Data:*
  - Surface Leak-Off Pressure (LOP),
  - Drilling mud weight at the test (drilling mud weight at LOT may be different from the current mud weight), and
  - Maximum Allowable Mud Weight (MAMW):

    The MAMW is the maximum weight/density that the drilling mud can have; otherwise, the formation is fractured/broken down:

$$\text{MAMW} = \text{MW}_{\text{LOT}} + \frac{\text{LOP}}{0.052 \times \text{TVD}_{\text{Casing}}} \tag{10.1}$$

As a common practice, round down MAMW to tenth (e.g., round down 13.991 ppg to 13.9 ppg). This is carried out as a margin of safety providing some further overbalance.

- Initial MAASP (Maximum Allowable Annular Surface Pressure):

$$\text{Initial MAASP} = 0.052 \times (\text{MAMW} - \text{MW}_{\text{Current}}) \times \text{TVD}_{\text{Casing}} \tag{10.2}$$

For the safety purpose, round down MAASP to an integer (e.g., 874.56 psi to 874 psi).

- *Pump-Related Data:*
  - Mud Pump Factor/Displacement

    The pump factor/displacement [bbl/stk] is the volume in bbl pumped by the pump in one stroke. Pumps with different liner sizes have different pump factors. For each pump, this factor should be found and recorded.
  - Slow Pump Rate Data (Dynamic Pressure Loss at Slow Circulating Rates, SCRs):

    For each pump, dynamic pressure losses should be found at several Slow Circulating Rates (e.g., 30 and 40 SPM). This is conducted by simply measuring the drill-pipe pressures at these SCRs.

- *Pre-Recorded Volume Data:*
  - Volumes

    The volumes to be calculated are generally the drill-string (DS) volume and the annular volume. Specifically, the volumes consist of drill-string volume, the annulus volume, total well system volume, and the total active mud volume.

$$\text{DS Volume bbl]} = \text{Volume}_{\text{DC}} + \text{Volume}_{\text{HWDP}} + \text{Volume}_{\text{DP}} \tag{10.3}$$

$$\begin{aligned}\text{Annulus Volume} \\ &= \text{Volume}_{\text{Ann,DC*OH}} + \text{Volume}_{\text{Ann,DP*OH}} \\ &\quad + \text{Volume}_{\text{Ann,DP*Casing}}\end{aligned} \tag{10.4}$$

Following calculation of volumes, keep only two decimal places (e.g., 187.283 bbl to 187.28 bbl).

Note: In deviated wells, the drill-pipe volume is calculated for several sections. This is conducted because later they will be used for a separate calculation of drill-pipe pressure (DPP) step-down. These sections are

a) from surface to KOP, b) from KOP to EOB, and c) from EOB to total depth (TD).

Note: Typical kill sheets are not predicted for tapered drill-string, i.e., with different sizes of drill-pipes.

Note: As a limitation of conventional kill sheets, the formulas for the calculation of annular volumes assume that all the DCs' length is located in the open-hole section (long enough open-hole section). In case part of the DCs' length is located in the cased-hole section and the rest is in the open-hole section, the user should not use kill sheet formula but should use his own calculations or computer based kill sheets to find the annular volume.

- Strokes

  The number of strokes for each calculated volume is found by dividing the mentioned calculated volumes by the pump displacement [bbl/stroke].

$$\text{No. of Strokes} = \frac{\text{Calculated Volume}\left[\text{bbl}\right]}{\text{Pump Displacement}\left[\dfrac{\text{bbl}}{\text{stk}}\right]} \tag{10.5}$$

The stroke number should be rounded up to an integer number (e.g., round up 1,718.24 to 1,719 stroke).

- Pumping Time:

  To find the pumping time of mud [minute], the number of strokes is divided by the Slow Circulating Rate, SCR [in stroke/minute].

$$\text{Pumping Time} = \frac{\text{No. of Strokes}\left[\text{stk}\right]}{\text{SCR}\left[\dfrac{\text{stk}}{\text{min}}\right]} \tag{10.6}$$

Write the pumping time while keeping only two decimal numbers (e.g., write 42.978 to 42.97 minutes). The pumping time is not calculated for active surface volume in the kill sheet, the reason being is that just the drill-sting and annulus volumes are required for circulating the kick out.

Note: An important reason to find the strokes and pumping time is that they are independent of the unit systems used (either SI or field units).

10.2.1.2 Subsea Well

The layout for subsea kill sheet is the same as for the surface kill sheet. However, further data is required to be filled out on the subsea kill sheet compared to the surface kill sheet. They are listed as given next:

Current Well Data:

- Marine riser length
- Choke-line length

Pump-Related Data:

- Dynamic Pressure Loss at SCRs through the Riser ($SCRP_{Riser}$) through the choke-line ($SCRP_{Choke\text{-}line}$) must be found.
- Choke-line Friction Pressure (CLFP) should be found. There are several methods, but one method is by subtracting $SCRP_{Choke\text{-}line}$ from $SCRP_{Riser}$:

$$CLFP = SCRP_{Choke\text{-}Line} - SCRP_{Riser} \tag{10.7}$$

Pre-Recorded Volume Data:

- Annular Volume:
  The Drill-pipe casing volume (which is part of the cased-hole volume) must be calculated from the BOP down:

$$\begin{gathered} DP \times Casing\,Volume[bbl] = \\ (MD_{Casing} - L_{Riser})[ft] \times Capacity_{Ann,DP*Casing}\left[\frac{bbl}{ft}\right] \end{gathered} \tag{10.8}$$

- choke-line Capacity and Volume:
  The choke-line volume which is part of the annular volume is found using the choke-line capacity and length:

$$CLFP = SCRP_{Choke\text{-}Line} - SCRP_{Riser} \tag{10.9}$$

$$\begin{gathered} Choke\text{-}Line\ Volume\ bbl] \\ = Length_{\ Choke\text{-}Line}[ft] \times Choke\text{-}Line\,Capacity\left[\frac{bbl}{ft}\right] \end{gathered} \tag{10.10}$$

The total annulus volume in a subsea well is:

$$\begin{aligned} &Total\ Annulus\ Volume \\ &\quad = OH\ Volume + \text{"}DP \times Casing\text{"}\ Volume \\ &\quad + Choke\text{-}Line\ Volume \end{aligned} \tag{10.11}$$

- Marine Riser drill-pipe volume:
  After the kick influx is circulated out of the hole using the choke-line (i.e., the well was killed), it is necessary to displace the annulus between riser and drill-pipe with kill mud as a pre-requisite before we can safely resume drilling. This volume is found from the marine riser length and marine riser drill-pipe capacity:

$$\text{“Riser} \times \text{DP” Volume bbl]} = \text{Length}_{\text{Riser}}[\text{ft}] \times \text{Capacity}_{\text{Riser*DP}}\left[\frac{\text{bbl}}{\text{ft}}\right] \quad (10.12)$$

10.2.2 Next Pages (Post-Kick Kill Sheet)

The next page(s) of the kill sheet pertain to post-kick. The post-kick kill sheet is only one page (except for the deviated wells which is two pages). This part consists of:

- *(Recorded) Kick Data:*
  - Shut-In Drill-Pipe Pressure (SIDPP)
  - Shut-In Casing Pressure (SICP)
  - Pit Gain (Volume) which indicates the taken kick size.
- *Well Kill Data (Found Using Pre-Kick and Kick Data):*
  - Kill Mud Weight/Density:

$$MW_{\text{Kill}} = MW_{\text{Current}} + \frac{\text{SIDPP}}{0.052 \times \text{TVD}} \quad (10.13)$$

  - Initial Circulating Pressure (ICP):

$$\text{ICP} = \text{SCRP} + \text{SIDPP} \quad (10.14)$$

  - Final Circulating Pressure (FCP):

$$\text{FCP} = \text{SCRP} \times \frac{MW_{\text{Kill}}}{MW_{\text{Original}}} \quad (10.15)$$

  - Initial dynamic casing pressure at Kill Pump Rate (for Subsea Wells):
    For the dynamic casing pressure, the pressure loss (due to friction) through the choke-line must be subtracted from the SICP:

$$\text{Initial Dynamic CP} = \text{SICP} - \text{CLFP} \quad (10.16)$$

  - Drill-Pipe Pressure (DPP) Step-Down Chart and Graph:
    This parameter is important for the second circulation in the Driller's method or for the Wait and Weight method in order to control and manage the drill-pipe pressure to maintain constant bottomhole pressure.

For vertical wells, it is found simply as:

$$\text{DPP StepDown}\left[\frac{\text{psi}}{100\,\text{stk}}\right] = \frac{(\text{ICP} - \text{FCP})}{\text{DS Strokes}} \times 100 \tag{10.17}$$

For deviated wells, there are different DPP Step-Down values for the several geometries that exist in the well (surface to KOP, KOP to EOB, EOB to TD). To find them, first, the circulating pressure at KOP and EOB should be respectively found when the kill mud reaches KOP and EOB. Then, using them, three DPP step-down values are found:

a) From Surface to KOP

$$\text{DPP StepDown}_1 = \frac{(\text{ICP} - \text{KOP CP})}{\text{Surface to KOP Strokes}} \times 100 \tag{10.18}$$

b) From KOP to EOB

$$\text{DPP StepDown}_2 = \frac{(\text{KOP ICP} - \text{EOB CP})}{\text{KOP to EOB Strokes}} \times 100 \tag{10.19}$$

c) From EOB to TD

$$\text{DPP StepDown}_3 = \frac{(\text{EOB ICP} - \text{FCP})}{\text{EOB to TD Strokes}} \times 100 \tag{10.20}$$

To become familiar with the process of completing kill sheets, three examples will be presented next.

10.3 SURFACE-BOP VERTICAL WELLS

10.3.1 Introduction

First, Figure 10.1 gives a general overview of different parts of the kill sheet.

10.3.2 Exercise 1

Using the following data (in Table 10.1) for a surface vertical well, complete the kill sheet and answer the questions.

10.3.2.1 Questions

Q1) What is the Maximum Allowable Mud Weight based on the leak-off-test data?

Q2) What is the Maximum Allowable Annular Surface Pressure (MAASP) with the well shut in and the pressures stable?

Q3) What is the safety margin at the shoe with the well shut in?

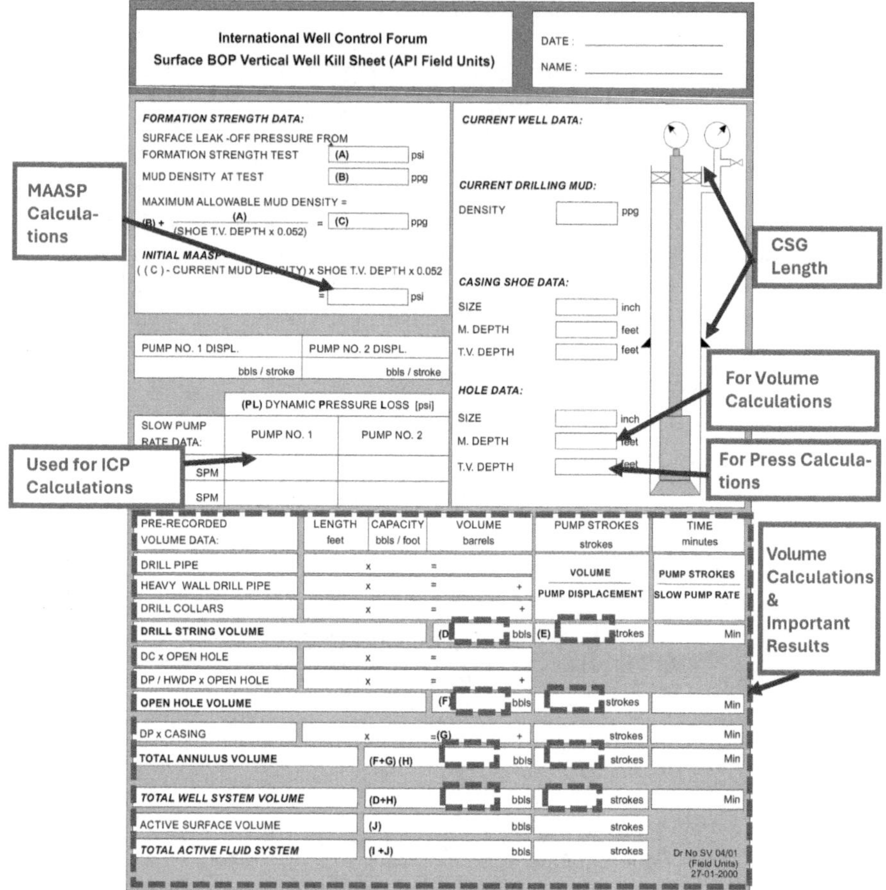

FIGURE 10.1 Overview of important points in a surface-BOP – vertical kill sheet (pre-kick sheet/first page).

Q4) How many strokes are needed to get kill mud from pump to bit?
Q5) What is the volume of open hole?
Q6) What is the total annulus volume?
Q7) Calculate the formation pressure based on the shut-in data.
Q8) What kill mud is required to balance formation pressure?
Q9) What will the Initial Circulating Pressure (ICP) be at 30 SPM?
Q10) What will the Final Circulating Pressure (FCP) be at 30 SPM?
Q11) After reaching FCP, it is decided to increase the pump rate to 40 SPM. What would happen to bottomhole pressure if the choke operator holds casing pressure constant at whatever it is reading at the time as the pump operator speed is increased?

a. Increase
b. Decrease
c. Remain Constant

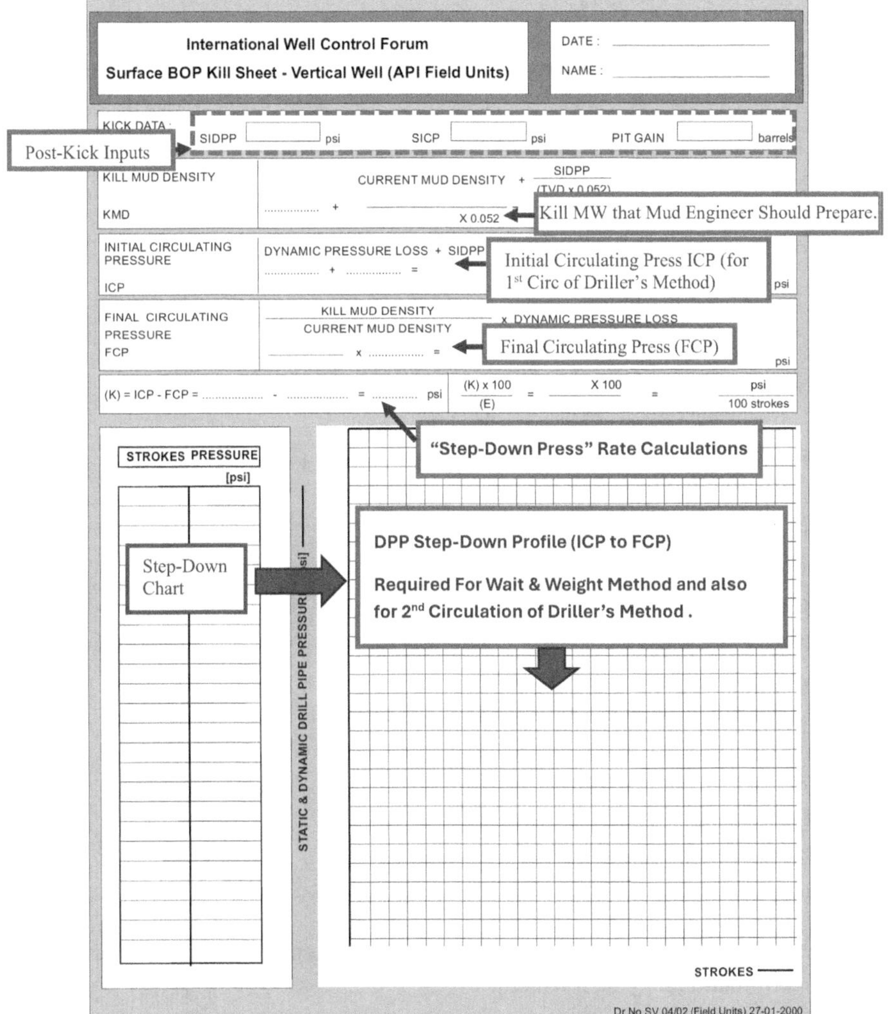

FIGURE 10.1 (Continued)

Q12) How many strokes to go from ICP to FCP?
Q13) What would the new MAASP be once the well has been killed?
Q14) What would be the drill-pipe pressure step-down per 100 strokes of kill mud pumped down the drill-string?

Answer:

Figure 10.2 provides the completed kill sheet for Exercise 1.

Q1) What is the Maximum Allowable Mud Weight based on the leak-off-test data?

TABLE 10.1
Given Data for Surface Vertical Well Kill Sheet and Questions for Exercise 1

Pit Gain	9
SIDPP	450
SICP	600
Hole Size	8 3/8-in
Hole Depth	11,095 ft (TVD/MD)
Casing	9 5/8-in (Shoe at 8,856 ft)
Drill-pipe (DP)	OD = 5-in, Capacity: 0.0178 bbl/ft
Heavy Weight Drill-pipe (HWDP)	OD = 5-in, ID = 3-in, Capacity: 0.0088 bbl/f (Length: 630 ft)
Drill-collar (DC)	OD = 6 ½-in, ID = 2 13/16-in, Capacity: 0.0077 bbl/ft (Length: 450 ft)
Mud Weight/Density	12 ppg
Annular Capacity of Open Hole/DC	0.0271 bbl/ft
Annular Capacity of Open Hole/HWDP or DP	0.0439 bbl/ft
Annular Capacity of Casing/DP	0.0493 bbl/ft
Leak-Off Pressure Using 10.3 ppg Mud	1,700 psi
Pump Factor/Output (for Pump 1 and 2):	0.109 bbl/stroke
Dynamic Pressure Loss: (at Slow Circulating Rate of 40 SPM)	450 psi
Surface Lines (between Pump and Kelly):	7 bbl
Surface Lines (between Pump and Kelly):	7 bbl

MAMW is the maximum weight that the drilling mud can have; otherwise, the formation may be fractured.

$$\text{MAMW} = \text{MW}_{\text{LOT}} + \frac{\text{LOP}}{0.052 \times \text{TVD}_{\text{Casing}}}$$

$$= 10.3 + \frac{1700}{0.052 \times 8856} = 13.991 \rightarrow 13.9 \text{ ppg}$$

MAMW is the maximum mud weight that we should not exceed. For safety, it is rounded down to one decimal place.

Q2) What is the Maximum Allowable Annular Surface Pressure (MAASP) with the well shut-in and the pressures stable?

$$\text{MAASP} = 0.052 \times (\text{MAMW} - \text{MW}_{\text{Current}}) \times \text{TVD}_{\text{CSG}}$$

$$= 0.052 \times (13.9 - 12) \times 8856 = 874.97 \text{ psi} \rightarrow 874 \text{ psi}$$

For safety, MAASP is rounded down to an integer number.

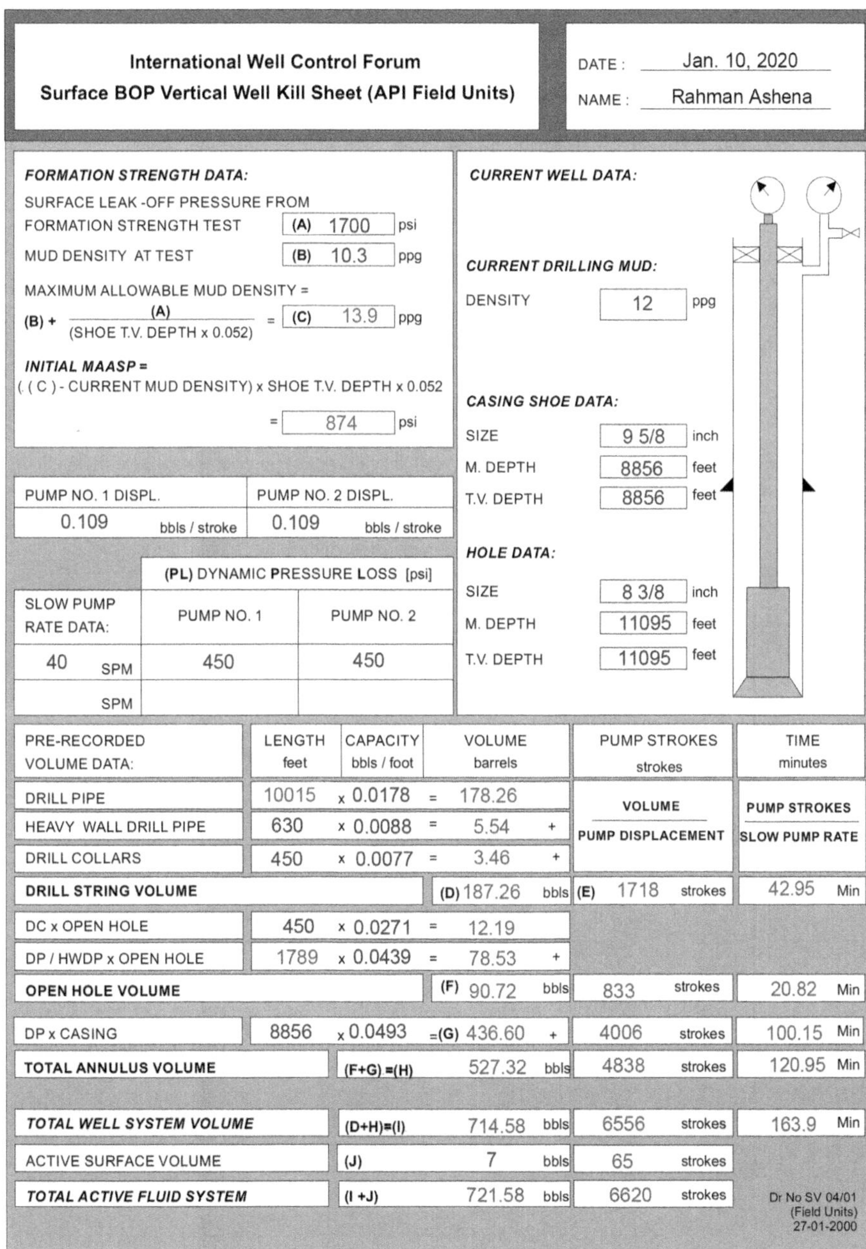

International Well Control Forum
Surface BOP Vertical Well Kill Sheet (API Field Units)

DATE : Jan. 10, 2020
NAME : Rahman Ashena

FORMATION STRENGTH DATA:

SURFACE LEAK -OFF PRESSURE FROM FORMATION STRENGTH TEST (A) 1700 psi

MUD DENSITY AT TEST (B) 10.3 ppg

MAXIMUM ALLOWABLE MUD DENSITY =

$(B) + \frac{(A)}{(\text{SHOE T.V. DEPTH} \times 0.052)}$ = (C) 13.9 ppg

INITIAL MAASP =
((C) - CURRENT MUD DENSITY) x SHOE T.V. DEPTH x 0.052
= 874 psi

PUMP NO. 1 DISPL.	PUMP NO. 2 DISPL.
0.109 bbls / stroke	0.109 bbls / stroke

SLOW PUMP RATE DATA:	(PL) DYNAMIC PRESSURE LOSS [psi] PUMP NO. 1	PUMP NO. 2
40 SPM	450	450
SPM		

CURRENT WELL DATA:

CURRENT DRILLING MUD:
DENSITY 12 ppg

CASING SHOE DATA:
SIZE 9 5/8 inch
M. DEPTH 8856 feet
T.V. DEPTH 8856 feet

HOLE DATA:
SIZE 8 3/8 inch
M. DEPTH 11095 feet
T.V. DEPTH 11095 feet

PRE-RECORDED VOLUME DATA:	LENGTH feet	CAPACITY bbls / foot	VOLUME barrels	PUMP STROKES strokes	TIME minutes
				VOLUME / PUMP DISPLACEMENT	PUMP STROKES / SLOW PUMP RATE
DRILL PIPE	10015	x 0.0178	= 178.26		
HEAVY WALL DRILL PIPE	630	x 0.0088	= 5.54 +		
DRILL COLLARS	450	x 0.0077	= 3.46 +		
DRILL STRING VOLUME			(D) 187.26 bbls	(E) 1718 strokes	42.95 Min
DC x OPEN HOLE	450	x 0.0271	= 12.19		
DP / HWDP x OPEN HOLE	1789	x 0.0439	= 78.53 +		
OPEN HOLE VOLUME			(F) 90.72 bbls	833 strokes	20.82 Min
DP x CASING	8856	x 0.0493	=(G) 436.60 +	4006 strokes	100.15 Min
TOTAL ANNULUS VOLUME		(F+G) =(H)	527.32 bbls	4838 strokes	120.95 Min
TOTAL WELL SYSTEM VOLUME		(D+H)=(I)	714.58 bbls	6556 strokes	163.9 Min
ACTIVE SURFACE VOLUME		(J)	7 bbls	65 strokes	
TOTAL ACTIVE FLUID SYSTEM		(I +J)	721.58 bbls	6620 strokes	

Dr No SV 04/01
(Field Units)
27-01-2000

FIGURE 10.2 The first page of completed (pre-kick) kill sheet of surface vertical well exercise. The collected data are added to the kill sheet in blue color (or gray color for black and white print). The calculated data are inserted into the kill sheet in red color.

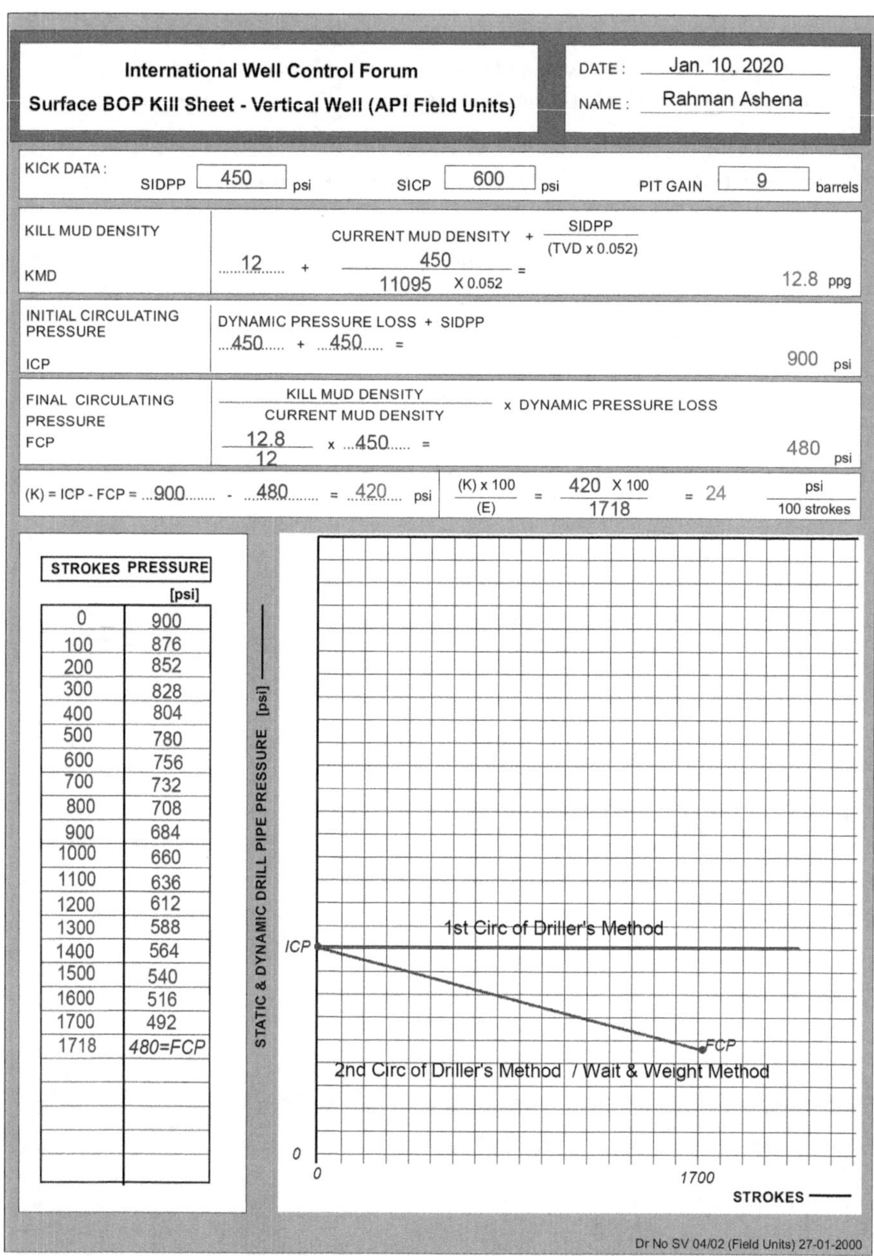

International Well Control Forum
Surface BOP Kill Sheet - Vertical Well (API Field Units)

DATE: Jan. 10, 2020
NAME: Rahman Ashena

KICK DATA: SIDPP 450 psi SICP 600 psi PIT GAIN 9 barrels

KILL MUD DENSITY
KMD
CURRENT MUD DENSITY + SIDPP / (TVD x 0.052)
12 + 450 / (11095 X 0.052) = 12.8 ppg

INITIAL CIRCULATING PRESSURE
ICP
DYNAMIC PRESSURE LOSS + SIDPP
450 + 450 = 900 psi

FINAL CIRCULATING PRESSURE
FCP
KILL MUD DENSITY / CURRENT MUD DENSITY x DYNAMIC PRESSURE LOSS
12.8 / 12 x 450 = 480 psi

(K) = ICP - FCP = 900 - 480 = 420 psi
(K) x 100 / (E) = 420 X 100 / 1718 = 24 psi / 100 strokes

STROKES	PRESSURE [psi]
0	900
100	876
200	852
300	828
400	804
500	780
600	756
700	732
800	708
900	684
1000	660
1100	636
1200	612
1300	588
1400	564
1500	540
1600	516
1700	492
1718	480=FCP

FIGURE 10.2 Continued: The second page (post-kick) of the kill sheet of surface vertical well exercise.

Q3) What is the safety margin at the shoe with the well shut-in?

Safety Margin is defined as the difference between MASSP and CP. This can be in static or dynamic conditions. As the questions asks for that at initial shut-in time (static), SICP is considered. Therefore:

$$Safety\ Margin = MAASP - SICP = 874 - 600 = 274\ psi$$

Q4) How many strokes to get kill mud from pump to bit?

This is equal to the drill-string volume plus the surface volumes. As the surface line volume is not given in this example, we ignore it. The calculation is done as follows::

The length of the drill-pipe (DP) is:

$$\text{Length}_{\text{DP}} = \text{MD}_{\text{TD}} - \text{Length}_{\text{DC}} - \text{Length}_{\text{HWDP}}$$
$$\text{Length}_{\text{DP}} = 11{,}095 - 450 - 630 = 10{,}015 \text{ ft}$$

The Drill-string (DS) Volume is:

$$\begin{aligned}\text{DS Volume} &= \text{Volume}_{\text{DC}} + \text{Volume}_{\text{HWDP}} + \text{Volume}_{\text{DP}} \\ &= \text{Length}_{\text{DC}} \times \text{Cap}_{\text{DC}} + \text{Length}_{\text{HWDP}} \times \text{Cap}_{\text{HWDP}} + \text{Length}_{\text{DP}} \times \text{Cap}_{\text{DP}} \\ &= 0.0077 \times 450 + 0.0088 \times 630 + 0.0178 \times 10{,}015 \\ &= 3.46 + 5.54 + 178.26 \\ &= 187.26 \text{ bbl}\end{aligned}$$

Therefore, the number of strokes for drill-string volume is:

$$\begin{aligned}\text{DS Strokes} &= \frac{\text{DS Volume}}{\text{Pump Displacement}} \\ &= \frac{187.26}{0.109} = 1717.98 \rightarrow 1718 \text{ stk}\end{aligned}$$

Q5) What is the volume of open hole?

The length of DP or HWDP in the annulus is:

$$\text{Length}_{\text{DP}\times\text{OH}} = \text{Length}_{\text{OH}} - \text{Length}_{\text{DC}} = (11095 - 8856) - 450 = 1789 \text{ ft}$$

$$\begin{aligned}\text{Annular OH Volume} &= \text{Length}_{\text{Ann,DC*OH}} \times \text{Cap}_{\text{Ann,DC*OH}} \\ &\quad + \text{Length}_{\text{Ann,DP*OH}} \times \text{Cap}_{\text{Ann,DP*OH}} \\ &= 450 \times 0.0271 + 1789 \times 0.0439 = 90.72 \text{ bbl}\end{aligned}$$

Q6) What is the total annulus volume?

The Cased Volume is:

$$\begin{aligned}\text{Cased Volume} &= \text{Length}_{\text{Ann,DP*Casing}} \times \text{Cap}_{\text{Ann,DP*Casing}} \\ &= 8856 \times 0.0493 = 436.6 \text{ bbl}\end{aligned}$$

The total Annulus Volume is:

$$\text{Total Annulus Volume} = \text{Annular OH Volume} + \text{Cased Volume}$$
$$= 90.72 + 436.6 = 527.32 \text{ bbl}$$

Q7) Calculate the formation pressure based on the shut-in data.

Formation Pressure can be found by adding SIDPP to the (uniform) mud hydrostatic pressure in the drill-string:

$$\text{Formation Press} = \text{SIDPP} + 0.052 \times \text{MW} \times \text{TVD}_{\text{TD}}$$
$$= 450 + 0.052 \times 12 \times 11095 = 7374 \text{ psi}$$

Q8) What kill mud is required to balance formation pressure?

The required kill mud is the equivalent mud weight to balance the formation pressure. Thus, it is found using the following relation:

$$\text{MW}_{\text{Kill}} = \text{MW}_{\text{Current}} + \frac{\text{SIDPP}}{0.052 \times \text{TVD}} = 12 + \frac{450}{0.052 \times 11095}$$
$$= 12.779 \rightarrow 12.8 \text{ ppg}$$

Note: Always round up the kill mud weight to one decimal place. This is done for safety reasons (to create some overbalance pressure).

Q9) What will the ICP be at 30 SPM?

ICP is the drill-pipe pressure, DPP, which is maintained during the first circulation in the Driller's method. It is found by adding the SIDPP to the pressure loss at SCR (also called SCRP):

$$\text{ICP} = \text{SCRP} + \text{SIDPP} = 450 + 450 = 900 \text{ psi}$$

Q10 What will the Final Circulating Pressure (FCP) be at 30 SPM?

Final Circulating Pressure (FCP) is the DPP which is obtained when the kill mud reaches the bit (either in Driller's method or in Wait and Weight method). It is found by converting the pressure loss at SCR (also called SCRP) to that of the kill mud weight:

$$\text{FCP} = \text{SCRP} \times \frac{\text{MW}_{\text{Kill}}}{\text{MW}_{\text{Original}}} = 450 \times \frac{12.8}{12} = 480 \text{ psi}$$

Note: Here, the FCP was calculated as an integer. However, always round up FCP to an integer for safety reasons (to create some overbalance above formation pressure).

Q11) After reaching FCP, it is decided to increase the pump rate to 40 SPM. What would happen to bottomhole pressure if the choke operator holds casing pressure constant at whatever it is reading at the time when the pump operator speed is increased?

a. Increase
b. Decrease
c. Remain constant

"Choice-b" is correct.

During any bring-up of the pump rate, it is required to keep casing pressure (CP) constant so that the bottomhole pressure remains constant. It is expected that during bring-up the drill-pipe pressure, DPP increases (due to its rate increase). If we try to keep DPP (rather than CP) constant during bring-up, it means we are opening the choke which causes a decrease in casing pressure and thus bottomhole pressure.

Q12) How many strokes to go from ICP to FCP?

The number of strokes from ICP to FCP is that of the drill-string volume which is (based on the kill sheet volumetric calculations):

$$\text{DS Strokes} = 1718\,\text{stk}$$

Q13) What would the new MAASP be once the well has been killed?

MAASP depends on mud weight. At the end of kill operations, we have the kill mud in the hole, and it should be used for the calculations:

$$\begin{aligned}\text{MAASP} &= 0.052 \times \left(\text{MAMW} - \text{MW}_{\text{Current}}\right) \times \text{TVD}_{\text{Casing}} \\ &= 0.052 \times (13.9 - 12.8) \times 8856 = 506.56 \text{ psi} \rightarrow 506 \text{ psi}\end{aligned}$$

Q14) What would be the drill-pipe pressure step-down per 100 strokes of kill mud pumped down the drill-string?

The pressure step-down is found as the pressure drop (from ICP to FCP) for 100 strokes pumped [psi per 100 strokes:

$$\begin{aligned}\text{DPP StepDown} &= \frac{\text{ICP} - \text{FCP}}{\text{DS Strokes}} \times 100 = \frac{900 - 480}{1718} \times 100 \\ &= 24.44 \rightarrow 24 \left[\frac{\text{psi}}{100\,\text{stk}}\right]\end{aligned}$$

Note: Always round down the pressure step-down to an integer. This is done for safety reasons (to create some overbalance above formation pressure).

10.4 SURFACE-BOP – DEVIATED WELL

10.4.1 Introduction

First, Figure 10.3 gives a general overview of different parts of this type of kill sheet.

10.4.2 Exercise 2

Using the following data (in Table 10.2) for a surface deviated well, complete the kill sheet and answer the questions.

10.4.2.1 Questions

Q1) What is the Maximum Allowable Mud Weight based on the leak-off-test data?

Q2) What is the Maximum Allowable Annular Surface Pressure (MAASP) with the well shut-in and the pressures stable?

Q3) What is the volume to get kill mud from pump to bit? Divide this volume by a) from the surface to the KOP, b) from KOP to the EOB, and c) from EOB to BHA, d) in HWDPs, and e) in DCs?

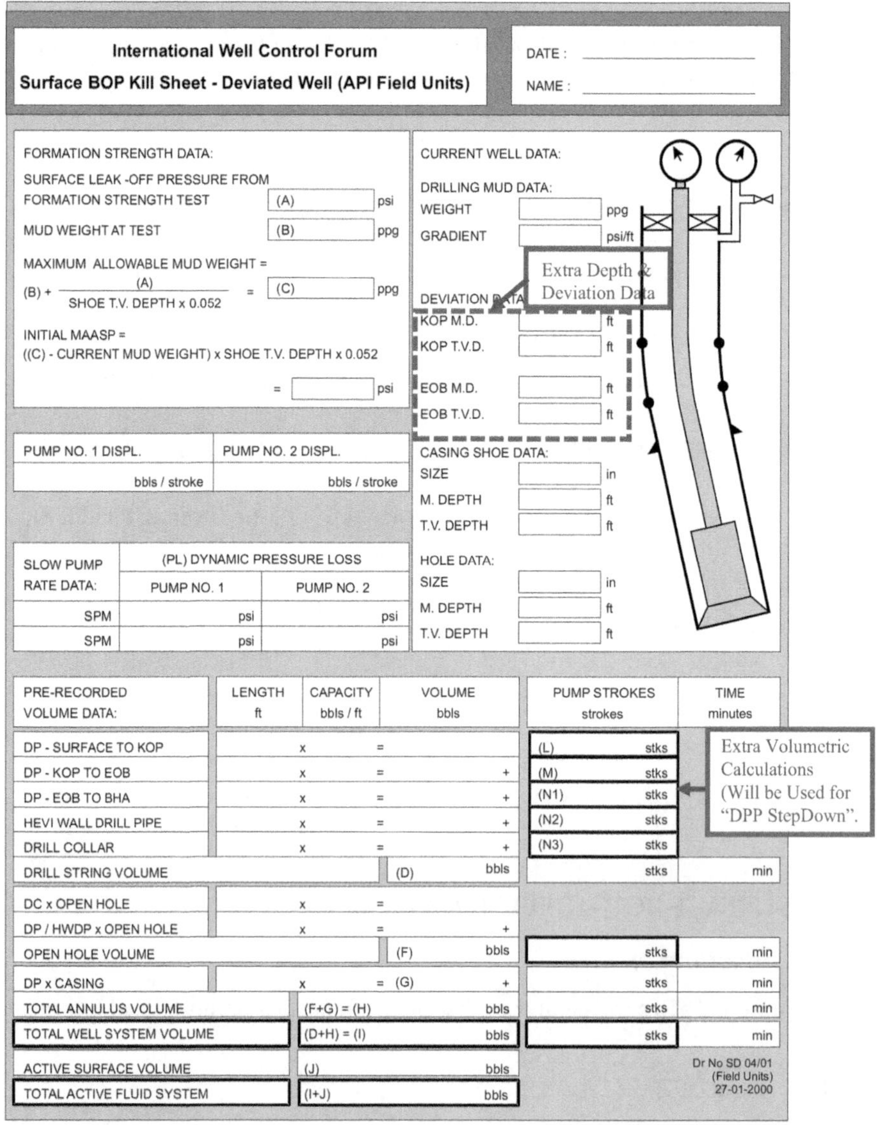

International Well Control Forum

Surface BOP Kill Sheet - Deviated Well (API Field Units)

DATE : ______

NAME : ______

FORMATION STRENGTH DATA:

SURFACE LEAK -OFF PRESSURE FROM FORMATION STRENGTH TEST (A) psi

MUD WEIGHT AT TEST (B) ppg

MAXIMUM ALLOWABLE MUD WEIGHT =

$(B) + \frac{(A)}{\text{SHOE T.V. DEPTH} \times 0.052} = (C)$ ppg

INITIAL MAASP =

((C) - CURRENT MUD WEIGHT) x SHOE T.V. DEPTH x 0.052

= psi

CURRENT WELL DATA:

DRILLING MUD DATA:

WEIGHT ppg

GRADIENT psi/ft

DEVIATION DATA

KOP M.D. ft

KOP T.V.D. ft

EOB M.D. ft

EOB T.V.D. ft

CASING SHOE DATA:

SIZE in

M. DEPTH ft

T.V. DEPTH ft

HOLE DATA:

SIZE in

M. DEPTH ft

T.V. DEPTH ft

PUMP NO. 1 DISPL.	PUMP NO. 2 DISPL.
bbls / stroke	bbls / stroke

SLOW PUMP RATE DATA:	(PL) DYNAMIC PRESSURE LOSS PUMP NO. 1	PUMP NO. 2
SPM	psi	psi
SPM	psi	psi

PRE-RECORDED VOLUME DATA:	LENGTH ft	CAPACITY bbls / ft	VOLUME bbls	PUMP STROKES strokes	TIME minutes
DP - SURFACE TO KOP		x	=	(L) stks	
DP - KOP TO EOB		x	= +	(M) stks	
DP - EOB TO BHA		x	= +	(N1) stks	
HEVI WALL DRILL PIPE		x	= +	(N2) stks	
DRILL COLLAR		x	= +	(N3) stks	
DRILL STRING VOLUME			(D) bbls	stks	min
DC x OPEN HOLE		x	=		
DP / HWDP x OPEN HOLE		x	= +		
OPEN HOLE VOLUME			(F) bbls	stks	min
DP x CASING		x	= (G) +	stks	min
TOTAL ANNULUS VOLUME		(F+G) = (H)	bbls	stks	min
TOTAL WELL SYSTEM VOLUME		(D+H) = (I)	bbls	stks	min
ACTIVE SURFACE VOLUME		(J)	bbls		
TOTAL ACTIVE FLUID SYSTEM		(I+J)	bbls		

Dr No SD 04/01 (Field Units) 27-01-2000

s

FIGURE 10.3 Overview of important points in a surface-BOP – deviated kill sheet (pre-kick kill sheet/first page).

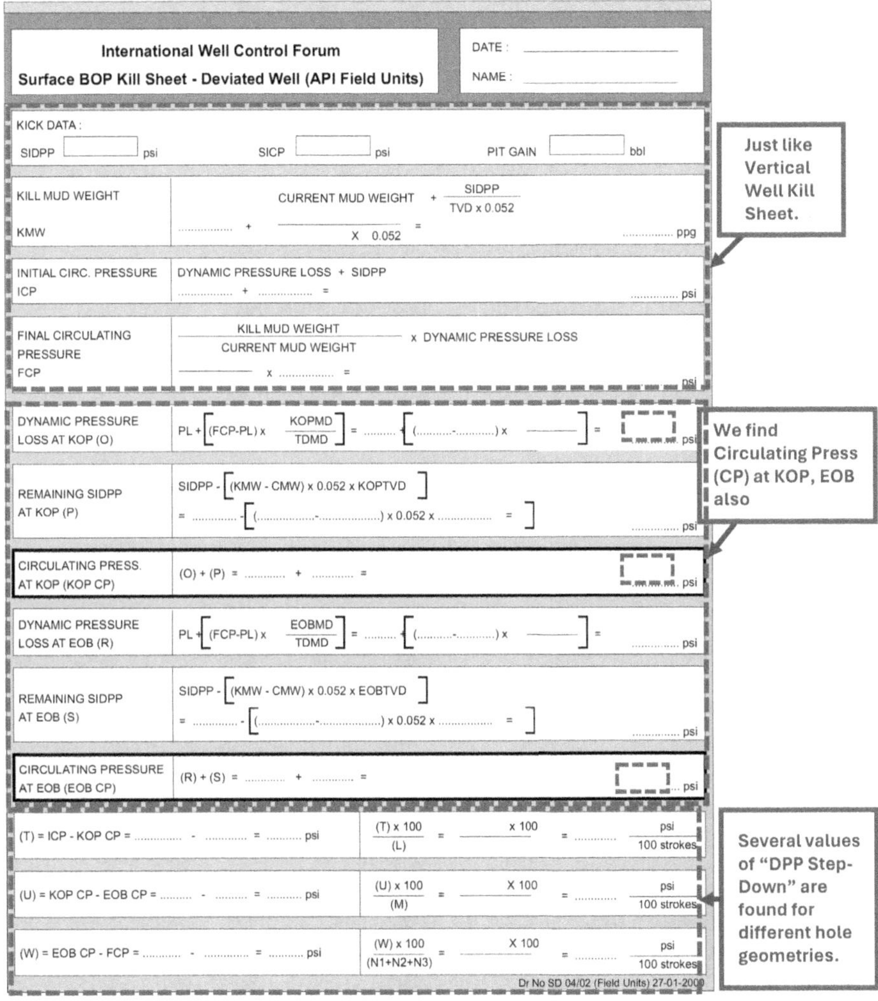

International Well Control Forum
Surface BOP Kill Sheet - Deviated Well (API Field Units)

DATE : ____________
NAME : ____________

KICK DATA :
SIDPP [ ] psi SICP [ ] psi PIT GAIN [ ] bbl

KILL MUD WEIGHT KMW	CURRENT MUD WEIGHT + SIDPP / (TVD x 0.052) .... + ________ / (X 0.052) = ... ppg
INITIAL CIRC. PRESSURE ICP	DYNAMIC PRESSURE LOSS + SIDPP .... + ... = ... psi
FINAL CIRCULATING PRESSURE FCP	KILL MUD WEIGHT / CURRENT MUD WEIGHT x DYNAMIC PRESSURE LOSS ________ x .... = ... psi
DYNAMIC PRESSURE LOSS AT KOP (O)	PL + [(FCP-PL) x KOPMD / TDMD] = ... + [(..........-..........) x ________] = ... psi
REMAINING SIDPP AT KOP (P)	SIDPP - [(KMW - CMW) x 0.052 x KOPTVD] = ... - [(..........-..........) x 0.052 x .... =] .... psi
CIRCULATING PRESS. AT KOP (KOP CP)	(O) + (P) = ... + ... = ... psi
DYNAMIC PRESSURE LOSS AT EOB (R)	PL + [(FCP-PL) x EOBMD / TDMD] = ... + [(..........-..........) x ________] = ... psi
REMAINING SIDPP AT EOB (S)	SIDPP - [(KMW - CMW) x 0.052 x EOBTVD] = ... - [(..........-..........) x 0.052 x .... =] .... psi
CIRCULATING PRESSURE AT EOB (EOB CP)	(R) + (S) = ... + ... = ... psi

(T) = ICP - KOP CP = ... - .... = ... psi	(T) x 100 / (L) = ________ x 100 / ________ = ... psi / 100 strokes
(U) = KOP CP - EOB CP = ... - .... = ... psi	(U) x 100 / (M) = ________ X 100 / ________ = ... psi / 100 strokes
(W) = EOB CP - FCP = ... - .... = ... psi	(W) x 100 / (N1+N2+N3) = ________ X 100 / ________ = ... psi / 100 strokes

Dr No SD 04/02 (Field Units) 27-01-2000

s

FIGURE 10.3 Continued: Overview of important points in a surface-BOP – deviated kill sheet (second page).

Q4) Calculate the formation pressure based on the shut-in data.

Q5) What kill mud weight is required to balance formation pressure?

Q6) What will the ICP be at 30 SPM?

Q7) What will the FCP be at 30 SPM?

Q8) What would be the drill-pipe pressure step-down per 100 strokes of kill mud pumped down the drill-string? Divide this into three geometries of a) surface to KOP, b) KOP to EOB, and c) EOB to TD.

Answers for Exercise:

Figure 10.4 and Table 10.3 provide the completed kill sheet for Exercise 2.

Q1) What is the Maximum Allowable Mud Weight based on the leak-off-test data?

In this example, the surface leak-off pressure is not given. Instead, the leak-off or fracture pressure gradient of 0.791 psi/ft is given. Therefore, first, we find the leak-off/fracture pressure at the casing shoe:

$$P_{Frac} = 0.791 \times 5007 = 3960.53 \rightarrow 3960\,\text{psi}$$

TABLE 10.2
Given Data for Surface Deviated Well Kill Sheet and Questions for Exercise 2

Pit Gain	6 bbl
SIDPP	300 psi
SICP	350 psi
Hole Size	8 ½ - in
Hole Depth and Inclination Data: [Including Kill-Off-Point (KOP), End Of Build (EOB), Total Depth (TD)]	KOP = 2585.42 ft Build - Up Rate (BUR) = 2.5 deg / 90 ft Final Inclination Angle = 90 deg . MD_{EOB} = 6,010 ft TVD_{EOB} = 5,007 ft MD_{TD} = 14,835 ft TVD_{TD} = 5,007 ft
Casing	OD = 9 5 / 8 - in Shoe at MD = 6,562 ft/TVD = 5,007 ft
Drill-pipe (DP)	OD = 5 in, Capacity = 0.01727 bbl/ft
Heavy Weight Drill-pipe (HWDP)	OD = 5 in, Capacity = 0.00870 bbl/ft (Length : 902 ft)
Drill-collar (DC)	OD = 6 ¼ in, Capacity = 0.00767 bbl/ft (Length : 262 ft)
Mud Weight/Density	10 ppg
Annular Capacity Open Hole/DC	0.03069 bbl / ft
Annular Capacity Open Hole/HWDP or DP	0.04590 bbl / ft
Annular Capacity Casing/DP or HWDP	0.04989 bbl / ft
Leak-Off (Fracture) Pressure Gradient	0.791 psi / ft
Pump Factor/Output (for Pump 1 and 2)	0.0986 bbl / stroke
Dynamic Pressure at Slow Circulating Rate of 40 SPM:	1000 psi
Active Surface Mud Volume	437 bbl
Surface Lines (between Pump and Kelly)	13 bbl

Therefore, MAMW is found:

$$\text{MAMW} = \frac{P_{\text{Frac}}}{0.052 \times \text{TVD}_{\text{Casing}}} = \frac{3960}{0.052 \times 5007} = 15.209 \rightarrow 15.2\,\text{ppg}$$

For safety (to prevent formation fracture/breakdown), it is rounded down to the tenth.

TABLE 10.3
Drill-Pipe Pressure (DPP) Step-Down Chart Which Is Used for Plotting the Step-Down Graph (for the Second Circulation of Driller's Method or the Wait and Weight Method) – Exercise 2

Stroke	Pressure [psi]	Pressure Step-Down [psi/100 stk]
0	1,300 = ICP (Kill Mud at Surface)	31
100	1,269	31
200	1,238	31
300	1,207	31
400	1,176 (~KOP)	0.5 × 31 + 0.5 × 16 = 24
500	1,152	16
600	1,136	16
700	1,120	16
800	1,104	16
900	1,088	16
1,000	1,072 (~EOB)	0.5 × 16 + 0.5(−4) = 6
1,100	1,066	−4
1,200	1,070	−4
1,300	1,074	−4
1,400	1,078	−4
1,500	1,082	−4
1,600	1,086	−4
1,700	1,092	−4
1,800	1,098	−4
1,900	1,102	−4
2,000	1,108	−4
2,100	1,112	−4
2,200	1,116	−4
2,300	1,122	−4
2,400	1,128	−4
2,495	1,120 = FCP (Kill Mud at Bit)	

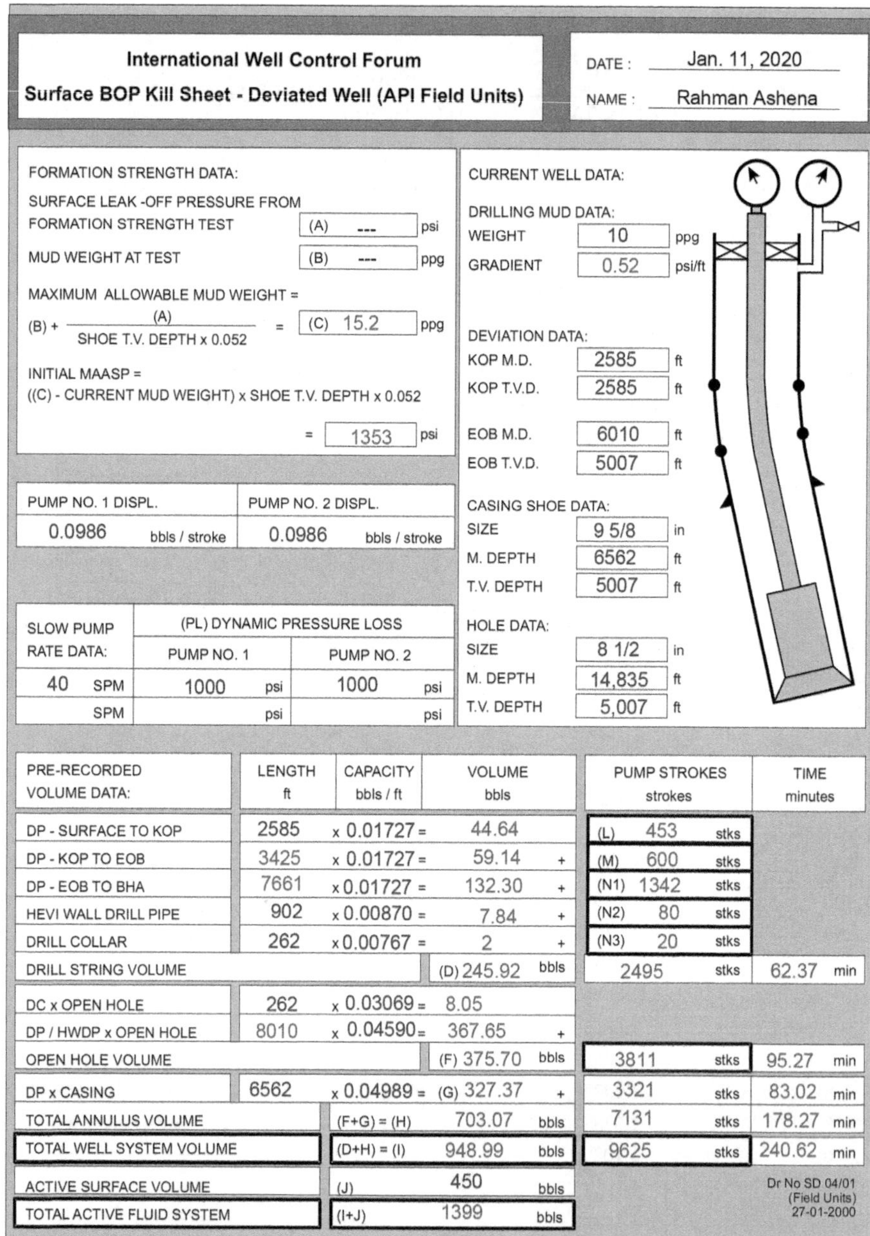

International Well Control Forum

Surface BOP Kill Sheet - Deviated Well (API Field Units)

DATE : Jan. 11, 2020

NAME : Rahman Ashena

FORMATION STRENGTH DATA:

SURFACE LEAK -OFF PRESSURE FROM FORMATION STRENGTH TEST (A) --- psi

MUD WEIGHT AT TEST (B) --- ppg

MAXIMUM ALLOWABLE MUD WEIGHT =

(B) + (A) / (SHOE T.V. DEPTH x 0.052) = (C) 15.2 ppg

INITIAL MAASP =

((C) - CURRENT MUD WEIGHT) x SHOE T.V. DEPTH x 0.052

= 1353 psi

CURRENT WELL DATA:

DRILLING MUD DATA:

WEIGHT 10 ppg

GRADIENT 0.52 psi/ft

DEVIATION DATA:

KOP M.D. 2585 ft

KOP T.V.D. 2585 ft

EOB M.D. 6010 ft

EOB T.V.D. 5007 ft

CASING SHOE DATA:

SIZE 9 5/8 in

M. DEPTH 6562 ft

T.V. DEPTH 5007 ft

HOLE DATA:

SIZE 8 1/2 in

M. DEPTH 14,835 ft

T.V. DEPTH 5,007 ft

PUMP NO. 1 DISPL.	PUMP NO. 2 DISPL.
0.0986 bbls / stroke	0.0986 bbls / stroke

SLOW PUMP RATE DATA:	(PL) DYNAMIC PRESSURE LOSS	
	PUMP NO. 1	PUMP NO. 2
40 SPM	1000 psi	1000 psi
SPM	psi	psi

PRE-RECORDED VOLUME DATA:	LENGTH ft	CAPACITY bbls / ft	VOLUME bbls	PUMP STROKES strokes	TIME minutes
DP - SURFACE TO KOP	2585	x 0.01727 =	44.64	(L) 453 stks	
DP - KOP TO EOB	3425	x 0.01727 =	59.14 +	(M) 600 stks	
DP - EOB TO BHA	7661	x 0.01727 =	132.30 +	(N1) 1342 stks	
HEVI WALL DRILL PIPE	902	x 0.00870 =	7.84 +	(N2) 80 stks	
DRILL COLLAR	262	x 0.00767 =	2 +	(N3) 20 stks	
DRILL STRING VOLUME			(D) 245.92 bbls	2495 stks	62.37 min
DC x OPEN HOLE	262	x 0.03069 =	8.05		
DP / HWDP x OPEN HOLE	8010	x 0.04590 =	367.65 +		
OPEN HOLE VOLUME			(F) 375.70 bbls	3811 stks	95.27 min
DP x CASING	6562	x 0.04989 =	(G) 327.37 +	3321 stks	83.02 min
TOTAL ANNULUS VOLUME		(F+G) = (H)	703.07 bbls	7131 stks	178.27 min
TOTAL WELL SYSTEM VOLUME		(D+H) = (I)	948.99 bbls	9625 stks	240.62 min
ACTIVE SURFACE VOLUME		(J)	450 bbls		
TOTAL ACTIVE FLUID SYSTEM		(I+J)	1399 bbls		

Dr No SD 04/01 (Field Units) 27-01-2000

FIGURE 10.4 a) The first page (pre-kick kill sheet) of the kill sheet of surface deviated well for Exercise 2.

International Well Control Forum Surface BOP Kill Sheet - Deviated Well (API Field Units)	DATE : Jan. 11, 2020 NAME : Rahman Ashena

KICK DATA :		
SIDPP 300 psi	SICP 350 psi	PIT GAIN 6 bbl

KILL MUD WEIGHT KMW	CURRENT MUD WEIGHT + $\frac{\text{SIDPP}}{\text{TVD} \times 0.052}$ 10 + $\frac{300}{5007 \times 0.052}$ =	11.2 ppg
INITIAL CIRC. PRESSURE ICP	DYNAMIC PRESSURE LOSS + SIDPP 1000 + 300 =	1300 psi
FINAL CIRCULATING PRESSURE FCP	$\frac{\text{KILL MUD WEIGHT}}{\text{CURRENT MUD WEIGHT}}$ x DYNAMIC PRESSURE LOSS $\frac{11.2}{10}$ x 1000 =	1120 psi
DYNAMIC PRESSURE LOSS AT KOP (O)	PL + [(FCP-PL) x $\frac{\text{KOPMD}}{\text{TDMD}}$] = 1000 + [(1120 - 1000) x $\frac{2585}{14835}$] =	1021 psi
REMAINING SIDPP AT KOP (P)	SIDPP - [(KMW - CMW) x 0.052 x KOPTVD] = 300 - [(11.2 - 10) x 0.052 x 2585 =]	138 psi
CIRCULATING PRESS. AT KOP (KOP CP)	(O) + (P) = 1021 + 138 =	1159 psi
DYNAMIC PRESSURE LOSS AT EOB (R)	PL + [(FCP-PL) x $\frac{\text{EOBMD}}{\text{TDMD}}$] = 1000 + [(1120 - 1000) x $\frac{6010}{14835}$] =	1048 psi
REMAINING SIDPP AT EOB (S)	SIDPP - [(KMW - CMW) x 0.052 x EOBTVD] = 300 - [(11.2 - 10) x 0.052 x 5007 =]	13 psi
CIRCULATING PRESSURE AT EOB (EOB CP)	(R) + (S) = 1048 + 13 =	1061 psi

(T) = ICP - KOP CP = 1300 - 1159 = 141 psi	$\frac{(T) \times 100}{(L)}$ = $\frac{141 \times 100}{453}$	= 31	$\frac{\text{psi}}{\text{100 strokes}}$
(U) = KOP CP - EOB CP = 1159 - 1061 = 98 psi	$\frac{(U) \times 100}{(M)}$ = $\frac{98 \times 100}{600}$	= 16	$\frac{\text{psi}}{\text{100 strokes}}$
(W) = EOB CP - FCP = 1061 - 1120 = -59 psi	$\frac{(W) \times 100}{(N1+N2+N3)}$ = $\frac{-59 \times 100}{1442}$	= -4	$\frac{\text{psi}}{\text{100 strokes}}$

Dr No SD 04/02 (Field Units) 27-01-2000

FIGURE 10.4 b) The second page (post-kick) of kill sheet of surface deviated well for Exercise 2. (Continued)

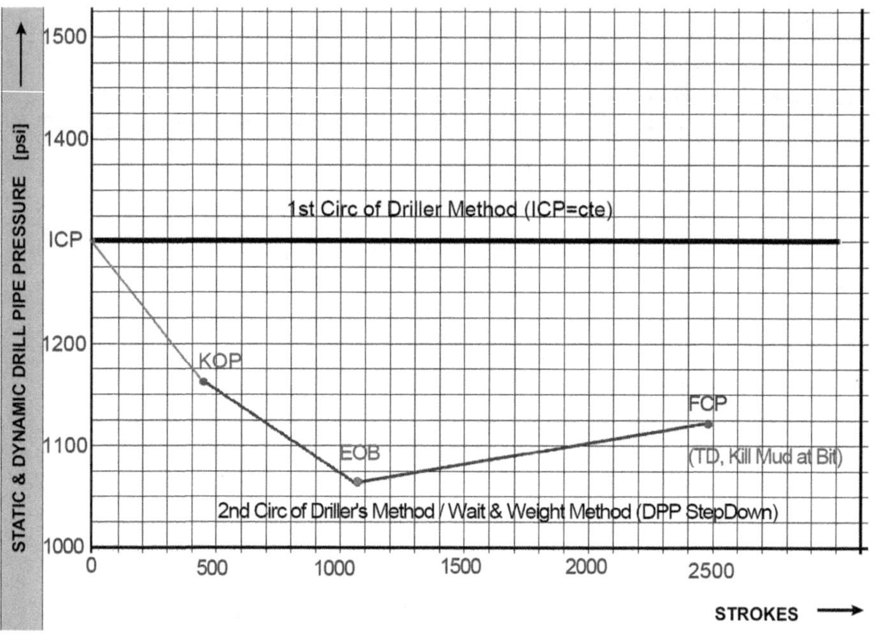

FIGURE 10.4 c) Continued: The third page of the kill sheet of surface deviated well (DPP Step-Down Profile) for Exercise 2.

Q2) What is the Maximum Allowable Annular Surface Pressure (MAASP) with the well shut-in and the pressures stable?

The initial Maximum Allowable Annular Surface Pressure (MAASP) is:

$$\text{MAASP} = 0.052 \times (15.2 - 10) \times 5007 = 1{,}353.89 \rightarrow 1{,}353\,\text{psi}$$

Q3) What is the number of strokes to get kill mud from pump to bit? Divide the strokes to: a) from the surface to the KOP, b) from KOP to the EOB, and c) from EOB to BHA, d) in HWDPs and e) in DCs?

In deviated wells, the drill-pipe volume is divided into the following three sections of a) surface to KOP, b) KOP to EOB, and c) EOB to the total depth (TD) or top of BHA. The whole drill-string volume is:

$$\text{Volume}_{\text{DS}} = \left[\text{Volume}_{\text{surf to KOP}} + \text{Volume}_{\text{KOP to EOB}} + \text{Volume}_{\text{EOB to BHA}}\right] + \text{Volume}_{\text{HWDP}} + \text{Volume}_{\text{DC}}$$

The volume of drill-pipe from the surface to KOP is:

$$\text{Volume}_{\text{Surf to KOP}} = 2585 \times 0.01727 = 44.64\,\text{bbl}$$

The corresponding number of strokes is:

$$\text{Strokes}_{\text{Surf to KOP}} = \frac{44.64}{0.0986} = 453\,\text{stk}$$

The volume of KOP to EOB is:

$$\text{Volume}_{\text{KOP to EOB}} = (6010 - 2585) \times 0.01727 = 3425 \times 0.01727 = 59.14\,\text{bbl}$$

The corresponding number of strokes is:

$$\text{Strokes}_{\text{KOP to EOB}} = \frac{59.14}{0.0986} = 600\,\text{stk}$$

The volume of EOB to TD is:

$$\text{Volume}_{\text{EOB to BHA}} = (14835 - 3425 - 2585 - 902 - 262) \times 0.01727 = 7661 \times 0.01727$$
$$= 132.3\text{ bbl}$$

The corresponding number of strokes is:

$$\text{Strokes}_{\text{KOP to EOB}} = \frac{132.3}{0.0986} = 1342\,\text{stk}$$

The HWDP volume is:

$$\text{Volume}_{\text{HWDP}} = 902 \times 0.0087 = 7.84\,\text{bbl}$$

The corresponding number of strokes is:

$$\text{Strokes}_{\text{KOP to EOB}} = \frac{7.84}{0.0986} = 80\,\text{stk}$$

The DC volume is:

$$\text{Volume}_{\text{DC}} = 262 \times 0.00767 = 2\,\text{bbl}$$

The corresponding number of strokes is:

$$\text{Strokes}_{\text{KOP to EOB}} = \frac{2}{0.0986} = 21\,\text{stk}$$

Therefore, the drill-string volume is:

$$\text{Volume}_{\text{DS}} = [44.64 + 59.14 + 132.3] + 7.84 + 2 = 245.92\,\text{bbl}$$

The corresponding number of strokes is:

$$\text{Strokes}_{\text{KOP to EOB}} = \frac{245.92}{0.0986} = 2495\,\text{stk}$$

Q4) Calculate the formation pressure based on the shut-in data.
The formation pressure is:

$$\begin{aligned}\text{Formation Press} &= \text{SIDPP} + 0.052 \times \text{MW} \times \text{TVD} \\ &= 300 + 0.052 \times 10 \times 5007 = 2904\,\text{psi}\end{aligned}$$

Q-5) What kill mud weight is required to balance formation pressure?

$$\begin{aligned}\text{MW}_{\text{Kill}} &= \text{MW}_{\text{Current}} + \frac{\text{SIDPP}}{0.052 \times \text{TVD}} \\ &= 10 + \frac{300}{0.052 \times 5007} = 11.152 \rightarrow 11.2\ \text{ppg}\end{aligned}$$

Q-6) What will the ICP be at 30 SPM?

$$\text{ICP} = \text{SCRP} + \text{SIDPP} = 1000 + 300 = 1300\,\text{psi}$$

Q-7) What will the FCP be at 30 SPM?

$$\text{FCP} = \text{SCRP} \times \frac{\text{MW}_{\text{Kill}}}{\text{MW}_{\text{Original}}} = 1000 \times \frac{11.2}{10} = 1120\,\text{psi}$$

Q-8) What would be the drill-pipe pressure step-down per 100 strokes of kill mud pumped down the drill-string? Divide this into three geometries of a) surface to KOP, b) KOP to EOB, and c) EOB to TD.

First, we need to evaluate the circulating pressures when kill mud reaches KOP, EOB, and TD.

–KOP CP:

$$\begin{aligned}\text{Dynamic Pressure Loss at KOP} &= \text{SCRP} + \left[(\text{FCP} - \text{SCRP}) \times \frac{\text{MD}_{\text{KOP}}}{\text{MD}_{\text{TD}}}\right] \\ &= 1000 + \left[(1120 - 1000) \times \frac{2585}{14835}\right] = 1021\,\text{psi}\end{aligned}$$

$$\begin{aligned}\text{Remaining SIDPP at KOP} &= \text{SIDPP} - \left[0.052 \times (\text{MW}_{\text{Kill}} - \text{MW}_{\text{Current}}) \times \text{TVD}_{\text{KOP}}\right] \\ &= 300 - \left[0.052 \times (11.2 - 10) \times 2585\right] = 138\,\text{psi}\end{aligned}$$

Therefore, Circulating Pressure at KOP (KOP CP) is found:

$$\text{KOP CP} = \text{Dynamic Pressure Loss at KOP} + \text{Remaining SIDPP at KOP}$$
$$= 1021 + 138 = 1159\,\text{psi}$$

–EOB CP:

$$\text{Dynamic Pressure Loss at EOB} = \text{SCRP} + \left[(\text{FCP} - \text{SCRP}) \times \frac{\text{MD}_{\text{EOB}}}{\text{MD}_{\text{TD}}}\right]$$
$$= 1000 + \left[(1120 - 1000) \times \frac{6010}{14835}\right] = 1048\,\text{psi}$$

$$\text{Remaining SIDPP at KOP} = \text{SIDPP} - \left[0.052 \times (\text{MW}_{\text{Kill}} - \text{MW}_{\text{Current}}) \times \text{TVD}_{\text{EOB}}\right]$$
$$= 300 - \left[0.052 \times (11.2 - 10) \times 5007\right] = 13\,\text{psi}$$

Therefore, Circulating Pressure at EOB (EOB CP) is found:

$$\text{EOB CP} = \text{Dynamic Pressure Loss at EOB} + \text{Remaining SIDPP at EOB}$$
$$= 1048 + 13 = 1061\,\text{psi}$$

a) *From Surface to KOP*

$$\text{DPP StepDown}_1 = \frac{(\text{ICP} - \text{KOB CP})}{\text{Surface to KOP Strokes}} \times 100 = \frac{(1300 - 1159)}{453} \times 100$$
$$= 31.12 \rightarrow 31\ \text{psi} / 100\,\text{stk}$$

b) *From KOP to EOB*

$$\text{DPP StepDown}_2 = \frac{(\text{KOP ICP} - \text{EOB CP})}{\text{KOP to EOB Strokes}} \times 100 = \frac{(1159 - 1061)}{600} \times 100$$
$$= 16.33 \rightarrow 16\ \text{psi} / 100\,\text{stk}$$

c) *From EOB to TD*

$$\text{DPP StepDown}_3 = \frac{(1061 - 1120)}{1442} \times 100 = -4\ \text{psi} / 100\text{stk}$$

10.5 SUBSEA-BOP VERTICAL WELLS

10.5.1 Introduction

First, Figure 10.5 illustrates a general overview of different parts of this type of kill sheet.

10.5.2 Exercises 3 and 4

Using the following data (in Table 10.4) for a subsea vertical well, complete the kill sheet and answer the questions.

Questions:

Q1) Calculate the strokes to pump down inside the drill-string from the surface to the bit.

Q2) Calculate the strokes to pump from the bit to the shoe.

Q3) Calculate the strokes to pump through the total annulus.

Q4) Calculate the time in minutes for calculating the total well system volume at 45 SPM.

Q5) Calculate the strokes to pump to displace the marine riser to kill fluid before opening the BOP.

Q6) Calculate the pressure safety margin at the casing shoe at a) initial shut-in/static and b) dynamic conditions.

Q7) Calculate the pressure safety margin at the casing shoe in static conditions, assuming the top of the kick is below the casing shoe.

Q8) Calculate the kill mud density.

Q9) Calculate the Initial Circulating Pressure (ICP).

Q10) Calculate the Final Circulating Pressure (FCP).

Q11) Calculate the initial dynamic casing pressure at kill pump rate.

Q12) Calculate the MAASP after circulation of kill fluid.

Q13) Calculate the pressure drop per 100 strokes kill fluid pumped inside the drill-string.

Answers for Exercise:

Figure 10.6 provides the completed kill sheet for Exercise 3.

Q1) Calculate the strokes to pump down inside the drill-string from the surface to the bit.

The length of the Drill-pipe (DP) is:

$$\text{Length}_{DP} = \text{MD}_{TD} - \text{Length}_{DC} - \text{Length}_{HWDP}$$
$$\text{Length}_{DP} = 10{,}450 - 912 - 723 = 8815 \text{ ft}$$

The drill-string (DS) volume is:

$$\begin{aligned}\text{DS Volume} &= \text{Length}_{DC} \times \text{Cap}_{DC} + \text{Length}_{HWDP} \times \text{Cap}_{HWDP} + \text{Length}_{DP} \times \text{Cap}_{DP} \\ &= 0.0077 \times 912 + 0.0088 \times 723 + 0.0172 \times 8815 = 151.61 + 6.36 + 7.02 \\ &= 165 \text{ bbl}\end{aligned}$$

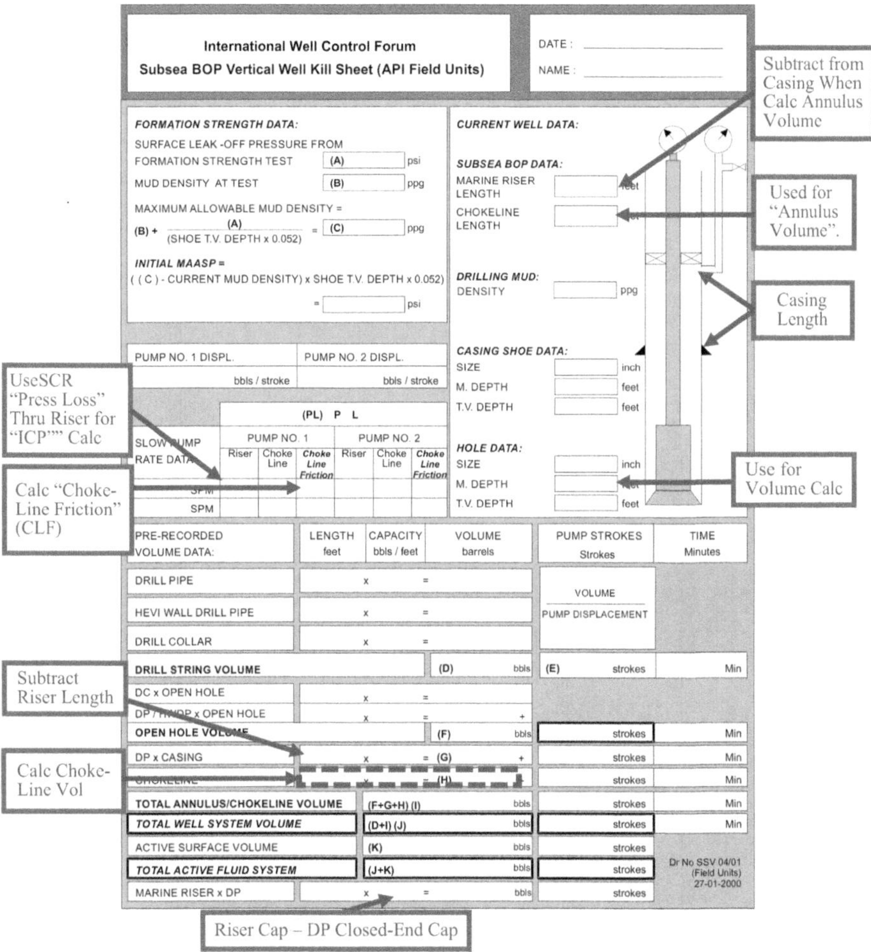
International Well Control Forum
Subsea BOP Vertical Well Kill Sheet (API Field Units)

DATE :
NAME :

FORMATION STRENGTH DATA:
SURFACE LEAK -OFF PRESSURE FROM FORMATION STRENGTH TEST (A) psi
MUD DENSITY AT TEST (B) ppg
MAXIMUM ALLOWABLE MUD DENSITY =
(B) + (A) / (SHOE T.V. DEPTH x 0.052) = (C) ppg
INITIAL MAASP =
((C) - CURRENT MUD DENSITY) x SHOE T.V. DEPTH x 0.052)
= psi

CURRENT WELL DATA:
SUBSEA BOP DATA:
MARINE RISER LENGTH feet
CHOKELINE LENGTH feet
DRILLING MUD:
DENSITY ppg
CASING SHOE DATA:
SIZE inch
M. DEPTH feet
T.V. DEPTH feet
HOLE DATA:
SIZE inch
M. DEPTH feet
T.V. DEPTH feet

PUMP NO. 1 DISPL.	PUMP NO. 2 DISPL.
bbls / stroke	bbls / stroke

SLOW PUMP RATE DATA	(PL) P L					
	PUMP NO. 1			PUMP NO. 2		
	Riser	Choke Line	Choke Line Friction	Riser	Choke Line	Choke Line Friction
SPM						
SPM						

PRE-RECORDED VOLUME DATA:	LENGTH feet	CAPACITY bbls / feet	VOLUME barrels	PUMP STROKES Strokes	TIME Minutes
DRILL PIPE		x	=	VOLUME / PUMP DISPLACEMENT	
HEVI WALL DRILL PIPE		x	=		
DRILL COLLAR		x	=		
DRILL STRING VOLUME			(D) bbls	(E) strokes	Min
DC x OPEN HOLE		x	=		
DP / HWDP x OPEN HOLE		x	= +		
OPEN HOLE VOLUME			(F) bbls	strokes	Min
DP x CASING		x	= (G) +	strokes	Min
CHOKELINE		x	= (H)	strokes	Min
TOTAL ANNULUS/CHOKELINE VOLUME			(F+G+H) (I) bbls	strokes	Min
TOTAL WELL SYSTEM VOLUME			(D+I) (J) bbls	strokes	Min
ACTIVE SURFACE VOLUME			(K) bbls	strokes	
TOTAL ACTIVE FLUID SYSTEM			(J+K) bbls	strokes	
MARINE RISER x DP		x	= bbls	strokes	

Dr No SSV 04/01 (Field Units) 27-01-2000

FIGURE 10.5 a) Overview of important points in a surface-BOP – deviated kill sheet (pre-kick kill sheet/first page). (Continued)

Therefore, the number of strokes for drill-string volume is:

$$\text{DS Strokes} = \frac{\text{DS Volume}}{\text{Pump Displacement}}$$

$$= \frac{165}{0.12} = 1375\,\text{stk}$$

Q2) Calculate the strokes to pump from the bit to the shoe.

The length of DP or HWDP in the annulus is:

$$\text{Length}_{\text{DP}\times\text{OH}} = \text{Length}_{\text{OH}} - \text{Length}_{\text{DC}} = (10450 - 7800) - 912 = 1738\,\text{ft}$$

$$\text{Annular OH Volume} = \text{Length}_{\text{Ann,DC*OH}} \times \text{Cap}_{\text{Ann,DC*OH}} + \text{Length}_{\text{Ann,DP*OH}} \times \text{Cap}_{\text{Ann,DP*OH}}$$

$$= 912 \times 0.0292 + 1738 \times 0.0447 = 104.31\,\text{bbl}$$

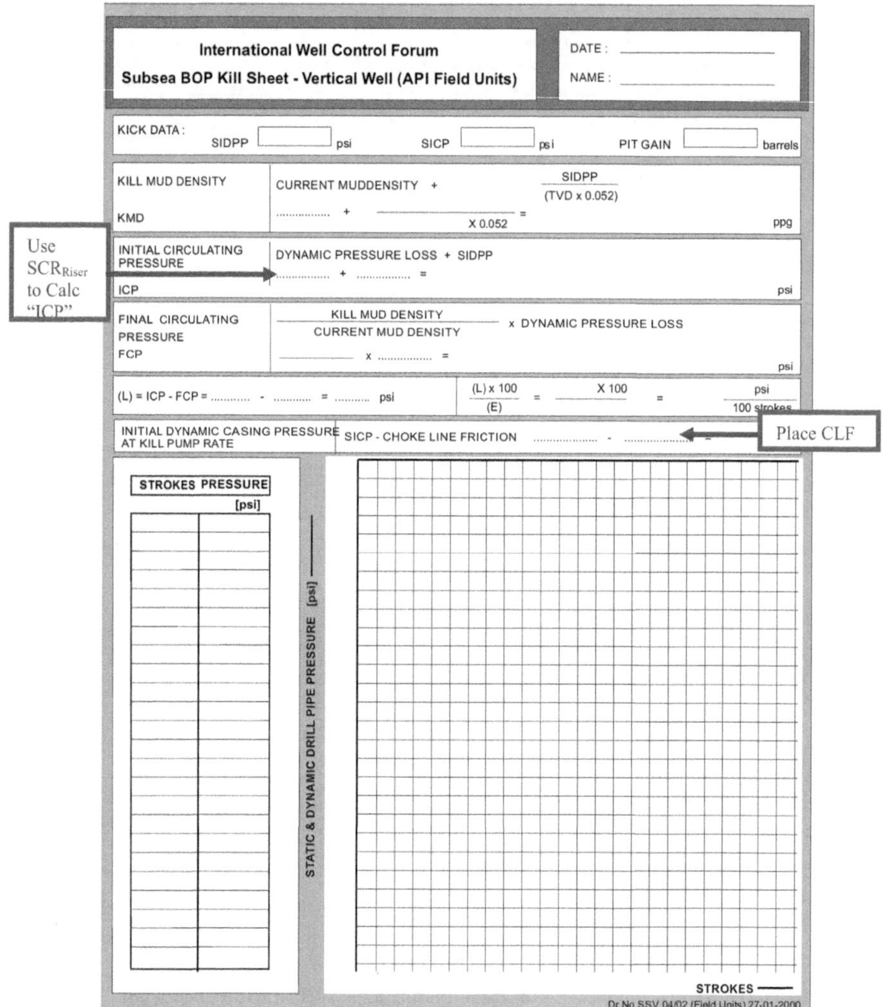

FIGURE 10.5 Continued: b) Overview of important points in a Surface-BOP – deviated kill sheet (second page).

Therefore:

$$\text{Strokes for Annular OH} = \frac{\text{OH Volume}}{\text{Pump Displacement}}$$

$$= \frac{104.31}{0.12} = 869.25 \rightarrow 870\,\text{stk}$$

TABLE 10.4
Given Data for Subsea Vertical Well Kill Sheet and Questions for Exercise 3

Pit Gain	10 bbl
SIDPP	550 psi
SICP	820 psi
Marine Riser	OD = 23 in/ID = 20 in Length : 400 ft, Capacity : 0.3892 bbl / ft)
Choke-Line	ID = 2.5 in, Capacity : 0.0061 bbl/ft Length: 415 ft
Hole Size	8 ½ in
Hole Depth	10,450 ft (TVD / MD)
Casing	9 5 / 8 - in (Shoe at 7,800 ft)
Drill-pipe (DP)	5 - in (Capacity : 0.0172 bbl / ft) (Closed-end distance = 0.0254 bbl/ft)
Heavy Weight Drill-pipe (HWDP:	OD = 5 in, ID = 3 in, Capacity: 0.0088 bbl/ft (Length : 723 ft)
Drill-collar (DC)	OD = 6.5 in, ID = 2 13/16 in, Capacity: 0.0077 bbl/ft (Length : 912 ft)
Mud Weight/Density	11.5 ppg
Annular Capacity Open Hole/DC	0.0292 bbl / ft
Annular Capacity Open Hole/HWDP or DP	0.0447 bbl / ft
Annular Capacity Casing/DP	0.0478 bbl / ft
Annular Capacity Riser/DP	0.3638 bbl / ft
Leak-Off Pressure Using 11 ppg Mud:	1,900 psi
Pump Factor/Output (for Pump 1 and 2)	0.1224 bbl / stroke (at 100% Efficiency) Pump Efficiency = 98%
Dynamic Pressure Losses	Through riser (at 45 SPM) = 780 psi Through choke-line at 45 SPM = 900 psi
Active Surface Mud Volume	480 bbl

Q3) Calculate the strokes to pump through the total annulus.

The OH Annulus Volume was already calculated equal to 104.31 bbl.

The DP length is found by subtracting the riser length from the casing measured depth:

$$\text{DP} \times \text{Casing Volume} = (7800 - 400) \times 0.0478 = 353.72 \text{ bbl}$$

The Choke-Line Volume is:

$$\begin{aligned}\text{Choke-Line Volume} &= \text{Choke-Line Length} \times \text{Capacity}_{\text{Choke-Line}} \\ &= 415 \times 0.0061 = 2.53 \text{ bbl}\end{aligned}$$

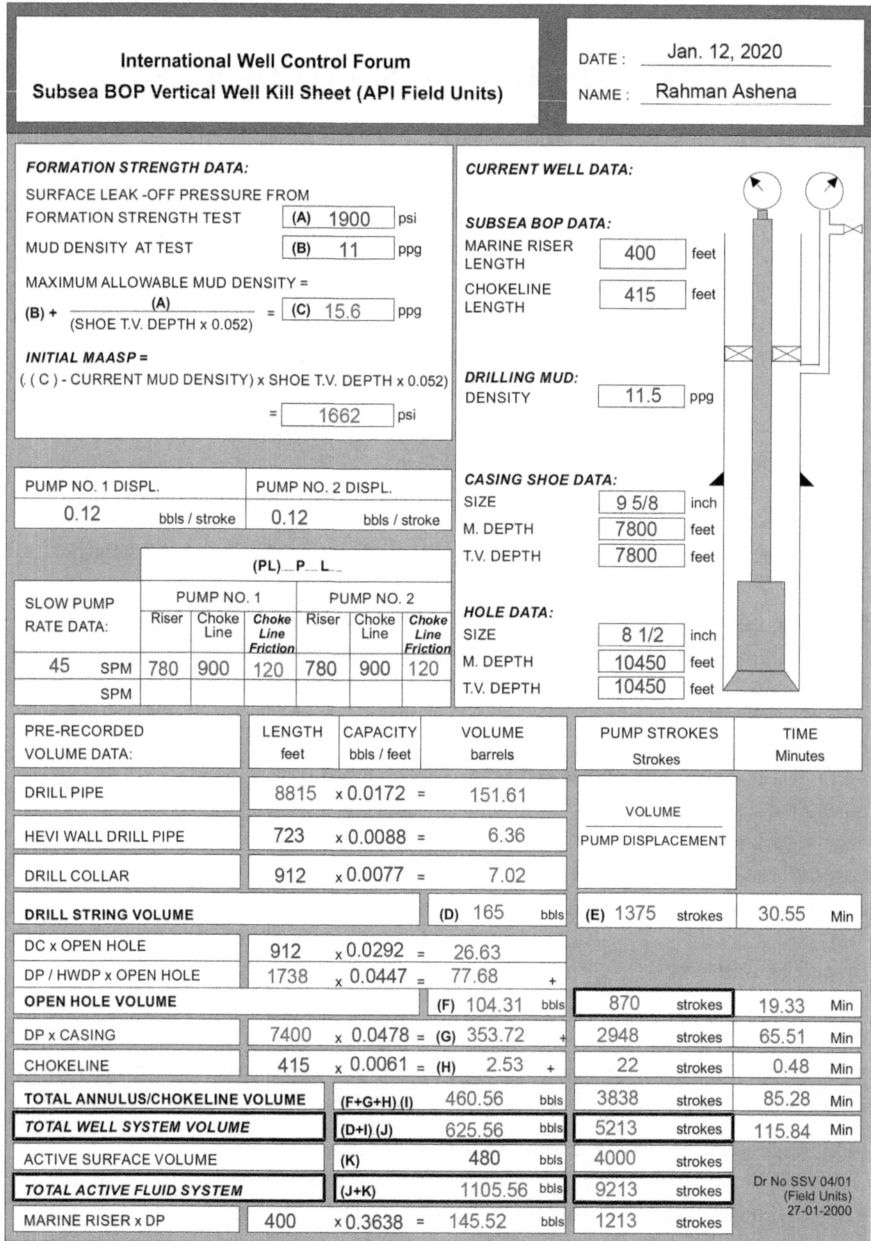

International Well Control Forum

Subsea BOP Vertical Well Kill Sheet (API Field Units)

DATE : Jan. 12, 2020

NAME : Rahman Ashena

FORMATION STRENGTH DATA:

SURFACE LEAK-OFF PRESSURE FROM FORMATION STRENGTH TEST (A) 1900 psi

MUD DENSITY AT TEST (B) 11 ppg

MAXIMUM ALLOWABLE MUD DENSITY =

(B) + (A) / (SHOE T.V. DEPTH x 0.052) = (C) 15.6 ppg

INITIAL MAASP =

((C) - CURRENT MUD DENSITY) x SHOE T.V. DEPTH x 0.052) = 1662 psi

CURRENT WELL DATA:

SUBSEA BOP DATA:

MARINE RISER LENGTH 400 feet

CHOKELINE LENGTH 415 feet

DRILLING MUD:

DENSITY 11.5 ppg

CASING SHOE DATA:

SIZE 9 5/8 inch

M. DEPTH 7800 feet

T.V. DEPTH 7800 feet

HOLE DATA:

SIZE 8 1/2 inch

M. DEPTH 10450 feet

T.V. DEPTH 10450 feet

PUMP NO. 1 DISPL.	PUMP NO. 2 DISPL.
0.12 bbls / stroke	0.12 bbls / stroke

SLOW PUMP RATE DATA:	(PL) P L					
	PUMP NO. 1			PUMP NO. 2		
	Riser	Choke Line	*Choke Line Friction*	Riser	Choke Line	*Choke Line Friction*
45 SPM	780	900	120	780	900	120
SPM						

PRE-RECORDED VOLUME DATA:	LENGTH feet	CAPACITY bbls / feet	VOLUME barrels	PUMP STROKES Strokes	TIME Minutes
DRILL PIPE	8815	x 0.0172 =	151.61	VOLUME / PUMP DISPLACEMENT	
HEVI WALL DRILL PIPE	723	x 0.0088 =	6.36		
DRILL COLLAR	912	x 0.0077 =	7.02		
DRILL STRING VOLUME			(D) 165 bbls	(E) 1375 strokes	30.55 Min
DC x OPEN HOLE	912	x 0.0292 =	26.63		
DP / HWDP x OPEN HOLE	1738	x 0.0447 =	77.68 +		
OPEN HOLE VOLUME			(F) 104.31 bbls	870 strokes	19.33 Min
DP x CASING	7400	x 0.0478 =	(G) 353.72 +	2948 strokes	65.51 Min
CHOKELINE	415	x 0.0061 =	(H) 2.53 +	22 strokes	0.48 Min
TOTAL ANNULUS/CHOKELINE VOLUME		(F+G+H) (I)	460.56 bbls	3838 strokes	85.28 Min
TOTAL WELL SYSTEM VOLUME		(D+I) (J)	625.56 bbls	5213 strokes	115.84 Min
ACTIVE SURFACE VOLUME		(K)	480 bbls	4000 strokes	
TOTAL ACTIVE FLUID SYSTEM		(J+K)	1105.56 bbls	9213 strokes	
MARINE RISER x DP	400	x 0.3638 =	145.52 bbls	1213 strokes	

Dr No SSV 04/01 (Field Units) 27-01-2000

FIGURE 10.6 a) The first page (pre-kick sheet) of the kill sheet of subsea vertical well for Exercise 3.

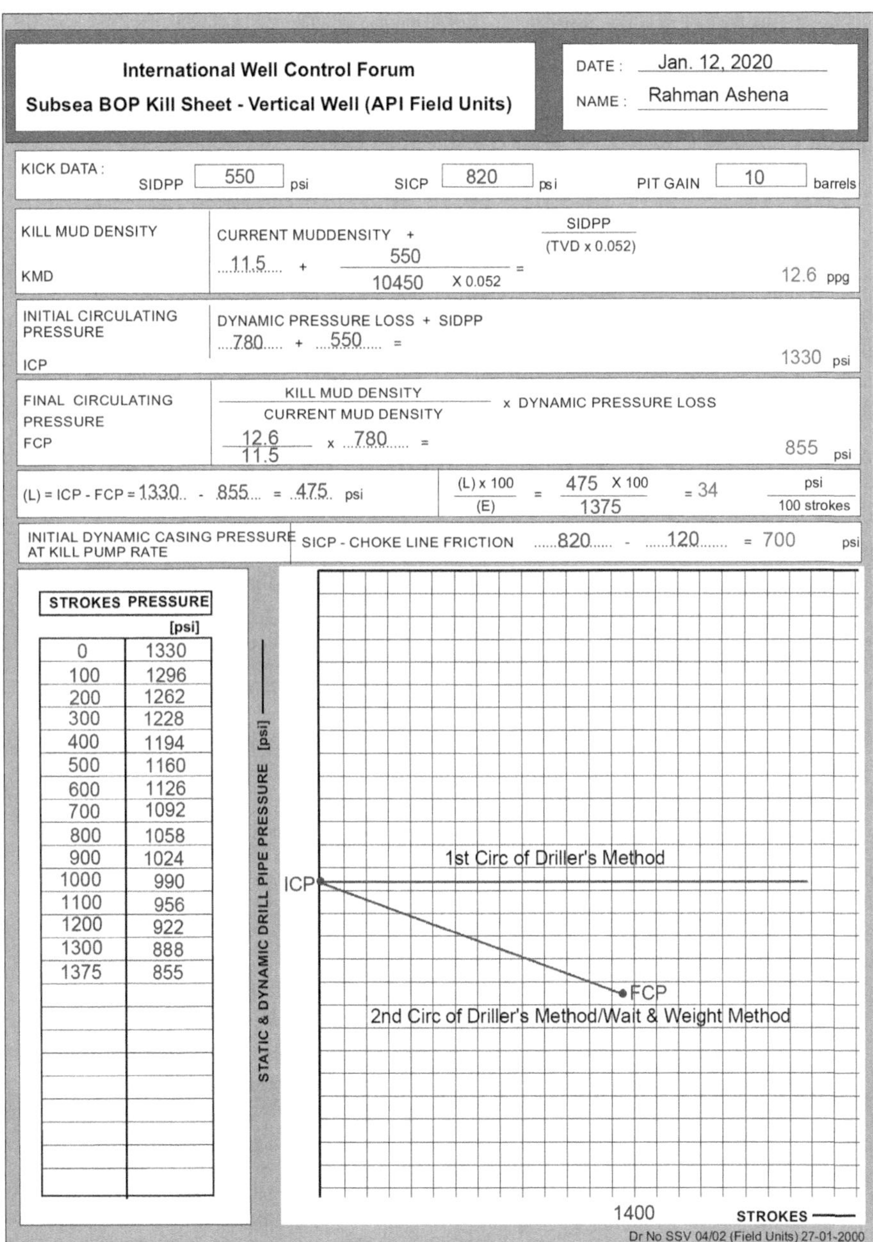

International Well Control Forum

Subsea BOP Kill Sheet - Vertical Well (API Field Units)

DATE : Jan. 12, 2020

NAME : Rahman Ashena

KICK DATA : SIDPP 550 psi SICP 820 psi PIT GAIN 10 barrels

KILL MUD DENSITY / KMD: CURRENT MUDDENSITY + SIDPP / (TVD x 0.052)

11.5 + 550 / (10450 X 0.052) = 12.6 ppg

INITIAL CIRCULATING PRESSURE / ICP: DYNAMIC PRESSURE LOSS + SIDPP

780 + 550 = 1330 psi

FINAL CIRCULATING PRESSURE / FCP: KILL MUD DENSITY / CURRENT MUD DENSITY x DYNAMIC PRESSURE LOSS

12.6 / 11.5 x 780 = 855 psi

(L) = ICP - FCP = 1330 - 855 = 475 psi

(L) x 100 / (E) = 475 X 100 / 1375 = 34 psi / 100 strokes

INITIAL DYNAMIC CASING PRESSURE AT KILL PUMP RATE: SICP - CHOKE LINE FRICTION 820 - 120 = 700 psi

STROKES	PRESSURE [psi]
0	1330
100	1296
200	1262
300	1228
400	1194
500	1160
600	1126
700	1092
800	1058
900	1024
1000	990
1100	956
1200	922
1300	888
1375	855

FIGURE 10.6 Continued: b) The second page (post-kick) of kill sheet of subsea vertical well, Exercise 3.

Therefore, the total annulus volume is:

$$\begin{aligned}\text{Total Annulus Volume}\\ &= \text{Annular OH Volume} + \text{“DP} \times \text{Casing” Volume} + \text{Choke-Line Volume}\\ &= 104.31 + 353.72 + 2.53 = 460.56\ \text{bbl}\end{aligned}$$

The total annulus strokes is:

$$\text{Total Annulus Strokes} = \frac{\text{Total Annulus Volume}}{\text{Pump Displacement}} = \frac{450.56}{0.12} = 3838\ \text{stk}$$

Q4) Calculate the time in minutes for calculating the total well system volume at 45 SPM.

Total well system volume is found by adding the drill-string volume and the annulus volume:

$$\begin{aligned}\text{Total Well System Volume} &= \text{Drill} - \text{String Volume } + \text{Total Annulus Volume}\\ &= 165 + 460.56 = 625.56\ \text{bbl}\end{aligned}$$

$$\begin{aligned}\text{Strokes for “Total Well System”} &= \frac{\text{Total Well System Volume}}{\text{Pump Displacement}}\\ &= \frac{625.56}{0.12} = 5213\ \text{stk}\end{aligned}$$

$$\text{Pumping Time} = \frac{\text{Strokes}}{\text{SCR}\ [\text{SPM}]} = \frac{5213}{45} = 115.84\ \text{min}$$

Q5) Calculate the strokes to pump to displace the marine riser to kill fluid before opening the BOP.

After the kick influx is circulated out of the hole using choke-line (i.e., well is killed), we need to displace the annulus between riser and drill-pipe with kill mud as a prerequisite before we can safely resume drilling. The volume to displace the annulus between the riser and the drill-pipe:

$$\text{“Riser} \times \text{DP” Volume bbl]} = \text{Length}_{\text{Riser}}\,[\text{ft}]\text{"} \times \text{Capacity}_{\text{Riser*DP}}\left[\frac{\text{bbl}}{\text{ft}}\right]$$

$$\text{Marine Riser} \times \text{DP Annulus} = \text{Riser Length} \times \text{Cap}_{\text{Ann},\frac{\text{Riser}}{\text{DP}}} = 400 \times 0.3638 = 145.52\ \text{bbl}$$

The number of strokes is:

$$\begin{aligned}\text{Strokes for Riser} \times \text{DP Displacement} &= \frac{\text{“Riser} \times \text{DP” Volume}}{\text{Pump Displacement}}\\ &= \frac{145.52}{0.12} = 1212.66 \rightarrow 1213\ \text{stk}\end{aligned}$$

Q6) Calculate the pressure safety margin at the casing shoe at a) initial shut-in (static) and b) at dynamic conditions.

To find the safety margin, we need to calculate some other parameters first.

Maximum Allowable Mud Weight (MAMW) is:

$$\text{MAMW} = \text{MW}_{\text{LOT}} + \frac{\text{LOP}}{0.052 \times \text{TVD}_{\text{Casing}}} = 11 + \frac{1900}{0.052 \times 7800} = 15.684 \rightarrow 15.6 \text{ ppg}$$

Note: For safety reason (providing some overbalance), MAMW is rounded up to one decimal part.

Maximum Allowable Annular Surface Pressure (MAASP) is:

$$\text{MAASP} = 0.052 \times (\text{MAMW} - \text{MW}) \times \text{TVD}_{\text{Casing}} = 0.052 \times (15.6 - 11.5) \times 7800$$
$$= 1662.96 \text{ psi} \rightarrow 1662 \text{ psi}$$

Note: For safety, MAASP is rounded down to an integer number.

a) *Static Safety Margin is:*

$$\textit{Safety Margin} = \textit{MAASP} - \textit{SICP} = 1662 - 820 = 842 \textit{ psi}$$

b) *Dynamic Safety Margin is:*

$$\textit{Dynamic Safety Margin} = (\textit{MAASP} - \textit{CLFP}) - (\textit{SICP} - \textit{CLFP})$$
$$= 1662 - 820 = 842 \textit{ psi}$$

Therefore, the static and dynamic safety margins have the same value.

Q7) Calculate the pressure safety margin at the casing shoe in static conditions, assuming the top of the kick is below the casing shoe.

As the question mentions "static", it means we have stopped circulating and shut in the well. The question also mentions that the kick influx has been circulated such that its top is located at the shoe. As well control method has been applied correctly, the bottomhole pressure has been maintained constant; thus, after stopping the circulation and well shut-in, the SICP is the same as the beginning. Therefore, the safety margin remains constant:

$$\textit{Safety Margin} = \textit{MAASP} - \textit{SICP} = 1662 - 820 = 842 \textit{ psi}$$

Q8) Calculate the kill mud density.

The required kill mud is the equivalent mud weight to balance the formation pressure. Thus, it is found using the following relation:

$$\text{MW}_{\text{Kill}} = \text{MW}_{\text{Current}} + \frac{\text{SIDPP}}{0.052 \times \text{TVD}} = 11.5 + \frac{550}{0.052 \times 10450} = 12.512 \rightarrow 12.6 \text{ ppg}$$

Note: Always round up the kill mud weight to one decimal place. This is done for safety reasons (to create some overbalance pressure).

Q9) Calculate the ICP.

The ICP in the subsea case is found as the summation of the SIDPP to the pressure loss at SCR (also called SCRP) through the riser:

$$\text{ICP} = \text{SCRP} + \text{SIDPP} = 780 + 550 = 1330 \text{ psi}$$

Q10) Calculate the FCP.

Final Circulating Pressure (FCP) is the drill-pipe pressure(DPP) which is obtained when the kill mud reaches the bit (either in Driller's method or in Wait and Weight method). It is found by converting the pressure loss at SCR (also called SCRP) to that of the kill mud weight:

$$\text{FCP} = \text{SCRP} \times \frac{\text{MW}_{\text{kill}}}{\text{MW}_{\text{original}}} = 780 \times \frac{12.6}{11.5} = 854.608 \rightarrow 855 \text{ psi}$$

Note: FCP is rounded up to an integer for safety (to create some overbalance above formation pressure).

Q11) Calculate the initial dynamic casing pressure at kill pump rate.

Initial dynamic casing pressure is a subsea term which is found by subtracting CLFP (Choke-line Friction Pressure) from SICP:

$$\text{Initial Dynamic CP} = \text{SICP} - \text{CLFP} = 820 - 120 = 700 \text{ psi}$$

Q12) Calculate the MAASP after the circulation of kill fluid.

MAASP depends on mud weight. At the end of kill operations, we have the kill mud in the hole, and it should be used for the calculations:

$$\begin{aligned}\text{MAASP} &= 0.052 \times \left(\text{MAMW} - \text{MW}_{\text{Current}}\right) \times \text{TVD}_{\text{CSG}} \\ &= 0.052 \times \left(15.6 - 12.6\right) \times 7800 = 1216.8 \text{ psi} \rightarrow 1216 \text{ psi}\end{aligned}$$

Q13) Calculate the drill-pipe pressure drop per 100 strokes kill fluid pumped inside the drill-string.

The drill-pipe pressure step-down (DPP Step-Down) value is found as the pressure drop (from ICP to FCP) for 100 strokes pumped [psi per 100 strokes]:

$$\begin{aligned}\text{DPP StepDown} &= \frac{\text{ICP} - \text{FCP}}{\text{DS Strokes}} \times 100 \\ &= \frac{1330 - 855}{1375} \times 100 = 34.54 \rightarrow 34 \left[\frac{\text{psi}}{100 \text{ stk}}\right]\end{aligned}$$

Note: Always round down the pressure step-down to an integer. This is done for safety reasons (to create some overbalance above formation pressure).

Exercise 4:

When killing a well with a horizontal section, what will happen if you use a vertical kill sheet to circulate out the kick?

a. The strokes to the bit will be wrong.
b. You will be applying too little pressure to the well.
c. The FCP is more difficult to calculate.
d. You will be applying too much pressure to the well.

Answer – d:

Assume a vertical kill sheet is mistakenly used for a horizontal well while regarding the difference between measured depth (MD) and true vertical depth (TVD).

- In this case, choice a is wrong because the number of strokes to the bit is not wrong as the length of the drill-string is the same.
- Choice c is wrong because FCP depends on Slow Circulating Rate Pressure (SCP) and kill mud weight; KMW = MW = SIDPP/(0.052 × TVD), which in turn depends on shut-in drill-pipe pressure (SIDPP) and vertical depth (TVD), and the user would simply use the same values that the kill sheet asks if a vertical kill sheet is used in lieu of a deviated one.
- However, as we already know, the bottomhole pressure equals static hydrostatic mud pressure plus annular pressure loss.
- When we assume a vertical wellbore than a horizontal one, more annular pressure losses would be exerted to the bottomhole pressure as compared to a deviated or horizontal wellbore.
- Therefore, the bottomhole pressure is greater in a vertical well than in a horizontal well. Therefore, choice d is correct.
- Another way to answer this question is by comparing Figure 10.2 and Figure 10.4 by which it is understandable that the dynamic well pressures applied from the surface in a vertical well are greater than those applied in a horizontal well. This is because around the end of build (EOB) section, lower pressure than FCP is exerted at the surface; thus, the wellbore pressure would be expected to be lower as well.

11 Well Control Methods

Rahman Ashena

11.1 INTRODUCTION

After a successful well shut-in, the kill sheet is completed using recorded post shut-in data. Following that, the well must be killed or taken under control using an appropriate well control or kill method. Prior to killing the well, the first action is holding a pre-kill planning meeting. Next, different well kill methods should be considered for applications.

This chapter provides an overview of well kill methods and techniques for controlling wells kicked. Well control methods consist of using mud hydrostatic pressure, using Blowout Preventers (BOPs) to seal the wellbore, and circulate the kick influx either safely to the surface, or squeeze it back into the formation. The main kill methods that are discussed in detail in this chapter consist of Driller's method, Wait and Weight methods, Concurrent methods, and volumetric method which are based on maintaining constant bottomhole pressure. These methods are categorized as influx-out methods whereas the Bullheading method is the influx-in method which is based on pumping or squeezing the kick influx fluids in the well to push them back to the formation and thus kill the well. Next, to select the appropriate kill method, the advantages and disadvantages of each method are discussed (which allow comparison and delineation of their selection criteria). In addition, their step-by-step procedures are pinpointed, and the pressure curves during operations are analyzed. The well kill method in wireline operations is also covered. Next, possible kill problems or equipment malfunctions and the right measures to control the situation are covered. Following successful kill, the well can be safely opened with some necessary activities to be explained.

11.2 PRE-KILL PLANNING MEETING

After completing the kill sheet and before proceeding to kill the well, a pre-kill planning meeting must be held by the drilling supervisor. The following points are mentioned in this meeting:

- *To Mud Engineer:*
  Ask the mud engineer to:
  - Make enough volume of the kill mud (with kill mud weight).
  - Prepare and keep the kill mud in a separate reserve tank and not to transfer it to the active tanks (so that kill mud does not mix with the original mud).

DOI: 10.1201/9781003473770-11

    - Take/measure the mud levels frequently (e.g., every ten minutes) to make sure about possible mud volume changes. Explain how the mud level is expected to change during the operations. Due to gas influx expansion, as expected, some increase in return mud and thus the mud level is observed until the gas influx exits the choke.
- *To Driller and Assistant Driller:*
    - Briefly explain to the driller about the entire kill operations.
    - Tell them that they should make "zero" the stroke-counter at the beginning of the well kill.
    - Tell them to coordinate the pump bring-up with the supervisor. Tell them that they should raise the pump rate slowly (e.g., in five stroke/minute) to reach the kill rate.
    - Ask them to communicate with the drilling supervisor properly.
    - Ask them to monitor the shut-in pressures and alert the supervisor/tool pusher about any changes in the drill-pipe and casing pressures during the operations.
    - Tell the driller that they should shut down the well at the end of the operations and
- Announce the time you plan to start the kill operations. Let the crew know the time you plan to start the operations.

11.3 WELL KILL METHODS

Following the kick detection and the well shut-in, there are several methods used to kill the well and take it under control. The methods of well kill are categorized into influx-out and influx-in methods. Because of their more safety, the influx-out methods are the main kill methods under attention by international organizations (such as International Association of Drilling Contractors, IADC, and International Well Control Forum, IWCF), which are based on the principle of maintaining constant bottomhole pressure. In these methods, the bottomhole pressure is equal or slightly greater than the formation pore pressure so that no formation influx can enter the wellbore again. The influx-out methods consist of the *Driller's* method, *Wait and Weight* method, C*oncurrent* method (combination of Driller's and Wait and Weight methods), and the *Volumetric and Bleed* method. In this manual, the Driller's, Wait and Weight, and Concurrent methods are first discussed in detail (which apply when the drill-string is on-bottom); and then the volumetric method is discussed (which applies when the drill-string is off-bottom or out of the hole. Finally, the other category of well kill methods (i.e., *influx-in*) is discussed, which consists of the *bullheading* method. In this method, the kick influx is squeezed back to the formation.

11.4 DRILLER'S METHOD

11.4.1 Description

The Driller's method is considered the basic well kill method which takes the well under control via two circulations. In the first circulation, the original mud is pumped to circulate out the kick influx, e.g., gas. At the end of the first circulation, there is no kick fluid in the well. In the second circulation, a heavy enough mud (i.e., kill mud) is circulated through the well to replace the original mud. Therefore, at the end of the second circulation, the final bottomhole pressure would be equal or slightly greater than the formation pressure, and thus no new kick influx can enter the well again; therefore, the well can be safely opened for continuing the well operations (e.g., drilling).

The fundamental basis for all influx-out well control methods is that the bottomhole pressure (BHP) must be always maintained constant. Therefore, in all these methods, during well control operations, all the opening and closing of chokes should be controlled for this purpose.

It is noted that in drilling wells with fixed rigs and surface-BOPs, kick circulation is easier than in subsea-BOPs because the kill and choke-lines are short which reduces circulation frictional pressure losses. Thus, slightly higher pump rate can frequently be used in surface-BOP cases.

11.4.2 Advantages and Disadvantages

11.4.2.1 Advantages

Generally, the longer the time elapses after the kick influx entrance and well shut-in, the more difficult it would be to kill and control the well (due to gas migration). The origin of advantages of the Driller's method lies in the fact that circulating the kick out of the hole can be immediately started (without any waste of time). Therefore, some advantages of this method are:

- Killing the well can be proceeded quickly after well shut-in.
- The kill sheet calculations (for density, pressures, etc.) are quite simple, and the implementation of the method is also simple.
- In this method, less safety risks are involved as the well control operations can be started with less chance of gas migration. Gas migration makes shut-in pressures increase; thus, controlling the well condition can become more difficult.
- It is the recommended well control method for severe or special cases which are subject to hole problems that can follow the kick – for example, for horizontal and highly inclined wells which are prone to additional hole problems such as stuck pipe.

11.4.2.2 Disadvantages

The origins of the disadvantages of the Driller's method lie in the realization of two facts: a) safe well control conditions will be established after two subsequent mud circulations (rather than one) and b) the original mud weight (which is lower than the formation pressure's equivalent density) is used for the first mud circulation. Therefore, a few disadvantages of the method are:

- During the kill operations, high annular pressures are exerted on the wellbore walls including the casing shoe and formation for a long time, which may cause unprecedented formation fracturing at the shoe. This is because the mud weight used for the first circulation is the original one (i.e., low mud weight).

- During the kill operations, the surface equipment, such as choke and BOP, need to stand high pressures. Thus, the possibility of failure of the surface equipment is higher than the other kill methods. Again, this is because the mud weight used for the first circulation is the original one (i.e., low mud weight).
- The period of time that the surface equipment need to stand pressures is the higher than that in the Wait and Weight method. Therefore, the possibility of failure of the equipment is higher than the other methods. This is because killing operations are conducted via two mud circulations (more time needed).

11.4.3 Procedure

The procedure for implementing the Driller's method to kill the well is divided into three parts of a) preparation, b) first circulation, and c) second circulation. In all these phases, close communication between all the engaged drilling personnel (particularly the driller and the choke-person/supervisor) is extremely important for the success of the operation.

11.4.3.1 Preparation

Prior to kick occurrence, make sure that the pre-kick data has been already gathered in pre-kick kill sheet. For example, ensure that the Slow Circulating Rate Pressures (SCRPs) corresponding to the Slow Circulating Rate (SCR), e.g., 30 or 40 gpm, have been taken and recorded – in better words, ensure that the pre-kick kill sheet has been filled out.

Following kick detection, make sure that the well has safely been shut-in and secured. Make a safety meeting with the drilling crew to review the kill operation. Wait until the Shut-In Drill-pipe Pressure (SIDPP) and Shut-In Casing Pressures (SICP) are already stabilized on the pressure gauges. As an example, in Figure 11.1, the SICP and SIDPP are respectively equal to 950 and 630 psi. For more information on the required calculations in the kill sheet, please refer to Section 7.7 and Chapter 10.

11.4.3.2 First Circulation

The first mud circulation in the Driller's method is considered the first stage of the well kill operation which by itself consists of two sub-stages. For better understanding, these stages should be followed using Figure 11.1.

11.4.3.2.1 Circ 1–1: Bringing Up the Pump to SCR

First, make the "stroke-counter" zero. During the bring-up stage or raising the pump rate to slow circulation rate (SCR), it is essential to maintain the bottomhole pressure during bring-up. This is done by operating the choke in a controlled manner to keep the casing pressure constant (in surface-BOP-stacks) or the kill-line pressure

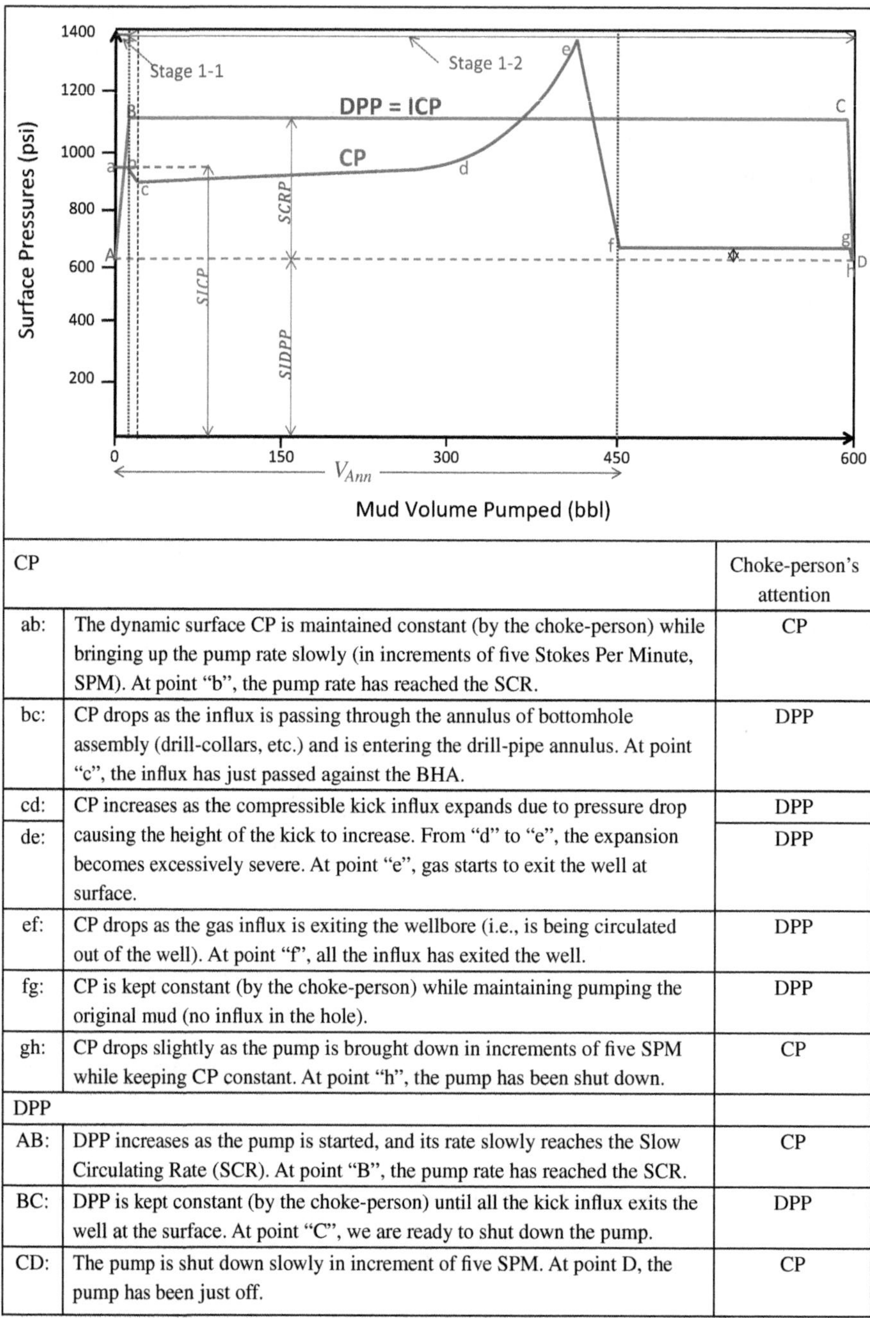

CP		Choke-person's attention
ab:	The dynamic surface CP is maintained constant (by the choke-person) while bringing up the pump rate slowly (in increments of five Stokes Per Minute, SPM). At point "b", the pump rate has reached the SCR.	CP
bc:	CP drops as the influx is passing through the annulus of bottomhole assembly (drill-collars, etc.) and is entering the drill-pipe annulus. At point "c", the influx has just passed against the BHA.	DPP
cd:	CP increases as the compressible kick influx expands due to pressure drop	DPP
de:	causing the height of the kick to increase. From "d" to "e", the expansion becomes excessively severe. At point "e", gas starts to exit the well at surface.	DPP
ef:	CP drops as the gas influx is exiting the wellbore (i.e., is being circulated out of the well). At point "f", all the influx has exited the well.	DPP
fg:	CP is kept constant (by the choke-person) while maintaining pumping the original mud (no influx in the hole).	DPP
gh:	CP drops slightly as the pump is brought down in increments of five SPM while keeping CP constant. At point "h", the pump has been shut down.	CP
DPP		
AB:	DPP increases as the pump is started, and its rate slowly reaches the Slow Circulating Rate (SCR). At point "B", the pump rate has reached the SCR.	CP
BC:	DPP is kept constant (by the choke-person) until all the kick influx exits the well at the surface. At point "C", we are ready to shut down the pump.	DPP
CD:	The pump is shut down slowly in increment of five SPM. At point D, the pump has been just off.	CP

FIGURE 11.1 Surface casing pressure (CP) and drill-pipe pressure (DPP) curves versus mud volume pumped (*the first circulation of the Driller's method*).

constant (in subsea stacks). Start one pump slowly with the pump rate of five Strokes Per Minute (SPM). Monitor the casing pressure (CP) (in surface stacks)/kill-line pressure (KLP) (in subsea stacks) until a slight rise from the SICP value is observed. At this instant, slightly open the choke slowly to reduce the CP/KLP and wait for confirmation from the crew that the well is flowing. Continue increasing the pump rate slowly (in increments of five SPM) while maintaining the CP (surface stacks)/ KLP (subsea stacks) constant (look at *CP curve*, line *a-b* in Figure 11.1). Continue this process until the SCR of, e.g., 30 or 40 SPM, is reached.

Note 1: At this stage, the CP (surface)/KLP (subsea) must be kept constant in order to keep the bottomhole pressure constant. By no means, the CP/KLP should be lower than the SICP. If so, it means that the choke has not been adjusted properly and thus further kick influx has already entered the well.

Note 2: At this stage, the DPP is not a good basis for being kept constant because the drill-pipe frictional pressure loss continuously increases as the pump rate is increased. Thus, keeping the DPP constant has not only no logic behind, but also it causes the bottomhole pressure to drop.

Note 3: When the pump rate has reached the SCR, record the stabilized DPP as the measured Initial Circulating Pressure, ICP (i.e., point "B" in Figure 11.1). Compare the measured ICP with the expected calculated ICP in the kill sheet. The measured ICP should be equal or slightly greater than the calculated ICP.

11.4.3.2.2 Circ 1–2: Circulating the Kick Influx Out of the Hole

After the pump rate has reached the SCR and the DPP has stabilized, the choke-person is required to transfer attention from the CP/KLP to the DPP. In other words, thereafter, he/she must adjust the choke in a way that the DPP can be maintained constant at ICP.

As the original mud is being pumped, the kick influx is circulated up in the annulus. As the influx is passing through the annulus of bottomhole assembly (drill-collars, etc.) and is entering the drill-pipe annulus, CP/KLP drops (*CP curve, line b-c* in Figure 11.1). This pressure drop occurs as the height of the influx in the annulus of the drill-pipe is lower than the height of the influx in the annulus of the drill-collar. After the kick influx has entered the annulus of the drill-pipe and is raised up, it starts to expand as gas influx is compressible (look at *CP curve, line c-e* in the same figure).

The influx expansion causes the height of the kick influx to increase and thus resulting in greater surface annular pressure or CP/KLP. Adjusting the choke to keep the DPP constant is not easy and indeed requires great tact of the choke-person.

Immediately after the kick influx reaches the surface and starts to exit the well, the CP/KLP starts to decrease (look at *point "e"* in Figure 11.1). When all the gas influx has been completely circulated out of the well, the decrease in the CP stops (*point "f"* in the same figure). At this time, shut down the pump. As the pump is brought down to zero, the choke-person needs to adjust the choke to keep the CP/ KLP constant to maintain the bottomhole pressure.

11.4.3.3 Second Circulation

Following the kick occurrence and well shut-in and recording the stabilized shut-in pressures, the second page of the kill sheet is already completed. The required volume of the kill mud is prepared. At this time, the second stage (consisting of three substages) of the well kill can be started.

11.4.3.3.1 Circ 2–1: Bringing Up the Pump to SCR with Kill Mud

Having prepared the heavy weight mud, connect the pump line to the kill mud tank. Pump slowly until the kill mud has passed the surface lines (based on the kill sheet). Make "stroke-counter" zero at this moment.

Bring up the pump (rate) to an SCR in increments of five SPM. During the bring-up stage, the bottomhole pressure must essentially be maintained constant by operating the choke in a controlled manner. In surface-BOP-stacks, this is done by maintaining the CP constant. In subsea stacks, we must maintain the KLP constant. Monitor the CP/KLP gauge until a slight rise from the new SICP[1] is observed. At this instant, open the choke slowly and wait for the confirmation from the crew that the well is flowing. Continue this process until the desired SCR, e.g., 30 or 40 SPM, is reached. At the SCR, DPP reaches the ICP again (look at *DPP curve, line D-E in* Figure 11.2).

11.4.3.3.2 Circ 2–2: Pumping the Kill Mud until Reaching Bit Nozzles

After the pumped mud rate has reached the SCR, the choke-person continues to monitor the CP gauge until the kill mud reaches the bit nozzles. In other words, thereafter, he/she would continue to adjust the choke in the way that the CP is maintained constant at the new SICP. As the kill mud is pumped down through the drill-string, the surface DPP is being decreased from Initial Circulating Pressure (ICP) to Final Circulating Pressure (FCP) (look at *DPP curve, line E-F* in Figure 11.2). This is because the hydrostatic head pressure is increased in the pipe. The FCP is equal to the frictional pressure loss in drill-pipe when the kill mud has filled the drill-string and is pumped at a Slow Circulating Rate.

11.4.3.3.3 Circ 2–3: Kill Mud Travels from Bit Nozzles to Surface

When the kill mud reaches the bit nozzles and is about to rise through the annulus to the surface, the choke-person must transfer attention to the DPP until the kill mud reaches the surface. In other words, thereafter, he/she would continue to adjust the choke in the way that the DPP is maintained constant at FCP. When the kill mud is displacing the original mud in the annulus, CP/KLP decreases from the new SICP to zero after the kill mud reaches the surface (look at *CP curve, line j-k in* Figure 11.2). This drop happens because the hydrostatic head pressure of the annulus increases as the kill mud is moving up until it reaches the surface. As a compensation, the CP/KLP drops. At this time, the well, i.e., the BOP can be safely opened, and drilling operations can be resumed.

- Refer to Figure 11.2 to notice which pressure (CP or DPP) the choke-person should pay attention or try to maintain during the second circulation.

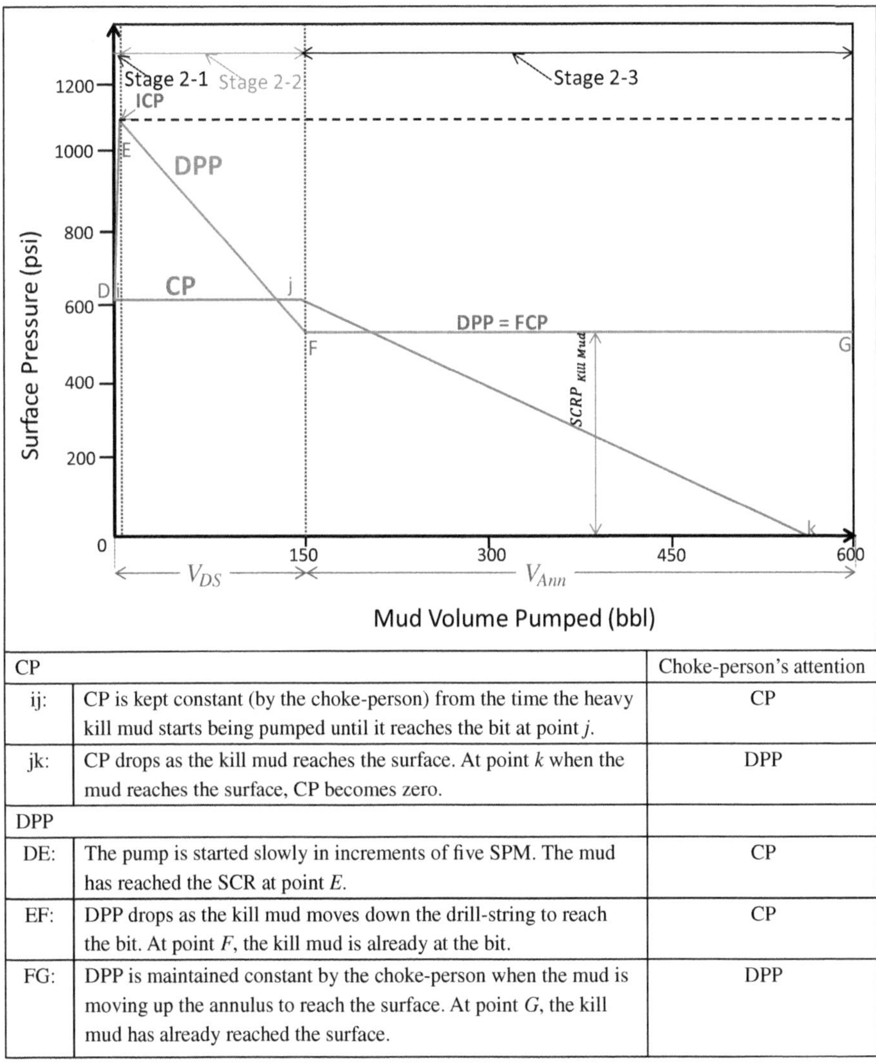

CP		Choke-person's attention
ij:	CP is kept constant (by the choke-person) from the time the heavy kill mud starts being pumped until it reaches the bit at point *j*.	CP
jk:	CP drops as the kill mud reaches the surface. At point *k* when the mud reaches the surface, CP becomes zero.	DPP
DPP		
DE:	The pump is started slowly in increments of five SPM. The mud has reached the SCR at point *E*.	CP
EF:	DPP drops as the kill mud moves down the drill-string to reach the bit. At point *F*, the kill mud is already at the bit.	CP
FG:	DPP is maintained constant by the choke-person when the mud is moving up the annulus to reach the surface. At point *G*, the kill mud has already reached the surface.	DPP

FIGURE 11.2 Surface CP and DPP curves versus mud volume pumped (*the second circulation of the Driller's method*).

11.4.4 Pressure Profiles and Interpretations

This section illustrates and explains different pressure profiles corresponding to the Driller's method including surface CP (surface stacks)/KLP (subsea stacks) and shoe pressure curves. As the worst-case scenario, it is assumed that a gas kick influx has entered the well and the curves are corresponding to this scenario. This helps readers to pinpoint any variation of the parameters. These curves have been already referred in the text corresponding to precudure, as given in Section 11.4.3.

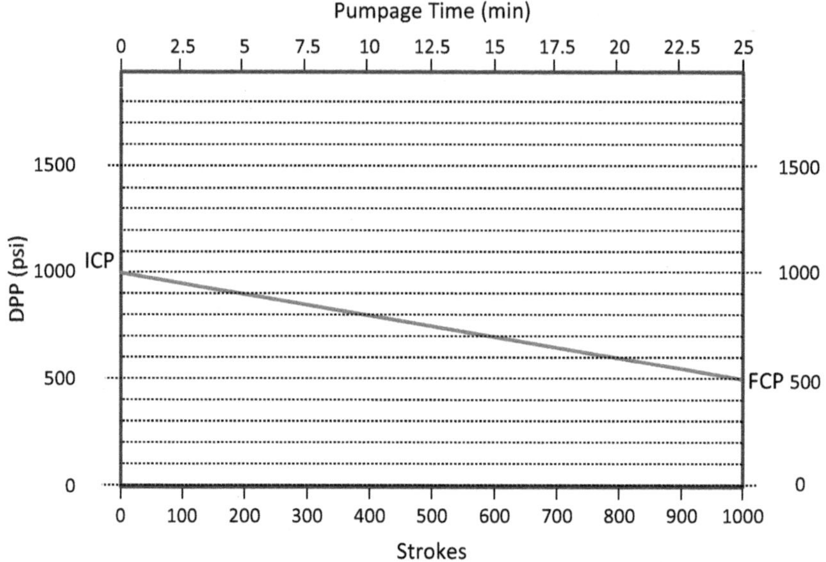

FIGURE 11.3 DPP step-down curve from Initial Circulating Pressure (ICP) to Final Circulating Pressure (FCP) (the *second circulation of the Driller's method).*

11.4.4.1 CP and DPP

During the killing operations, it is important to know the variations of the CP/KLP and DPP as they are expected to be. This is crucial for the correct implementation of the kill procedure. To maintain CP/KLP or DPP constant, it is the responsibility of the choke-person to keep the pressures constant by operating the chokes. Figure 11.1 and Figure 11.2 respectively show the CP/KLP and DPP profiles versus the mud volume pumped. Figure 11.3 shows that the DPP drop from ICP to FCP as the heavy kill mud is pumped from the surface down the drill-string until it reaches the drill bit.

11.4.4.2 Casing Shoe Pressure

Casing shoe is generally considered as the weakest point of the well in terms of the possibility of formation fracturing. Thus, it is important to be aware of its variations during the kill operations. To prevent formation fracture at shoe, CP/KLP should be kept below MAASP. In the first circulation, casing shoe pressure profile looks more similar to the CP/KLP profile than the DPP profile (see Figure 11.4). However, in the second circulation, shoe pressure profile looks more like that of the DPP (see Figure 11.5).

11.4.5 Impacting Parameters and Interpretation

11.4.5.1 Kick Size

A most important factor impacting wellbore pressures during well control operations is the size or volume of the kick influx. It is expected that the larger the kick size, the greater the casing pressures and shoe pressures. In accordance with the expectation,

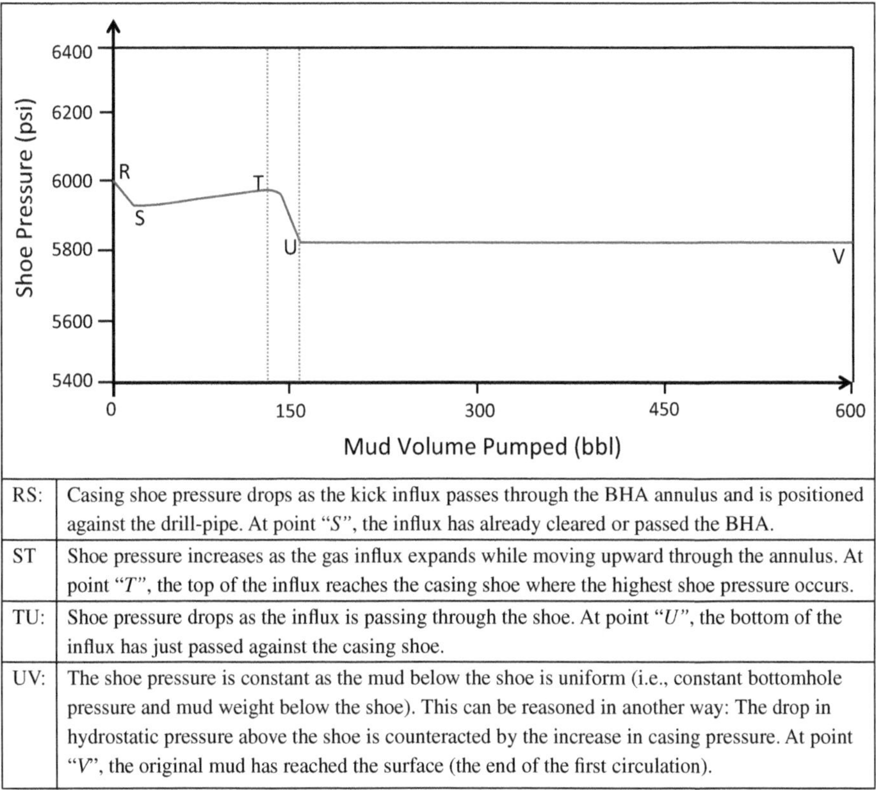

RS:	Casing shoe pressure drops as the kick influx passes through the BHA annulus and is positioned against the drill-pipe. At point "*S*", the influx has already cleared or passed the BHA.
ST	Shoe pressure increases as the gas influx expands while moving upward through the annulus. At point "*T*", the top of the influx reaches the casing shoe where the highest shoe pressure occurs.
TU:	Shoe pressure drops as the influx is passing through the shoe. At point "*U*", the bottom of the influx has just passed against the casing shoe.
UV:	The shoe pressure is constant as the mud below the shoe is uniform (i.e., constant bottomhole pressure and mud weight below the shoe). This can be reasoned in another way: The drop in hydrostatic pressure above the shoe is counteracted by the increase in casing pressure. At point "*V*", the original mud has reached the surface (the end of the first circulation).

FIGURE 11.4 Casing shoe pressure profile (*the first circulation of the Driller's method*).

Figure 11.6 and Figure 11.7 compare the CP and shoe pressure profiles for three typical kick sizes of 30, 50, and 70 barrels.

11.4.5.2 Kick Intensity

Kick intensity is defined as the difference between the equivalent mud weight corresponding to pore pressure of the kicking formation and the original mud weight. In other words, the kick intensity expresses the magnitude of the anticipated underbalance in equivalent mud weight. It is defined as:

$$\text{Kick Intensity} = \frac{\text{Anticipated } P_{\text{pore}}}{0.052 \times \text{TVD}} - \text{MW} \tag{11.1}$$

As an example, assume drilling at the depth of 10,000 ft with the mud weight of 10 ppg, where an abnormally pressured formation is encountered with the pore pressure (Ppore or Pf) of 5,500 psi. In this example, the kick intensity is found:

$$\text{Kick Intensity} = \frac{5{,}500}{0.052 \times 10{,}000} - 10 = 0.57 \text{ ppg}$$

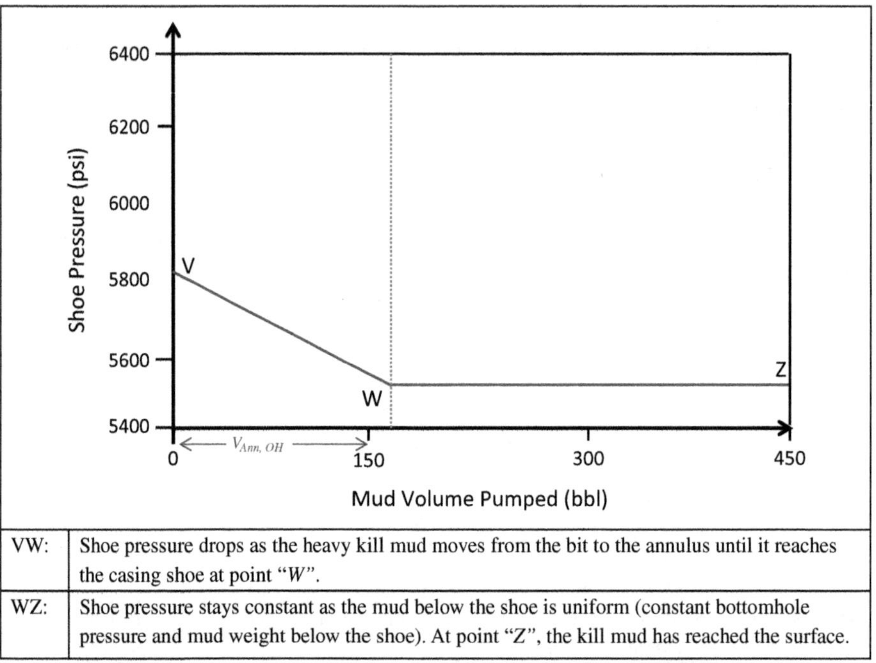

VW:	Shoe pressure drops as the heavy kill mud moves from the bit to the annulus until it reaches the casing shoe at point "*W*".
WZ:	Shoe pressure stays constant as the mud below the shoe is uniform (constant bottomhole pressure and mud weight below the shoe). At point "Z", the kill mud has reached the surface.

FIGURE 11.5 Casing shoe pressure profile/curve (*the second circulation of the Driller's method*).

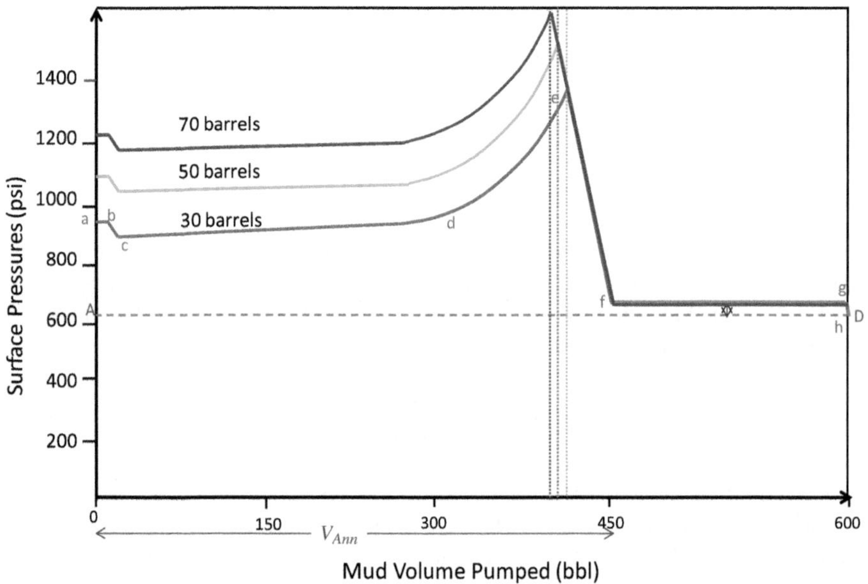

FIGURE 11.6 Effect of kick size on surface CP profile (the *first circulation of the Driller's method*).

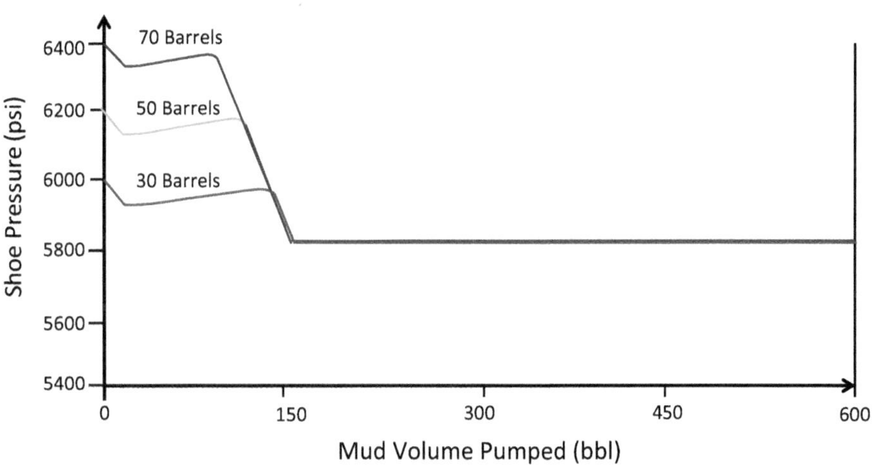

FIGURE 11.7 Effect of kick size on surface CP profile (the *first circulation of the Driller's method*).

It is expected that the greater the kick intensity is, the greater the wellbore pressures encountered during well kill operations. Figure 11.8 and Figure 11.9 show that a minor increase in kick intensity causes considerable increases in surface CPs and shoe pressures.

11.5 WAIT AND WEIGHT METHOD

11.5.1 Description

The "Wait and Weight" method takes the well under control by only one circulation. The title "Wait and Weight" comes from the fact that the well kill operation does not start immediately after well shut-in, unlike Driller's method. Instead, we should "Wait" for a while to "Weight" the current mud to kill mud, or first prepare the heavy kill mud.

In critical cases with low formation fracture pressures or long open holes, this method is preferred and recommended for taking the well under control following a kick. Since the lowest annular pressures and unsafe time are experienced during the circulation in this method, it is considered a safe well kill method in long open holes to prevent formation fracture at the shoe.

The fundamental basic for all well control methods is that the Bottomhole Pressure (BHP) must be always maintained constant. Therefore, in Wait and Weight method, similar to the Driller's method, this will be done by appropriately opening or closing the choke during the well kill.

11.5.2 Advantages and Disadvantages

A few advantages and disadvantages of the Wait and Weight method are listed in the next sections.

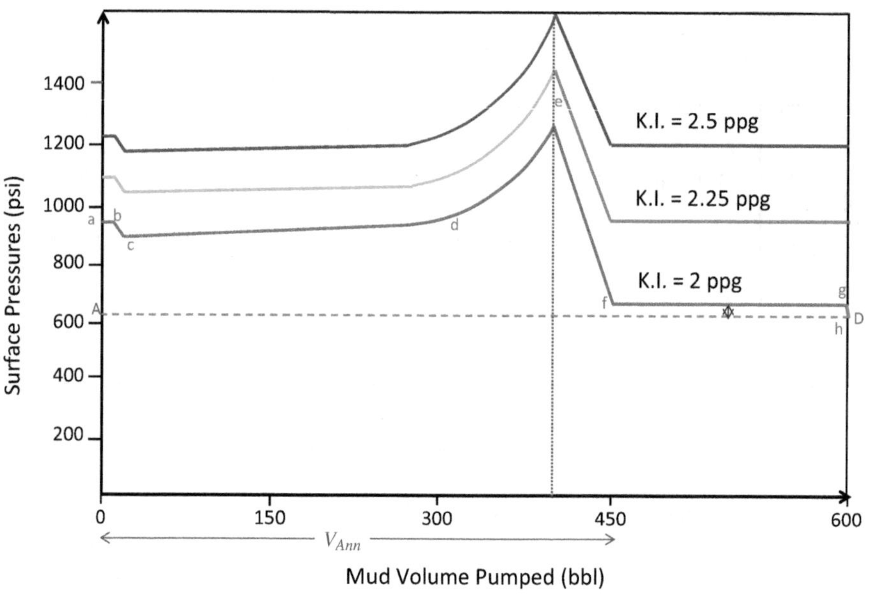

FIGURE 11.8 Effect of kick intensity on surface CP profile (the *first circulation of the Driller's method*).

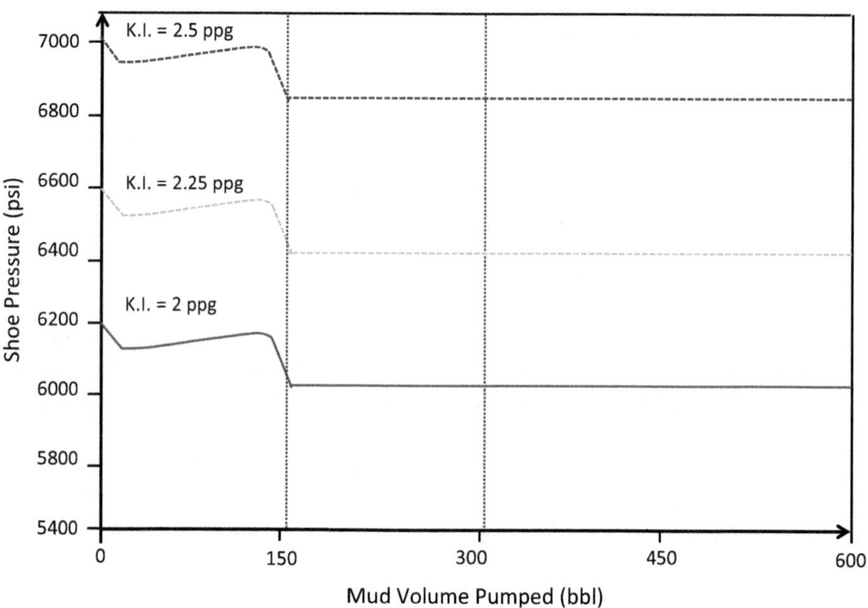

FIGURE 11.9 Effect of kick intensity on the shoe pressure profile (the *first circulation of the Driller's method*).

11.5.2.1 Advantages

- Generally, the Wait and Weight method results in significantly lower shoe and wellbore pressures than the Driller's method (refer to the Section 11.5.5.2). This is more significant in wells with long open-hole sections, or with low formation fracture pressures, or when the kick size is large.
- The period for circulating through choke is short.
- In general, the surface and bottomhole equipment undergo the least pressures.

11.5.2.2 Disadvantages

- Preparing the heavy weight mud (by weighing the original mud) requires some time which causes the kick to migrate. If considerable time passes, considerable stresses may be imposed on the downhole and surface equipment.
- If a large increase from original mud weight is required to make the kill mud, it is not possible to perform this task uniformly in one step. This means it needs some steps and may be a timely process.
- Because of the time required for weighing up, there is a possibility for cuttings' settlement in the wellbore, and in worst case even the annulus can pack off.
- Kill sheet calculations are a bit more difficult than the Driller's method.
- Manipulating the choke during well control operations is more challenging than the Driller's method. In fact, until the kill mud reaches the bit nozzles, the trend of pressure drop from ICP to FCP should be strictly followed.

11.5.3 Procedure

Following the well shut-in and stabilization of shut-in pressures, the kill sheet is completed. Using the kill sheet calculations, the required kill mud weight and volume are calculated. Then, the kill mud is prepared in the mud tanks. Following that, the well kill operation is executed.

11.5.3.1 Stage 1: Start Pumping Kill Mud by Bring-Up the Pump to SCR

Connect the pump line to the kill mud tank. Pump slowly until the kill mud has already passed through the surface lines. Make the "stroke-counter" zero at this moment.

Bring up the pump rate to the SCR with rate steps of five SPM. Monitor the (surface) CP gauge until a rise from the new value of SICP is observed. It is noted that SICP might be slightly greater than the initial one (called new SICP) because of gas migration during the time taken for the kill mud to get prepared. At this instant, open the choke slowly and wait for confirmation from the crew that the well is flowing. Increase the pump rate slowly (in increments of five (5) SPM) while maintaining the CP constant so that the bottomhole pressure is maintained constant. This is done by slowly opening and adjusting the choke size in a controlled manner. Continue this process until the SCR, e.g., 30 or 40 SPM, is reached. At the SCR, the DPP reaches the ICP again (look at *Point "B" in the DPP curve, or point "b" in the CP curve*, in

Figure 11.10). Please note that the kill mud is still in the surface lines and thus has not yet entered the drill-pipe.

11.5.3.2 Stage 2: Pumping Kill Mud until It Reaches Bit Nozzles

After the pumped mud rate has reached the SCR, continue pumping while the choke-person should pay attention to allow the DPP drop from Initial Circulating Pressure (ICP) to Final Circulating Pressure (FCP) based on the trend line in the kill sheet. Therefore, for each 100 strokes pumped, just a specified pressure increment is slowed to drop in order to maintain the bottomhole pressure constant. As the kill mud is pumped down through the drill-string, the surface DPP is being decreased from ICP to FCP (DPP curve, line B-C in Figure 11.10). At point "C" in the same figure, the DPP is equal to FCP. This is because the hydrostatic pressure in the drill-string is being increased. FCP is equal to the frictional pressure loss in drill-pipe when the kill mud is pumped at the Slow Circulating Rate. For more information on the corresponding calculations, refer to Chapter 10.

11.5.3.3 Stage 3: Kill Mud Travels from Bit Nozzles to Surface

When the kill mud reaches the bit nozzles and is just about to rise up through the annulus to the surface, the choke-person has to pay attention to maintaining the DPP at FCP by adjusting the choke.

Different phases of displacing the kick influx and the original mud in the annulus (by the kill mud) can be divided into the following:

11.5.3.3.1 The Kick Influx is Displaced Until its Top Reaches the Casing Shoe

When the kill mud is pumped, the gas kick influx moves up in the annulus: It passes through the annulus of the drill-collars and the drill-pipes until its top reaches the casing shoe (point *e* in Figure 11.10). At this moment, the casing shoe experiences the highest pressure and that is the most critical time in terms of fracture (Point "T" in Figure 11.11). The surface CP at this time must be maintained lower than the MAASP so that formation fracture pressure is not exceeded.

11.5.3.3.2 The Kick Influx is Displaced Until its Top Reaches the Surface

Pumping the kill mud is continued, and the gas kick influx enters the cased section of the hole. As the influx moves up the casing, the shoe pressure drops further until its top reaches the surface. At this moment, the highest CP/KLP is experienced (point "*f*" in Figure 11.10). Until this time, the choke-person is responsible for keeping the DPP constant by adjusting (closing) the choke.

11.5.3.3.3 The Gas Influx Exits the Well from the Surface

After the kick influx has reached the surface, it starts to exit the well through the choke. As it is exiting, the DPP suddenly drops due to U-Tubing (much lighter fluid in the annulus than in the drill-string). The choke-person is expected to close the choke a bit in order to keep the DPP constant until all the influx exits the choke (point "g" *in* Figure 11.10).

11.5.3.3.4 The Original Mud is Displaced Out of the Well

After all the gas influx has been circulated out of the hole, the original mud is also displaced out of the hole by the kill mud. This makes the current mud in the hole heavy enough so that the hydrostatic bottomhole pressure is equal or slightly greater than the formation pore pressure. According to the constant bottomhole pressure principle in well control, the surface pressure must become zero at this point (point "h" *in* Figure 11.10).

11.5.4 Pressure Profiles and Interpretations

Figure 11.10 and Figure 11.11 illustrate and explain drill-pipe pressure (DPP) and casing pressure (CP) profiles of the Wait and Weight method. Detailed interpretation for variations of pressures at different stages is given in the figures.

11.5.4.1 CP and DPP

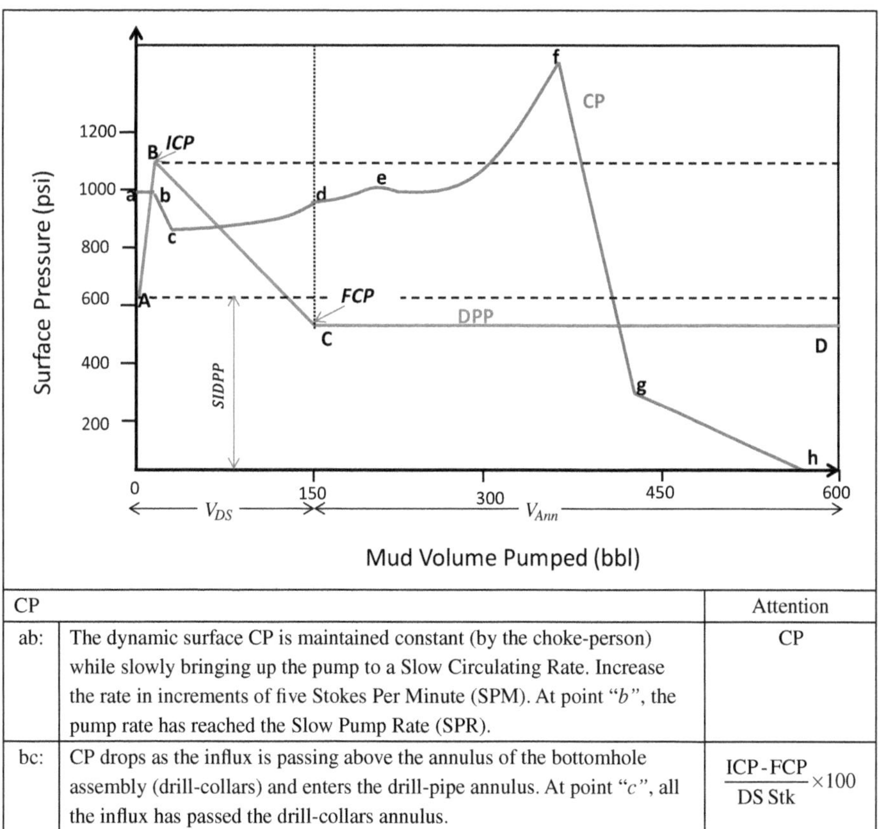

CP		Attention
ab:	The dynamic surface CP is maintained constant (by the choke-person) while slowly bringing up the pump to a Slow Circulating Rate. Increase the rate in increments of five Stokes Per Minute (SPM). At point "*b*", the pump rate has reached the Slow Pump Rate (SPR).	CP
bc:	CP drops as the influx is passing above the annulus of the bottomhole assembly (drill-collars) and enters the drill-pipe annulus. At point "*c*", all the influx has passed the drill-collars annulus.	$\frac{\text{ICP-FCP}}{\text{DS Stk}} \times 100$

FIGURE 11.10 Drill-pipe pressure (DPP) and surface casing pressure (CP) curves during the Wait and Weight Circulation Method. (Continued)

cd:	CP shows an increase because of gas influx expansion (due to its compressible characteristic) or increase in the influx height. At point "*d*", the kill mud reaches the bit; therefore, the rate of CP increase would become lower than before.	$\frac{\text{ICP-FCP}}{\text{DS Stk}} \times 100$
de:	Between "d" and "e", the top of gas influx enters the cased-hole section (which is not exactly recognizable using the CP curve). As the capacity of the cased-hole is greater than that of the open hole, entrance of the influx in the casing results in lower height of the influx and lower CP. Around point "e", almost all the gas influx is in the cased-hole.	DPP
ef:	With the influx being in the cased section, CP suddenly increases due to abrupt expansion near the surface. Point "*f*" represents the moment when the kick influx reaches the surface and starts exiting from the choke.	DPP
fg:	CP drops as the influx is exiting the well through the choke. At point "*g*", the influx is voided out from the well through the choke.	DPP
gh:	CP decreases further as the kill mud replaces the original mud. At point "*h*", all the original mud has exited the well at the surface.	DPP
DPP		
AB:	DPP is increased as the pump is started and slowly reaches the ICP at the Slow Circulating Rate (SCR).	CP
BC:	DPP drops as the kill mud moves down the drill-string to reach the bit.	$\frac{\text{ICP-FCP}}{\text{DS Stk}} \times 100$
CD:	DPP reaches the FCP and stays constant thereafter.	DPP

FIGURE 11.10 (*Continued*)

11.5.4.2 Shoe Pressure

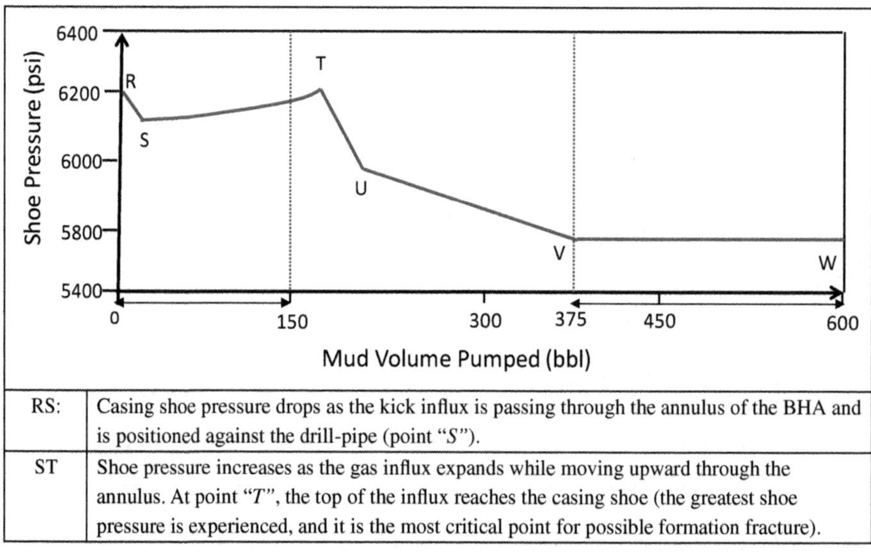

RS:	Casing shoe pressure drops as the kick influx is passing through the annulus of the BHA and is positioned against the drill-pipe (point "*S*").
ST	Shoe pressure increases as the gas influx expands while moving upward through the annulus. At point "*T*", the top of the influx reaches the casing shoe (the greatest shoe pressure is experienced, and it is the most critical point for possible formation fracture).

FIGURE 11.11 Casing shoe pressure during circulation (the Wait and Weight method).

TU:	Shoe pressure drops steeply as the influx passes through the shoe. At point "*U*", the bottom of the influx has just passed against/cleared the casing shoe already. Now, the top of the original mud has reached the shoe at point "*U*". The kick size at the shoe can be found between points "T" and "U" which is around 40 bbl.
UV:	Shoe pressure continues dropping (but less steeply) as the kill mud is displacing all the original mud above the shoe. At point "*V*", the kill mud reaches the casing shoe.
VW:	Shoe pressure stays constant because the influx and original mud have already passed the shoe, and the mud below the shoe is uniform. At point "W", the kill mud reaches the surface when the circulation can be safely stopped.

FIGURE 11.11 *(Continued)*

11.5.5 Impacting Factors and Interpretation

11.5.5.1 Kick Size

Figure 11.12 and Figure 11.13 respectively compare the (surface) CP and the shoe pressure profiles for three typical kick size cases of 30, 50, and 70 barrels. The figures show that the greater the kick size, the greater the CPs and shoe pressures.

11.5.5.2 Kick Intensity

Kick intensity is defined as the difference between the formation pore pressure's equivalent mud weight and the original mud weight. The relation for this parameter was provided in Section 11.4.5.2. Therefore, the kick intensity expresses the magnitude of the underbalance in equivalent mud weight.

It is expected that the greater the kick intensity is, the greater the wellbore pressures encountered during well kill operations.

Figure 11.14 shows that a minor increase in kick intensity brings about a considerable increase in surface CP in the Wait and Weight method.

Figure 11.15 shows that a minor increase in kick intensity brings about a considerable increase in shoe pressure in both the Wait and Weight method and first circulation of the Driller's method.

11.6 CONCURRENT METHOD

The Concurrent method is not a commonly used well kill method as it is complicated and involves weighing up the active drilling mud system as it is being pumped in the well at the kill pump rate.

The concurrent method amounts to a combination of, or a compromise between the Driller's and the Wait and Weight methods. In this method, immediately after the determination of the stabilized shut-in pressures, we proceed to circulating the well with the original weight mud. As we are circulating, the mud weight is increased in stages, and the drill-pipe circulating pressures are adjusted accordingly to keep the bottomhole pressure almost constant. This process is continued until the kick influx is circulated out which may require some bottoms-up to ensure the well is completely safe.

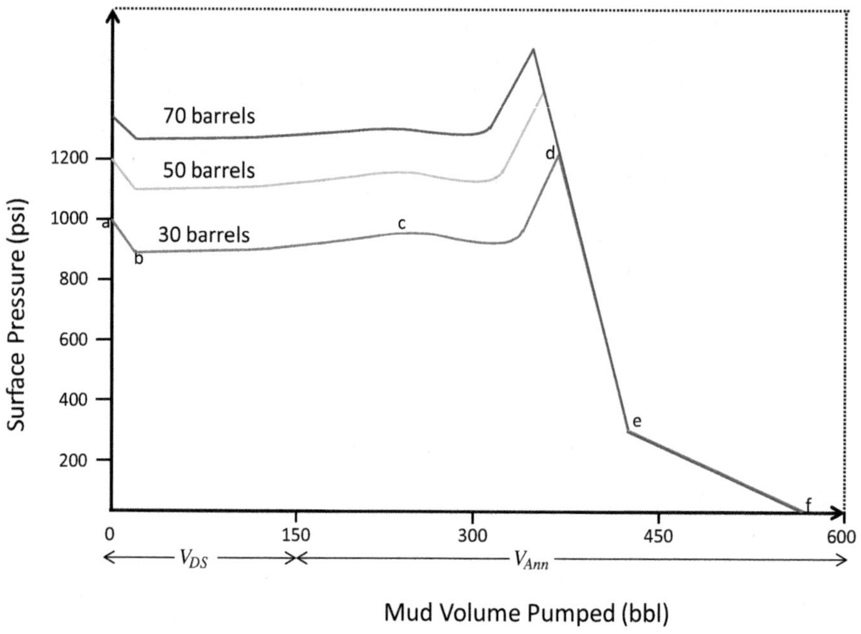

FIGURE 11.12 Effect of kick size on surface CP profile (the Wait and Weight method).

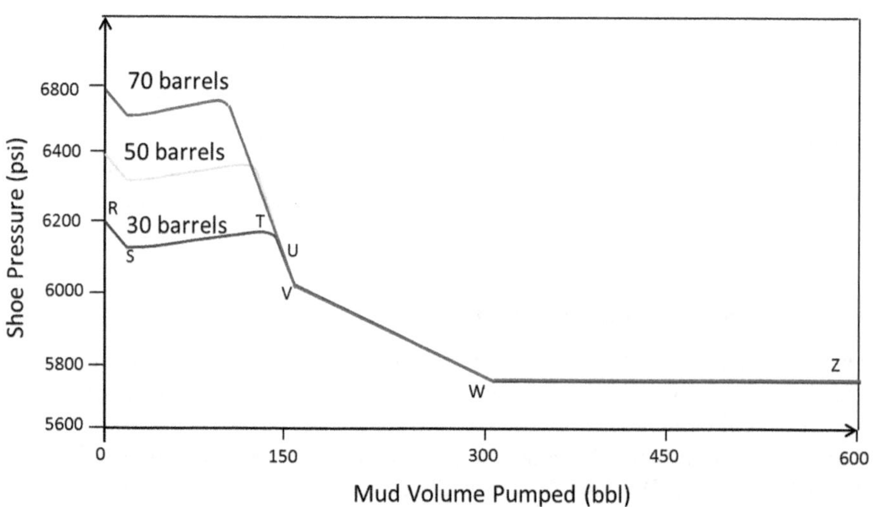

FIGURE 11.13 Effect of kick size on shoe pressure profile (the Wait and Weight method).

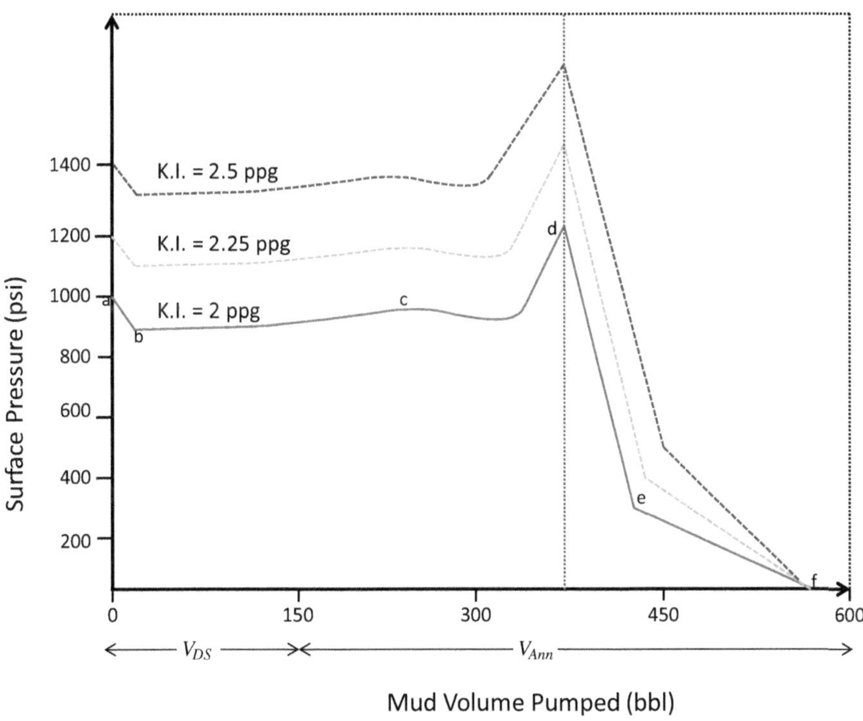

FIGURE 11.14 Effect of kick intensity on surface CP profile (the Wait and Weight method).

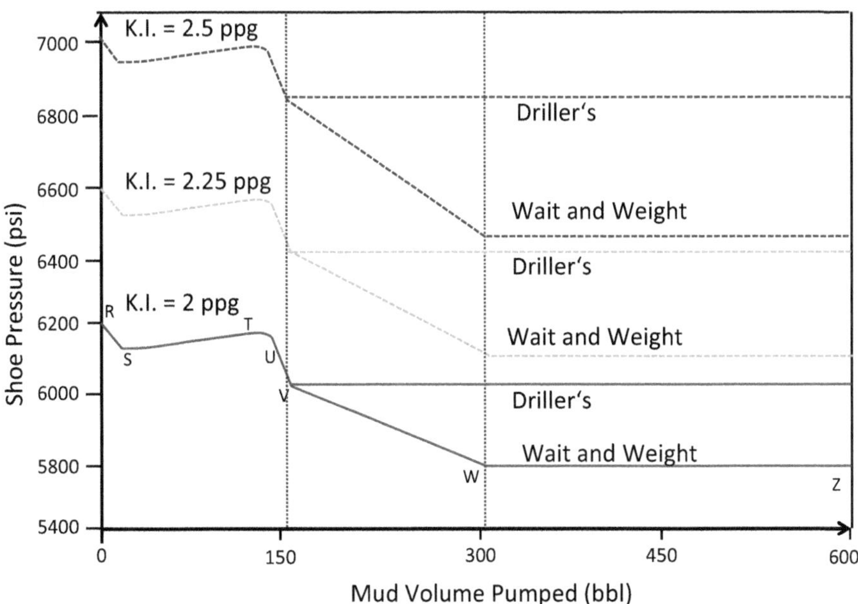

FIGURE 11.15 Effect of kick intensity on casing shoe profile (the Wait and Weight method and the Driller's first circulation).

As the mud weight is increased, the required drill-pipe pressures must be calculated, and a schedule is prepared for the drill-pipe pressure, to be followed by manipulating the choke. The calculations are generally complex, but there are some similarities with the Driller's and Wait and Weight method:

- The calculation of ICP is done in the same manner as the previous two methods.
- When the final heavy kill mud reaches the bit, it is expected to reach the FCP as in the second circulation of the Driller's method or the Wait and Weight method. Please refer to Figure 11.3.

The advantages of the method (compared with Driller's and Wait and Weight method) are:

- The well kill operations can start immediately after completing the kill sheet with minimum waiting time.
- It allows gradual and smooth increasing of the mud weight (especially when a large weight increase is required) without impacting the mud properties.

The disadvantages of the method are:

- It includes complex calculation.
- It is a complex method in terms of its management, and care should be taken in not making a mistake during the operation.
- The well is circulated under pressure for a longer time due to several bottoms-up required.

11.7 VOLUMETRIC METHOD

The well control methods discussed so far, i.e., the Driller's, the Wait and Weight methods, and Concurrent method, cannot be implemented in some situations. These situations are as follows:

- *Pumps broken down:* In case mud pumps have broken down, it is impossible to circulate the kick and take the kick influx out of the hole.
- *The drill-string is out of the hole or is off-bottom:* When the drill-string or the bit is at the bottom (of the hole), the drill-string can be used to pump the mud and circulate the kick influx out of the hole. However, there are periods when the bit is not at the bottom or is off-bottom (during dummy or real tripping operations) or when the drill-string is out of the hole (i.e., it has already been pulled out of the hole for bit change, etc.). In such times, it is not possible to circulate the mud.
- *The bit is at the bottom, but preparing the heavy kill mud requires considerable time:* The longer the elapsed time for preparing the heavy/kill mud,

the farther the influx migrates up and the greater the casing and bottomhole pressures. Therefore, before the surface casing pressure approaches to reach MAASP, it must be bled off.

- *The drill-string or bit nozzles are plugged:* Kicks may occur when the drill-string or the bit nozzles are plugged/blocked. In this case, mud circulation to take the influx out of the hole is impossible.

In all the mentioned situations above, the drilling mud cannot be circulated to take the kick influx out of the hole. In addition, the surface drill-pipe pressure (DPP) cannot be monitored, and thus the formation pore pressure is not known exactly for performing a proper well kill operation. Therefore, the volumetric method is alternatively implemented as a special well control method to control the well. In this method, the volume of gas influx will be allowed to migrate, and thus the casing pressure will increase until a certain figure and then a specific amount of mud will be bled off to compensate the increase in casing pressure. The volumetric method will allow the kick influx to rise to the surface safely with the bottomhole pressure being maintained almost constant.

The volumetric methods fall into partial and complete ones, which are explained next.

11.7.1 Partial Volumetric Method

There are scenarios where mud circulation may not be temporarily possible, but it can take a considerable time. The reasons could be due to the equipment failure, the necessity of considerable weighing of the mud (for several pounds per gallon), or the bit is off-bottom. In all these situations, due to considerable time to take, gas migration occurs, resulting in considerable increase in shut-in pressures. The gas migration may potentially jeopardize the surface CP to reach the MAASP. In such cases, while waiting for preparations (kill mud preparation, repairing the mud pump, or stripping[2] the drill string back to the bottom), it can be decided to simultaneously apply the partial volumetric method. In this method, an increase in CP must be essentially bled off to prevent reaching the MAASP. To do this correctly, as the CP is increased for a specified magnitude (e.g., 200 psi), the choke is opened to allow bleeding-off of some mud volume out to the trip tank. The bled-off mud volume should have a hydrostatic pressure equal to the increase in bottomhole pressure. This provides the expansion of the gas influx and thus prevents further increase in the CP at each step (see Figure 11.15). In other words, the increase in CP due to the gas migration is counteracted by the hydrostatic mud pressure drop (by allowing mud bleed-off), resulting in rather constant bottomhole pressure and successful maintenance of the well conditions in safe conditions. Next, after preparations are made, a circulation-based method can be safely implemented to circulate the kick influx out of the well.

The partial version of volumetric method is not considered a well control method because it does not extract the kick influx out and does not counteract the formation pore pressure. However, it alleviates the situation by preventing excessive or sudden rise in CP to reach MAASP. The bottomhole pressure is maintained rather constant by maintaining the Shut-In Drill-Pipe Pressure (SIDPP) constant (in case the bit nozzles are not plugged). In turn, it provides enough time for preparing to start the well kill operations by a circulation-based well control method (such as Driller's method).

11.7.2 Complete Volumetric Method

The volumetric method can also be used to let all the influx vent out of the hole. This is called the "complete volumetric method". This method requires a rather long period of time (compared with circulation-based methods). The regular choke opening required in this method according to gas expansion process prevents an excessive increase in CP due to gas migration up the annulus (Figure 11.16a). When the gas top is reaching the surface, it can later be vented completely by applying the volumetric method. At this time, for each time step, the emptied space in the hole must be replaced/filled with mud; in other words, the bottomhole pressure is kept rather constant by adding additional mud volume or hydrostatic pressure (Figure 11.16b). An example of volumetric method application is presented in exercises in Section 11.10.

The volumetric method is not called a kill method as it just vents the gas out of the well, thus the well is still underbalanced. But rather, it can be called a well control

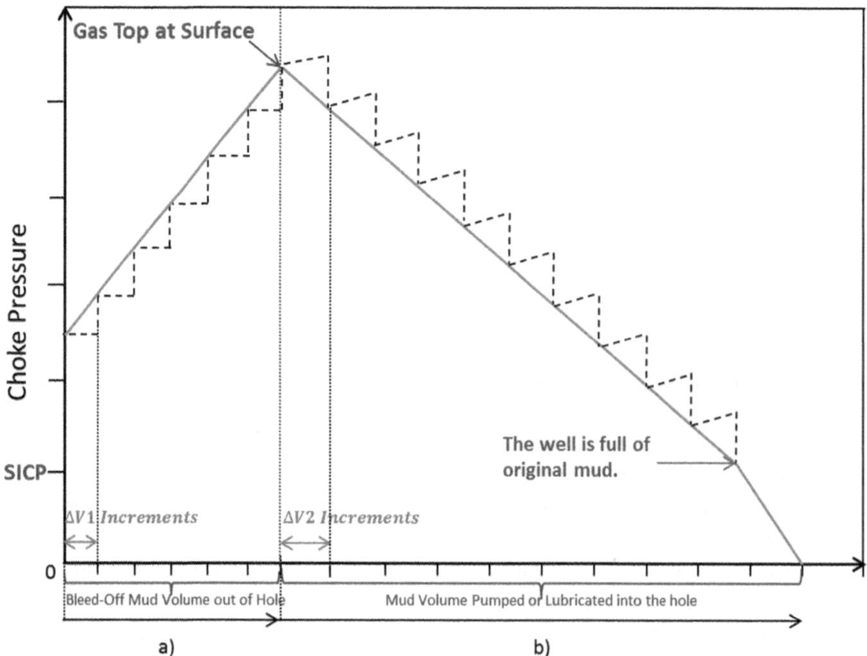

FIGURE 11.16 Volumetric well control. a) The volume of mud bled-off out of the hole in pressure stages and b) the volume of mud pumped into the hole since the gas top reaches the surface.

method. Subsequently, to kill the well, we still need to apply the lubricate method by which the kill mud is pumped through the annulus.

11.7.3 Advantages and Disadvantages

11.7.3.1 Advantages

In all the four situations mentioned in Section 11.7, the volumetric method contributes to preventing an increase in SICP as well as allowing gas influx to vent out of the well. Thus, the bottomhole pressure is maintained constant. In case the drill-string or nozzles are plugged up or the string is out of the hole, the volumetric method (combined with the lubricate and bleed-off) is the only well control method.

11.7.3.2 Disadvantages

- It takes a considerable period of time to let all the gas influx get out the well and thus control the well. It is by far longer than the circulation-based methods.
- The influx volume, which is necessary for calculation, is not known with great accuracy.
- The volumetric method does not kill the well. In partial volumetric methods, it allows more time to get prepared (without reaching MAASP) and then apply circulation-based killing methods. In the complete volumetric method, only influx is bled off. However, the well is still underbalanced. To kill the well, the lubricate and bleed-off method should be essentially applied.

11.7.4 Calculations

Using Boyle's Law (see Appendix), the mud volume to be bled off in each step, $\Delta V_{m,Bleed}$ [bbl], is found as follows:

$$\Delta V_{m,Bleed} = \frac{\Delta P_s \times \text{Influx Vol}}{P_{pore} - \Delta P_s} \tag{11.2}$$

where ΔP_s is the increase in surface casing pressure and P_{pore} is the formation pore pressure.

11.8 KICK CIRCULATION COMPLICATIONS AND MALFUNCTIONS

During well kill operations, some downhole or surface problems may happen to complicate the operation. The crew should be vigilant enough to carefully monitor the pressures for detecting possible problems or malfunctions immediately. In case the problem is not detected and thus the right action is not taken, it may aggravate the situation and a really critical situation is resulted.

With good monitoring of well control operations, problems or malfunctions are detected and proper actions can be taken to resolve the problems. It is essential that there is good communication between all parties involved in the well kill (especially the driller and the drilling supervisor/tool pusher) at all stages of the well control operation.

11.8.1 Detection

Maintaining constant bottomhole pressure is considered the fundamental principle and the key to success in all the influx-out kill methods. This is guaranteed by manipulating the choke valve according to the calculations and pattern/procedure in the completed kill sheet. The choke valve is opened or closed to either keep the drill-pipe constant, or keep the casing pressure constant, or allow a scheduled drop.

If a kill problem occurs, it undoubtedly shows itself on pressure gauges or logs, so any trends far away from the expected ones can be observed, indicating that the bottomhole pressure is deviated from being constant. Thus, it is important to observe the indications or changes in parameters and have a reference guide to detect the exact kill problem/malfunction.

Table 11.1 shows a brief reference guide for the detection. A kill problem may originate from an equipment malfunction or a human error. It is noted that choke plugged-up and gas hydrate formation have almost the same symptoms.

Gas hydrates are ice-like solids formed when gases are flowing in the presence of small quantities of water vapor. In deepwater subsea operations (seawater

TABLE 11.1
Detection of Kill Problems/Equipment Malfunctions Using Indications/Observations

		Indications			
Equipment	**Kill Problem /Malfunction**	**Drill-pipe Pressure (DPP)**	**Casing pressure, CP (Surface-BOP)/ kill-line pressure, KLP (subsea)**	**Hook load (held drill-string weight)**	**Mud pit level**
Bit Nozzles	Plugged-up:	↑	–	–	–
	Washout:	↓	–	–	–
Drill-string	Washout:	↓	–	–	–
Hole/Annulus	Downhole Pack-off:	↑	↓	↓	↓
Choke	Plugged-up:	↑	↑	–	–
	Washout:	↓	↓	–	–
Mud Loss (Underground Blowout):		↓	↓	↑	↓
Pump	Failure:	↓	↓	–	–
	Leakage:	↓	↓	–	–
Gas hydrate		↑	↑	–	–

Note: ↑ Shows "Increase" and ↓ Shows "Decrease".

> 600 m/1,969 ft) or cold climates, hydrates are normally formed in choke-line through failsafe valves or at the choke where a considerable pressure and temperature drop occurs. The implications are first an increased upstream pressure, and then complete blockage of the choke-line, or even the choke manifold. Usually, hydrate formation does not pose a problem with oil-based muds. In brief, for gas hydrate formation, the following conditions should essentially exist:

- Free-water presence in the gas influx,
- Gas at or below its dew point,
- Low temperature, and
- High pressure.

11.8.2 Actions

11.8.2.1 First Action

When a kill problem is detected or doubted, the first action is to stop the operation. Stop circulation/pumping, and then shut-in the choke/well safely. To do this safely, the bottomhole pressure should be maintained constant. In surface BOP-stacks, as we bring down the pump (reduce the pump rate to zero), the casing pressure (CP) must be maintained constant by closing the choke in a controlled manner. In subsea-BOP-stacks, the kill-line pressure (KLP) is maintained constant as we bring down. It is important to keep the closed-in (stoppage) period as short as possible and, thus, better to solve the problems quickly. If there is an issue with the choke (choke washout), instead, the High Closing Ratio (HCR) valve or another manual valve upstream the choke must be closed.

11.8.2.2 Next Actions

For each kill problems mentioned in Table 11.1, after well kill stoppage and shut-in, some actions should be taken to resolve the issues:

- *Bit nozzles*

  - *Plugged-up:* In well kill operation while circulating the at a constant pump rate, bit nozzles may plug up (see Table 11.1). In this case, the drill-pipe pressure shows an increase, whereas the casing pressure CP (surface stacks)/ Kill Rate Pressure KLP (subsea stacks) usually remains unchanged as long as the circulation rate remains constant. Generally, if the drill-pipe pressure goes greater than the pressure schedule (especially when it must be maintained), the choke-person should open the choke in order to bleed off pressure. After opening the choke, if the drill-pipe pressure is still greater than expected and instead the choke/annulus pressure drops, it is inferred that the bit nozzles may be partially plugged, or there is a restriction in the annulus.

Following the detection, the choke is closed (thus the well is closed as well). Next, select the current drill-pipe pressure as the newly adjusted pressure for using it during the upcoming well kill operations. For example, during the first circulation of the Driller's method, if nozzles plug up, consider the current drill-pipe pressure as the ICP (which is greater than the initial ICP) and continue the kill operation. It should be noted that sometimes nozzles may *unplug* again after a while, and graphs should be re-adjusted accordingly.

- *Washout:* During circulating the mud at a constant rate, bit nozzles may wash out. In this case, the drill-pipe pressure (DPP) shows a decrease (Table 11.1), whereas the casing pressure CP (surface stacks)/Kill Rate Pressure KLP (subsea stacks) usually remains unchanged as long as the circulation rate remains constant. Generally, if DPP goes lower than the pressure schedule (especially when it must be maintained), the choke-person should close the choke in order to raise the pressure. After closing the choke, if the drill-pipe pressure still remains lower than expected but the choke/annulus pressure (CP) rises, it can be inferred that either the bit nozzles may be washed out (or come-off from their places), or it is possible that the drill-pipe is washed out (hole inside the drill-pipes). First, assume that washout of bit nozzles has occurred.

Following the detection, the choke/well is shut-in. To resolve the issue when we have a uniform mud (weight) in the string, usually select the current drill-pipe pressure as the newly adjusted pressure for use during further well kill operations. For example, during the first circulation of the Driller's method, if nozzles washout occurs and are detected, consider the current pressure as ICP (which is lower than the schedule's ICP) and continue. However, if this problem occurs when there is no uniform mud in the string (e.g., kill mud is being circulated to the bit in the wait and weight method), the procedure to go ahead is not as straightforward as when we have uniform mud in the string. In such a case, we must do some calculations to revise the step-down pressure schedule, then switch on the pump and continue. For better understanding how to revise the schedule, an example is given in Figure 11.17 and Figure 11.18. Based on the calculations, it is briefly suggested that if a nozzle washout occurred with the pressure drop of 100 psi, just reduce 100 psi from the step-down pressure schedule and continue.

- *String Washout (Hole):* In case of a drill-string washout, the same indications are observed as in the case with the bit nozzles' washout. In both cases, only the drill-pipe pressure decreases, as seen in Table 11.1. In most cases, it is not possible to exactly identify which issue has exactly happened. However, when the kill mud is being pumped through the drill-string, and suddenly the DPP drops and the heavy mud reaches the surface, the issue is definitely the drill-string washout (see Figure 11.19).

In facing the problem, the first action is to stop circulation/pumps and then shut-in the well by closing the choke. Next, if the drill-string washout is confirmed, it is advised to essentially employ the volumetric method of well control. When the kick

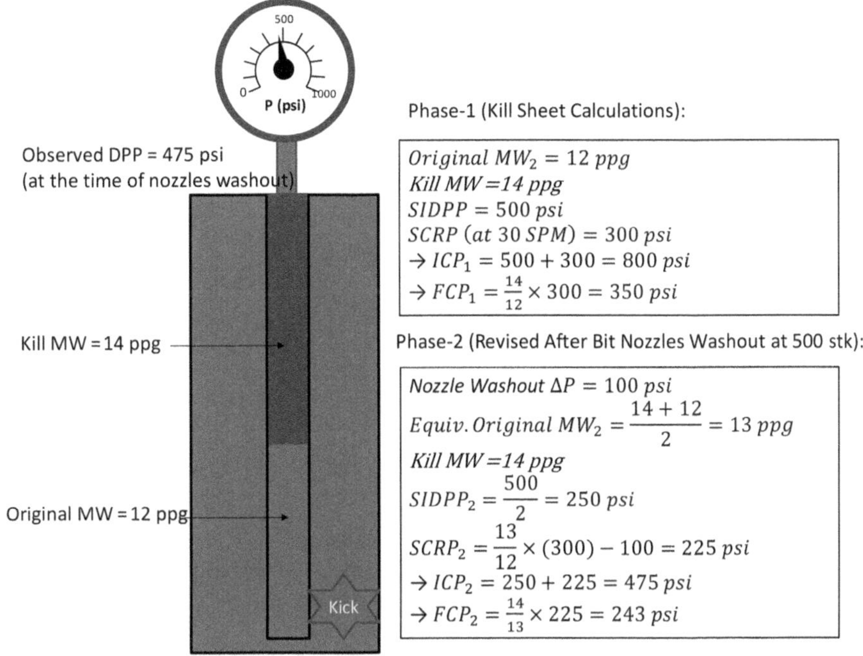

FIGURE 11.17 An example of bit nozzles' washout for the two phases of kill sheet calculations (Phase-1) and following detection of bit nozzles' washout where we revised the calculations to find the second initial and final circulating pressures (ICP_2 and FCP_2). Refer to Figure 11.17 to find the revised plot of the step-down chart.

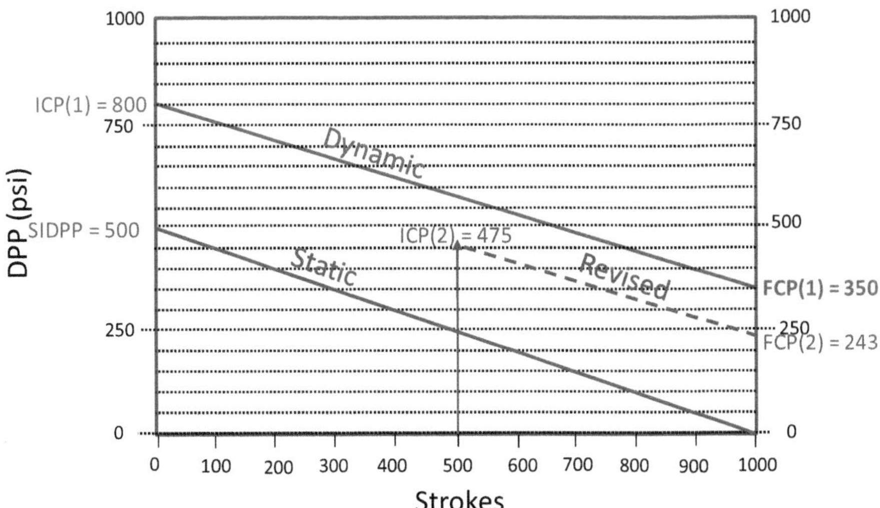

FIGURE 11.18 The initial and revised drill-pipe pressure (DPP) step-down graphs for the example and calculations provided (in Figure 11.16) for bit nozzle washout as the kill problem.

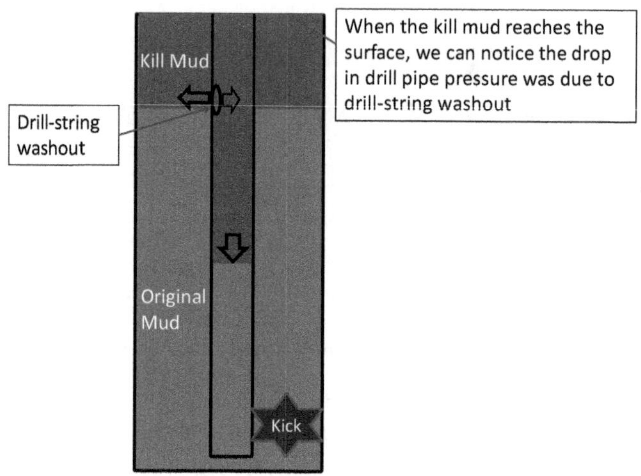

FIGURE 11.19 Drill-string washout (shallow) when the kill mud is pumped through the drill-string. The observation of drop in the drill-pipe pressure can be attributed to either the nozzle washout or the drill-string washout (but none of them can be specified), unless when we are pumping the kill mud with a shallow drill-string washout.

influx reaches above the washout depth (if the washout depth is deep enough), it is possible to reestablish the mud circulation (with a lower rate) to displace the influx out of the well. However, this may not be recommended as it comes with certain risks including the parting of the drill-string with continued circulation, even with a lower circulation rate.

- *Downhole Pack-Off/Blockage:* From the time the well is shut-in till beginning of the well kill operations, the well conditions are almost static (there is no mud circulating in the hole with pumps off). In this period, there is a possibility for cuttings to settle down around the bottomhole assembly (BHA) which can cause stuck pipe or even pack-off/downhole blockage. This condition is more critical in high-angle wells or when formations are unconsolidated. In case of a downhole pack-off/blockage, there are some observations: a) While circulating the mud, the DPP shows an increase (because the mud cannot flow and reach the annulus due to the downhole blockage); b) in contrast, the CP shows a decrease (as mud flows out of the hole reduces and perishes); 3) in addition, since the string is stuck from outside, particularly at the BHA, part of its weight would be held by the bottomhole. Therefore, the hook load (the drill-string weight held on the hook) drops; and 4) since there is no mud flow feed to the annulus, the level of the mud pit drops.

In this situation, to kill the well, the only possibility is to bullhead the well from the drill-string (to squeeze the kick influx back into the formation) because we still have hydraulic communication with the bottomhole. In case there is a kick influx in the annulus above the pack-off point, the only well control method would be the Volumetric and Bleed Method. As such, the bullhead through the drill-string and volumetric through the annulus may be carried out simultaneously.

- *Choke*

  - *Plugged-up:* When the annulus is loaded with cuttings, particularly loose formation particles, the choke may be plugged up. In this case, the CP shows an increase first, and, shortly after, the DPP will increase by an equal amount of pressure increase. When the said indications are observed, the choke-person would try opening the choke. If it did not have any effect to maintain pressures, it is confirmed that choke plugging has occurred.

Following the problem detection, it is advised to shut-in the choke/well immediately to stop over-pressuring the well. Next, we must switch to an alternate choke. Then, bring up the pump and displace with a lower circulation rate/Slow Circulating Rate (SCR) (e.g., continue by reducing SCR from 40 to 30 stroke/minute).

Note: When there is a large volume of cuttings in the annulus or the drilled formation was loosened, select a low SCR. After choke plugging, to prevent plugging of the second choke, it is more important to select a lower SCR than before.

- *Washout:* Chokes may wear over time. In case of choke washout, the CP shows a drop, and, shortly after, the DPP will increase with an equal amount of pressure increase (with the mud pit level remaining almost constant). When the said indications are observed, the choke-person would try closing the choke. If it did not have any effect to maintain the pressures, it is confirmed that choke washout has occurred. Following the problem detection, it is advised to shut-in the choke/well immediately to stop down-pressuring the well. Next, we must switch to an alternate choke. Then, bring up the pump and displace the well.

Note: The pressure observations of choke washout are the same as mud loss (underground blowout). To distinguish between the two, we should pay attention to the mud pit levels.

- Pump

  - *Failure:* If the pump fails during well kill/circulation, it does not show itself immediately on the pressure gauges. When the pump fails, the derrickhand can notice that, and the driller can immediately observe *stroke-counter* going to zero soon. Following that, both the drill-pipe and casing/annulus pressures (DPP and CP) would show a decrease. Following the detection, immediately proceed to well shut-in. Next, switch to another pump. Then, bring up the pump to the kill rate while maintaining the casing/annulus pressure constant.

- *Leakage:* If the pump leaks during well kill/circulation, the drill-pipe and casing/annulus pressure (DPP and CP) would show a decrease; thus, the driller should observe and notice this. Following the detection, holding the CP constant, slow down the pump gradually to a stop (shut down). Shut-in the well. Next, change to another pump. Then, bring up the pump to the kill rate while maintaining the CP constant.

Note 1: It is noted that the circulating pressure of the second pump may be different from the first one.

Note 2: If the second pump is unavailable (for either case of failure or leakage), it is essential to apply the partial volumetric method while the pump is repaired.

- *Mud Loss (Underground Blowout):* There are two possible cases when mud loss can occur together with kick: 1) The kick occurs first and then the mud loss occurs at an overlying (shallower) layer; and this case is the most relevant to our discussion in this section. 2) Mud loss occurs during drilling (the loss zone), then later an over-pressured formation is drilled when kick occurs.

Case 1: Imagine kick indicators are observed at the drilling depth of 10,000 ft. The well is shut-in, and shut-in pressure and data are read and recorded. Next, the kill sheet is completed. Then, we proceed to well kill. During mud circulation, both shut-in pressures suddenly show a decrease, and a drop in the mud pit level is also observed (Figure 11.20). These observations indicate that due to gas migration after well shut-in and the resultant too high downhole pressures, formation has fractured at the shoe, and mud loss has occurred. The mud loss has definitely occurred at a weak zone above the kick zone. It is seen from the figure that well shut-in after mud loss (see “d”) occurrence just causes further drop in surface and downhole pressures, making the well prone for another kick influx. In this case, to reduce the mud loss, circulation/pump rate should be reduced so that the annular pressure loss (APL) reduces, and thus the bottomhole pressure slightly decreases. For example, reduce the circulation rate/SCR from 40 to 30 stroke/minute, as well as try opening the choke further.

Case 2: Another case of mud loss, which is more difficult to handle, is when the mud loss had occurred first. Imagine we drilled to the depth of 8,000 ft where 40 bbl/hour mud loss occurred. Then, drilling was continued while partially controlling the loss (e.g., using Lost Circulation Material (LCM) pills) to the depth of 10,000 ft where a kick flow occurs. In this case, most of the kick influx may penetrate the loss zone, and thus just a partial effect of the real kick may reach to the surface (as a pit gain). Therefore, the apparent/recorded pit gain and later the shut-in pressures are

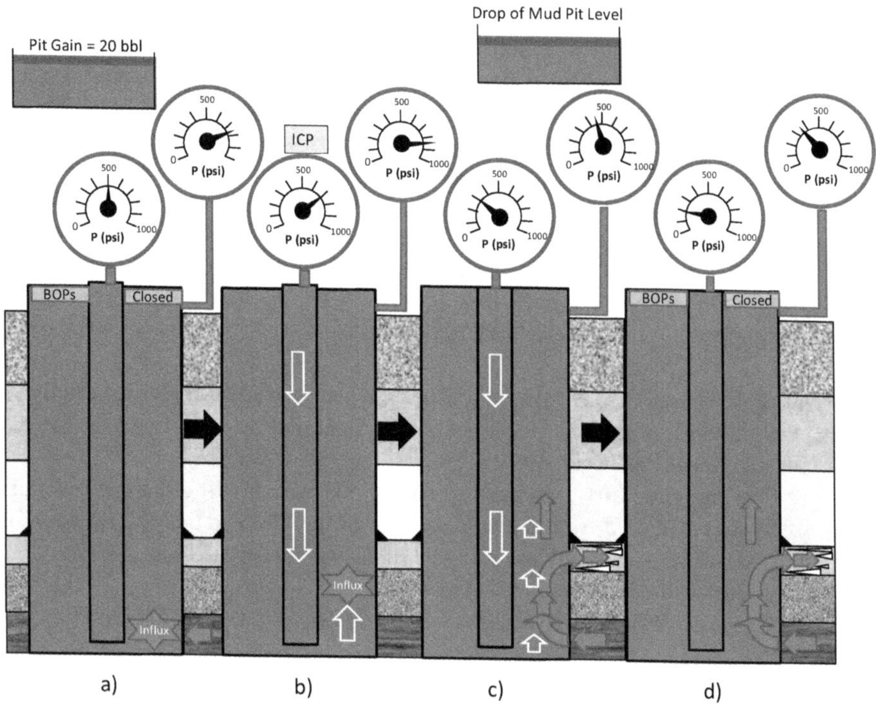

FIGURE 11.20 Detection of an underground blowout when the kick occurred first (Case 1). a) Initial well shut-in after kick detection; b) well shut-in while gas is migrating; c) mud loss occurred causing both shut-in pressures to drop: the drill-pipe pressure (DPP) dropped from 700 to 300 psi, the casing/annulus pressure (CP) dropped from 850 to 450 psi, and a drop in the mud pit level was observed; d) we can also see that if the well is shut-in, pressures after mud loss (d) are lower than the initial shut-in pressures (a) and lower than after mud loss (c).

not representative of (i.e., less than) the real kick. In such a case, we have to complete the kill sheet using the apparent shut-in data (though they are not representative) and proceed to killing the well/circulation. Similar to the previous mud loss case (when kick occurred first), it is recommended to control the loss by keeping the APL low by selecting a low SCR and trying further opening the choke. As long as there is mud return back, it is ensured that both circulating the kick influx and managing the loss are under progress. If the kick was not controlled, gradually increase the kill mud weight and continue the circulation. The process may take a long time and requires patience and tact to succeed.

- *Mud Gas Separator:* Mud gas separator, MGS (also called poor-boy degasser), is placed downstream of the choke manifold, which is useful in well control/kill situations when large volume of gas must be separated.

Poor-boy degasser has two outlets (vent line and mud line). In offshore operations, the vent line reaches above the mast where the gas would vent. In onshore wells, the vent line is extended to the burn pit located, e.g., 1,000 ft, away from the rig site. The other outlet is the U-shaped mud line out of which the cleaned mud (gas-separated) would pour into the possum-belly (shaker box/divider) or to the return line. The U-shape or the siphon of the mud line acts as a means of preventing gas from penetrating into the exiting mud to the possum-belly. The mud would then pour into the shaker pit/sand-trap pit. Next, it would go to the middle tanks where probably a vacuum degasser may be installed to further separate the gas and clean the mud in the middle tank.

If mud gas separator overloads the shaker (gas is observed at the possum-belly), it means the gas has replaced the liquid seal in the siphon of the mud line (i.e., there is no liquid seal in the mud line).

To resolve the issue, first, we need to shut-in the well holding the CP constant. Next, we inject clean mud from the *hot mud line* of the MGS so that the *mud seal* is created. Next, we open the well and bring up to a lower circulation rate (SCR) so that gas overloading would not occur any longer.

For more information on MGS and standards of operation, see Chapter 17.

- *BOP Failure:*

  - *Gas Hydrate Formation:* It is a fact that prevention is a great deal simpler than removal of hydrates. Thus, in areas where there is possibility of hydrate formation, it is essential to control at least one of the four (4) conditions necessary for hydrate formation, mentioned before in Section 11.8.1. Hydrates usually form in deepwater drilling or in drilling in very cold regions like in Alaska or Siberia. A common practice on surface choke manifolds is to inject glycol at a rate of 0.5–1 gallon/minute on the upstream side of the choke-line to suppress hydrate formation. With a water-based mud in the hole, it is recommended to keep a high chloride concentration (i.e., high salt concentration) in the mud to lower its freezing point.

Removal of hydrates is challenging as it requires considerable heat to begin hydrates' decomposition. Such decomposition can be explosive with the possibility of human fatalities because once the process begins, it will continue extremely quickly. Therefore, circulation of a hot brine solution is the best method of decomposing hydrates. Methanol is also recommended to decompose hydrate formations.

11.9 BULLHEADING

The other category of well kill methods is "influx-in method" which mainly consists of *bullheading*. In bullheading, the kick influx is squeezed to be pushed back into the formation. The circulation rate should definitely exceed the migration rate of the possible gas influx. Due to the surface applied pressure, this may or may not cause formation fracturing/breakdown. Specifically, if considerations for geomechanics and formation strength are made, it is expected that fracturing does not occur.

11.9.1 Advantages and Disadvantages

Bullheading is not considered a routine well control method. However, it is commonly considered particularly in workover operations and if influx-out kill methods are not possible or result in critical well conditions. Sometimes, allowing to apply bullheading in such cases must be predicted in the drilling plan. Generally, bullheading is recommended in the following well control conditions:

- When the formation (and thus the influx) contains H_2S or a high-pressure gas which cannot be safely handled at the surface by the equipment and rig crew. Such an influx would pose considerable risk for the crew. This well kill method must be predicted in the well plans of H_2S-bearing or high-pressure, high-temperature (HPHT) wells.
- A combined kick and loss situation (or an underground blowout) is experienced. The downhole annulus bullhead rate should exceed the gas migration rate to ensure the situation does not deteriorate further.
- When the casing pressure is expected or probable to exceed MAASP during conventional well control methods (e.g., Driller's method) according to the kill sheet calculations.
- A possible case that the casing pressure may exceed MAASP is when a very large influx has been taken.
- To push at least a part of the gas influx into the formation so that the application of conventional well control methods would be possible (without later exceeding MAASP).
- Not only normal mud circulation is not possible (like in volumetric method), but also stripping the drill-pipes to the hole is not possible because.
    - The drill-string is out of the hole, or it has been sheared.
    - The drill-string is off-bottom.
    - The drill-string or bit nozzles are plugged/blocked.
    - The drill-string is washed out or parted.

On the other hand, bullheading is not sometimes an appropriate method for killing the well. These cases are:

- *Not high enough working pressure ratings of the surface equipment:* Bullheading is not an option for taking the well under control when the required applied pressure exceeds the working pressure ratings of the surface facilities (such as the BOPs and the wellheads) or the burst pressure of the casings. Particularly in case of bullheading in workover (old) wells, case should be taken as such the casings are worn, and their burst strength may be exceeded by the applied surface pressure of bullheading.

- *High pumping rate is not possible in large-diameter wells:* If a gas influx is suspected (shut-in pressures continue to rise indicating that gas is migrating in the hole), the pumping rate for bullheading should be great enough to exceed the rate of gas migration. During bullheading, if pump pressures increase (instead of decreasing), it is an indication that the pumping rate is too low to be successful. In large-diameter holes, we may fail to reach high enough pump rates, which jeopardizes the success of bullheading.
- *In long open holes:* In wells with long-hole sections, bullheading may require applying surface pressures exceeding formation fracture pressure. In such a case, when the drilling mud is pumped for bullheading, it may cause fracturing of the formation at the casing shoe, which is the weakest zone exposed in the open hole. Therefore, bullheading is not recommended with long open-hole sections. Otherwise, the possible consequence is formation fracturing causing an underground blowout.
- *Position of the influx is far off-bottom:* If the position of the gas influx is far off-bottom (e.g., 1,500 m/4,922 ft up the bottom of hole), the rate of success of the bullheading is low.
- *Low rock permeability:* Bullheading cannot squeeze or inject the drilling mud into the formation without fracturing it when the formation rock permeability is very low (tight rocks). Fracturing the rock may cause losses or an underground blowout. Conversely, if the formation permeability is high enough, bullheading can be conducted successfully without fracturing.
- *Possibility of Partial Success:* It is possible that an apparently successful bullheading job has not killed the well completely (the influx reverts back to the wellbore). Thus, it may be essential to displace the influx out and then displace the original drilling mud with the heavy kill mud.

11.9.2 Procedure

The detailed procedure of bullheading is as follows:

1. Find the maximum allowable surface pressure that can be applied in bullheading (MASP) by considering the working pressure ratings of equipment at surface and the downhole like casings, as well as using the leak-off test results, mud weight, and casing shoe depth.
2. Select an appropriate injection pressure and make sure it does not exceed the rated working pressures of the equipment (wellhead or BOPs).
3. Make sure that enough drilling mud volume is in the tanks to displace the entire open-hole volume by 50% excess.
4. Inject at a slow rate (but higher than the gas migration rate) to establish the intended injection pressure. Try to keep a constant rate and plot the injection pressure versus the volume.

5. Continue pumping until minimum volume displaced is influx volume plus 50%. In some cases, the entire open-hole volume should be displaced plus a margin.
6. Stop pumping and observe the well. The shut-in drill-pipe and casing/annulus pressures should be approximately the same.
7. Increase the original mud weight to the kill mud weight (if necessary) and circulate using the Wait and Weight method until the annulus is clear of influx, and thus the well is killed.

Depending on whether the drill-string is in the hole or not, bullheading may be conducted in two scenarios:

- *Pipes in the hole:* Close a set of pipe rams and inject the mud (or apply enough surface pressure) through the drill-string to squeeze the influx back into the kicking formation.
- *No pipes in the hole:* Close the blind rams and inject the mud (apply enough surface pressure) through a kill-line to squeeze the influx back into the kicking formation.

11.9.3 Well Kill in Wireline Operations

Following the well shut-in, a quick option to kill the well in a wireline case is bullheading. To do this, we can simply apply mud pressure to squeeze/push the influx back into the formation. When the rig BOP stack is still in place, this bullheading pressure is applied via a kill-line (with the BOP above the kill-line closed). During production, when the wellhead is already installed and the wireline lubricator is mounted on that (see type C of wireline setup in Chapter 12), the mud circulation is applied through the 2-inch gate valve of the swept bent. The applied pressure should be within the working pressure ratings of the surface equipment and does not cause burst of the casings.

Bullheading may not be technically possible due to limitations in working pressure ratings of the casings and surface equipment, or the company's well kill methods may be the displacement/influx-out ones. In such conditions, it is necessary to strip the drill-pipes in the hole (using the snubbing unit or by closing the annular preventer) and simultaneously bleed off mud volume until we reach the bottom. At this moment, well kill operations can be started.

11.9.4 End of Well Kill

After successfully killing the well and regaining its control:

- The well can be safely opened (by opening the BOPs).
- The line-up including that of the choke manifold must be reverted to that of drilling.

- As the kill mud weight used is the lowest possible weight for killing the well, it is usually increased slightly at this stage. Thus, an additional trip margin can be provided for the overbalance pressure. For greater safety, this step can be taken before opening the BOPs.

11.10 EXERCISES

Exercise 1:

What is the danger when a gas kick is circulated through the choke manifold?

a. The increased volume can overload the mud gas separator.
b. The gas will change to a liquid and increase pit level.
c. The increased volume of gas at the mud gas separator will increase the bottomhole pressure.
d. The gas will cause a temperature increase at the remote choke and damage rubber seals.

Answer – a:

Choice a is correct. As gas flows downstream the choke, it reaches the mud gas separator (MGS) and that can overload it depending on its capacity. All the other choices do not make sense are wrong.

Exercise 2:

Kill mud is being circulated down the drill-string in a horizontal well. The pumps are stopped, and the well is shut-in as the kill mud reaches the start of the horizontal section (2,000 feet long).

What would you expect the shut-in Drill-pipe Pressure to be?

a. Zero.
b. The same as the original Shut-In Drill-Pipe Pressure.
c. The same as the Shut-In Casing Pressure.
d. Original SIDPP less than the hydrostatic pressure of 2,000 feet of mud.

Answer – a:

When the kill mud reaches the start of the horizontal section, the hydrostatic mud pressure equals the formation pore pressure; therefore, shut-in drill-pipe pressure (SIDPP) should become zero.

Exercise 3:

You are circulating kill mud to the bit with the Weight and Wait method. The drill-string is tapered. Drill-pipe length is 10,000 feet with:

From surface (0) to 5,000 feet, there is 6 5/8-inch drill-pipe
From 5,000 to 10,000 feet, there is 5-inch drill-pipe
Total strokes to bit = 2,100 strokes
ICP = 800 psi
FCP = 425 psi

If the pressure step-down graph/schedule is made on the basis of an average of 18 psi per 100 strokes pumped, what would be the effect on bottomhole pressure after pumping kill mud down to the top of the 5-inch drill-pipe?

a. Bottomhole pressure will be too high.
b. Bottomhole pressure will be too low.
c. Bottomhole pressure will be correct.

Answer – c:

This is a very tricky question. First, it is noted that the step-down chart is just important for the Wait and Weight method. Second, in each step of 100 strokes pumping kill mud in the drill-string, the calculated drill-pipe pressure is found on the basis of the assumption of same drill-pipe size (inside diameter), *not* tapered drill-pipe. If this were the case, with 18 psi pressure step (as calculated below) the bottomhole pressure would remain constant. However, since here it is a tapered drill-pipe with the size of half of the pipe length larger near the surface, 6 5/8 in > 5 in (which is usually the case), 18 psi is not sufficient, and it would experience an underbalance.

$$\frac{\Delta P}{100\text{ stk}} = \frac{800 - 425}{2100} \times 100 = 17.85\text{ psi} \rightarrow 18\text{ psi}$$

Exercise 4:

A kick is taken in a horizontal well, where SIDPP = SICP. The influx is circulated out using the Driller's method. Why does the casing pressure increase quite rapidly when the influx is circulated out of the horizontal section and into the vertical section?

a. ECD is much greater in the horizontal section.
b. This is because kill mud was not pumped from the start of the kill.
c. This is normal for any well when Driller's method is used.
d. There is no loss of hydrostatic head until gas arrives at the inclined or vertical section of the well.

Answer – d:

When the gas influx is in the horizontal section, it is at the same TVD (same pressure), and therefore it does not expand. However, when it enters the deviated/inclined or vertical section, it undergoes lower pressures and thus it expands. This would show itself as greater casing pressure at the surface.

Exercise 5:

The objective of Volumetric Well Control is to maintain BHP constant as gas migrates up the well to the surface. To accomplish this, it is required to maintain:

a. Pressure in the influx constant as the gas rises up the well.
b. The sum of mud hydrostatic pressure and SICP constant as the gas migrates.
c. SIDPP constant.
d. The pressure at the shoe constant as the gas migrates up the open hole.

Answer – c:

It is required to keep the SIDPP constant in order to maintain the BHP constant (equal to the formation pressure when the kick influx originally occurred).

Exercise 6:

A 13-bbl kick is taken and the well is shut-in, and the following recorded data were recorded:

SIDPP = 500 psi
SICP = 750 psi
Current Mud Weight = 12.7 ppg
Well TVD = 11,750 ft

After 30 minutes, the SIDPP and SICP have both increased by 150 psi. How much mud would you need to bleed off after one hour of gas migration to get hole pressure back to the original stabilized shut-in value?

Answer:

The pressure increase in one hour is 2 × 150 = 300 psi. Therefore, the first mud bleeding-off should be done at the casing pressure of 1,050 psi.

Next, the formation pressure is calculated:

$$\text{Formation Press} = 500 + 0.052 \times 11{,}750 \times 12.7 = 7{,}760 \text{ psi}$$

The mud volume to be bled off is:

$$\Delta V_{m,Bleed} = \frac{\Delta P_s \times \text{Influx Vol}}{P_{pore} - \Delta P_s} = \frac{300 \times 13}{7{,}760 - 300} = 0.52 \text{ bbl}$$

Therefore, each time the SICP increases for 300 psi, 0.52 bbl mud volume should be bled off at the choke so that BHP can be maintained constant.

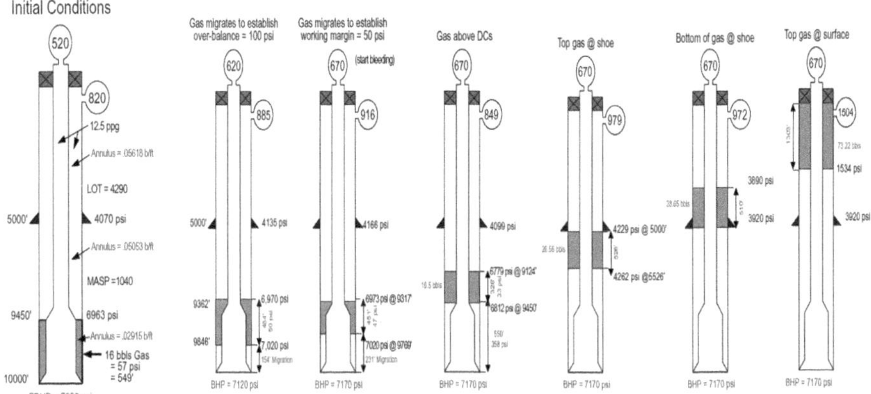

FIGURE 11.21 a) The volume of mud bled-off out of the hole in pressure stages (volumetric method). It is noted that the SIDPP is maintained constant to ensure maintaining BHP constant (from Well Control School, Well Control Manual, 2002).

Exercise 7:

During drilling at the true vertical depth of 9,000 ft (see Figure 11.21), pit gain was observed by the mud engineer. Following well shut-in, the shut-in pressures were recorded as follows.

Shut-In Casing Pressure = 600 psi
Shut-in drill-pipe pressure = 0
Formation pore pressure (SIDPP) = 4,100 psi

a) Identify why SIDPP is zero?
b) Based on Figure 11.22, the casing pressure has risen for 200 psi, and 0.3 bbl was bled off to revert the pressure back to 600 psi. How much is the influx volume?

Answer:

a) Since the bit nozzles are plugged as seen in Figure 11.21, the pressure is not felt by the standpipe pipe gauge and thus the SIDPP is zero.
b) Using the following calculations, the influx volume is 5.85 bbl.

$$\Delta V_{m,Bleed} = \frac{\Delta P_s \times \text{Influx Vol}}{P_{pore} - \Delta P_s} \rightarrow$$

$$0.3 = \frac{200 \times \text{Influx Vol}}{P_{4,100} - 200} \rightarrow \text{Influx Vol} = 5.85 \text{ bbl}$$

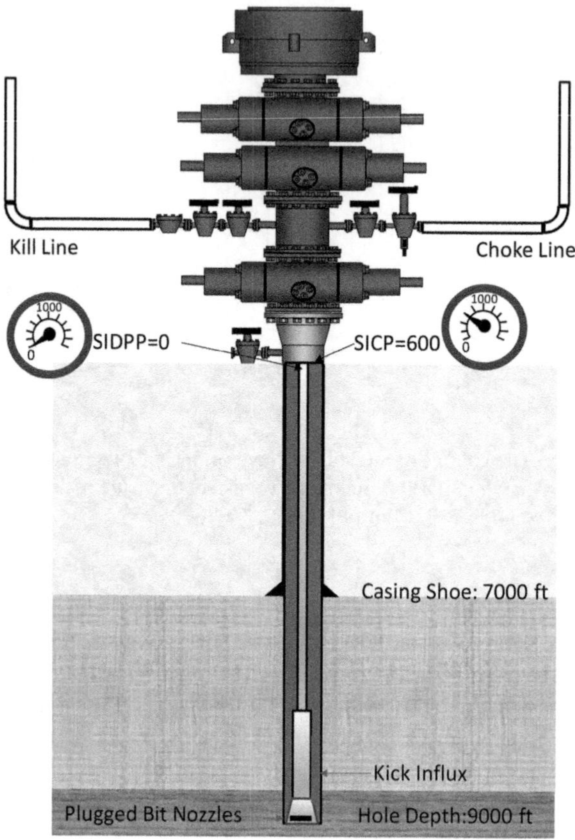

FIGURE 11.22 Gas migration and casing pressure increase from 500 psi (SICP) to 600 psi (Exercise 7). At this time, some mud must be bled-off to maintain bottomhole pressure. It is noted that due to bit nozzles' plugging, SIDPP is read zero. Only the volumetric method can be applied to control the well.

Exercise 8:

What are "hydrates"?

a. Hydrates are a solid, frozen mixture of oil and water.
b. Hydrates are a solid, frozen mixture of water and gas.
c. Hydrates are a solid, frozen mixture of oil and gas.
d. Hydrates are a solid, frozen mixture of methanol and water.

Answer – b:

Gas hydrate is considered a frozen and solid mixture of gas and water.

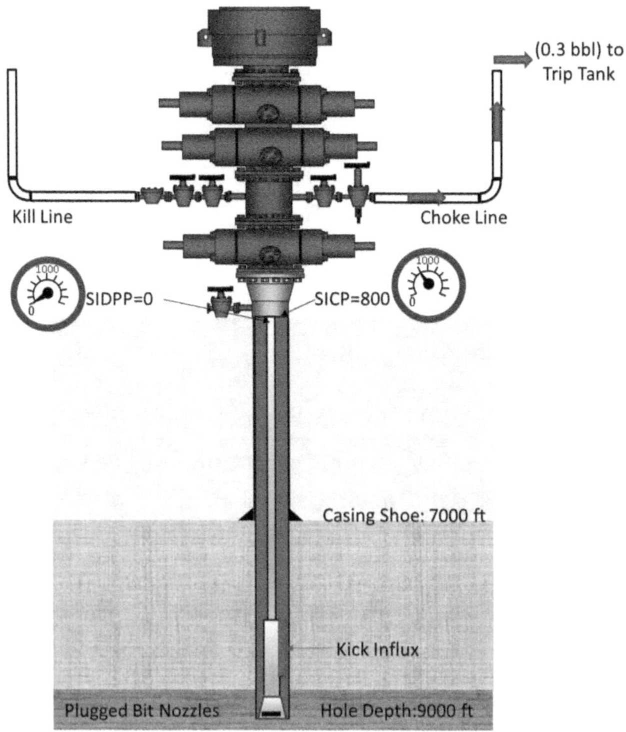

FIGURE 11.23 When SICP reaches 800 Psi, the choke was opened to allow bleeding-off of 0.3 bbl mud to the trip tank (exercise 7).

11.A APPENDIX

11.A.1 Proof of "Equation 10.2" for Volumetric Method

When the gas influx is in the wellbore, with pressure being initially P_1 which is equal to the formation pressure, its volume is also equal to the pit gain. Based on Boyle's Law, ignoring gas Z-factor effect, the following relation holds between the initial gas state (subscript "1") and the second state (subscript "2"):

$$P_2V_2 = P_1V_1 \tag{11.A.1}$$

As in state "2", the gas has expanded for the magnitude of ΔV_m (i.e., volume increase due to gas expansion), therefore:

$$V_2 = V_1 + \Delta V_m \tag{11.A.2}$$

Combining Equations (11.A.1) and (11.A.2) gives:

$$P_2\left(\Delta V_m + V_1\right) = P_1V_1 \tag{11.A.3}$$

Thus, rearrangement for ΔV_m gives:

$$\Delta V_m = \frac{P_1 V_1 - P_2 V_1}{P_2} \tag{11.A.4}$$

Knowing P_2 can be written as "$P_1 - (P_1 - P_2)$", Equation (11.A.4) becomes:

$$\Delta V_m = \frac{(P_1 - P_2) V_1}{P_1 - (P_1 - P_2)} \tag{11.A.5}$$

Therefore:

$$\Delta V_m = \frac{\text{Inc in Surf Press} \times \text{Influx Vol}}{\text{Formation Press} - \text{Inc in Surf Press}} \tag{11.A.6}$$

This is the same as Eq. 11.2 in the text.

NOTES

1 The new value of SICP is lower than the initial value of SICP because the kick influx has already been extracted/removed from the annulus.

2 Stripping is running the drill string back under pressure to the bottom while the drill-string is packed-off/sealed by the BOPs.

12 Wireline Operations and Well Control Setup

Rahman Ashena

12.1 INTRODUCTION

Wireline is a strong, thin length of wire cable mounted on a powered reel, which is used to carry out many oilfield operations. Wireline operations may be needed during drilling, completion, production, and workover operations. Field operators can use wireline to take downhole directional survey, open-hole formation evaluation well logs, mechanical back-off of stuck pipe, carry out cased-hole logging, install or remove plugs in/from the completion string, and perforate casings.

This chapter covers the well control setup and aspects required for wireline operations, based on different situations and scenarios possible to be encountered in wells.

Regardless of the purpose of the wireline operations, wireline operations fall into some categories depending on the following situations:

1) No pipes in the hole,
2) With pipes in the hole, and
3) With Christmas tree on the well (during production).

12.2 NO PIPES IN THE HOLE

With no pipes in the hole, wireline jobs are conducted using two common methods as are explained next.

12.2.1 Without Pressure Control

Sometimes, when wireline jobs are conducted, the well is not exposed to underground or reservoir formations, and thus there would be very little/no risk of kicks or getting underbalanced. Some examples of such conditions are when a cement plug is placed in the open hole and it is intended to conduct cased-hole well logging (such as a cement bond logging/CBL or taking a corrosion log from the casing), or it is planned to perform a wireline job through the completion string while the well is not exposed the least to the reservoir formation (because the reservoir is not yet perforated).

DOI: 10.1201/9781003473770-12

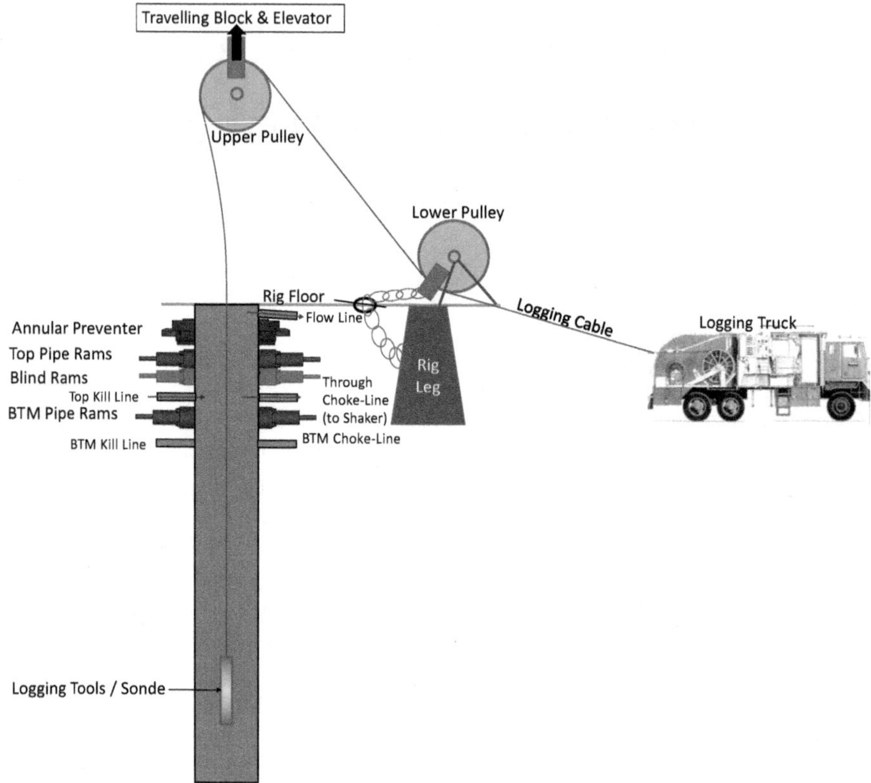

FIGURE 12.1 Schematic of surface installations for a wireline job (A-1 setup) without any pipes in the hole (in the presence of a rig). The surface setup is without pressure control as it relies on the drilling mud as the primary barrier and the rig annular BOP as the secondary barrier. Refer to "A-1" type setup in the text.

In such conditions, the wireline jobs may be performed without strict pressure control. The "without pressure control" wireline job means that we rely only on the drilling mud as the primary barrier while using only rig BOPs as the secondary barrier. Figure 12.1 shows a typical setup for such a case. For simplicity, it is abbreviated as A-1 type setup.

In case of a kick flow, the rig annular BOP can be used to shut-in and secure the well. In this case, it is important that the rig annular BOP is designed for "Complete Shut-Off" (CSO). If this is the case, the relevant manufacturer's tables should be referred to establish the correct closing pressure. In this case, establish a cross-wellhead mud circulation and monitor the well continuously (to detect any possible kick flows using flowmeter sensors or pit level sensors confirmed by the mud engineer's checks of mud level). The disadvantage of this method is the possibility of fluid leakage through the annular preventer packing and possible damage of the annular packing by the wireline. It must be noted that the rig annular BOPs would leak and cannot seal properly in case of using "braided" wireline.

12.2.2 Using Shooting Nipple and Wireline BOPs (with Pressure Control)

On the other hand, when there is some risk of underbalance or using braided wireline, wireline jobs are carried out with pressure control (using wireline BOPs as the secondary barrier). Wireline BOPs are more reliable to seal around the wireline than

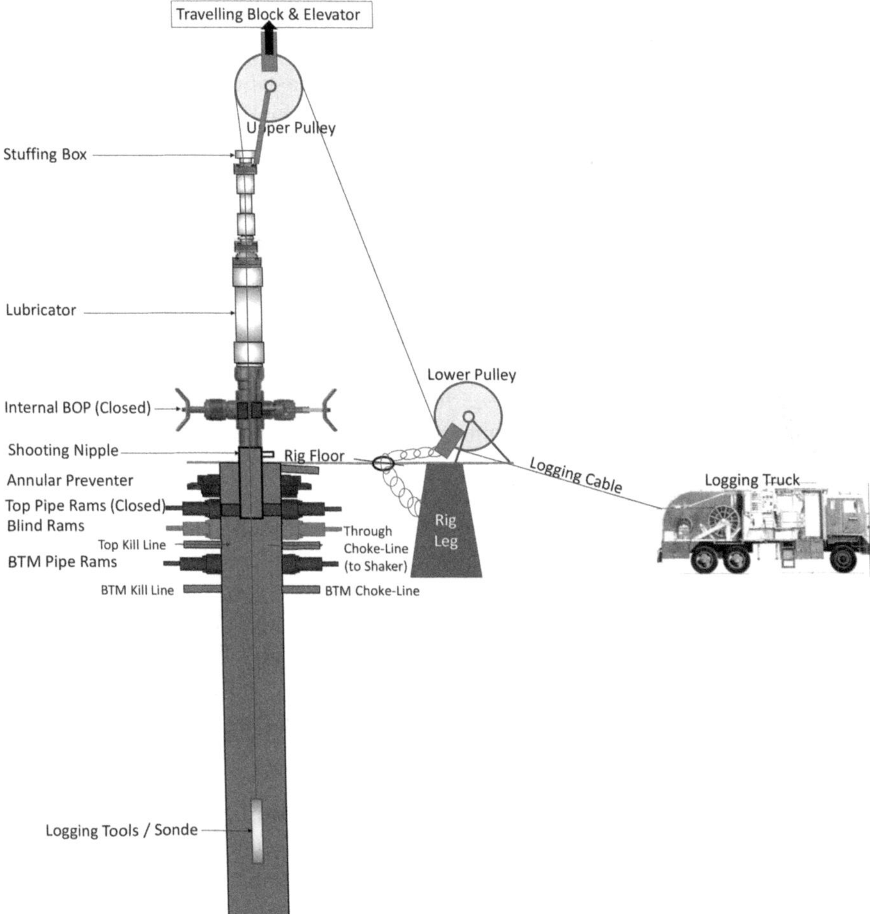

FIGURE 12.2 Schematic of surface installations for a wireline job (A-2 type setup using shooting nipple and wireline BOPs) with no pipes in the hole (in the presence of a rig or Hydraulic Control Unit). The surface setup includes the shooting nipple above which wireline BOPs (internal BOP/IBOP) are used to seal around the wireline. In case of more considerations for safety, it is also possible to install a lubricator or the stuffing box above the IBOP. Other surface installations include the logging truck, logging cable, shooting nipple (kept in its place by closing the top pipe rams around it), the lower pulley/sheave (which is connected to the rig leg using chains), the upper pulley/sheave (kept in its place using the traveling block and elevator), etc. A cross-wellhead mud circulation is maintained by pumping mud from the top kill-line back through the choke-line to the shaker tank.

the rig annular BOP and are recommended as the preferred method. Therefore, first, a special pipe joint called *shooting nipple* should be used so that the wireline BOPs (including an *Internal BOP or IBOP*, sometimes with a *stuffing box*) can be installed on it. Figure 12.2 shows a typical schematic of the setup. For simplicity, it is abbreviated as A-2 type setup. The shooting nipple is a single joint pipe of about 7 inches with a circular base at around three feet below its upper end, which makes it a proper basis or structure for the IBOP to be mounted on that. After running the shooting nipple in the hole, it should immediately be tested by closing the top pipe rams and blind rams and applying for example 1,000 psi pressure through the fill-up line to the shooting nipple's specific test port. Next, the IBOP is installed, tested, and closed to seal around the wireline during the operations. During wireline operations, to detect any possible signs of kick flow or mud loss, the crew should establish a so-called cross-wellhead mud circulation (pumping mud for example from the top kill-line and back through the choke-line and choke manifold to the shaker tank, see the figure). In case a partial kick flow occurs (e.g., less than 5 bbl), while maintaining the so-called cross-wellhead mud circulation, slightly close the adjustable manual choke so that some back-pressure is maintained (say 200–300 psi). Then, continue logging and quickly pull out of the hole. After the logging tool reaches the surface, quickly proceed to closing the blind or shear rams. If the kick flow is severe or is approaching loss of well control/a blowout, proceed to cutting the wireline and close the blind-shear rams. In critical cases, it is recommended to use blind-shear rams as they cut the wireline and shield the well immediately. It is noted that the stuffing box may not be used for conventional logging. However, it may be installed above the IBOP in some special cases such as under-pressure cement bond logging (CBL) jobs. It is noted that CBL jobs may be performed under surface applied pressure to eliminate the negative effect of micro-annulus from the log quality.

In periods when the wireline is out of the hole, the crew must always close the blind rams to seal and shield the well (for prevention of any possible flow out of the well).

12.2.3 Using Connection, Spacer Spool on Rig BOPs and Wireline BOPs (with Pressure Control)

In special cases such as well logging with coiled tubing in high-angle well (called E-Line) or cases with high mud loss, it is suggested to use another setup to guarantee greater safety. Therefore, as the second method of performing wireline jobs with no pipes in the hole, the annular preventer is nippled down first. Next, using a connector (e.g., the Christmas Cap) and a spacer spool, the lubricator can be installed on the drilling BOPs to ensure safe operations. Figure 12.3 shows a typical schematic of the setup. For simplicity, it is abbreviated as A-3 type setup.

A lubricator consists of:

- Stuffing box or pack-off head
- Grease injectors
- Lubricator joints or tube bodies
- Wireline Blowout Preventers (BOPs)

A bleed or pump-in valve is used for additional safety in special/risky cases.

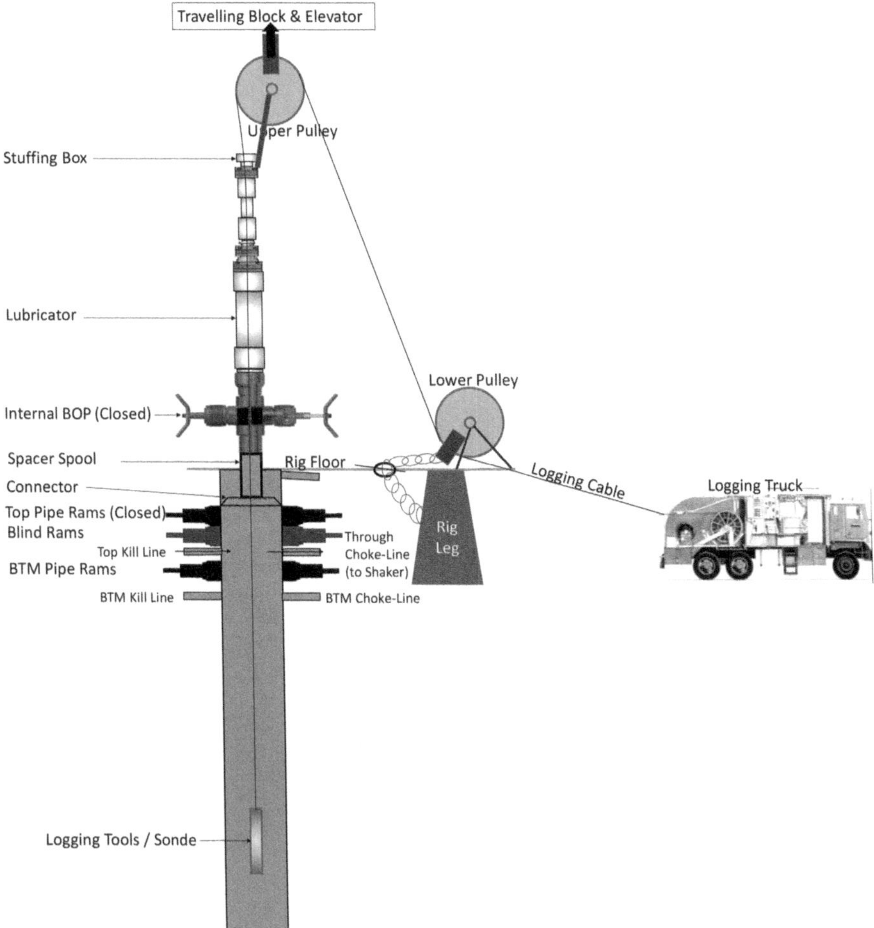

FIGURE 12.3 Schematic of surface installations for a wireline job (A-3 type setup using spacer spool and wireline BOPs) without any pipes in the hole (in the presence of a rig or Hydraulic Control Unit). In this setup, the annular preventer is uninstalled, a connector (e.g., a Christmas Cap) is connected, and a spacer spool is used to reach above the floor. Next, an internal bop and lubricator (with stuffing box) are made up as wireline BOPs.

12.3 PIPE IN THE HOLE

With pipes in the hole, there are cases where wireline is essential to perform some tasks. There are two main cases: First, in special well conditions (particularly high inclination wells), well logging may be conducted with wireline passing through the drill-pipes (Tough Logging Conditions, TLC). Second, wireline jobs in the well completion phase are done through the completion string: During running the completion string into the hole, there are the plug and prong in the Sliding Side Door (SSD). When the completion string reaches the desired depth, the crew need to prepare to set the packer. Setting the packer requires sealing/closing the bottom-end of the string below the packer (at the No-Go Nipple). Next, by applying pump pressure, the packer can be set/installed. To seal/close the bottom-end of the string, it is essential to run in the hole with wireline (five runs) to:

1) remove the prong from the SSD,
2) remove the plug from the SSD,
3) convert the SSD port's position from open to close,
4) place a different plug in the No-Go Nipple, and
5) place a different prong in the No-Go Nipple.

To conduct the wireline jobs with pipes in the hole, the shooting nipple is not required to establish a basis for the wireline surface setup. In any case, to provide safety for the wireline jobs, we need some safety setup. First, it is necessary to make up a crossover/

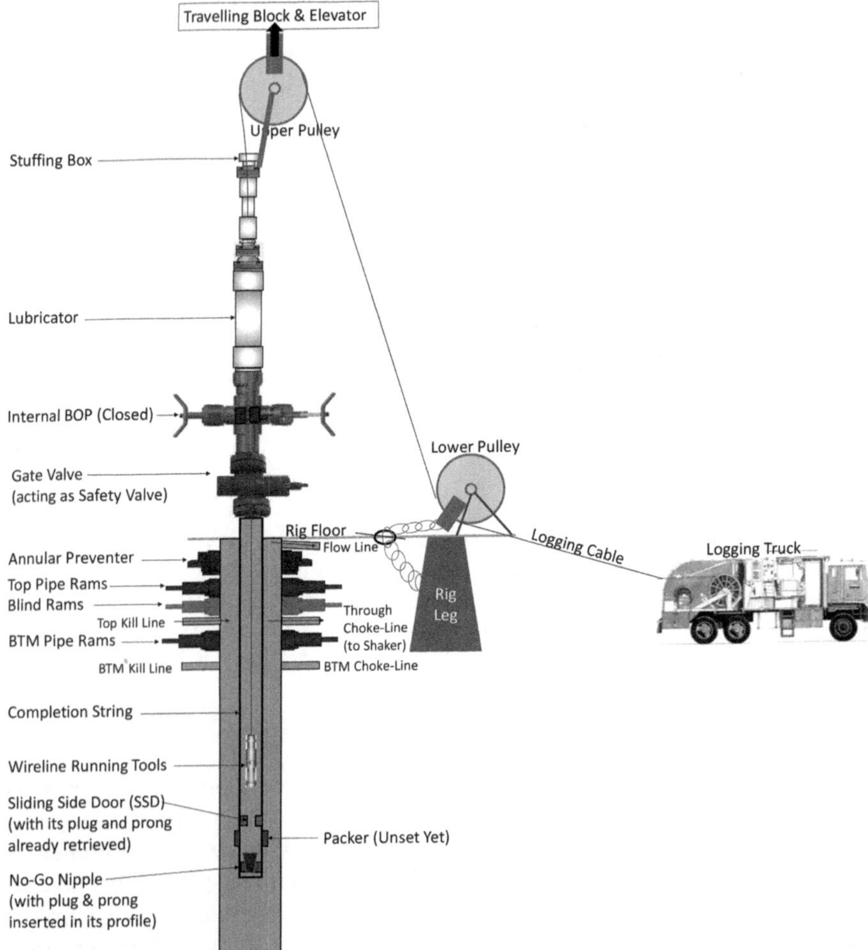

FIGURE 12.4 Schematic of surface installations for a wireline job with pipe in the hole (in the presence of a rig or Hydraulic Control Unit). Refer to "B" type setup in the text.

converter (of thread to flange) on the landing joint/last joint of the completion string. Next, a gate valve is connected to the flange to be closed for sealing the well when there is no wireline in the hole. Sealing the well by the gate valve in such periods prevents any U-tubing flow or a kick flow through the completion string to the surface, so it acts just like a Drill-Pipe Safety Valve (FOSV). The role of the gate valve when there is pipe in the hole is analogous to the role of blind rams when there is no pipe in the hole. In addition, a lubricator (equipped with a stuffing box) is connected above the gate valve to seal around the wireline when the wireline is in the hole. If the gate valve and the lubricator were not mounted on the completion string, some mud in the annulus would flow from the annulus to the completion string, back to the surface. This usually occurs due to the U-tube effect (possible existence of heavier mud in the annulus than in the completion string) which is considerable in long completion strings. Without using the safety setup, the possible U-tube effect causes a mud level drop in the annulus. If the annulus mud level drop is considerable, it can lead to a kick flow situation during the wireline operation.

Next, the rest of wireline surface equipment (including the pulleys/sheaves and the wireline trucks) are installed to complete the required surface setup. At this point, the wireline operations can safely commence. Figure 12.4 shows a typical schematic of the setup. For simplicity, it is abbreviated as B setup.

12.4 CHRISTMAS TREE ON THE WELL (DURING PRODUCTION)

During production, for wireline jobs with an already-installed Christmas tree, a lubricator should be installed. When the full wellhead (with Christmas tree) is already installed, the lubricator is simply connected to its Christmas Cap. Figure 12.5 shows a typical schematic of the setup. For simplicity, it is abbreviated as type C setup.

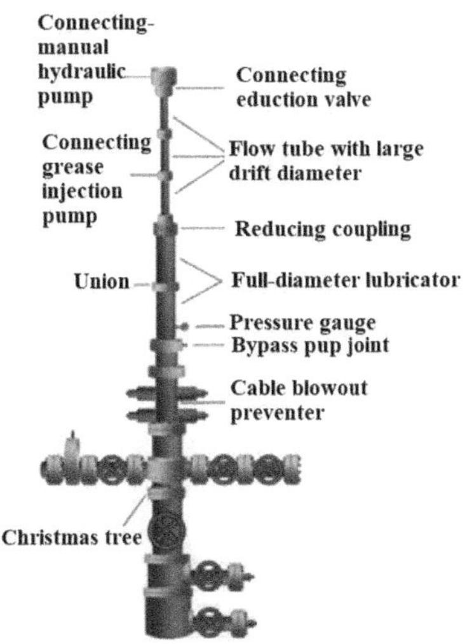

FIGURE 12.5 Schematic of surface installations for a wireline well logging job without any pipes in the hole with wireline BOPs on an already-installed wellhead. Refer to "C" type setup in the text.

13 Early Kick Prediction by Data-Driven AI/ML Techniques

Rahman Ashena, Elliot H. Kurnia and Terrence Josiah

13.1 INTRODUCTION

Kicks are often encountered during drilling petroleum or geothermal wells, particularly in the exploration phase. The implications are irreversible reservoir formation damage, lost time causing low drilling efficiency and considerable cost, possible equipment failure, casualties, and tremendous damage to the ecological environment. The earlier the kick detection, the sooner the crew can take action to control the well and prevent or lessen the consequences. In case kicks are not controlled and lead to blowouts, the damage will be immensely more severe. Therefore, the development of early kick prediction and detection systems is of great benefit for the petroleum and geothermal industries.

Before kicks can be detected on-site, some surface and downhole parameters usually undergo minor changes, which cannot usually be detected by rig personnel or field equipment. The advantage of using surface parameters is that they can be easily measured or continuously collected by surface sensors. However, downhole or subsurface parameters associated with logging while drilling (LWD) and measurement while drilling (MWD) require data telemetry such as mud pulse or electromagnetic telemetry. It is noted that such data can be used for post-drilling analysis in terms of kicks statistics (see two examples by Holand, 2017; Ashena et al., 2023).

Traditional alarms are designed in a way that when certain (monitored) parameters (usually pit tank volume and outflow rate) exceed the operational limits, the alarm will trigger. The issue with this method is that mud volume and outflow rate (outflow percentage by paddles) are expected to fluctuate when standard tasks are carried out such as making connections, drilling the hole deeper, mud transfers, and increasing or decreasing pump rates. The traditional detection is carried out by rig crew (drilling supervisor, tool pusher, or driller) using mud logging sensors measuring mud tanks levels and outflow rate. The mud level measurement is a definite kick indicator, but there may be an inaccuracy of the sensor or crew mistake, and there is usually a detection delay (i.e., a lag time between the actual initial kick occurrence time and the detection time); the traditional outflow rate/percentage is not accurate and cannot be due to possible malfunction of the paddle-type flowmeter. Coriolis

DOI: 10.1201/9781003473770-13

mass flowmeters are preferred to traditional meters, but they have their own disadvantages including high cost and possible inaccuracies, and obviously they cannot predict forthcoming kicks. Due to possible fluctuations of the mentioned parameters during different tasks and the inaccuracy of particularly the outflow meter, many false alarms may occur. It is a matter of fact that accurate, early, and true detection of influx and loss during drilling is of great significance for drilling operations. Therefore, to overcome the mentioned shortcoming, ML methods can detect such minor changes and judge the occurrence of gas kicks.

Kick prediction is possible via two approaches of the physical method and the ML method based on data analysis. The physical or analytical method is straightforward giving a flowchart with few conditions or if-statements to determine when kicks may occur. The ML approach takes in the whole data, chews it together using, and then provides a model using AI with the results expressed as "kick probability". For prediction purposes, data-driven methods employing ML may be effective for real-time kick prediction.

An artificial neural network model, when properly trained and tuned for a specified dataset with high accuracy, can be plugged in with real-time drilling data to predict and detect kicks which may be earlier than traditional methods. To verify this, in this research work, following a literature review, several ML models were developed and applied to some measured kick data from the Middle East to find the best performing model. The objective was to use the tremendous amount of measured drilling data available for the development of the best performing model to predict kicks and prevent or mitigate the consequences.

13.2 LITERATURE REVIEW

Real-time kick prediction was found to be effectively possible by using ML methods (Yang et al., 2019). Particularly in deepwater cases (Yin et al., 2019), an ML model can act as an early kick predicter to prevent or predict kicks much earlier. It is noted that deep-water occurs when seawater exceeds 600 m/1,969 ft.

Shi et al. (2019) discussed the use of ML to predict kick influx and loss in an oil well. The models used were Random Forest (RF) and Super Vector Machine (SVM). Both models were trained using a real-time drilling data during kick occurrence, with the data being preprocessed and then split to train and test the SVM and RF models. Both models had a dedicated line graph indicating when approximately a kick happened. The RF was able to detect kicks on average 27 minutes earlier than the pit volume method. However, there were four instances where "false normal detection" occurred in the middle of a kick which caused the detection accuracy to drop, but the RF error rate was still low. When the data was plugged in using SVM model, again the kick influxes were predicted 27 minutes earlier than pit volume method, and the model had only one "false kick detection" which dropped the prediction slightly. Since the SVM model had much less error compared to the RF model, it can be considered more suitable for kick prediction in their study case.

In addition, Shi et al. investigated mud loss detection using the same two models developed for kick detection. The RF model predicted losses on average 20 minutes earlier than the traditional pit volume decrease method, but it came with a "false detection" before the beginning of the loss which made the accuracy rate slightly imperfect. Meanwhile, the SVM prediction outcomes for the loss detection came out the same as the RF model but without the "false detection". Collectively, for the detection of kick and loss cases, both models were observed to have some "false detection" going into the accuracy testing, but, overall, both models were still very accurate in prediction, hitting 93.72% accuracy for SVM and 92.23% accuracy for the RF model. Both models also had 20–25 minutes of earlier kick and loss prediction periods compared to the pit volume method.

Yang et al. (2019) used ANN (artificial neural network) models which had the capacities to learn and develop themselves for kick detection. The way a neural network worked is that the model itself would determine what was the most important variable within the dataset. Therefore, initially, the parameters consisted of total depth, weight of hook (WOH), weight of bit (WOB), revolutions per minute (RPM), torque, rate of penetration (ROP), standpipe pressure (SPP), DC exponent (D_{exp}), and flow rates (FLW), Pump1 and Pump2, and Total Gas (T_{gas}). Following using the Principal Component Analysis (PCA) method, the contributing factors in the variables were eliminated to seven consisting of WOH, ROP, SPP, RPM, FLWpmps, Pump1, and Pump2. The authors did not mention which parameters of pumps 1 and 2 they referred.

After gathering and preprocessing of all the data and running the ANN model, Yang et al. (2019) found the accuracies of 97.3% and 96.2% for training set and testing set, respectively,\ which are quite high. The model was able to identify gas kicks at 12–38 time points faster than on-field equipment does; however, the authors did not mention the duration of each time point. The researchers did mention about the necessity of considering probability as a challenge when using the ANN models, that is why running the same dataset for testing may result in several different results regarding the gas kick time frame.

Yin et al. (2019) developed an ANN-based kick warning model using big data for an ultra-deepwater high-pressure, high-temperature (HPHT) case well of up to 1,688 m water depth and 176°C bottomhole temperature. The data consisted of mud-log data (18 parameters) and well-log data (11 parameters) from 12 kick incidents of 5 drilled wells. The tuned ANN was then plugged in with various sensors in the rig, constantly feeding it with data so that when an early overflow appeared due to an emerging kick, the model would detect it and warn the rig crew about it. Using the model, the overflow in the flow line system was detected 20 seconds after the initial symptoms, whereas when the mud-log was solely used, it took up to 160 seconds into the overflow to detect that a kick had happened which was when the overflow had almost reached its maximum value (see Figure 13.1).

Unrau et al. (2017) mentioned that by incorporating ML into the kick alarm system, AI would learn to adapt the threshold by learning the pattern of the inputted data. This type of alarm was called Adaptive Alarm by ML. As a result, the number of the false alarms could be reduced, kicks could be detected earlier, and corrective action could also be taken earlier. The ML model was trained and tested using 33 kicks, 20 loss events, 80,000 hours (9+ years) of rig data, 465,000 meters drilled, 80,000 pump off/on events, and 1,500 trips in order for the model to be able to learn the data pattern and be used accurately. With the adaptive alarm, false alarms were eliminated, see an example in Figure 13.2. Therefore, for future drilling, using such

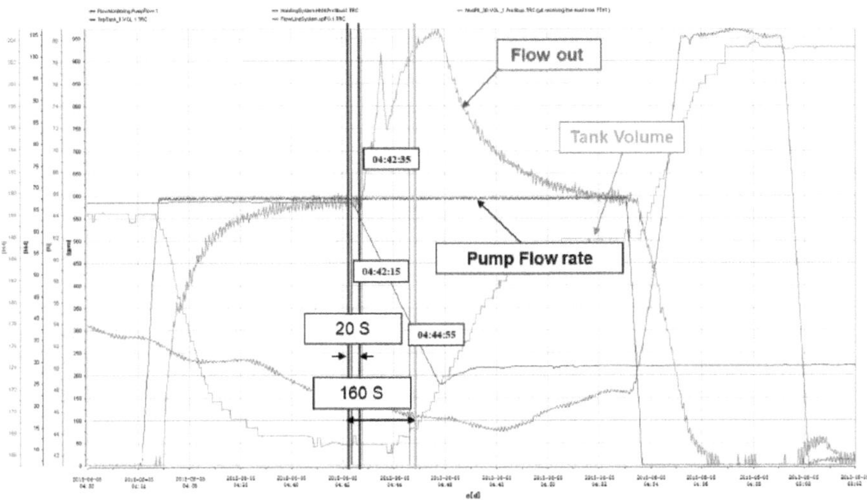

FIGURE 13.1 Kick detection time by mud logging unit and the ANN model (Yin et al., 2019).

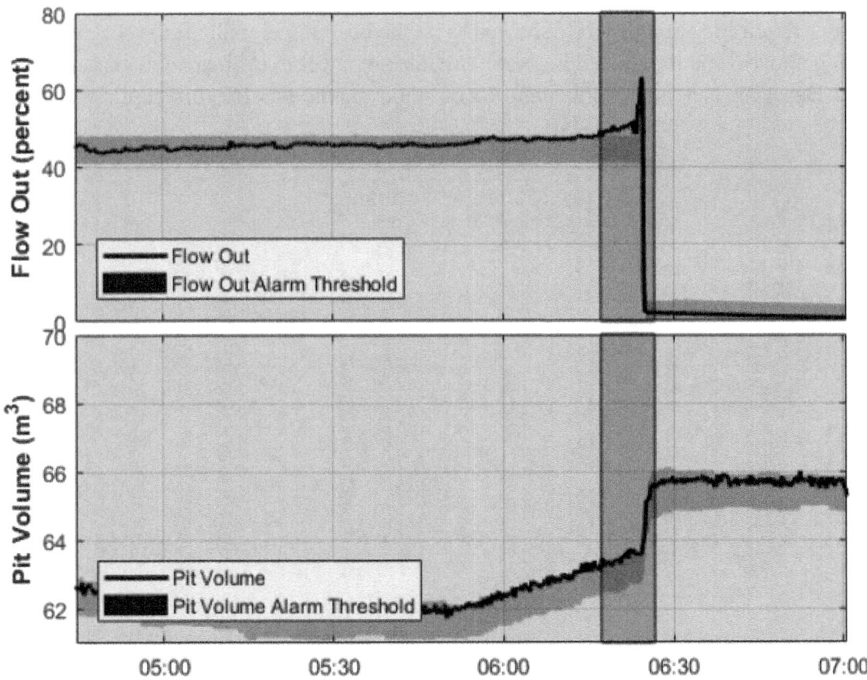

FIGURE 13.2 An example of kick detection (Unrau et al., 2017).

adaptive alarms, it would be possible to reduce false alarms to an extremely low number while maintaining a very tight alarm threshold.

Alouhali et al. (2018) found the ML approach an appropriate approach to predicting kicks real-time using five ML models consisting of Artificial Neural Network (ANN), Sequential Minimal Optimization (SMO), Decision Tree (DT), K-Nearest Neighbors (KNN), and Bayesian Network. As for the data, they used over 1 million initial data sets or instances which were measured with high frequency of one instance per 5 seconds (0.2 Hz). They classified the data into two groups of Kick (1) and No Kick (0), and thus the ML models used are of classification type. The researchers said that just few kicks were among the data, but they refrained from mentioning the number. In order to delete outliers and reduce the noise, preprocessing and quality control of the initial data were carried out, and this process is also called the dataset condensation or instance reduction stage, leaving just above 122,000 data sets for final utilization.

Next, in order to find the best performing model, each of the five models was tested using seven evaluation criteria or performance matrix applied to the test data. The ratio of train and test data was not mentioned. The data consisted of Precision, Recall, F-Measure, Mattews Correlation Coefficient (MCC), Receiver Operator Characteristic (ROC) Area, Precision Recall Curves (PRC) Area, and Kappa Statistic. Figure 13.3 shows that the Decision Tree was found to be the highest performing model because of its high accuracy and relatively lower computation time/power required. Though KNN showed to be the best model with all the seven criteria being above 98%, it required higher computation power to evaluate new inputs. ANN and Bayesian Network were found to be good candidates for this application with

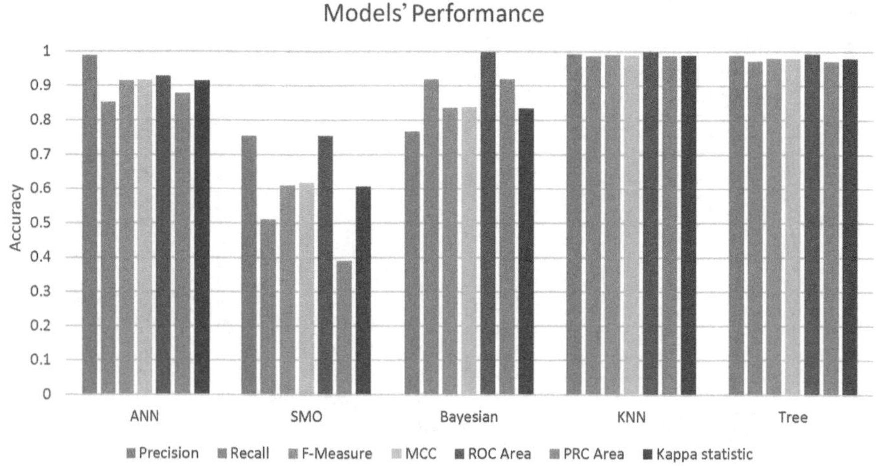

FIGURE 13.3 Comparison of performance matrix/evaluation criteria of the different models used by Alouhali et al. (2018).

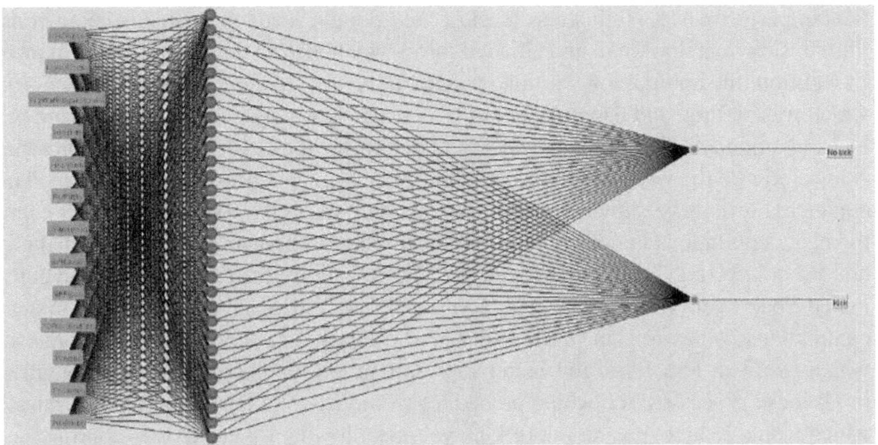

FIGURE 13.4 The ANN model structure used by Alouhali et al. (2018).

slightly less accuracy than Decision Tree and KNN. SMO may be used for classification problems, but it is more suitable for regression analysis problems, thus it is not proposed in this case. It is noted that the researchers did not specify the number of hidden layers used for the ANN model, whereas only one layer was shown in Figure 13.4. No information about computation times/powers of the ANN and other models was mentioned. This is an important factor to be investigated in another work.

Yalamarty et al.'s (2022) 9th pdf: *Early Detection of Well Control Kick Events by Applying Data Analytics on Real Time Drilling Data* inferred that ML models can be used to predict kicks earlier, lowering the risk of conversion to a blowout. The variables that they considered for ML consisted of depth, pump pressure, gain in active system, gain in trip tank, and increase in the outflow. They trained the model with 20 historical well test data for the purpose of parametric tuning, then they provided a good description of how the model can be applied in field practice. After training the model, it would be incorporated with the sensors and into the dashboard to create an alarm system. When an alarm pops, it would create a warning with a type of abnormalities it has detected into the dashboard. An SME (Subject Matter Expert) will then check the error to determine whether it is a false alarm or not. After the initial check by SME, the user will then give feedback to label the abnormalities which in this case it was triggered when engineer was transferring a connection thus, the current feedback is produced. These algorithms had been developed through a thorough iterative process of development and testing with collaboration with SMEs from an operator. This approach helped ensure that the algorithms incorporate practical concepts and operations that was common in the oilfield.

Curina et al. (2022) inferred that by using automatic kick detection, human error could be reduced, and kicks can be detected earlier with its symptoms. Their automation process has some features one of which is by measuring the operator stress level, based on the temperature of their bodies. When high bodies temperature was detected in the operators, it indicates high stress level which were directly correlated to the human performance. High stress level of the operator leads to mistakes being committed, slow reaction time, and clouded mind which affects the overall performance. In addition, the automation was also trained to be able to detect different processes, which was drilling and tripping. The system uses the weight on hook (WOH) and the block position and movement to accurately determine the depth of each tool joint that is lowered into the well, which is important for subsequent possible well shut-in. The conducted tests have shown a prediction accuracy of 98.1%. This allows the system to keep a continuous log of the tool joint positions and update it as the drill-string is hoisted or lowered. Other functions of the ML include BOP layout function and automated Space-out. With the incorporation of the ML models, the automated system could choose when to shut in the well based on pure data feeding from the sensors which result in less error and better kick mitigation. By incorporating automation in all these processes, the whole procedure would become faster and more accurate which would reduce the amount of kick volume allowing for easier well controls.

13.3 DATASET PREPARATION

In this work, each data instance represents one time step. Attributes or features refer to some parameters measured at the surface during drilling. Each dataset or instance had seven input attributes which were surface drilling parameters, consisting of drilled depth, suction pit volume, active pit volume totalizer (APVT), drag force, surface torque, pump pressure, and mud circulation rate (or flow-in). The last output attribute, called kick probability, was generated by the ML models. Due to the high frequency of data measurement (1 Hz/one instance per second), there were many initial drilling data sets. Among the initial drilling data sets, there were only five kick occurrences. Following the preprocessing stage, also called dataset condensation, to reduce outliers and noise, finally 21,833 instances remained, using which the ML models were created or developed.

13.4 METHODS

Five types of ML models were applied to the data consisting of Static ANN (SNN), Dynamic ANN (DNN), XGBRF, KNN Regressor, and Gaussian Naive Bayes (NB).

13.4.1 Static ANN (SNN)

Most ANN structures used so far in engineering have been the static feed-forward type because they are appropriate for applications which are time independent. These ANNs, called static neural networks (SNNs), respond instantaneously to inputs. A static ANN model structurally consists of some neurons linked to each other, which cooperate to transform inputs to outputs in the best possible manner. Specifically, an ANN structure is in turn made up of an input layer consisting of input parameters, one or several hidden layers as wells as an output layer. In each layer, there are one or several neurons. It is noted that there exist only one input layer and only one output layer all the time. No computation is made in the neurons of the input layer – they just receive the input parameters data from the user.

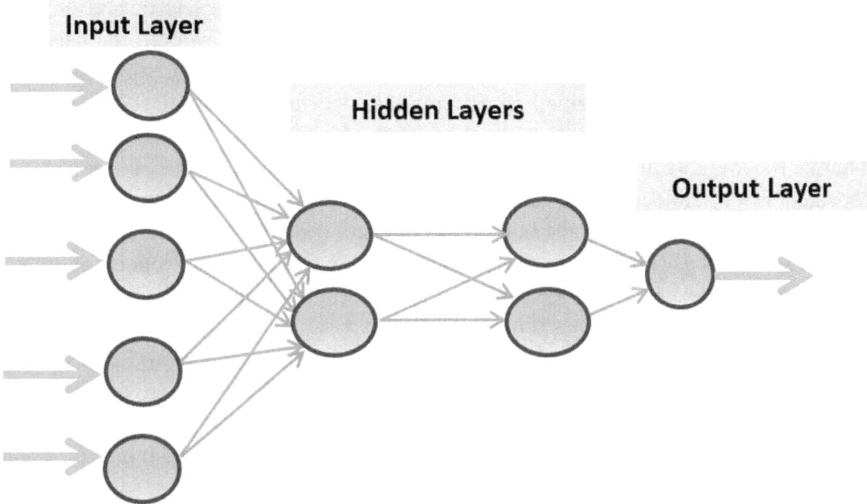

FIGURE 13.5 The typical structure of a feed-forward artificial neural network (Ashena and Thonhauser, 2015).

Since there are seven surface parameters or attributes in this work, the number of neurons in the input layer is seven as well. The number of hidden layers depends on the complexity of the problem; thus, a deep ANN has several hidden layers, but usually two or three layers are sufficient. The neurons in the hidden layers have no communication with the outside world, they only deal with the input and output neurons. The output neurons are responsible for transferring outputs with the outside world. Since the output parameter is kick probability in this work, there is only one output neuron. Since most SNNs that are used in many engineering applications are feed-forward propagation type (e.g., Ashena et al., 2010, 2021), feed-forward propagation was used in this work to result "kick probability". Figure 13.5 shows the structure of such a typical ANN.

As the main drawback of the model, an SNN does not have the capability to capture temporal dependencies which are the "time delays" or the past values of parameters which could be used to help the model perform better in time-series modeling. As the other drawbacks, an SNN cannot learn properly from complex data sequence, irregularities, and noisy observations that may exist in time-series data.

The static neuron is not dynamic in nature as it does not consider the time delays impacting the dynamics of the system.

13.5 DYNAMIC ANN

It is noted that time delays are the intrinsic characteristics of biological neurons contributing to dynamic learning process of humans. As the kick problem is considered a time-dependent one, an SNN model may not be an appropriate algorithm to be practically applicable.

Therefore, unlike an SNN model, a DNN has a dynamic characteristic and employs extensive feedback between the neurons of a layer and/or the layers of the network. This feedback implies that the network has local memory characteristics. The node equations in DNN are described by differential or difference equations. Because of feedback paths from their outputs to the inputs, the response of DNN is recursive. That is, the weights are adjusted, the output is then recalculated, and the process is repeated. For a stable network, successive iterations produce smaller and smaller output changes until eventually the outputs become constant (Sinha et al., 2000).

Dynamic ANNs (DNNs) are essentially considered an upgraded version of SNNs. In the SNN model, there is a feed-forward connection between all the layers, that is, information only travels forward in which it does not have a chance to adjust all the neurons based on the error. Due to this point, the performance of an SNN will be less accurate than that of a DNN, but due to only having a one-way passing of information, the time it takes to run an SNN is shorter than for the dynamic case. A type of DNN has also feed-forward connections, but the second type which has recurrent connections is more common or preferred because recurrent-dynamic networks typically have a longer response than feedforward-dynamic networks. Recurrent DNNs need longer computation time and computer power than SNNs since information travels to the end of the hidden layer and passes back to the other way which is the series of hidden layer to correct its error (i.e., the two-way connection). This means essentially that DNNs can better adjust all the parameters it had in order to perform the best calculation and finally be sent to the output layer.

Owing to their advantages, dynamic ANNs are generally more powerful than static ANNs although it is more difficult to train them. Because DNNs have memory, they can be trained to learn sequential or time-varying patterns. This has applications in such disparate areas as prediction in financial markets (Roman and Jameel, 1996), channel equalization in communication systems (Feng et al., 2003), phase detection in power systems, sorting, fault detection, speech recognition, and even the prediction of protein structure in genetics. One principal application of dynamic neural networks is in control systems.

It is noted that dynamic neural network is used for the time-series regression. When time is involved in an ML, the "shuffle" must be set to false, which effectively turns off the data shuffling. This is important because if it is turned on, the order of the training data will be shuffled, thus making the time factor useless in the time-series regression analysis (the time factor is not being considered a parameter if we shuffle the order of the data). After undergoing various testing using various parameters, the "Relu" activation and "Adamax" optimizer have been proven to perform the best as they give a good rise in probability before the actual event happening. By performing Hyperparametric Tuning using Bayesian Optimization, we were able to conclude in this work that 128 neurons with 1 hidden layer and 100-time delay were the most well-performing hyperparameter settings for this specific case study. We could get our current results by combining the hyperparameters along with the previous activation and optimizer.

In this work, the long strong term memory (LSTM) specifically was used, which is one of the best models commonly used for time-series regression. LSTM allows us to capture the long-term dependencies of well using its built-in mechanism, which overcomes the limitation of traditional neural networks. LSTM is also able to utilize the complexity of several components such as memory cells, input gates, and output gates, which allows LSTM to manage the given information effectively well for prediction. LSTM has also been shown to be capable of capturing and modeling long sequences and time-series data with complex patterns and irregularities, which can be very suitable for our work.

In lieu of LSTM, it is possible to use the Focused Time-Delay Neural Network (FTDNN); however, FTDNN does not have the built-in mechanism to handle long-term dependencies. Moreover, FTDNN also is not as flexible and effective in handling delayed inputs and complex sequences, resulting in lower performance than by LSTM in performing time-series regression. Therefore, FTDNN was not considered in this research.

13.6 XGBRF/KNN/GAUSSIAN NB

For sake of performance comparison with SNN and DNN, some other ML models were used in this work consisting of Extreme Gradient Boosting Random Forest Regressor (XGBRF), K-Nearest Neighbor (KNN), as well as Gaussian Naive Bayes (Gaussian NB). These models do not have the time factor, which is one of the most essential parameters in our analysis. Some of these models are designed to shuffle the order of the training data, which help the models learn the pattern and subtle changes in variables, which is useful for normal regression and classification tasks. These models do not either consider past variables (lack of temporal dependency), meaning that these models treat each data point as independent, which, as mentioned before, make these models not suitable for time-series analysis.

Although it is worth noting that XGBRF may be used in some time-series tasks, it requires complex data preprocessing, feature engineering, and major adjustments since XGBRF is not specifically designed for these kinds of tasks. The KNN and Gaussian NB do not have internal memory to retain information about past observations. Moreover, these models have some assumptions that must be fulfilled by the data for these models to perform well, which requires much more data preparation and preprocessing. Despite the limitations of these models making them not worth to be explored deeper due to time and cost limits, their performance was checked.

13.7 RESULTS AND DISCUSSION

Having applied different ML models to the data, Table 13.1 shows the performance results of different ML models used in this work. Figures 13.6–13.15 show time-based kick probability graphs of different models used for a test example.

As for the SNN model, good performance was found when one hidden layer was considered with 128 neurons. The activation was performed by the hyperbolic function "tanh", and the optimizer was "adam". Figure 13.6 shows that kick prediction is not reliable due to many inconsistencies in kick probability (y axis) or unusual rise in probability of kick, way before the kick happens and the sudden rise in kick probability to 100% without any warning, which all indicate false predictions. Therefore, it is concluded that Static Neural Network is not a suitable model for use in our case.

TABLE 13.1
Results of Performance of Different ML Models

Model	No. of Neuron	Lag Time	Activation Function	Optimizer	RMSE (0.001)	False Warning	Gradual Spike (Symptoms)	Computation Time
SNN	128	–	Tanh	Adam	153	Yes	No 50 minutes early	18.7 sec
DNN-1	256	100	Relu	Adamax	43	0	Yes 56.91 minutes early	235.18 minutes
DNN-2 (Best Model)	**128**	**100**	**Relu**	**Adamax**	**40**	**0**	**Yes** **48.38 minutes early**	**43 minutes**
DNN-3	128	100	Relu	Adam	64	2	No 48 minutes early	42.63 minutes
DNN-4	128	100	Relu	Adam	94	7	No 48 minutes early	27.78 minutes
DNN-5	128	140	Sigmoid (worst)	Adamax	410	1	NA NA	90.76 minutes
XGBRF	–	–	–	–	26	0	No 50 minutes early	0.8 sec
KNN	–	–	–	–	22	0	Yes 40 minutes early	0.4 sec
Gaussian NB	–	–	–	–	93	0	No 50 minutes early	0.1 sec

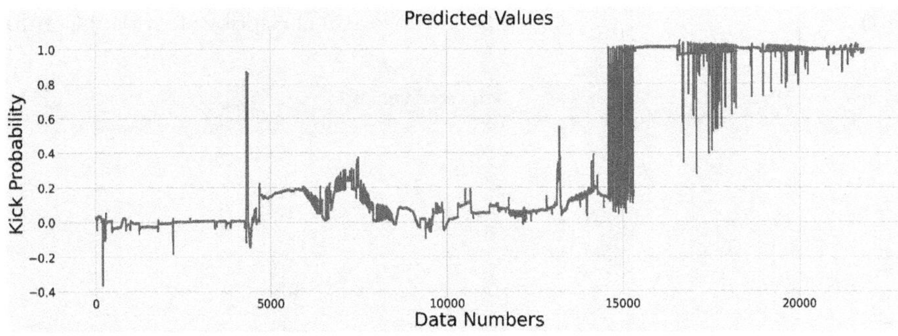

FIGURE 13.6 Modeling result by the SNN showing kick probability versus time.

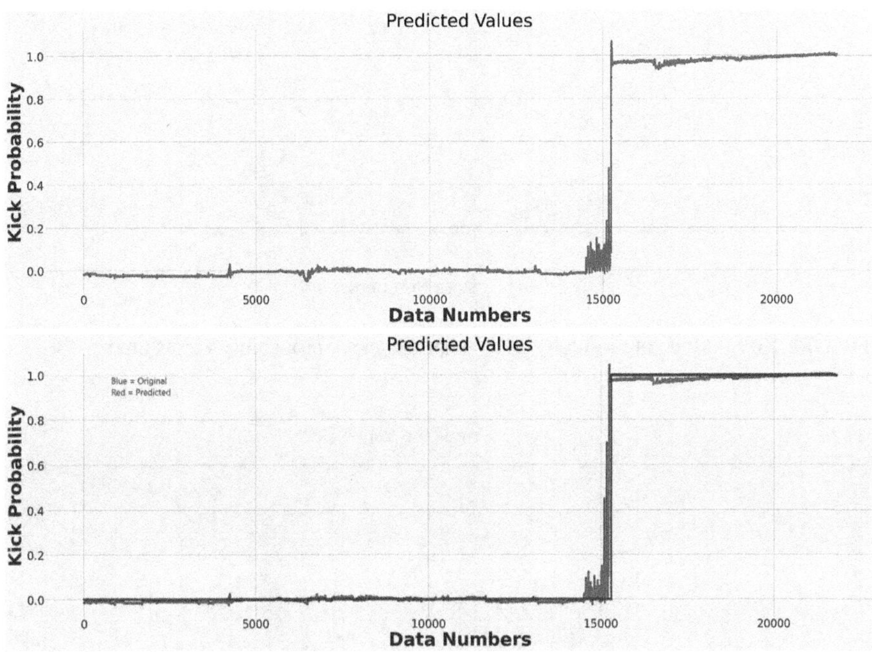

FIGURE 13.7 Modeling result by DNN-1 showing kick probability versus time.

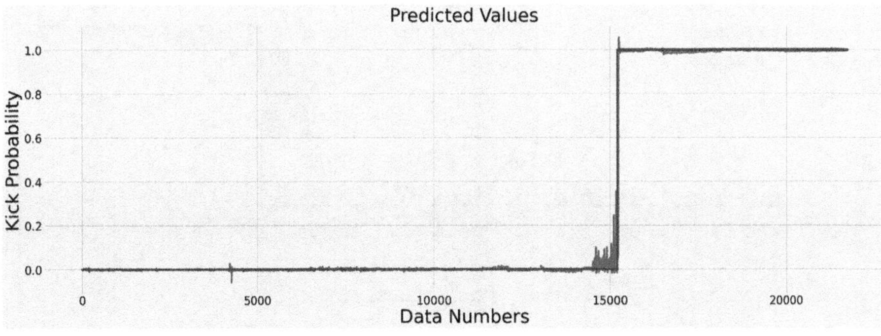

FIGURE 13.8 Modeling result by DNN-2 showing kick probability versus time.

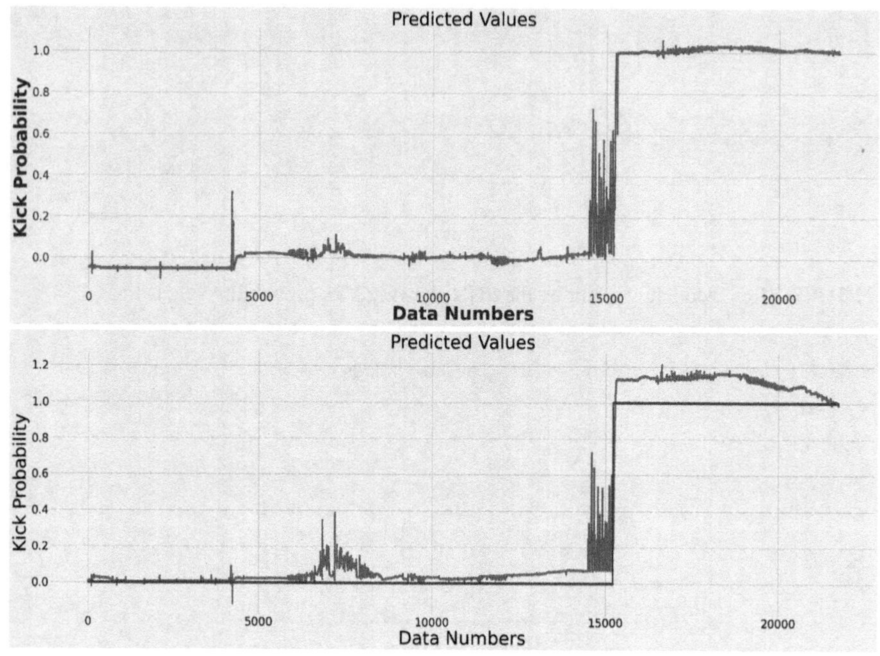

FIGURE 13.9 Modeling result by DNN-3 showing kick probability versus time.

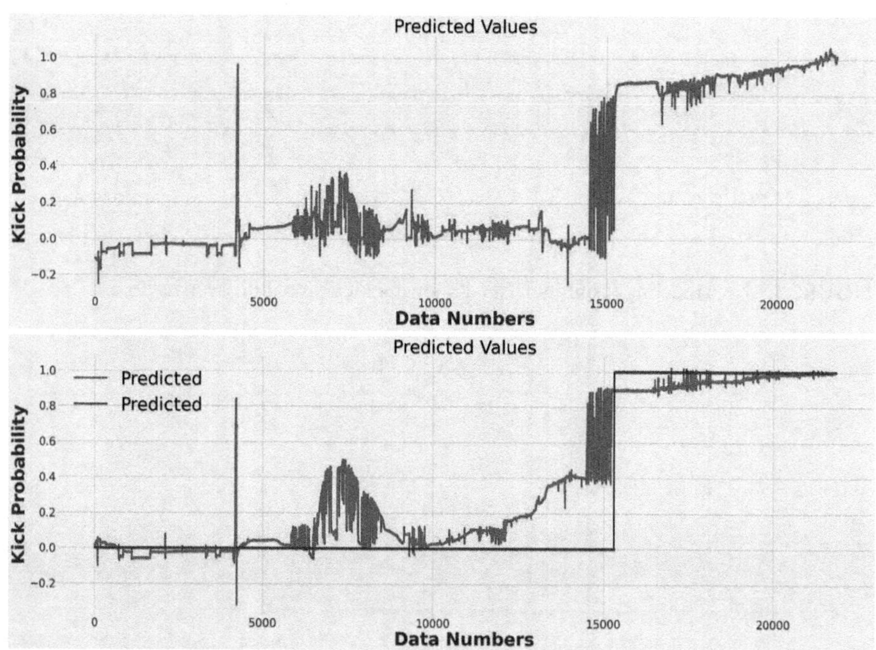

FIGURE 13.10 Modeling result by DNN-4 showing kick probability versus time.

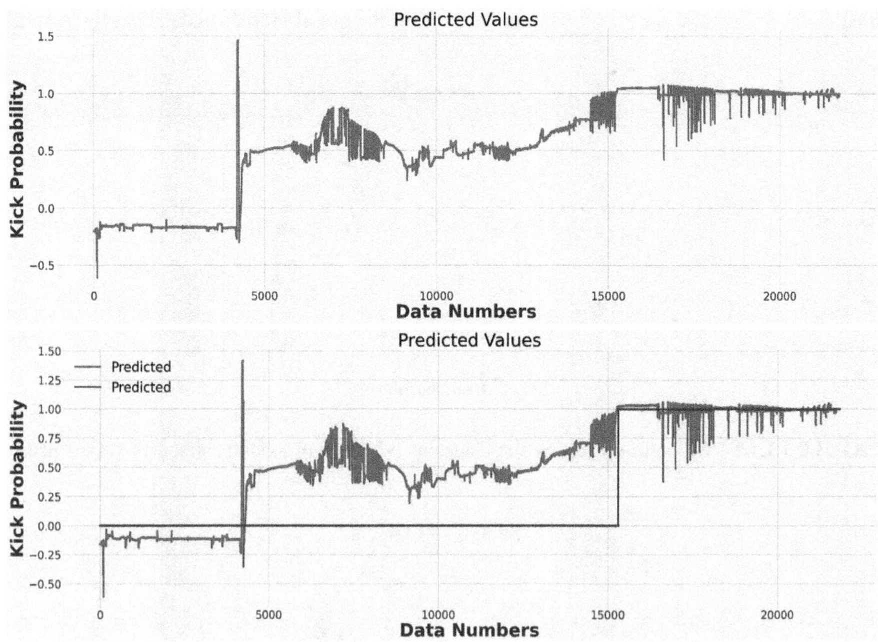

FIGURE 13.11 Modeling result by DNN-5 showing kick probability versus time.

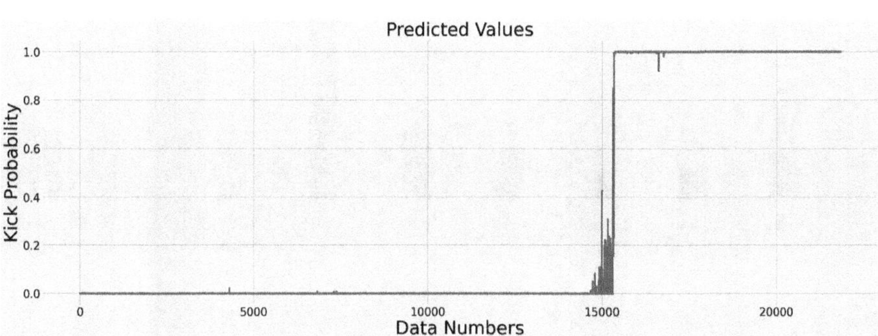

FIGURE 13.12 Modeling result by the XGBRF showing kick probability versus time.

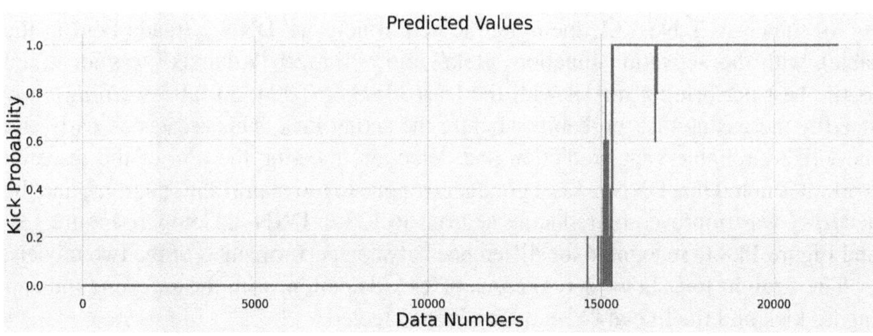

FIGURE 13.13 Modeling result by the KNN Regressor showing kick probability versus time.

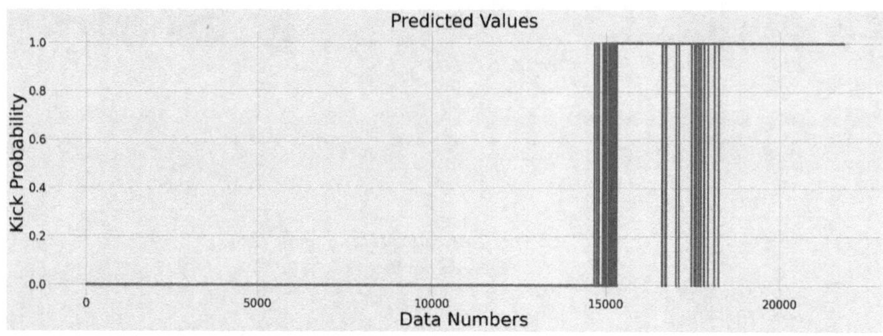

FIGURE 13.14 Modeling result by the Gaussian NB showing kick probability versus time.

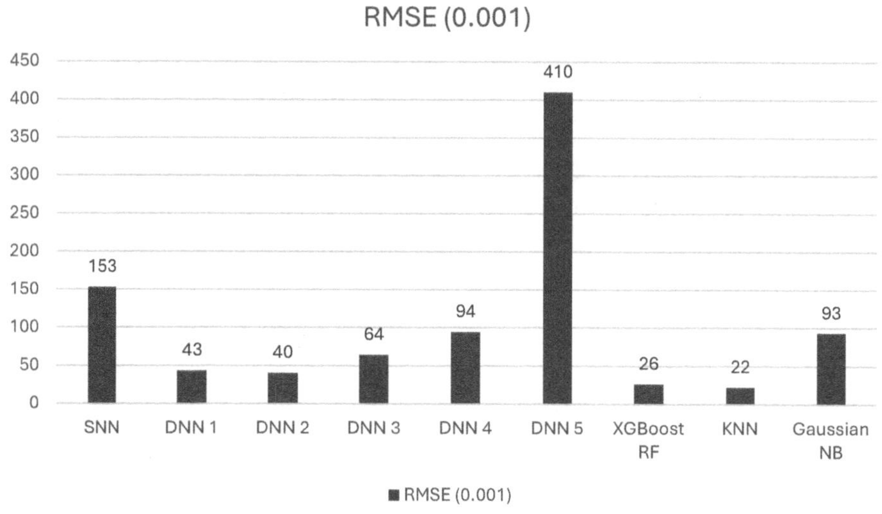

FIGURE 13.15 Comparison of RMSE (errors) of different AI models used for kick prediction and detection.

As shown in Table 13.1, the dynamic neural network DNN-2 (made bold in the table), with the activation function "Relu" and optimizer "Adamax" was identified as the best performing model with the least MSE of 0.004, no false warnings, and steadily increasing kick probability before the actual kick. This creates an early and most likely reliable kick prediction and detection, meeting the aim of the research work. It is noted that DNN-1 has a greater error due to too many (256) neurons, and the accuracy was improved by reducing neurons to 128 in DNN-2. Compare Figure 13.7 and Figure 13.8 to recognize the difference between performances of the two models.

The straight line shows actual cases of kick detection, with the zero line indicating no kick and the line at "1" showing kick detected.

The straight line shows actual cases of kick detection, with the zero line indicating no kick and the line at "1" showing kick detected.

The straight line shows actual cases of kick detection, with the zero line indicating no kick and the line at "1" showing kick detected.

13.8 CONCLUSIONS AND RECOMMENDATIONS

Five types of AI models were used for kick prediction and detection, consisting of static and dynamic artificial neural networks (SNN and DNN), extreme gradient boosting random forest (XGBRF), K Nearest Neighbor (KNN), and Gaussian Naive Bayes (NB). The following conclusions were drawn:

1. All the models predicted the symptoms of forthcoming kicks from 40 to 57 minutes earlier than the actual kick occurrence; however, only in the DNN model, a realistically steady gradual increase in kick probability was observed as it was approaching the actual kick time.
2. Having dynamic characteristic, the DNN model showed good performance with the highest accuracy or lowest RMSE error of 0.004, contributing to a gradual and steady rise of probability nearing the actual kick event.
3. The computation time of the DNN model is 43 minutes using a conventional PC, which is considerably larger than the static type models. This would jeopardize the practicality of the model for field use unless using stronger computers or cloud-based computation is viable.
4. Due to the limited number of kicks in the data sets, some overfitting was observed. Therefore, it is recommended in future works to include drilling data with more kick occurrences.

REFERENCES

Alouhali, R., Aljubran, M., & Gharbi, S., 2018. Drilling Through Data: Automated Kick Detection Using Data Mining. Paper presented at the *SPE International Heavy Oil Conference and Exhibition*, Kuwait City, Kuwait, December. https://doi.org/10.2118/193687-MS

Ashena, R., Bahreini, H., & Lotfi, M., 2023. Statistical Analysis of Past Kicks and Blowouts Occurred in a Middle Eastern Oilfield. *Journal of Petroleum Exploration and Production Technology*. https://doi.org/10.1007/s13202-023-01664-9

Ashena, R., Moghadasi, J., & Ghalambor, A., 2010. Neural Networks in BHCP Prediction Performed Much Better Than Mechanistic Models. Paper presented at the *International Oil and Gas Conference and Exhibition in China*, Beijing, China. https://doi.org/10.2118/130095-MS

Ashena, R., Rabiei, M., Rasouli, V., et al., 2021, May. Drilling Parameters Optimization Using an Innovative Artificial Intelligence Model. *ASME. Journal of Energy Resources Technology*, 143 (5), 052110. https://doi.org/10.1115/1.4050050

Ashena, R., & Thonhauser, G., 2015. Application of Artificial Neural Networks in Geoscience and Petroleum Industry. In Cranganu, C., Luchian, H., & Breaban, M. (eds) *Artificial Intelligent Approaches in Petroleum Geosciences*. Cham: Springer. https://doi.org/10.1007/978-3-319-16531-8_4

Curina, F., Abdo, E., & Mustapha, H., 2022. Reducing Human Error Through Automatic Kick Detection, Space Out and Dynamic Well Monitoring. Paper presented at the *IADC/SPE International Drilling Conference and Exhibition*, Galveston, TX, March. https://doi.org/10.2118/208784-MS

Feng, J., Tse, C. K., & Lau, F. C. M., 2003. A Neural-Network-Based Channel-Equalization Strategy for Chaos-Based Communication Systems. *IEEE Transactions on Circuits and Systems I: Fundamental Theory and Applications*, 50 (7), 954–957.

Holand, 2017. *Loss of Well Control Occurrence and Size Estimators, Phase I and II.* Report No. ES201471/2. Office/Division Program, TAP, Project No. 765. Category, Deepwater. https://www.bsee.gov/research-record/loss-well-control-occurrence-and-size-estimators

Roman, J., & Jameel, A., 1996. Backpropagation and Recurrent Neural Networks in Financial Analysis of Multiple Stock Market Returns. *Proceedings of the Twenty-Ninth Hawaii International Conference on System Sciences*, 2, 454–460.

Shi, X., Zhou, Y., Zhao, Q., et al., 2019. A New Method to Detect Influx and Loss During Drilling Based on ML. Paper presented at the *International Petroleum Technology Conference*, Beijing, China. https://doi.org/10.2523/IPTC-19489-MS

Sinha, N. K., Gupta, M. M., & Rao, D. H., 2000. Dynamic Neural Networks: An Overview. *Proceedings of IEEE International Conference on Industrial Technology 2000*, vol. 2, pp. 491–496 (IEEE Cat. No. 00TH8482), Goa, India. https://doi.org/10.1109/ICIT.2000.854201

Unrau, S., Torrione, P., Hibbard, M., et al., 2017. ML Algorithms Applied to Detection of Well Control Events. Paper presented at the *SPE Kingdom of Saudi Arabia Annual Technical Symposium and Exhibition*, Dammam, Saudi Arabia. https://doi.org/10.2118/188104-MS

Yalamarty, S. S., Singh, K., & Kamyab, M., 2022. Early Detection of Well Control Kick Events by Applying Data Analytics on Real Time Drilling Data. Paper presented at the *IADC/SPE International Drilling Conference and Exhibition*, Galveston, TX, March. https://doi.org/10.2118/208770-MS

Yang, J., Sun, T., Zhao, Y., et al., 2019. Advanced Real-Time Gas Kick Detection Using ML Technology. Paper presented at the *29th International Ocean and Polar Engineering Conference*, Honolulu, HI. https://onepetro.org/ISOPEIOPEC/proceedings-abstract/ISOPE19/All-ISOPE19/ISOPE-I-19-602/21663

Yin, Q., Yang, J., Borujeni, A. T., et al., 2019. Intelligent Early Kick Detection in Ultra-Deepwater High-Temperature High-Pressure (HPHT) Wells Based on Big Data Technology. Paper presented at the *29th International Ocean and Polar Engineering Conference*, Honolulu, HI.

14 IoT Infrastructures for Well Control

Stefan Erakovic and Asad Elmgerbi

ABBREVIATIONS

AI	Artificial Intelligence
AP	Annular Preventer
BHA	Bottomhole Assembly
BOP	Blowout Preventer
DL	Deep Learning
E&P	Exploration & Production
IIoT	Industrial Internet of Things
IoT	Internet of Things
IoT-ECVSs	Internet of Things Edge Computer Vision Systems
ML	Machine Learning
RAM	Ram Preventers
4IR	Fourth Industrial Revolution
LWD	Logging While Drilling
MWD	Monitoring While Drilling
PPE	Personal Protective Equipment
WC-SOS	Well Control Space-out System

14.1 INTRODUCTION

Internet of Things (IoT) uses a system of interconnected smart sensors and actuators that enable sharing, processing, and analysis of data within the system and over the Internet to improve manufacturing and industrial processes. Communication between multiple sensors is ensured by embedding the system with special network streams. IoT influences various information tasks and promotes the technological development of industries (Temizel et al., 2021).

IoT implies that the used devices and the information they process serve the purposes of an industrial entity, not the end consumer. Industry Internet Consortium, a well-known noncommercial organization, defines IoT as "the execution of intelligent industrial operations using data analytics to achieve transformative business results" (Canvan, 2022). Broadly speaking, IoT refers to the automated and interconnected use of devices, machines, and sensors in

DOI: 10.1201/9781003473770-14

industrial applications. Through big data processing and continuous machine learning, IoT can improve the efficiency and reliability of operations in various industries without explicit human–machine interaction. This type of analytics also aims to create new business models and revenue streams based on big data analytics.

Typical industrial sectors are characterized by data analysis with predictions occurring after big data is uploaded to centralized servers. However, data transfer delays and network pressures create server-based systems and bandwidth problems. The solution to the limitations of server-based systems has been edge computing. Edge computing is different from a centralized system because data analysis and storage take place at the edge, that is, near the data sources. The use of edge computing and IoT aims to solve the problems of traditional infrastructures, including data storage capacity, data transfer time delays, and processing power; thereby, edge computing improves the performance of IoT platforms (Yu et al., 2017). Thus, edge computing is advantageous from a security point of view since sensitive data analysis takes place directly on-site without the need to transmit all data over the net. At the same time, if certain sensitive information needs to be stored on a central server, the IoT edge platform can selectively send only the necessary items. This approach prevents unexpected data leakage and reduces the bandwidth required for data transmission (Alsheikh et al., 2021; A. Magana-Mora et al., 2021).

IoT also uses special systems that analyze images or video recordings and thereby create computer vision. This field is actively developing and is about the computer analysis of images or videos to look for patterns or sequences of information in order to form a description of what is being seen. Because computer vision technology deals with large amounts of information, it requires high computing power. Through the use of IoT computing edge platforms, computer vision performs analysis at higher speeds and is used in various industrial applications. Thus, camera-based computer vision has found application in video surveillance, communications, public safety, customer analysis, traffic surveillance, etc. (Andriulo et al 2024).

Cameras are also used on drilling rigs – however, in most cases for manual surveillance. The main working area on a drilling rig is the drill floor, where the driller and crew monitor drilling operations and assemble pipe connections, bottomhole assembly, and drill bit. Cameras are needed to monitor rig site operations and ensure the safety of workers in red zones and areas performing critical and hazardous rig operations. These areas include monkey board, mud pumps, cranes, and boat operations on offshore rigs. However, there are cameras not only for monitoring the drilling site but also as a data source for an intelligent analysis of drilling operations and processes. The mentioned IoT edge platform was installed on drilling rigs to provide real-time processed information to the driller and other interested parties on the rig. In this way, possible well control problems, lost circulation, and stuck pipe incidents can be alerted in advance so that the order of work processes can be changed in advance. Manual or automatic control and actuation of drilling dynamics and rheology systems can be carried out to optimize and automate drilling operations (Alsheikh et al., 2021).

14.1.1 The Power of 1%

IoT has the potential to reduce operating costs. On the Industrial Internet, there is what is known as the power of the 1%. This means that significant savings in operating costs require savings from the industrial Internet of only 1% in many industries. By comparison, an aviation fuel savings of 1% per year mean savings of $30 billion. Likewise, fuel savings for gas generators in power plants of 1% yield operational savings of $66 billion. In addition, a 1% annual reduction in capital expenditures on equipment would yield about $90 billion in the oil and gas industry. The same is valid for agriculture, transportation, and health care. Thus, in most industries, a modest improvement of 1% would significantly increase the return on capital and operating costs incurred in implementing the Industrial Internet (Gilchrist, 2016).

14.2 CURRENT DATA SYSTEM VERSUS IOT EDGE DATA SYSTEM ON DRILLING RIGS

14.2.1 Current Drilling Rig Data System

Figure 14.1 shows a typical rig's data system. To facilitate the driller's job, real-time sensors are used both at the surface and in the downhole to capture data and provide up-to-date information about the well being drilled. Surface sensors are wired in real time, while downhole sensors read changes in fluid pressure pulses used to transmit data from the bottomhole to the surface while drilling. The collected data is transmitted to a central server via satellite link to the company's headquarters, where geologists, drilling engineers, and reservoir engineers can use and evaluate the data. While the infrastructure in use is well established, there are several problems with satellite systems, including the high cost of satellite infrastructure and the ability to transmit only low-frequency data (typically less than 1 Hz), which leads to transmission delays and interruptions. In addition, this infrastructure does not currently support intelligent algorithms such as artificial intelligence (AI), machine learning

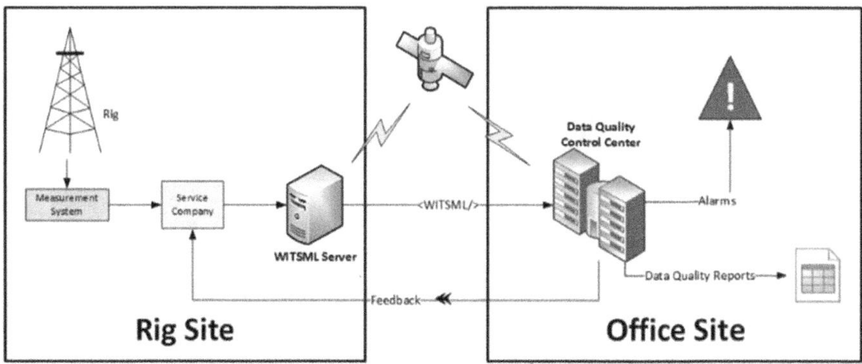

FIGURE 14.1 Current rig data system (Alsheikh et al., 2013).

(ML), and deep learning (DL) (Mohammad Arghad 2013). Implementing new sensors is essential for improving measurement accuracy as well as generating new ideas and measurements in well analytics. However, rapid and seamless integration of new sensors to improve drilling efficiency is not always easy, as introducing new sensors into a system with outdated standards becomes an obstacle to developing more accurate algorithms and predicting wellbore problems (Alsheikh et al., 2021; Macpherson et al 2013).

14.3 IOT EDGE DRILLING RIG DATA SYSTEM

Using edge analytics with an IoT platform leads to a change in the rig system in terms of real-time analysis of downhole data. Figure 14.2 shows that, in an IoT edge platform, sensors transmit data using wired or wireless communication protocols. The data is then collected at the edge server, where it is all grouped, filtered, and processed. Data analysis is performed on the edge server based on ML, AI, and DL models to identify patterns and relationships between the raw data. This process transforms the raw data into useful information for understanding and prediction. Thus, the driller will be able to make real-time adjustments to the drilling process and make decisions. At the same time, key drilling metrics, such as efficiency, critical data, and any other relevant information, are accumulated and updated in the cloud computing system. Likewise, the data from the edge server and the cloud computing system are stored at the company's headquarters, where they are stored in a database. The analyzed data can also be used by other E&P sectors to update workflows and models. Using edge analytics with an IoT platform ultimately enables data to be used in an integrated and real-time manner rather than in a single cloud computing system (Shi et al., 2016).

For sensors to accurately analyze an input signal, they must be pre-calibrated for type, resolution, range, and other expected parameters. This process becomes more complicated as the number of input signals increases, but pluggable sensors solve this problem with an automatic self-configuration process. The integration of pluggable sensors and data collection platforms at the IoT edge allows useful information to be obtained during data collection (Alsheikh et al., 2021; Meziou et al., 2020).

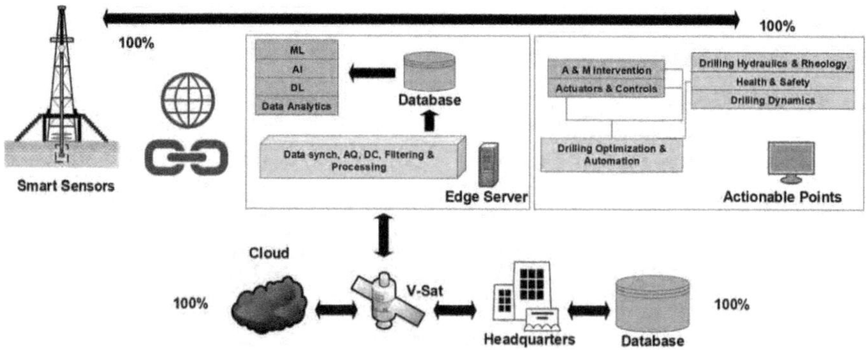

FIGURE 14.2 IoT edge rig data system (Alsheikh et al., 2021).

14.4 IOT-ECVS FOR WELL CONTROL

14.4.1 General

Well control is a well-planned process and includes the following steps:

1. Stopping the rotation of the drill-string assembly.
2. Pulling the drill bit away from the bottom of the formation and taking it aside.
3. Stopping the drilling mud pumps.
4. Well shut-in or closure.
5. Killing the well by adding weighting additives to the drilling fluid in some severe cases.

Spacing-out is aimed at pulling the drill-string assembly to place the thickest part (tool joint or pipe joint) several feet above the rig floor. In most cases, the distance between the drill-pipe's tool joint at the surface and the tool joint in the hole is greater than the length of the Blowout Preventer (BOP), as shown in Figure 14.3. The spacing-out ensures that the rams in the BOP will close around the body of the pipe which has the smallest diameter in the drill-string (Alsheikh et al., 2021).

During a well control incident, the driller's responsibility is to ensure that the BOPs are engaged and sealed properly (Han, 2015). The primary use of BOPs is to prevent mud or hydrocarbon flow through the annulus section (between the drill-pipes and casing/borehole) to the surface. The BOP stack consists of an annular preventer (AP) on the top with an underlying set of ram preventers (equipped with plungers) stacked on top of each other, see Figure 14.3. An AP, which is a flexible donut-shaped rubber seal, allows the drill-pipes to rotate and move without breaking the seal by installing the APs on the drill-pipes of different sizes. Alternatively, to seal the well, ram BOPs use plungers which are also rubber but coated with steel. Rams can either seal around the drill-pipe (pipe rams) or even cut through it (shear rams' type).

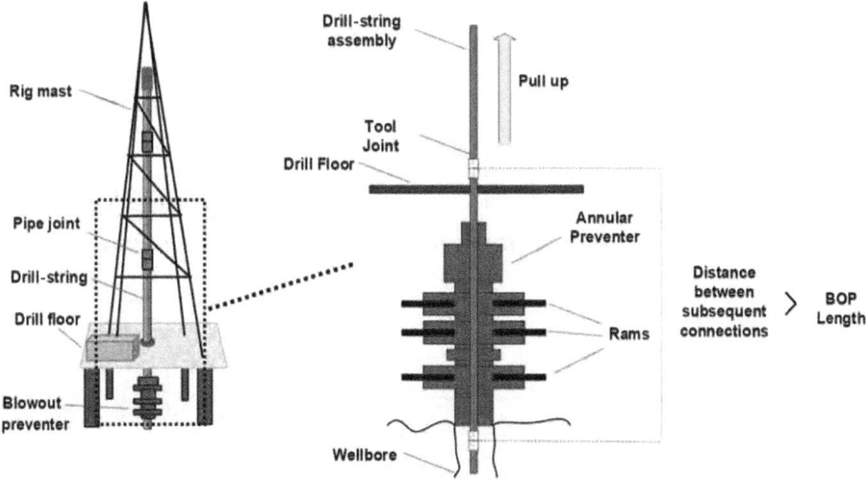

FIGURE 14.3 Drill-pipe space-out illustration in the blowout preventer (Alsheikh et al., 2021).

14.4.2 Well Control Space-Out System (WC-SOS)

Proper drill-pipe spacing-out is one of the key parameters for ensuring accurate well closure during an incident and maintaining BOP integrity. However, if the borehole is closed at the tool joint (i.e., in the thickest part of the drill-string), proper borehole sealing cannot be guaranteed. When a well control incident occurs, the rig crew must assess the situation to take further specific actions to resolve the problem as quickly as possible. It is important to note that the potential for human error arises because the rig crew is forced to follow a manual, time-consuming process while a well control incident is critical and can escalate quickly. These conditions create the potential for improper well sealing. To eliminate human error, a solution should be presented. WC-SOS is a pilot IoT-ECVS consisting of smart, waterproof, high-resolution wireless cameras for image/video capture and fog computing hardware. WC-SOS can be used as a part of rapid well-closure solution by using sensors and system AI/ML-based software for image/video processing and intelligent analytics.

Figure 14.4 shows that the camera should be mounted at some distance from the rig floor and pointed toward the drill-string assembly. The recorded data from the camera is sent to an edge server located in the foreman's office and/or in the driller's booth. The data is preprocessed using image-processing techniques to enhance and trim the images. This data is then fed into a DL model, which performs data analysis on the edge server to precisely locate the pipe joint on the drill-string. Thus, the post-processing logic is combined with DL models to detect pipe connections to predict the location of the next pipe connection within the drill-string. The advantage of WC-SOS is the continuous collection, validation, and enrichment of recorded data

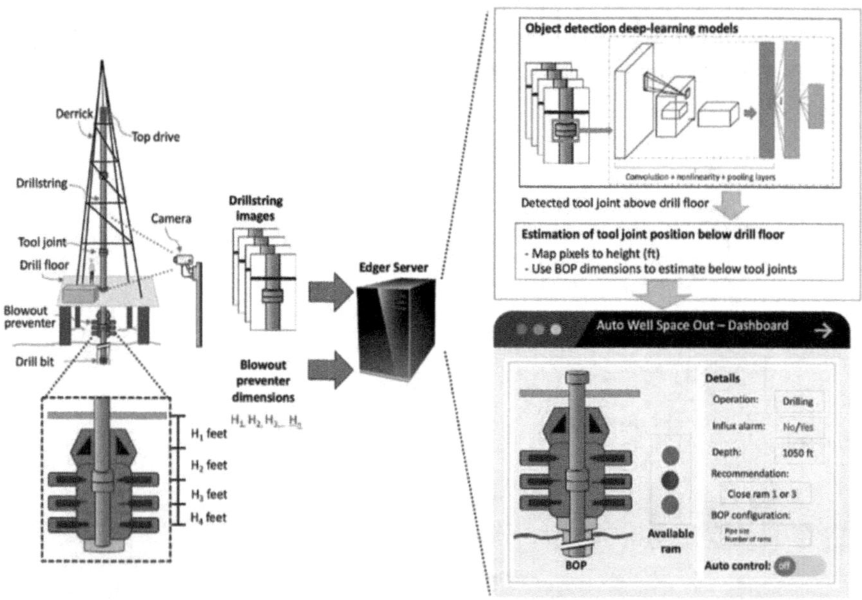

FIGURE 14.4 The Well Control Space-Out System (WC-SOS) (A. Magana-Mora et al., 2021).

to identify patterns and create models. The collected patterns show the position of the tool joints relative to the BOP in real time on the visual analysis panel so that the BOP rams do not close against the pipe joint.

For instance, Figure 14.4 illustrates that the pipe connection is located at RAM-1 of the Blowout Preventer (BOP). In the event of a kick flow incident, the driller has the option to activate either RAM-2 or RAM-3 to control the situation and prevent further escalation. However, WC-SOS, which is already being tested in the field, can help the driller make more accurate, safer, and faster decisions during an incident and under time pressure. Thus, WC-SOS plays a vital role in significantly reducing the likelihood of well control escalation by:

1. Eliminating well control space-out errors by increasing well shut-in probability.
2. Eliminating extra time to space-out as well as hardware (BOP or pipe) damage through incorrect BOP closure.

By automatically detecting tubing connections during drilling and tripping operations, WC-SOS also provides real-time, concurrent drilling/stripping rate, number of pipe joints in the wellbore, and drilling/stripping/stripping operations (Alsheikh et al., 2021; A. Magana-Mora et al., 2021).

14.4.3 DL Models

Efficient deep learning (DL) models are required for the WC-SOS technology. Figure 14.5 demonstrates the methodology used to develop and train the DL models. DL model development includes three (3) phases of training, testing, and validation. The training phase includes collecting and storing video recordings from the drill-string assembly under different drilling conditions and operating scenarios. These scenarios include weather conditions, run-in-hole and pull-out-of-hole of drill-string, and drilling at different speeds and different pipe diameters.

Then, image processing is performed, including cropping the original frames to increase the size of only the necessary data by scaling the pixels and image dimensions. Figure 14.6 shows the different image-processing methods. Next, the VGG image annotation tool is used to mark the pipe connections in the frame manually. This allows the introduction of a coordinate system (x, y) to determine the locations and frames of the tools. These coordinates are the input parameters for training the DL models (Dutta et al 2016:Alsheikh et al., 2021).

Despite the growing interest and progress in applying DL models in many fields and industries, DL model development is challenging due to a large number of tunable parameters. At the same time, it has been proven that an already trained DL model structure can be retrained for another similar problem while achieving accurate results (Kalkatawi et al., 2019).

Pretrained DL models are often used as reference models to provide faster development phase. Figure 14.7 compares different pretrained DL models that were tested on processed drill-string videos. The DL models were tuned to detect pipe connections. The available 7,000 frames were divided into 80% training and 20% for both testing and validation. Figure 14.7 shows the results by several DL algorithms (Alsheikh et al., 2021).

Figure 14.7 shows that "Faster RCNN-ResNet100" DL model had the best performance in average accuracy and recall. However, the "Faster R-CNN-ResNet50" model also showed similar results, although it has a much simpler model structure (50 layers instead of 100). Since the R-CNN-ResNet50 also provides sufficiently

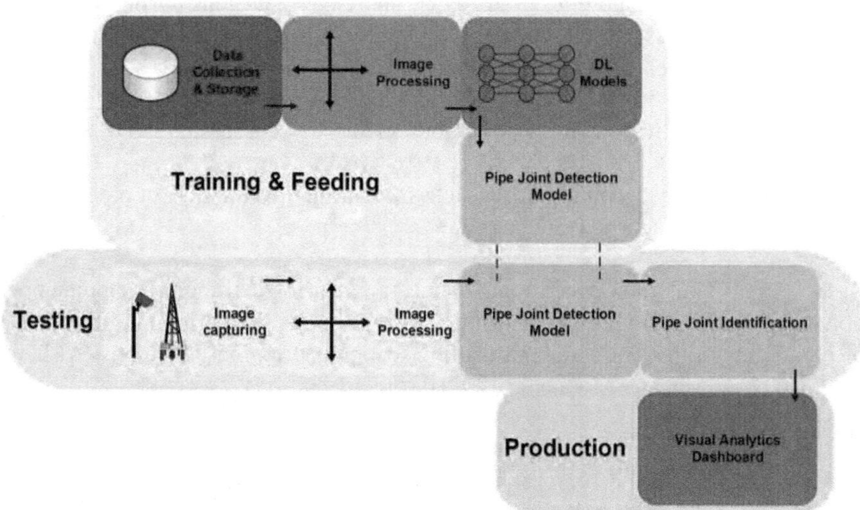

FIGURE 14.5 Methodology of developing and training the DL model used in the WC-SOS (Alsheikh et al., 2021).

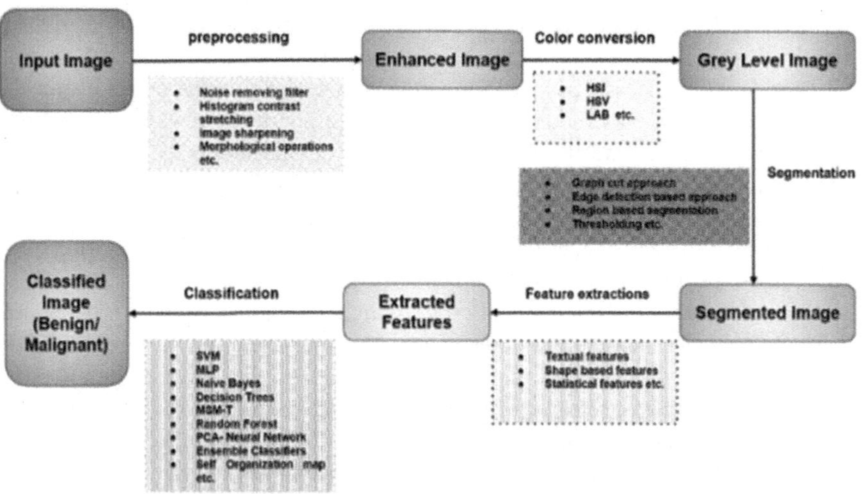

FIGURE 14.6 General steps involved in the image-processing segment of the DL model (Shyamali Mitra et al., 2022).

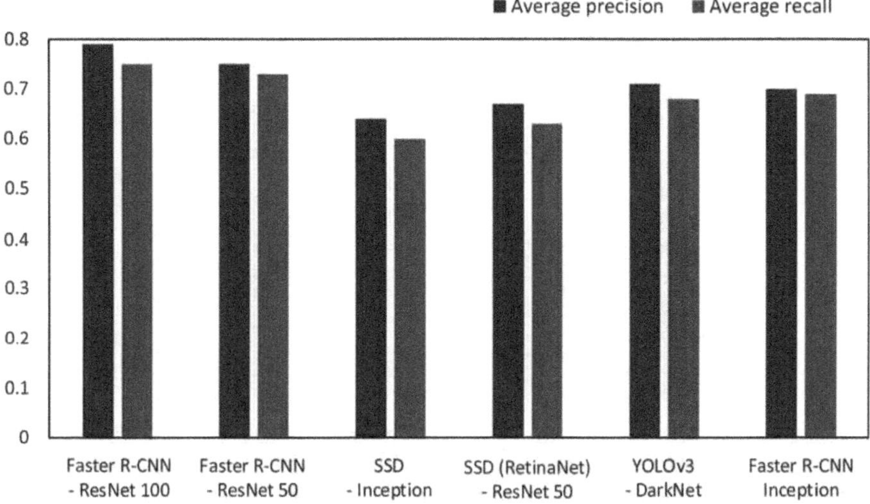

FIGURE 14.7 Average precision and recall results for DL models (Alsheikh et al., 2021).

fast data processing, it is suitable for real-time operation; therefore, it was chosen as the basis for model deployment. It is also worth noting that manual frame marking is required only in the training phase. In the next two stages (testing and validation/deployment), the model worked only with the processed and cropped frames and gave the coordinates of the predicted tool joint and the deviation value for each frame.

At the end, post-processing logics were also considered to further improve the accuracy of the model. The post-processing logics included the ability to visualize two pipe joints in one frame. An example of post-processing might be predicting the presence of a tool joint in the middle of the frame at time t and then predicting no tool joint in the middle of the frame at time t + 1. In this way, the discrepancy in the predictions may indicate a false positive or false negative result for time moments t and t + 1 (Alsheikh et al., 2021).

14.4.4 Shale Shaker Analysis

A warning sign of kicks is angular-shaped cuttings. Especially in development wells, kicks are also anticipated at some specific formations or lithologies. Thus, determining lithology during the drilling process is important, which depends directly on human (rig-site geologist) judgment.

Initially, the drilling fluid is pumped through the drill-string assembly during drilling and exits through the outlet pipe from the annulus into the shale shakers. The drilling fluid is filtered through a special screen mesh to remove solids and cuttings. After filtering in the shale shaker, the fluid is returned to the drill-string by pumps. However, the filtered cuttings are used to determine formation type to be drilled. Geologists manually determine cuttings lithology on the rig, which means that the final determination depends directly on the geologists' experience and skills.

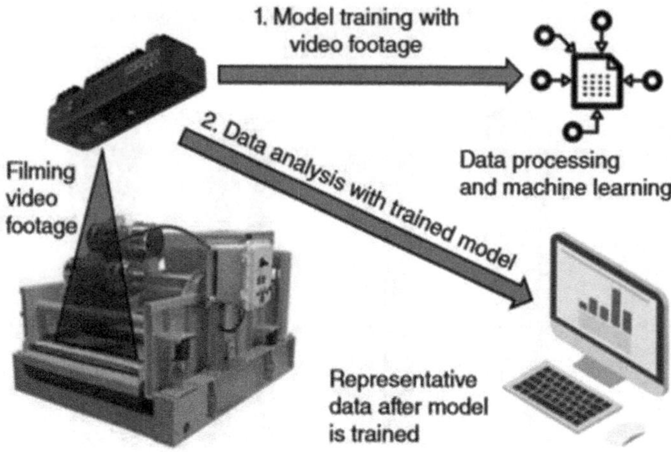

FIGURE 14.8 Analysis of fluid flowing through the shale shaker as segments from a camera view point (Carpenter, Chris 2023).

Thus, there may be human error in determining the lithology. This could happen due to the limited time required to analyze the data, which does not allow real-time decision-making. A potential solution to these problems can be the application of IoT-ECVS on shale shakers. IoT-ECVS can perform automated, real-time analytical operations on the mud coming out of the wellbore. Figure 14.8 shows that cameras can evaluate segments of fluid flowing through a shale shaker. The development of automated anomaly detection technology will, in turn, reveal patterns for evaluating the movement of fluids, solids, collapses, and rock debris coming out of the wellbore. DL models would use an image-recognition system to estimate flow rates and the shape of cutting segments (Ripperger et al 2022).

This approach can increase the likelihood of detection of stuck pipe which contributes to mitigating possible consequences of the incident, as well as detecting preliminary signs of lost circulation. In addition, IoT-ECVS would allow evaluating the effectiveness of well cleanup by comparing actual and expected results after the cuttings are removed from the surface (Alsheikh et al., 2021).

14.4.5 Crew Safety

Since a drilling rig is a place with many potential hazards, safety is always of paramount importance; thus, and the health and safety of workers should be constantly being evaluated. The IoT-ECVS can also be used to monitor threats to the health and life of workers on the drilling rig in real time, including warnings of possible dangers during work. The IoT-ECVS can also warn of potential incidents, accidents, and the safety of the rig crew in terms of personal protective equipment (PPE) and during well control events (Shirkhorshidi et al 2023).

One of the main benefits of using IoT-ECVS is tracking personnel in red zones, as shown in Figure 14.9, and assessing unsafe worker behavior. Infrared cameras can assess the body temperature and general well-being of rig personnel (Javay et al 2024).

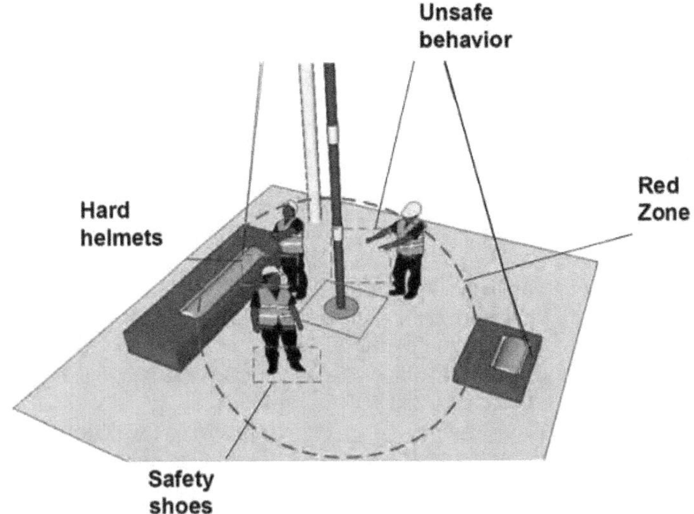

FIGURE 14.9 Monitoring of rig crew working on the rig floor inside the red zone. Classification of PPE and unsafe behavior of working personnel (Alsheikh et al., 2021).

14.5 CONCLUSION

1. The major goal of this chapter was to describe the broad potential and practical application of IoT in the drilling well control.
2. Utilizing edge analytics with an IoT platform leads to a change in the rig system in terms of real-time analysis of downhole data. WC-SOS can be used as a rapid space-out and well-closure solution. The system software is AI/ML-based for image/video processing and intelligent analytics. Efficient DL models have been developed for WC-SOS technology. DL model development included three phases of training, testing, and production.
3. IoT can be also applied for cuttings analysis and safety analysis.
4. Cameras can gather and analyze data at the edge to determine patterns and develop models to predict and prevent problems faced during drilling operations.
5. It is also required to note that IoT can help drilling companies save billions annually by providing an entirely new approach for data processing based on fast and real-time data analysis with minimum human involvement.

REFERENCES

Alsheikh, M., Gooneratne, C., Magana-Mora, A., et al., 2021. Internet of Things IoT Edge Computer Vision Systems on Drilling Rigs. In *Day 4 Wed*, p. 11282021, December 1. https://faculty.kaust.edu.sa/en/publications/internet-of-things-iot-edge-computer-vision-systems-on-drilling-rigs

Andriulo, F.C.; Fiore, M.; Mongiello, M.; Traversa, E.; Zizzo, V. Edge Computing and Cloud Computing for Internet of Things: A Review. Informatics 2024, 11, 71. https://doi.org/10.3390/informatics11040071

Arnaout, A., Zoellner, P., Johnstone, N., and G. Thonhauser. "Intelligent Data Quality Control of Real-time Rig Data." Paper presented at the SPE Middle East Intelligent Energy Conference and Exhibition, Manama, Bahrain, October 2013. doi: https://doi.org/10.2118/167437-MS

Canavan, L., 2022. *What Is IIoT? The Industrial Internet of Things Primer*. https://blog.iiconsortium.org/2019/09/what-is-iiot-the-industrial-Internet-of-things-primer.html (Last Accessed on May 12, 2022).

Carpenter, Chris. "AI-Driven Permanent Cuttings Monitoring Enables Safer and Faster Drilling." J Pet Technol 75 (2023): 65–67. doi: https://doi.org/10.2118/0223-0065-JPT

Dutta, Abhishek, Gupta, Ankush, and Zissermann, Andrew. 2016. VGG image annotator (VIA). URL: https://www.robots.ox.ac.uk/~vgg/software/via/

Gilchrist, A., 2016. *Industry 4.0: The Industrial Internet of Things/Alasdair Gilchrist*. Berkeley, CA: Apress; New York: Distributed to the Book Trade Worldwide by Springer.

Han, Z., 2015. *Stochastic Modelling for Condition Based Maintenance*. NTNU. https://ntnuopen.ntnu.no/ntnu-xmlui/handle/11250/2351227

Javay, Alexandre, Zaini, Muhamad Zaki, Almana, Abdullah, and Viktor Yurtaev. "Improving Rig Floor Safety by Leveraging Artificial Intelligence Capabilities." Paper presented at the International Petroleum Technology Conference, Dhahran, Saudi Arabia, February 2024. doi: https://doi.org/10.2523/IPTC-23482-EA

Kalkatawi, M., Magana-Mora, A., Jankovic, B., et al., 2019. DeepGSR: An Optimised Deep-Learning Structure for the Recognition of Genomic Signals and Regions. *Bioinformatics*, 35 (7), 1125–1132.

Macpherson, John D., de Wardt, John P., Florence, Fred, Chapman, Clinton D., Zamora, Mario, Laing, Moray L., and Fionn P. Iversen. "Drilling-Systems Automation: Current State, Initiatives, and Potential Impact." SPE Drill & Compl 28 (2013): 296–308. doi: https://doi.org/10.2118/166263-PA

Magana-Mora, A. Michael Affleck, Mohamad Ibrahim, Greg Makowski, Hitesh Kapoor, William Contreras Otalvora, MUSAB A. JAMEA, UMAIRIN, GUODONG ZHAN, CHINTHAKA P. "Well Control Space Out: A Deep-Learning Approach for the Optimization of Drilling Safety Operations," in IEEE Access, vol. 9, pp. 76479–76492, 2021, doi: 10.1109/ACCESS.2021.3082661

Meziou, Amine, Hattab, Oussama, and Jose Meraz. "Advancing BOP Reilability through IoT Technology." Paper presented at the Offshore Technology Conference, Houston, Texas, USA, May 2020. doi: https://doi.org/10.4043/30912-MS

Mohammad Arghad Arnaout, Doctoral Thesis, "Distributed Multi-sensor Fusion System for Drilling Rig State Detection", 2013, https://pureadmin.unileoben.ac.at/ws/portalfiles/portal/2400154/AC11652858n01.pdf

Ripperger, Georg, Peisker, Jörg, Oberschmidleitner, Patrick, Kucs, Richard Josef Werner, and Doris Winkler. "Safer and Faster Drilling through AI Driven Permanent Cuttings Monitoring – An Operator's Approach." Paper presented at the ADIPEC, Abu Dhabi, UAE, October 2022. doi: https://doi.org/10.2118/211759-MS

Shi, W., Cao, J., Zhang, Q., et al., 2016. Edge Computing: Vision and Challenges. *IEEE Internet of Things Journal*, 3 (5), 637–646. https://doi.org/10.1109/JIOT.2016.2579198

Shirkhorshidi, R., Norazman, N., Rosli, M. B., Arriffin, M., and M. Karbasian. "How to Comply Ai Generated Data with Wells Hse Performance Procedures and Requirement in Practice." Paper presented at the SPE/IADC Middle East Drilling Technology Conference and Exhibition, Abu Dhabi, UAE, May 2023. doi: https://doi.org/10.2118/214598-MS

Shyamali Mitra, Nibaran Das, Soumyajyoti Dey, Sukanta Chakraborty, Mita Nasipuri, and Mrinal Kanti Naskar. 2021. Cytology Image Analysis Techniques Toward Automation: Systematically Revisited. ACM Comput. Surv. 54, 3, Article 52 (April 2022), 41 pages. https://doi.org/10.1145/3447238

Temizel, Cenk, Canbaz, Celal Hakan, Aydin, Hakki, Hosgor, Bahar F., Kayhan, Deniz Yagmur, and Raul Moreno. "A Comprehensive Review of the Fourth Industrial Revolution IR 4.0 in Oil and Gas Industry." Paper presented at the SPE/IATMI Asia Pacific Oil & Gas Conference and Exhibition, Virtual, October 2021. doi: https://doi.org/10.2118/205772-MS

Yu, W., Liang, F., He, X., et al., 2017–2018. A Survey on the Edge Computing for the Internet of Things. *IEEE Access*, 6, 6900–6919. https://doi.org/10.1109/ACCESS.2017.2778504

15 Automated Well Control

Bryan Atchison and
Muhammad Abbas Bhatti

15.1 INTRODUCTION

The drilling of challenging wells forces the petroleum industry to come up with innovative solutions in order to explore the remaining reserves in a safe and efficient manner. The industry works in high-pressure, high-temperature (HPHT) regions and in deep-water and narrow pressure margins where the probability of major accident hazards is high. Well control is one of the most important factors during the planning and execution of drilling operations. An uncontrolled kick can result in a loss of well control (LOWC) event resulting in the loss of human lives, environment impact, and assets and reputational damage (Abimbola et al., 2015).

The LOWC is a combination of processes that include technical and human factors. These factors are noticeable in an emergency and high-stress situation (Thorogood et al., 2015). Despite the modern control systems, the driller is still required to identify an influx and manually function well control equipment appropriately.

During all phases of well control, human errors can occur, increasing the risk and cost of the operations. The cyclic nature of the oil and gas industry continues to deplete the industry of many of the experienced resources previously available. The combination of human factors' issues, stretched skills, and loss of resources places severe stress on the driller to perform flawlessly in a high-stress situation (Health and Safety Executive, 2013).

This fact is emphasized by the events of April 20th, 2010, in the Gulf of Mexico. A LOWC on the Deepwater Horizon resulted in 11 fatalities and an estimated 4.9 million barrels of spilt crude. The massive environmental and economic impact on the affected area, including the subsequent litigation and compensation, and the cost of the order of $65 bn brought into sharp focus the huge negative impact of well control mistakes. A series of in-depth investigations indicated that there were several contributing factors, including the operating team on the rig who were the "last barrier" and literally "in the firing line". This team were under stress, were making quick decisions, and were influenced by several extraneous and often competing factors. Poor or misguided decision-making was judged to be the controlling factor in the escalation of the incident (St. John, 2016). The response to this incident resulted in the revision of the US regulatory framework; revised standards for BOP equipment, the development of the Bureau of Safety and Environmental Enforcement (BSEE), and the Bureau of Ocean Energy Management (BOEM); and more focus on the research, technological solutions, and mitigations (National Commission on the BP Deepwater Horizon Oil Spill and Offshore Drilling, 2011). According to the study performed

DOI: 10.1201/9781003473770-15

by the International Association of Oil and Gas Producers (IOGP, 2021) on 172 well control incidents, 71% of the incidents were attributable to human factors.

Automated Well Control technology was developed to bypass traditional well control process and dramatically reduce the human factors risk elements. This is supported by a detailed analysis carried out in 2020, comparing the influence of human factors of traditional well control versus Automated Well Control, showing a reduction in the probability of human error by 94% for blowout and serious well control events (Atchison and Sarpangal, 2022). An independent analysis was performed by Safe Influx on nearly 3,700 incident descriptions involving the loss of primary well control which occurred between 1940 and 2021, derived from eight publicly available incident-sharing databases. The analysis indicates that 46% of the incidents were attributable to human factors, 33% were attributable to organisational issues, and the remaining 21% involved a failure related to technology issues (Gillard and Bhatti, 2022).

15.2 AUTOMATED WELL CONTROL DESCRIPTION

Automated Well Control is an emerging technology in the upstream well construction industry. It is a driller assistive tool and is designed to be installed on a drilling unit with a minimum footprint. The footprint on the rig is of the controller, which is the size of a filing cabinet and installed in a safe area on the drilling rig, and a Human Machine Interface (HMI) screen which sits in front of the driller at the Rig Floor. The controller and HMI screen are connected via a fibre optic cable.

The Automated Well Control system continuously monitors well flow out together with selected drilling and well control equipment I/O data through its Programmable Logic Controller (PLC) server. The topology of the system is illustrated in Figure 15.1 (Atchison, 2021; Atchison and Wuest, 2021).

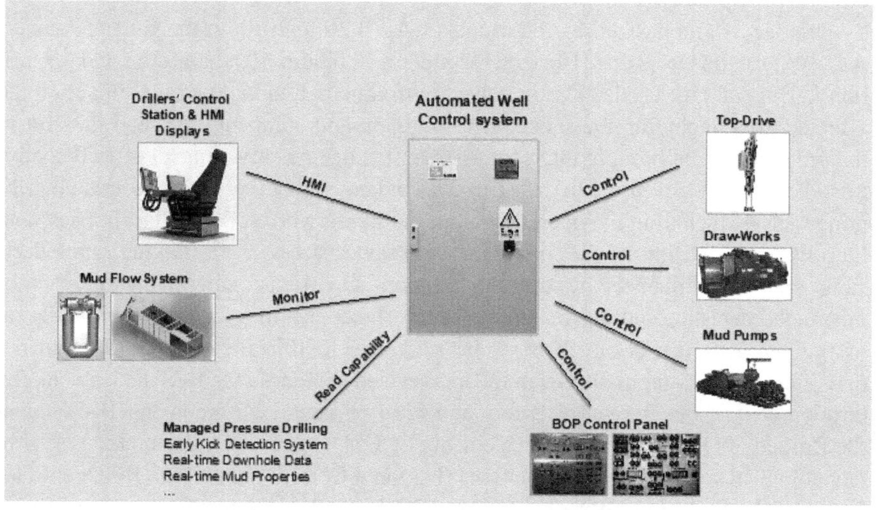

FIGURE 15.1 Automated Well Control Topology (Atchison, 2021).

Automated Well Control uses the outputs from the existing rig equipment and uses an automated process to perform a critical analysis of the appropriate signals. If a well control event is detected, within the predetermined parameters input by the driller, the system takes control of the rig equipment and makes the well safe efficiently.

Human factors cannot be totally eliminated. It is not convenient to eliminate the impact of human factors even with the high degree of automation. As according to Reason (1990), human beings design as well as operate the things (i.e., the system we are involved is man-made and man-run); therefore, human factor always lies in the workplace. Human factor causes can be reduced but are stubbornly resistant to new technology solutions.

When the Automated Well Control system identifies and confirms a self-sustained influx, a message appears on the driller's HMI screen indicating that an automated shut-in procedure is initiated. This prompts the driller to adopt a verification role based on the operation being conducted, while the system automatically performs the necessary sequence of operations with the rig equipment to safely shut in the well (Atchison, 2021; Atchison and Wuest, 2021; Atchison and Sarpangal, 2022).

A primary indication of an influx is, while pumping at a constant rate, that there is an increase in flow rate leaving the well. The increase in the return flow is one of the most reliable indicators of a kick. Another primary indication of an influx is an increase in volume of the circulating system mud tanks. An evaluation of the primary indicators of an influx, selecting the most sensitive, immediate information source, and least error prone indicator, is the increase in flow rate in the flow line. The flow rate down the flow line is a familiar indicator for the driller and was selected as the most important primary actionable indicator on which to base a decision-making capability. The rig flow rate sensors are continuously monitored to detect the increase in flow rate coming from the well.

In the event of a kick, the system warns the driller, automatically decides, spaces out the drill-string, stops the drilling equipment (top drive and mud pumps), and closes the preselected blowout preventer (BOP).

The driller has the power of veto throughout the automated sequence, reducing the frequency of errors caused by false positive indicators. An automated system reduces the opportunity for human error. This ensures that the volume of the influx is kept to a minimum (one of the key principles of well control) and reduces stress on equipment and personnel. This technology can be applied to all rigs including cyber rigs and, with additional minor modifications, to conventional rigs.

15.3 DESIGN AND MANUFACTURE PRINCIPLES

An Automated Well Control System operates under a Quality Management System which has been assessed and registered by the National Qualifications Authority (NQA) against the requirements of ISO 9001:2008. This Quality System includes procedures which cover all design, software, purchasing, manufacturing, installation, and testing activities. As an innovative product intended to be deployed to ensure enhanced safety and performance in well control operations, the Automated Well Control system has been designed and manufactured with product assurance as the key qualifier. Documented assurance measures have been executed during the development of the product as shown in Figure 15.2.

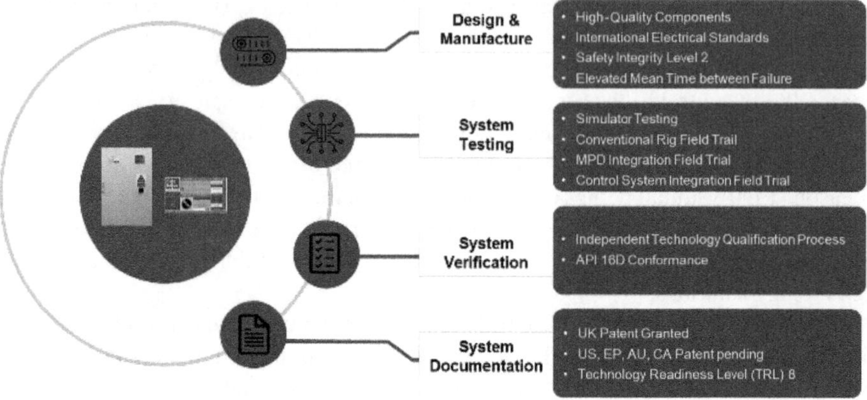

FIGURE 15.2 Technology assurance.

15.3.1 Equipment and Manufacturing Quality

The Automated Well Control System is built and programmed using high-specification Siemens PLC hardware and software. All equipment used in the system are designed such that their quality and reliability equal (or often exceeds) the quality and reliability of the connected equipment.

- PLC Control System – Siemens PLC offers high availability (seamless switchover between master and slave) and dual redundancy and has a proven record in several applications including rig controls.
- Power Supply(s) – The power supply is filtered to protect electronic componentry and incorporates an Uninterruptible Power Supply (UPS) battery power supply to ensure power availability for a minimum of two hours in accordance with API 16D.
- International Standard (IS) barriers – to protect HMI and other digital interfaces.
- HMI Screen – Atmosphere Explosible (ATEX) certified with a 19” Zone 1 touchscreen.
- Software – commonly used open-source Supervisory Control and Data Acquisition (SCADA) software.

15.3.2 International Electrical Standards

The design and implementation of the Automated Well Control Equipment, where applicable, conform to the following standards. The system is Conformité Européenne (CE) marked as self-declared by the manufacturer (Finesse Control Systems) to be compliant to these standards.

- BS 7671–2018 – IEE Wiring Regulations BS EN 60204–1 – Safety of Machines – Electrical Equipment
- IEC 62061 – Safety of Machines – Functional Safety

- BS EN 13849–1 – Safety of Machines – Safety-Related Parts of Control System
- IEC 60079–11 – ATEX – Intrinsically Safe Equipment
- IEC 61508–3 – Functional Safety – Software

15.3.3 Third-Party Systems Interfacing

It must be noted that this system has no capability on its own to either monitor instrumentation or control actuators. All monitoring and control are via existing rig sensors and third-party equipment control systems. A detailed Rig Site Survey document covers this issue.

Similarly, to achieve any Safety Integrity Level (SIL) rating on the shut-in functionality of the systems supplying the data and the systems activating the shut-in, the communication link between them will also need to be similarly SIL rated, if applicable. Protocol for interfacing will be determined for each individual system following a Rig Site Survey as it will depend on the existing equipment with which the system has to interface.

15.3.4 Safety Integrity Levels

A measure of the rate of unsafe failures is the safety integrity of the system, which is defined in Part 4 of International Electrotechnical Commission's (IEC) standard 61508 as "the likelihood of a safety-related system satisfactorily performing the required safety functions under all the stated conditions, within a stated period of time". The standard defines an SIL as "a discrete level (one of 4) for specifying the safety integrity requirements of safety functions".

The IEC 61508 defines SIL using requirements grouped into two main components: Functional component (hardware safety integrity) and systematic safety integrity. A device or system must meet the requirements for *both* categories to achieve a given SIL. The functional component is to "reduce the risk", and safety integrity component consists of SIL levels (between 1 and 4) (IEC 61508). The SIL requirements for hardware safety integrity are based on a probabilistic analysis of the device. To achieve a given SIL, the device must meet targets for the maximum probability of dangerous failure and a minimum risk reduction factor, both in continuous and "on demand" use. The Automated Well Control system has been manufactured using high quality components in compliance with the Safety Integrity Level 2 classification as detailed in Table 15.1 and Table 15.2.

TABLE 15.1
Safety Integration Level-2 (SIL 2) Compliance for Demand Use (after IEC 61508)

Probability of Dangerous Failure	Risk Reduction Factor
0.01–0.001	100–1000

TABLE 15.2
SIL 2 Compliance for Continuous Operation (after IEC 61508)

Probability of Dangerous Failure per hour	Risk Reduction Factor
0.000001–0.0000001	1,000,000–10,000,000

In most cases, the SIL classification of the Automated Well Control equipment exceeds the SIL classification of both the input and output phases of the rig's control system(s), meaning that the Automated Well Control System cannot be considered as the weakest link in the chain.

15.3.5 System Components – Mean Time between Failure

All components used in the manufacture of the Automated Well Control System have been selected with quality and reliability in mind. A measure of reliability is Mean Time Between Failure (MTBF), as quoted by the manufacturer, and a full system detail is summarized as:

- The minimum MTBF for the Control Panel is 90.3 years.
- The minimum MTBF for the HMI display is 7.99 years at 25°*C*.

15.4 TECHNOLOGY QUALIFICATION

The Automated Well Control system has been qualified by Lloyd's Register (LR) and comes with a Lloyds Qualification Certificate for both cyber and traditional rig applications. A Technology Qualification workshop for the Automated Well Control technology was held at LR, Aberdeen between 18th and 20th February 2019, in the presence of a range of relevant oil and gas industry experts. The outcome of the TQ workshop was an agreed set of technology goals, a system decomposition, technology maturity assessment, and a risk assessment.

The Technology Qualification (TQ) Process developed by Lloyds Register is a methodology to assess and help risks introduced by novel technology (Lloyds Register, 2022).

- TQ is a robust and systematic risk management process that demonstrates to interested parties that the uncertainties introduced by a novel technology, or new application of an existing technology, have been considered and that any associated technology risks have been mitigated.
- TQ is a risk-based process that uses the readiness level framework, a total system perspective and lifecycle approach to qualify innovative technologies, unconventional designs, and new ways of applying the existing technology.
- TQ is a methodology that provides assurance to owners, operators, suppliers, and investors at the distinct stages of novel technology development.

There are three main stages of the TQ process. The product of Stage 1 of the TQ process is the deliverable of a Technology and Risk Assessment report. Following the delivery of the report, Automated Well Control commences to Stage 2 of the TQ process by reviewing the recommendations made in the report and developing a TQ Plan. Upon the acceptance of the TQ Plan linked to the recommendations, LR issues the technology with a Statement of Endorsement. This document suggests that the technology under review is viable, if the recommendations made in the report and described in the TQ Plan were implemented in full. The final stage (Stage 3) of the process involves the implementation of the tests identified in the TQ Plan and a collection of all test results in the form of a dossier to support the assumption that the technology works in the intended environment under the conditions identified in the workshop.

A "Failure Modes, Effects, and Critical Analysis" (FMECA) was performed as a part of TQ examination process with Llyod's Register. A comparative human factors' study for the use of AUTOMATED WELL CONTROL against traditional well control methods was also performed. The human factors' analysis, completed by means of Human Reliability Assessment methodology, provides a qualitative and quantitative assessment of the human failure modes associated with the operation of the Automated Well Control compared with the equivalent associated with traditional methods.

The qualification process included a detailed Failure Mode Effect Analysis (FMEA), Factory Acceptance Test (FAT), and Site Acceptance Test (SAT). The SAT was performed as part of the test that was performed on a land drilling rig in Aberdeen, Scotland. The TQ plan covers the actual witnessed FAT of the control software and witnessed SAT of the technology on the RGU well-rig simulator and a traditional physical land rig.

The objective of the FAT and SAT for the simulator testing was to demonstrate prior to the delivery of equipment to the manufacturer that the integration of hardware and software has been correctly performed as per the approved MVP functional design specification.

15.5 RISKS ASSOCIATED WITH THE INTRODUCTION OF NEW TECHNOLOGY

There are several risks associated with the introduction of automation which several other industries have overcome. This section will discuss the prominent problems and how to deal with them.

15.5.1 Cybersecurity

The manufacturer recognizes the concern of operators and drilling contractors that essential well control equipment could be vulnerable to external hacking. At present, Automated Well Control technology does not have, nor does it rely upon, any connection to the internet or use of wireless technology thus significantly reducing the exposure to cyber-hacking.

15.5.2 Loss of Skills

Automated Well Control has redefined the drillers' role from manually and directly controlling all aspects of the well control operations to that of a system monitor and verifier. Other industries use extensive simulator training to ensure the necessary skill sets are still available if required.

15.5.3 Mistrust

Well Control is a safety critical process that requires confidence in the system by the user. A structured training program that has been developed that addresses the system theory and includes simulator training and rig floor training provided by coaches with an emphasis on repetition will develop confidence in the system. Easy-to-read process flow diagrams, using the swim lane format to portray parallel processes, have also been developed to enable comprehensive understanding of the system interactions.

15.5.4 Lack of Independence

Automated Well Control is a tool to assist the driller. The driller retains the responsibility for Well Control. The driller has the power of veto and can intervene at any moment. This principle is emphasized in training.

15.5.5 Lack of Familiarization with Systems

Lack of familiarity with automation technology can lead to serious consequences. A comprehensive training programme covering theory, simulator operations, practical training, and weekly drills will ensure a familiarity with the technology. Additional functionality of the system would require further training.

15.5.6 Performance Monitoring

The implementation of automation enables monitoring algorithms to be developed, providing the opportunity for data capture and analysis of each driller. Further training requirements could be identified, if required.

15.5.7 GAP Analysis versus Existing and Emerging Industry Guidance

As with any new evolving and emerging technologies, there is minimal instruction and guidance given to innovators on design and operational principles. Acknowledging that an automation of well control processes could encroach on statements in commonly used Standards and Procedures, there is a comprehensive GAP analysis of the overarching document – API Specification 16D, Control Systems for Drilling Well Control Equipment and Control Systems for Diverter Equipment (Third Edition, November 2018) and API Bulletin 16H, Automated Safety Instrumented Systems for Onshore Blowout Actuation (First Edition, February 2022) (API 16D and API 16H). It is important to mention here that the design and the TQ work of the Automated

Well Control occurred back in 2019 prior to the API bulletin 16H (published in 2022). Despite this, the GAP analysis did not identify any deficiencies in the work done to date. The GAP analysis is focused on documentation, procedures, and processes to ensure that the product has been designed, built, deployed, and operated within the criteria established by the industry.

15.5.8 API 16D Standard

It is recognized that most of the API 16D documentation refers to the hydraulic and electrohydraulic systems of a BOP control system and governs the standards and guidance for these controls (API 16D). It is therefore adopted as the minimum specifications for most BOP control systems. However, API 16D does recognize that sound business, scientific, engineering, and safety judgement should be used when employing the information in the specifications.

In line with the latter statement and because the Automated Well Control System forms a part of the control logic, a review of this specification was considered necessary to ensure that no specific conflicts with API 16D specification are noted. A total of 789 sections of API 16D were reviewed.

In simple terms, the operating criteria for the Automated Well Control system are to take over the activities of the driller (human) after detecting a pre-agreed volume of influx from a well and conduct the space-out, machinery shut-down, and shut-in duties without human intervention (unless chosen).

The vast majority (652) were not relevant to the Automated Well Control system. In the remaining 137 line items, no specific cases of non-compliance were identified. Of these 137 sections, 3 are relevant for the purchaser, 111 sections are relevant for the manufacturer, and remaining 23 of them are applicable to the vendor.

The Automated Well Control system cannot be classified as 'compliant' as these systems are not specifically mentioned in the specification. However, given that any API specifications are intended to 'facilitate the broad availability of proven sound engineering and operating practices,' a measure of conformance to API 16D can be applied.

15.5.9 API Bulletin 16H

A comprehensive GAP analysis of the document API Bulletin 16H, Automated Safety Instrumented Systems for Onshore Blowout Actuation (First Edition, February 2022) was performed to achieve a measure of compliance to the recommendations in the Bulletin.

The API Bulletin 16H provides information on existing and emerging technologies that could be integrated to bring a well to a safe state in the event other operational barriers fail. The bulletin discusses strategies to create an automated blowout preventer actuation system, the challenges and obstacles associated with these types of system, current existing technology, and the methods of achieving widespread implementation of such a system (API 16H).

An automated safety instrumented system is intended to bring the well to a safe state if the site personnel do not recognize a well influx or are unable to respond to one. The implementation of an automated safety instrumented system should consist of input from the lease operator, drilling contractor, and original equipment manufacturer.

The conclusion of this analysis is that the current Well Control System MVP and programmed additional modules are fully compliant with the requirements of API Bulletin 16H.

15.5.10 Simulator Testing and Rig Trials

Safe Influx have also established that there is very strong potential to collaborate with several other technologies to enhance the automated solution. The system can acquire data from different technologies to monitor for potentially dangerous situations and then initiate a shut-in process. These technologies include Managed Pressure Drilling, Early Kick Detection, Automated Mud Property Measurements, and Down Hole Data.

15.5.11 Simulator Testing at the Robert Gordon University (RGU) Facility

To ensure that the operation of the Automated Well Control system could be thoroughly evaluated in a benign environment, Safe Influx commissioned the compilation of an interface 'patch' to ensure effective and reliable data communications between the Automated Well Control system and a Drilling Systems DS6000 Simulator at the RGU facility.

This integration allowed several test scenarios to be run as follows:

a. To ensure that improved influx detection and reaction time were achieved by automation.
b. To ensure a number of fault- finding and user error scenarios could be recognized, documented, and, if applicable, incorporated into future upgrades to the system.

15.5.12 Rig Trial 1 – Test Land Rig – Bridge of Don, Aberdeen, UK

Following the various simulator tests in the DS6000 Drilling Simulator, the next stage in development was to demonstrate that the technology works reliably on a real working rig. In 2019, a field trial of Automated Well Control system was performed at the Weatherford test rig at Bridge of Don, Aberdeen. The objective of the field trial was to demonstrate and document that Automated Well Control technology reliably worked as designed on a 50-year-old traditional manually controlled drilling rig. This would also enable an extension of the existing technology qualification certificate, currently valid for cyber rigs, to include a manually drilling rig. The secondary objective, supporting the primary objective, was to test the system in multiple "additional", "normal", and "stress" scenarios to verify that the system works across a range of downhole scenarios.

The field trial objectives were fully met and exceeded the expectations of the team. The field trial was supported by the UK Oil & Gas Technology Centre (OGTC) and was witnessed by Lloyd's Register as the independent verifier and by personnel from independent and national oil companies.

A summary of the results is as follows:

1. All activities were incident free.
2. The Automated Well Control unit and interfaces were successfully and efficiently installed and commissioned at the rig.
3. A series of system tests and additional stress and normal tests, 15 in total were executed to demonstrate the system's functionality and capability in a range of downhole and surface situations.
4. The system was demonstrated successfully to Lloyd's Register (LR) and had achieved a Llyod's Register Technology Qualification certificate for both cyber and traditional rigs.
5. The system was successfully demonstrated to 4 VIP groups and a total of 33 people. This included senior staff from UK and overseas well operators, drilling contractors, IADC, OGTC, service companies, financial investors, and the press.

15.5.13 MPD Integrated Rig Trial 2 – Test Rig – Houston, USA

In March 2021, a rig trial and integration test of the service providers' Managed Pressure Drilling (MPD) and Automated Well Control system was performed on a test rig in Houston, USA. The objective of the integration test and rig trial was to witness and verify the integrity and functionality of the integrated MPD and Automated Well Control system (Atchison and Wuest, 2021).

The Automated Well Control equipment was required to be installed remotely by the service providers' MPD and Research and Development (R&D) team in Houston.

The complete design of the MPD and Automated Well Control technology link, the installation of the equipment, and the management of testing were operations that necessitated significant levels of videoconferencing between the teams over a period of many months.

Twenty-seven tests were performed. They were pre-agreed between MPD service provider, vendor, and manufacturer, which covered:

- Setting up of systems
- Independent system configuration and integrated commissioning
- Integrated contingency, communications, and comparison testing

For all tests, the expected and actual outcomes were documented and also whether the test was deemed successful. All tests were completed successfully and without major issues. All pre-agreed test criteria were met.

15.5.14 OEM's Drilling and BOP Operating Control Systems Simulator Testing, Arbroath, UK

As part of the agreement between Safe Influx and major Original Equipment Manufacturer (OEM), a programme of both simulator and rig interface testing was developed. The object of these tests was to prove that effective and reliable data communications could be achieved between Automated Well Control system and the OEM's Drilling and BOP Operating Control Systems.

A remote simulator test was performed in a virtual environment using a cloud-based rig simulator based in Norway, and the Remote BOP control system and Automated Well Control system were based in Scotland.

A simulator test was performed by the collaboration between the UK, Norway, and US teams. Two successful remote simulator tests were conducted to establish and test the interface arrangements to ensure that:

- The Automated Well Control system can be integrated with the OEM's Drilling and BOP Operating Control systems.
- The integrated system successfully functions and completes the automated shut-in sequence on the detection of an influx.

The test was conducted during two separate instances:

- **03 March 2022** – a full test team conducted six tests to prove connectivity and control of the OEM's systems. These were conducted successfully, but a PLC software error prevented the verification of the control sequence.
- **08 March 2022** – the software issue was identified, and a further nine tests were conducted with a small test team to prove the control sequence was valid across several scenarios. These tests were all successful.

For all tests, the expected and actual outcomes were documented. All integration tests were performed, and all were completed successfully without significant issues. All pre-agreed test criteria were met.

15.5.15 OEM's Drilling and BOP Operating Control Systems Rig Trial 3, Houston, USA

In April 2022, a rig trial and integration test between Automated Well Control system and the OEM's Amphion Rig Drilling and BOP Operating Control Systems was executed at STC Test Rig in Navasota, Texas, USA. This rig-based test was the second phase of the required Automated Well Control/OEM's integration after the successful evaluation of the Automated Well Control in the benign simulator environment. The test was performed to establish and test the interface arrangements to ensure that:

- The Automated Well Control system can be integrated with the OEM's Operating Control systems.
- The integrated system successfully functions and completes the automated shut-in sequence on the detection of an influx.

The test was conducted as detailed next:

- **01 Apr 2022** – the test team conducted five tests to prove the connectivity and control of the NOV systems.

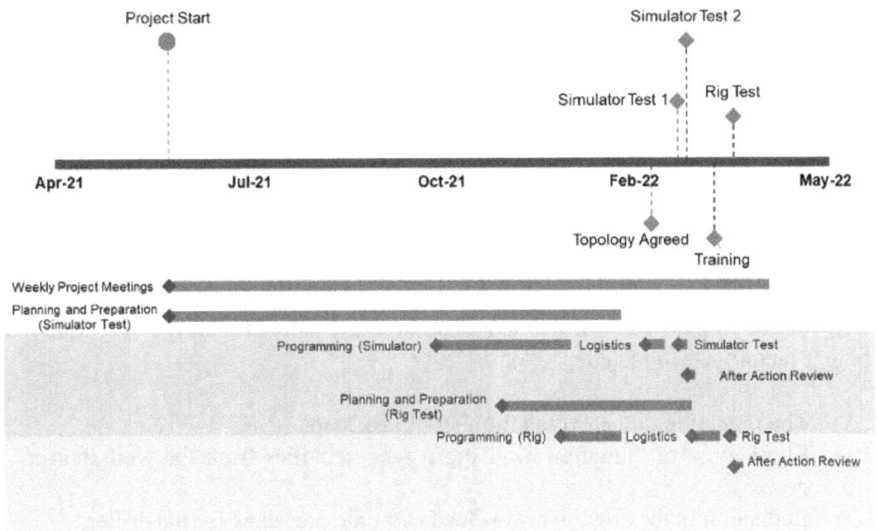

FIGURE 15.3 Simulator and rig test timeline.

The timeline of the project is shown in Figure 15.3.

The following highlights were recorded during the trial:

- Successful integration of Automated Well Control system with an OEM's Rig and BOP operating control system.
- Rig test achieved all the objectives.
- The Automated Well Control equipment was installed physically by the Safe Influx and OEM's team.
- Proper planning, experience, and competent personnel were the key factors that contributed to the success.
- This integration combines the functionally of both systems to provide a comprehensive Automated Well Control package to the upstream industry.
- All activities were incident free.

15.6 AUTOMATED WELL CONTROL DOCUMENTATION PATENT

Automated Well Control is patented technology capable of delivering the full well control protocol. To protect Intellectual Property (IP), the manufacturer applied for the UK Patent in April 2019.

The manufacturer was awarded UK Patent No GB2581586 on 27th January 2021 by the UK Intellectual Property Office, for Automated Well Control, together with an additional 50 potential modules to provide well control protection for all aspects of well operations.

A patent was applied for in the United States of America, and, using the Patent Cooperation Treaty (PCT), further patents have been applied for in Canada, Australia, and Europe.

15.7 COMPARATIVE HUMAN FACTOR ANALYSIS

Research and development is a vital part of the company work scope to retain its position as "Thought Leaders" in Automated Well Control. Given that a sizeable number of Well Control Incidents have been attributed to human factors (up to 71% have been reported by IOGP, 2021), Safe Influx commissioned a comprehensive study to compare the human factors' influence on traditional manual well control methods with the automated system (Atchison and Sarpangal, 2022).

The analysis provided a clear demonstration of the reduction in exposure to human factors' issues that Automated Well Control system brings compared to traditional well control methods. The dramatic reduction in exposure is achieved by a combination of factors which include:

- The reduction in the overall number of task steps.
- Elimination of 'situation evaluation' type activities from the well shut-in process.
- Reduction in the cognitive workload and time pressures for the driller.

The quantitative assessment demonstrated that:

- The human probability failures associated with the traditional well control method, which could lead to a potential loss of well control, would be reduced by 94% when the Automated Well Control system is used.
- The Automated Well Control could achieve 96% chance of reducing the shut-in influx volume (Marex Marine and Risk Consultancy, 2020).

15.8 TECHNOLOGY READINESS LEVEL

The concept of Technology Readiness levels was originally developed by *NASA* as a method of measuring the maturity of technology. Generic details of this process are included in Figure 15.4. This concept has been employed by Safe Influx to ensure that the maturity of the Automated Well Control Technology is documented, and it is considered that the automation system is categorized as

TABLE 15.3
Results Human Error Probabilities (after Marex Marine and Risk Consultancy, 2020)

	Failure to Shut in on an Influx	Delay in Shutting in on an Influx
Traditional Well Control	0.4411	0.5528
Automated Well Control	0.0250	0.0246
Reduction in HEP for Automated Well Control vs Traditional Well Control	94%	96%

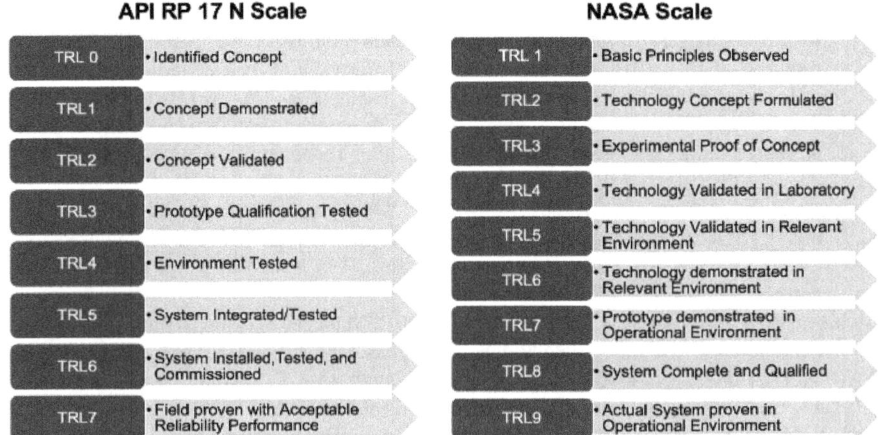

FIGURE 15.4 Technology Readiness Level (*NASA* and API RP 17N Scale).

NASA TRL-8 (System Completed and Qualified) and API RP 17 N TRL-6 (System Installed, Tested, and Commissioned) and is ready for field deployment (*NASA* and API RP 17N).

15.9 CONCLUDING REMARKS

Automated Well Control uses computer logic to identify the influx, take control of the rig equipment, and place the well into a secure position without the intervention of a human. In addition to the obvious benefits of prevention of fatalities, pollution, and reputational risk, there are several other benefits that will reduce well construction costs: Less time spent on recovery and remediation and potentially an enhanced well design leading to reduced well costs. Overall, the probability of successfully delivering all wells, especially technically challenging wells, within time and cost estimates will be increased.

ACKNOWLEDGEMENTS

The authors would like to thank John McGillivray and Richard Rose of Safe Influx; Fraser Dunphy of Finesse Control Systems Ltd; and Marex Marine and Risk Consultancy for assisting in preparation of this chapter.

REFERENCES

Abimbola, M., Khan, F., Garaniya, V., et al., 2015. Failure Analysis of the Tripping Operation and its Impact on Well Control. Vol. 1: Offshore Technology, Offshore Geotechnics. *Proceedings of the ASME 2015 34th International Conference on Ocean, Offshore and Arctic Engineering*, Newfoundland, Canada, May 31–June 5.

API Bulletin 16H, 2022. *Automated Safety Instrumented Systems for Onshore Blowout Actuation*, 1st edn. Washington, DC: API, February.

API RP 17N, 2015. *Recommended Practice Subsea Production System Reliability, Technical Risk, and Integrity Management*, 2nd edn. Washington, DC: API.

API Specification 16D, 2018. *Control Systems for Drilling Well Control Equipment and Control Systems for Diverter Equipment*, 3rd edn. Washington, DC: API, November.

Atchison, B. W., 2021. Automated Well Control: Automated Detection and Shut-In. Paper presented at the *SPE/IADC.* Middle East Drilling Technology, Abu Dhabi, UAE, May 25–27. SPE 202091-MS. https://doi.org/10.2118/202091-MS

Atchison, B. W., & Sarpangal, V., 2022. Automated Well Control Versus Traditional Well Control: A Human Error Comparison Process. Paper presented at the *IADC/SPE International Drilling Conference and Exhibition*, Galveston, TX, March 8–10. SPE-208786-MS. https://doi.org/10.2118/208786-MS

Atchison, B. W., & Wuest, C., 2021. The Integration of Managed Pressure Drilling (MPD) and Automated Well Control Technology. Paper presented at the *IADC/SPE Managed Pressure Drilling & Underbalanced Operations Conference& Exhibition, Virtual*, September 14–16. SPE-206385-MS. https://doi.org/10.2118/206385-MS

Gillard, M., & Bhatti, M. A., 2022. *Global Well Control Database Analysis*. Internal Report. Aberdeen, Scotland: Safe Influx Ltd.

Health and Safety Executive, 2013. *Human and Organisational Factors in Well Control: Multi-National Audit*. North Sea Offshore Authorities Forum. www.hse.gov.uk/offshore/nsoaf.pdf

IEC61508, 2010. *Functional Safety of Electrical/Electronic/Programmable Electronic Safety-Related Systems*, 2nd edn. International Electrotechnical Commission. https://webstore.iec.ch/publication/5515

International Oil and Gas Producers (IOGP), 2016. *Recommendations for Enhancements to Well Control.* https://www.iogp.org/bookstore/product/recommendations-for-enhancements-to-well-control-training-examination-and-certification/

Lloyds Register, 2022. *Guidance Notes for Technology Qualification. User Guide*. London, UK: LR. https://www.lr.org/en/guidance-notes-for-technology-qualification/

Marex Marine and Risk Consultancy, 2020. *Safe Influx Automated Well Control. Comparative Human Factors Analysis Report for Safe Influx*. https://www.safeinflux.com/wp-content/uploads/2020/09/MAREX-0283-01-v2.0-REDACTED_2020-09-07_11-28.pdf

NASA, 2012. *Technology Readiness Level.* https://www.nasa.gov/directorates/heo/scan/engineering/technology/technology_readiness_level (Last Accessed in January 16, 2023).

National Commission on the BP Deepwater Horizon Oil Spill and Offshore Drilling, 2011. *Deep Water: The Gulf Oil Disaster and the Future of Offshore Drilling*. Report to the President, USA.

Reason, J., 1990. *Human Error*. Cambridge University Press.

St. John, M. F., 2016. Macondo and Bardolino: Two Case Studies of the Human Factors of Kick Detection Prior to a Blowout. Paper presented at the *Offshore Technology Conference*, Houston, Texas, USA, May 2–5. OTC-26906-MS. https://doi.org/10.4043/26906-MS

Thorogood, J. L., Lauche, K., Crichton, M., et al., 2015. Getting to Grips with Human Factors in Drilling Operations. Paper presented at the *SPE/IADC Drilling Conference and Exhibition*, London, UK, March 17. SPE-173104-MS. https://doi.org/10.2118/173104-MS

16 Recent Advancements in Flowmeters for Kick Detection

Asad Elmgerbi and Rahman Ashena

16.1 INTRODUCTION

Flowmeters have become indispensable tools in modern drilling operations, significantly advancing well control capabilities. Precise measurement of fluid flow rates is paramount for effective well control during drilling activities. Modern flowmeters, equipped with advanced sensor technologies and data-processing capabilities, offer real-time insights into flow dynamics, enabling swift detection and mitigation of potential hazards such as kicks or fluid losses.

Recent advancements in flowmeter technology have further enhanced well control capabilities. Advanced metering technologies, including ultrasonic, electromagnetic, and Coriolis meters, excel in accurately measuring mud outflows. Therefore, this chapter covers these advancements which are particularly beneficial in challenging drilling environments, including those encountered in high-pressure high-temperature or deep-water cases.

16.2 PERFORMANCE COMPARISON OF COMMON FLOWMETERS

Detection of well control events in real-time and in accurate manner is extremely important to the integrity of the wellbore and safety of the rig site. Fundamentally, kicks occur when wellbore hydrostatic pressure falls below formation pore pressure, causing the well to become underbalanced; this allows unwanted fluids to enter the wellbore (Nas, 2011). In contrast, losses happen when mud weight is too high and the wellbore pressure exceeds formation fracture resistance, causing induced fractures and allowing drilling fluids to enter the formation. This can also happen when naturally fracture formations or very permeable formations are encountered (Lavrov, 2016). There are multiple causes for the existence of induced fractures, such as a high mud weight or high equivalent circulating density (ECD), surging, reservoir depletion, and formation pressures being lower than expected (Johnson et al., 2014). Conversely, for kicks, there are several reasons such as higher formation pressures than expected. These scenarios can become expensive in terms of both curing time and cost. Therefore, it is vital to find suitable ways to reliably monitor wellbore conditions. A method used to continuously monitor the wellbore is called the "delta flow". The delta flow method is simply monitoring and comparing the flow rate in and out

DOI: 10.1201/9781003473770-16

of the well to give direct and quick indication of the kick and loss events. Calculating delta flow is simply carried out by subtracting the flow into the well from the flow out of the well. A positive value indicates a kick flow, whereas a negative value indicates fluid loss to the formation (Cayeux and Daireaux, 2013)

The majority of inflow measurements today are still made by counting the strokes of the mud pumps. By knowing the displacement volume per stroke and the efficiency of the pump, the inflow volume can be determined. Other more accurate methods use Coriolis flowmeters applicable only to managed pressure drilling (MPD). The Coriolis flowmeter is installed downstream the mud pumps to measure the inflow rate. On the other hand, for MPD systems in which the return flow has pressure, the Coriolis flowmeter is installed in the return line as close to the bell nipple as possible to get the quickest measurement of fluid flow. When the flow rate is measured in the return line, the fluid is usually contaminated with cuttings from the drilling process as well as possibly small gas bubbles and wear material from the drill-string or the casing. Thus, the measured fluid can by no means be described as a pure fluid which brings some challenges in measuring its flow, hence application of some commercial flowmeters may not be feasible.

The type of measurement used by most available flowmeters can be split into two main categories based on their measurement output, mass flow, or volumetric flow. Mass flow is the amount of mass moving through an instrument over time, so the unit of measure is mass per amount of time, whereas volumetric flow is the measure of a substance moving through a device over time. The main advantage of the mass flowmeters over volumetric flowmeters is that mass or weight does not change depending on temperature or pressure (Bunyamin et al., 2010).

Figure 16.1 illustrates all flowmeters measuring either the volumetric or mass flow rate without creating obstructing elements in the flow path such as turbine or vortex flowmeters usually do.

The overall performance of any flowmeter is driven by multiple factors including:

- Effect of the pressure drop.
- Regular maintenance required.
- Effect of the fluid properties.

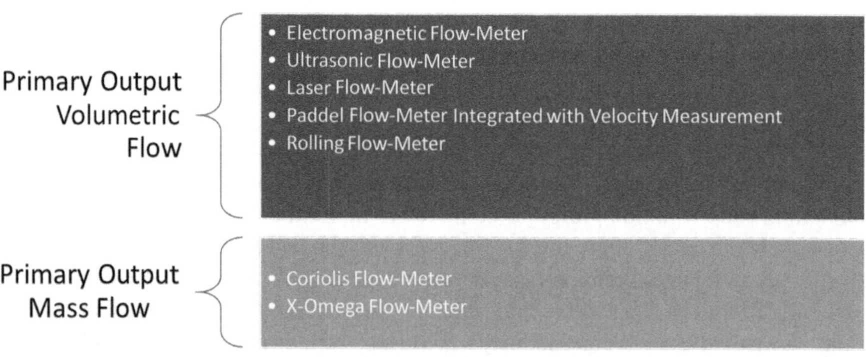

FIGURE 16.1 Typical flowmeter types used at the rig site. Modified from Roger (Roger, 2000).

- Cost.
- Type of drilling operation.
- Rig footprint changes required.

Thus, the criteria used to select the best flowmeter vary and depend on the environment of use. In the following section, the advantages and disadvantages of the flowmeters mentioned in Figure 16.1 are explained. The focus of the discussion is on their applications in the drilling industry.

16.2.1 Electromagnetic Flowmeter

Electromagnetic flowmeters use Faraday's Law of Electromagnetic Induction to define the flow of liquid in a pipe. A magnetic field is generated and channeled into the liquid flowing through the pipe. The flow of a conductive liquid through the magnetic field will cause a voltage signal which can be sensed by electrodes located on the flow tube walls. When the fluid moves quicker, more voltage is produced. The electronic transmitter processes the voltage signal to define the liquid flow. Currently, a conventional electromagnetic flowmeter can quantify the volumetric flow rate of a single-phase flow with an error as low as 0.05%, provided that the velocity profile is axisymmetric. Electromagnetic flowmeters can only function when the fluid is conductive. In multiphase flow which is of great importance to the oil industry, there is an upward inclined oil-in-water flow. Such flows are 'water continuous', and so the multiphase mixture is electrically conductive allowing the use of electromagnetic flowmeters (Leeungculsatien and Lucas, 2013). As for open channels, an electromagnetic flowmeter can be a good tool for measuring the flow rate when integrated with an ultrasonic level meter. In this case, the electromagnetic flowmeter measures the mean flow velocity, while the ultrasonic level sensor measures the flow area (Watral et al., 2015).

The advantages of this type of flowmeters are:

- Having minimum obstruction to the flow path, they provide no pressure drop
- They can measure forward and reverse flow with equal accuracy.
- They are capable of measuring large variations in flow rate.
- They need low maintenance cost as they have no moving parts.

Their disadvantages are:

- An electrically conductive fluid is required.
- They cannot be used to measure the outflow rate at the return flow line, due to the possible accumulation of the cuttings at the bottom of the line, even if it is integrated with the level sensor.
- At low flow rates, their accuracy gets extremely low.

16.2.2 Ultrasonic Flowmeter

Ultrasonic flowmeters utilize the sound waves' principle, which travel through the fluid to measure the velocity. In general, ultrasonic flowmeters can be classified into two categories: Doppler ultrasonic flowmeter and the transit-time meter. The Doppler-based ultrasonic flowmeters work better in dirty or aerated liquids like drilling fluids, whereas transit time-based ultrasonic flowmeters have a high accuracy when they are used to measure velocity of clean fluids like water or natural gas (Zhou et al., 2013)

The first successful application of using Doppler-based ultrasonic flowmeter to measure the flow out was published by Orban et al. (1987) and Orban et al. (1988). They used two sensors, with the first sensor measuring the level of the fluid and the second one measuring the local velocity. Both sensors are ultrasonic: One is ultrasonic level, whereas the second one is Doppler. The sensors are installed at the return flow line. Zhou et al. (2013) developed a test setup for testing the capability of a Doppler-based ultrasonic flowmeter to detect the gas kick. They concluded that the measurement accuracy is 6%, which is considered a good value for the drilling industry.

The advantages of these flowmeters are:

- They can handle high pressures, are repeatable and consistent, and can also handle extreme temperatures.
- They are non-obtrusive to pipework and can be clamped on.

Their disadvantages are:

- They have a relatively high cost.
- They have sensitivity to vibrations. In the environment of drilling rigs with huge vibrations, accuracy problems would be brought about.
- The accuracy of the flow measurement can be affected by the fluid chemistry and solid content.

16.2.3 Laser Flowmeters

Laser flowmeters use the Doppler shift to measure the velocity of a flowing fluid. The main uses currently for this technology are in open channel wastewater applications, and they can operate in both submerged and unsubmerged conditions. Also with the use of an ultrasonic level measurement, the laser flowmeter can measure both the velocity and flow height, and the laser flowmeter can measure at multiple points below the fluids surface, providing a more accurate mean flow measurement. It has a flow accuracy of plus or minus 5%, and a minimum depth of 25 mm is needed (Hummel, 2017).

The advantages of these flowmeters are given next:

- Portable and easy to deploy
- No contact with the fluid itself, so less risk of debris and silting up
- Can measure high and low velocities with a high accuracy
- Quick installation
- Can measure multiple points of flow at multiple depths, so work well in no uniform flows

The disadvantages of these flowmeters are as follows:

- Most applications are for open channel flows, and no proven performance in a closed flow line environment.
- Has not been used for the drilling industry

16.2.4 Paddle Flowmeter and Rolling Float Meter

The most used instrument in drilling rigs to monitor outflow rates is the flow paddle. Since it is installed on the flow line quite close to the bell nipple and before the mud pits, it presents a faster way to detect kick and loss events. The convention paddle flowmeter does not measure a real flow or a volume of drilling fluid, but rather it gives an indication of the mud level in the return flow line. Therefore, this kind of tool cannot be really reliable and accurate for detecting kick and loss events. Because the paddle flowmeter may be installed only at the return flow line which is an open channel, multiple factors can impact their accuracy. One of the main factors is the specific energy, where the level of the fluid inside the return flow line will be changeable based on the velocity (Lee et al., 2019). In recent years, some companies tried to improve the conventional paddle flowmeter by incorporating a velocity measurement into the paddle meter with a small paddle wheel at the end. A hall sensor measures the revolution per minute (RPM) of the wheel and can get a velocity reading from which it can be converted into flow rate (Matherne Instrumentation, 2024; Forerunner Technologies LLC, 2024).

The advantages of these flowmeters are:

- The relatively low price makes them attractive in a low oil price environment with tight budgets.
- The installation and maintenance are simple and can be done by one person compared to other flowmeters.

Their disadvantages are:

- The flow paddle gets easily worn by the solid material carried with the drilling fluid.
- Regular recalibration is required to maintain a reasonable accuracy.
- The accumulation of the cuttings at the bottom of the return flow line can lead to false alarms and can get the paddle to stuck in one position.
- The temperature rating of 135° for this flowmeter might become a problem for geothermal and high-temperature drilling applications.
- The paddle flowmeter position can be affected by the fluid rheology.
- The rolling float meter has not been tested in drilling rigs.

16.2.5 Coriolis Flowmeter

A Coriolis (mass) flowmeter uses the Coriolis effect to measure the amount of mass moving through the flowmeter element. The fluid to be measured runs through a U-shaped pipe that vibrates in an angular harmonic oscillation. Due to the Coriolis forces, the tubes will deform, and an additional vibration element will be added to the oscillation. This additional element causes a phase shift on some places of the tubes, which can be measured with sensors (Norman, 2011).

The Coriolis flowmeter offers a direct mass rate measurement. The introduction of Coriolis technology to obtain fluid returns' measurement has offered the ability to detect small changes in flow rate and fluid density caused by water or gas influx or losses that indicate change of formation zone or a possible loss of well control. The technology provides real-time mass, volume, and density data with improved accuracy and reliability. Coriolis meters can measure oil, water, and synthetic-based muds similarly well, regardless of the mud weight and/or chemical additives in the fluid and can also be used to continuously monitor drilling mud density to identify deviations in mud density at the surface which can signal or cause possible well control events.

The non-mechanical design and robustness of this flowmeter make them well suited to measuring particulate-containing fluids. In the sensor, the stiffness of the tubes is critical to measure consistency. If the tube stiffness changes as a result of corrosion, erosion, or over-pressure, measurement precision may be affected. Some sensors provide the ability to verify tube stiffness in situ without process interruption to detect any changes that may have occurred as a result of erosion or other damage, providing confidence in measurement (Casellas et al., 2016). Coriolis flowmeters can be used for both systems, the MPD and conventional open loop drilling system. They can increase the accuracy of detecting kick and loss events; consequently, more time will be available for the drilling crew to react before catastrophic events happen. Wellbore ballooning effects can sometimes cause false kick and loss alarms, or kicks can develop undetected when assumed to be the ballooning effect. Hence, another important application that is often mentioned is the identification and quantification of wellbore ballooning effects which are supported by their exceptional accuracy. In order to improve the measurement accuracy, it is important to install the device correctly. This makes it necessary to install it so that no cuttings and no gas bubbles can settle and accumulate at low flow velocities and consequently impair the measurement. In order to overcome the small pressure drop of the flowmeter on conventional drilling rigs, Micro Motion made an example calculation for such an installation. For example, if the flowmeter needs an input pressure of three psi at full flow rate, the minimum hydrostatic head of six ft would be required (Chris and Simons, 2013).

The advantages of this flowmeter are as follows:

- High accuracy and reliability,
- Capable of measuring difficult handling fluids,
- Independent of density changes, flow profile, and flow turbulence. Hence straight lengths are not required, and
- No routine maintenance is required since it has no moving parts.

Their disadvantages are as given next:

- Applicable to managed pressure drilling or underbalanced drilling cases where the flow returns under pressure.
- They are more expensive than other flowmeters.

- Erosion in the pipe can cause measurement inaccuracy.
- They can cause a pressure drop in hydraulic system.
- The rig footprint should be changed to allow for the Coriolis flowmeter to be installed, changing the hydraulic system pressure drops.
- They need to be regularly cleaned to prevent corrosion from taking place within the flowmeter and to keep the precision up to an acceptable level.
- Rig vibration may cause a reduction in the tool accuracy.
- The device does not work well in partially filled pipes in case of low outflow rates; therefore, if the entire return line is not rather full, it may not work properly, and
- Due to the pipe bending which is a part of the tool installation, the cuttings may accumulate at the bottom of the bend, causing malfunctioning the measurement.

16.2.6 X-Omega Flowmeter

The X-Omega flowmeter works by first measuring the fluid in a density module that contains a pressure sensor, next a wedge meter analyzes the flow velocity, and finally there is a second density module at the bottom-end that takes another pressure measurement. These simultaneous measurements allow the real-time data flow required in a managed pressure drilling (MPD) operation, and it can also be used in other drilling environments. This allows early kick and loss detection and also reduces the non-productive time (NPT). The X-Omega takes up a smaller amount of space than a Coriolis meter with a 4-ft height clearance compared to 14 ft of the Coriolis meter and has a 35% pressure drop reduction (Jacobs, 2015).

The advantages of these flowmeters are as follows:

- They can accommodate high flow rates without blockage.
- They have no moving parts.
- They are resistant to high pressures and temperatures.
- They can maintain accuracy with high levels of cuttings and gas in the drilling fluid.
- They do not require cleaning or flushing of the sensors, and they are resistant to abrasive wear.
- The pressure drop is roughly 35% less than typical Coriolis flowmeters.

Their disadvantages are as given next:

- The rig needs to be adapted to fit the flowmeter.
- They incur possible higher cost than other flowmeters.

16.3 DEVELOPED SYSTEMS AND TOOLS FOR IMPROVING KICKS AND LOSSES DETECTION

Hydrocarbon reservoirs are becoming more and more difficult to find and reach in terms of depth and remoteness of the drilling sites. Therefore, a trend has been observed to drill more intricate wells in the near future, which will make the drilling process more complex and require cutting-edge tools and technology, making the drilling cost reach double or treble higher than the current. Thus, to compensate for the extra cost, reducing the drilling time becomes a necessity, not an option (Elmgerbi et al., 2021). Therefore, in the recent decade, the focus and attention have been on the development and application of new measurement technologies in terms of downhole sensors and surface sensors to improve the real-time prediction and detection of kick and losses event. In this section, the limitations and related real-time application of the most recently disclosed and developed methods and tools aiming at improving the detection of the kick and loss events will be discussed.

16.4 STATE-OF-THE-ART APPROACHES

Gary Belcher (Belcher et al., 2008) disclosed a method to enhance kick and loss detection by permanently installing the relevant sensors at the tie-back casing string, which is a part of the wellhead. The proposed method is designed to be used for managed pressure drilling (MPD), where the wellbore pressure is controlled by a choke valve located at the choke manifold. Figure 16.2 depicts the proposed system.

The main featured challenges in his method are:

- Since all the sensors are installed below the Blowout Preventer (BOP), a special wellhead design is required to facilitate the sensor's transferring of data to the surface.
- Sensor maintenance is a big challenge, where tie-back casing must be completely retrieved to replace or maintain the sensors, which can require additional time and cost on the drilling operation.
- To avoid the return fluid entering the space between the tie-back casing and drill-pipe and divert it to the annular space between the tie-back casing and the conductor or surface casing, they suggest pumping heavy fluid into the annular space between the tie-back casing and drill-pipe, which can be a significant challenge when it comes to real application.

The acoustic velocity concept for detecting the kick was disclosed by Rocco DiFoggio in 2012 (DiFoggio and Blue, 2012). The gas influx within the disclosed method is detected by the measurement of the acoustic velocity of the borehole fluid and the temperature of the borehole fluid using a series of sensor subs installed along the drill-collar. The main challenges of their method from a real-time kick detection perspective can be mainly related to data transmission, where the only way of sending the downhole data is via the mud pulses (telemetry system). The second point which needed to be considered is that the measurement can be taken only if the drill-string is in static mode, with no rotation or movement.

Rogers (Rogers, 2015) as well as Knudsen and Vermeulen (2017) proposed a method for monitoring drilling fluid density, aiming to detect the kick and loss events. They suggested installing pressure and temperature sensors into the body of the bell nipple or raiser to monitor the return fluid density, which can be derived from the differential pressure between the two pressure sensors. The visible shortcoming of their methods are as follows: a) the impact of drill-string pipe movement and vibration on the sensor's measurement accuracy. Therefore, these methods can

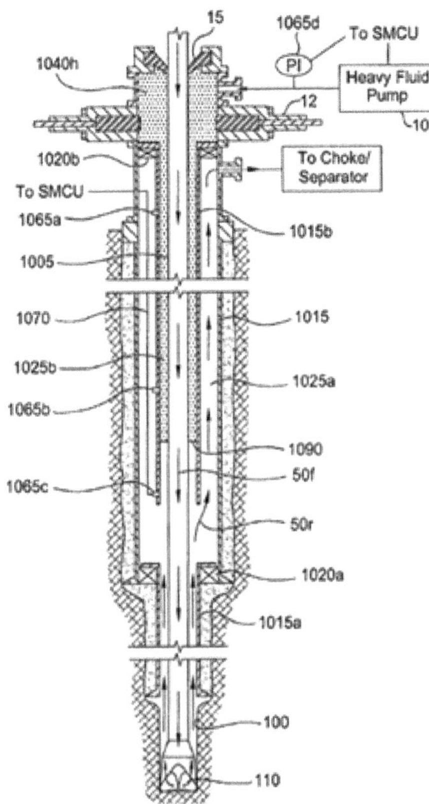

Legend
1. 1040h Heavy Fluid
2. 1005 Drill Pipe
3. 1015 Tie-back casing
4. 1065b and 1065c Sensors Location

FIGURE 16.2 Proposed configuration for the method disclosed by Belcher et al. (2008).

only be used at steady-state flow conditions without any substantial pipe movement or rotation, for instance, for coiled-tubing drilling or while sliding and b) if the kick detection will be based on the density changes, then it will be late to detect the kick, since all the measurement takes place at the bell nipple or the marine raiser.

Two researchers (Zhou et al., 2013; Fu et al., 2015) developed a test facility for evaluating the performance of Doppler ultrasonic meters in different environments. Their main considered factors were fluid properties, gas injection rate, and pump flow rate. Both reached similar conclusions, which can be summarized as following:

- Sensor accuracy is a function of drilling fluid density; as the density increases, the accuracy is reduced.
- Higher gas injection rate means less accuracy.
- The sensors work better under atmospheric conditions.
- The overall sensor error in terms of measurement is between 6% and 12% in the worst case.
- They did not study the impact of pipe movement and rotation on the sensor accuracy.

Pirir et al. (2017) proposed a new outflow measurement concept as an alternative solution to the Coriolis flowmeter. The method is based on installing Venturi flumes at the return flow line upstream of the shale shakers. In addition to the Venturi flumes, they suggested using two ultrasonic sensors to measure the fluid level at the beginning and the end of the Venturi flumes. The reading of the level sensors was used to estimate the flow rate. The proposed method was tested in the laboratory. The initial results show a good correlation between the flow rate measured by the Coriolis flowmeter and the new method. Nevertheless, several limitations can be concluded from their work:

- They assumed that the flow rate and density were stable to get good results. In reality, the flow rate and the density are prone to changes, especially the density, which is a function of cuttings' concentration, associated gas, pressure and temperature effects, etc.
- They did not consider the cuttings' bed, which can be accumulated at the bottom of the Venturi flumes.
- The recorded data while testing the system fluctuation. To overcome that, they applied an algorithm to process the data.

Another method that can be classified as thinking outside the box was discussed by Cayeux (Cayeux, 2020). He provided a conceptual design for an apparatus which is working on the vibration principle as the Coriolis flowmeter. In general, the tool consists of a rotating wheel and a vane located in the center of the wheel. To measure the mass flow, the fluid is pumped into the center of the rotating wheel, and as the fluid touches the vanes, they start to rotate. Due to the rotation of the wheel and the vanes, a friction force is created, which can be converted into a torque value. The measured torque value can be used to estimate the mass flow. Although the tool is still in the designing phase, the following points can be considered as challenges for the real application of the tool:

- The size of the tool must be big enough to facilitate high flow rate.
- Cuttings might accumulate at the bottom of the vanes, causing an error in the measurement.
- To convert mass flow to flow rate, he suggested using a small densitometer; in this case, only a fraction of the fluid will be used to measure the density. In other words, a percentage of the fluid will be used to evaluate the overall density of the return fluid.
- Considering the tool and the pumping facility of the densitometer, the expected footprint of the apparatus will be huge.

To evaluate the capability of the latest developed pressure sensors to detect the gas kick inside the marine riser, Zhou et al. (2021) conducted experimental work. The used test facility consists of a long vertical pipe (mimicking the marine riser in deepwater drilling) and 12 pressure sensors deployed along the pipe. Their work shows that the pressure sensors managed to identify and track the gas movement. Nevertheless, two main points can be very critical for such research, which they did not consider: First, the movement of the drill-pipe in their setup did not involve a pipe in the middle of the main pipe. The second point is regarding the simulated kick; in the conducted experiment, the kick was pumped as a slug; however, in reality, the gas kick can have multiple regimes such as interacting bubble flow, churn turbulent bubble flow, and clustered bubble flow (Gruber, 2016).

As an alternative to using pressure sensors for detecting the gas kick as it travels along the marine riser, Santos et al. (2021) assessed the use of fiber-optic sensing. They used PERTT Lab facilities at Louisiana State University to conduct the experimental work. They concluded that fiber-optic sensing technology could be used to detect gas kicks. However, data processing requires a specific algorithm, requiring time to be run.

16.6 CONVENTIONAL APPROACH

Despite the existence of advanced technology regarding detecting kick and loss events while drilling, most rigs today still rely on the conventional technique as the main approach to understanding the volumetric condition of the well for the detection. In the following section, the conventional approaches are briefly discussed.

16.6.1 Change in Total Volume in Active Mud Pits

Monitoring the active mud pit volume is the most used method for detecting loss and kick events. In practice, the level of the mud inside the mud pit is measured continuously by either a float level instrument or an ultrasonic level sensor, and the entire system is called Pit Volume Totalizer (PVT). Although this method is still the most commonly used, relying on this method alone during some drilling operations might be difficult to detect kicks and losses due to:

- A quite long response time because the mud pit is localized far from the wellhead.
- A large amount of mud might be buffered in the return flow lines, shakers, and sand traps due to the changes of the flow in, when mud pump switched off or when the flow rate changes.
- Transfer of mud from the active pit to another tank, or dumping the mud can be misinterpreted as a loss or kick.
- The change of the mud compressibility when moving from static to dynamic and vice versa.
- Thermal expansion of the drilling fluid.
- Transport and discharge the cuttings over the shale shakers.
- The retention capacity of some of mud transport and treatment equipment.

16.6.2 Delta Flow Method

As mentioned earlier, the delta flow method is simply monitoring and comparing the flow rate in and out of the well in order to give direct and quick indication of possible kick and loss events. Calculating delta flow is simply subtracting the flow into the well from the flow out of the well. A positive value indicates a kick flow whereas a negative value indicates a fluid loss to an underground formation. In steady-state theoretical conditions, this appears to be quite a simple solution. The main limitations of the delta flow method are as follows:

- This method requires relatively precise measurements of both inflow and outflow rates to obtain reliable results.
- It would generate false alarms in some cases such as may occur when starting and stopping the mud pumps or when reciprocating the drill-string.
- The other variations between inflow and outflow rates may also be induced by heave movement (offshore application).
- Detecting loss and kick events is not possible when the pump is off.

16.6.3 In and Out Pressure Variations' Monitoring Method

There is a method only applicable to the closed loop system where standpipe pressure (SPP) and Annular Discharge Pressure (ADP) are compared to identify loss and kick events. For example, when both SPP and ADP simultaneously decrease, it can be a reliable indication of a mud loss. In contrast, kick can be inferred when SPP decreases, but simultaneously ADP increases. Although this method improves the rapidity of the detection, it has several flaws and limitations:

- Some equipment pieces that must be installed on a rig and kept for maintenance.
- Steady-state flow conditions are required, otherwise the variation in SPP and ADP will exist, making the detection of the kick and losses very difficult.
- The changes in SPP and ADP can be caused by other downhole problems (plug nozzles, insufficient hole cleaning, and wellbore instability), which could mislead the interpretation of the data to identify the main causes of the variations.
- It can be used only for managed pressure drilling and underbalanced drilling methods.

16.7 SUMMARY

As mentioned earlier, depending on the drilling system (where open or closed), a different type of flowmeter may be the best option. Coriolis or a X-omega would be best suited to a closed system as they can form part of the flow system, and they need limited maintenance allowing for constant flow measurement including the mud density. Therefore, such meters are recommended be used in MPD (managed pressure drilling) and UBD (underbalanced drilling) applications because in these drilling systems there is an outflow pressure, and accurate inflow and outflow data as well as accurate drilling fluid density measurements are required. In contrast, for an open-system applications, other flowmeters will be preferred; for instance, the laser flowmeter and possibly the ultrasonic would be suitable because they both offer high accuracy and reliability compared to rolling flowmeter and paddle. However, an electromagnetic flowmeter can be an alternative of both in case only water-based mud is used. Table 16.1 shows a comprehensive comparison between the discussed flowmeters based on multiple criteria.

TABLE 16-1
Matrix for Flowmeter Performance Evaluation. Modified from Emerson Process Management (Truesdale, 2009)

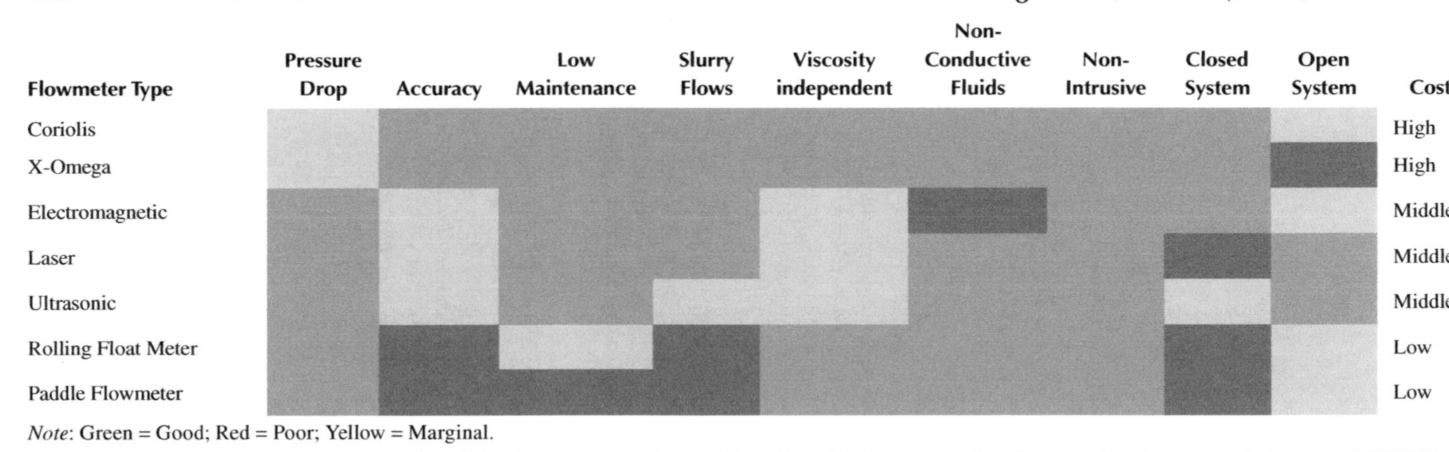

Flowmeter Type	Pressure Drop	Accuracy	Low Maintenance	Slurry Flows	Viscosity independent	Non-Conductive Fluids	Non-Intrusive	Closed System	Open System	Cost
Coriolis										High
X-Omega										High
Electromagnetic										Middle
Laser										Middle
Ultrasonic										Middle
Rolling Float Meter										Low
Paddle Flowmeter										Low

Note: Green = Good; Red = Poor; Yellow = Marginal.

In the second section of the chapter, both conventional and innovative systems developed to enhance kick and loss detection were thoroughly reviewed and analyzed.

REFERENCES

Belcher, G., Steiner, A., Schmigel, K., et al., 2008. *Annulus Pressure Control Drilling Systems and Methods.* European Patent Application, Patent No. EP1898044A3. https://patentimages.storage.googleapis.com/28/26/00/0c5037a489c1cb/EP1898044A3.pdf

Bunyamin, Husni, N. L., Basri, H., et al., 2010. Challenges in Turbine Flow Metering System: An Overview. *Journal of Physics: Conference Series*, 1198 (4). https://doi.org/10.1088/1742-6596/1198/4/042010

Casellas, H., Tonnessen, T. R., & Olsen, K. K., 2016. Step Changes in Drilling Fluid Measurement. *Oilfield Technology*, 23–26, July 1. https://pdf4pro.com/amp/view/step-changes-in-drilling-fluid-measurement-13493.html

Cayeux, E., 2020. Measurement of the Flowrate Out of a Well for Conventional Drilling Operations. Paper presented at the *IADC/SPE International Drilling Conference and Exhibition*, Galveston, TX. https://doi.org/10.2118/199661-MS

Cayeux, E., & Daireaux, B., 2013. Precise Gain and Loss Detection Using a Transient Hydraulic Model of the Return Flow to the Pit. Paper presented at the *SPE/IADC Middle East Drilling Technology Conference & Exhibition*, Dubai, UAE. https://doi.org/10.2118/166801-MS

Chris, R., & Simons, M., 2013. *Understanding and Selecting Coriolis Technology for Drilling Fluid Monitoring.* Micro Motion White Paper. www.emerson.com/documents/automation/white-paper-understanding-selecting-coriolis-technology-for-drilling-fluid-monitoring-micro-motion-en-64258.pdf

DiFoggio, R., & Blue, D. D., 2012. *Early Kick Detection in an Oil and Gas Well.* Patent No. US 2012/0170406 A1. https://patentimages.storage.googleapis.com/18/cf/b7/f1376ce558b164/US20120170406A1.pdf

Elmgerbi, A. M., Ettinger, C. P., Tekum, P. M., et al., 2021. Application of Machine Learning Techniques for Real Time Rate of Penetration Optimization. Paper presented at the *SPE/IADC Middle East Drilling Technology Conference and Exhibition*, Abu Dhabi, UAE. https://doi.org/10.2118/202184-MS

Forerunner Technologies LLC. *RFM-3600 Rolling Float Meter.* http://4runnertech.com/yahoo_site_admin/assets/docs/RFM-3600.103105606.pdf (Last Accessed in April 2024).

Fu, J., Su, Y., Jiang, W., et al., 2015. Development and Testing of Kick Detection System at Mud Line in Deepwater Drilling. *Journal of Petroleum Science and Engineering*, 135, 452–460. ISSN: 0920-4105. https://doi.org/10.1016/j.petrol.2015.10.013

Gruber, C. C., 2016. *Two-Phase Flow Investigations of Gas-Kick Scenarios.* Doctoral Thesis. Montanuniversität, Leoben, Austria.

Hummel, G., 2017. *Laser for Open Channel Flow Metering.* www.mi-wea.org/docs/Hummell%20Presentation.pdf (Last Accessed in August 2022).

Jacobs, T., 2015. Flow Sensor Technology Seeks to Replace the Coriolis Meter. *Journal of Petroleum Technology.* https://jpt.spe.org/flow-sensor-technology-seeks-replace-coriolis-meter

Johnson, A., Leuchtenberg, C., Petrie, S., et al., 2014. Advancing Deepwater Kick Detection. Paper presented at the *IADC/SPE Drilling Conference and Exhibition*, Fort Worth, TX, March. https://doi.org/10.2118/167990-MS

Knudsen, I., & Vermeulen, R., 2017. *A Drilling Rig and Method of Operating It.* Endress + Hauser Messtechnik, Patent No. US 2017/0145763 Al. https://patents.justia.com/patent/20170145763

Lavrov, A., 2016. Mechanisms and Diagnostics of Lost Circulation. In *Lost Circulation Mechanisms and Solutions Book.* Cambridge: Gulf Professional Publishing. ISBN: 978-0-12-803916-8.

Lee, J. S., Lee, S. O., Gray, D. D., et al., 2019. Visualization of Specific Energy for Open Channel Flow in Three Dimensions. *KSCE Journal of Civil Engineering*, 23, 2541–2549. https://doi.org/10.1007/s12205-019-2171-y

Leeungculsatien, T., & Lucas, G. P., 2013. Measurement of Velocity Profiles in Multiphase Flow Using a Multi-Electrode Electromagnetic Flow Meter. *Flow Measurement and Instrumentation*, 31, 86–95. ISSN: 0955-5986. https://doi.org/10.1016/j.flowmeasinst.2012.09.002

Matherne Instrumentation. *Mud Flow Sensors*. https://matherneis.com/product-item/mud-flow-sensors/ (Last Accessed in April 2024).

Nas, S., 2011. Kick Detection and Well Control in a Closed Wellbore. Paper presented at the *IADC/SPE Managed Pressure Drilling and Underbalanced Operations Conference & Exhibition*, Denver, CO, April. https://doi.org/10.2118/143099-MS

Norman, J., 2011. *Coriolis Sensors Open Lines to Real-Time Data*. www.drillingcontractor.org/coriolis-sensors-open-lines-to-real-time-data-10682 (Last Accessed in August 2022).

Orban, J. J., & Zanker, K. J., 1988. Accurate Flow-Out Measurements for Kick Detection, Actual Response to Controlled Gas Influxes. Paper presented at the *IADC/SPE Drilling Conference*, Dallas, TX. https://doi.org/10.2118/17229-MS

Orban, J. J., Zanner, K. J., & Orban, A. E., 1987. New Flowmeters for Kick and Loss Detection During Drilling. Paper presented at the *SPE Annual Technical Conference and Exhibition*, Dallas, TX. https://doi.org/10.2118/16665-MS

Pirir, I., Jinasena, A., & Sharma, R., 2017. Model Based Flow Measurement Using Venturi Flumes for Return Flow During Drilling. *Modeling, Identification and Control*, 38 (3), 135–142. https://doi.org/10.4173/mic.2017.3.3

Reason, J., 1990. *Human Error*. Cambridge: Cambridge University Press.

Roger, C. B., 2000. *Flow Measurement Handbook Industrial Designs, Operating Principles, Performance, and Applications*. Cambridge: Cambridge University Press (Verlag). ISBN: 978-0-521-48010-9. http://catdir.loc.gov/catdir/samples/cam032/99014190.pdf

Rogers, J. B., 2015. *Systems and Methods for Monitoring Drilling Fluid Conditions*. Patent No. US 2015/0211362 A1. CHEVRON U.S.A. Inc. https://patents.justia.com/patent/20150211362

Santos, O. L. A., Williams, W. C., Sharma, J., et al., 2021. Use of Fiber-Optic Information to Detect and Investigate the Gas-in-Riser Phenomenon. *SPE Drill & Completion*, 36, 798–815. https://doi.org/10.2118/204115-PA

St. John, M. F., 2016. Macondo and Bardolino: Two Case Studies of the Human Factors of Kick Detection Prior to a Blowout. Paper presented at the *Offshore Technology Conference*, Houston, TX, May 2–5. OTC-26906-MS. https://doi.org/10.4043/26906-MS

Thorogood, J. L., Lauche, K., Crichton, M., et al., 2015. Getting to Grips with Human Factors in Drilling Operations. Paper presented at the *SPE/IADC Drilling Conference and Exhibition*, London, UK, March 17. SPE-173104-MS. https://doi.org/10.2118/173104-MS

Truesdale, P., 2009. *Selecting Technology Required*. www.slideshare.net/JimCahill/reconciling-mass-and-energy-balances-in-an-ethylenecomplex (Last Accessed in August 2022).

Watral, Z., Jakubowski, J., & Michalski, A., 2015. Electromagnetic Flow Meters for Open Channels: Current State and Development Prospects. *Flow Measurement and Instrumentation*, 42, 16–25. https://doi.org/10.1016/j.flowmeasinst.2015.01.003

Zhou, G., Leach, C., Denduluri, V. S., et al., 2021. Pressure-Difference Method for Gas-Kick Detection in Risers. *SPE Journal*, 26, 2479–2497. https://doi.org/10.2118/205362-PA

Zhou, Q., Zhao, H., Zhan, H., et al., 2013. The Application of Ultrasonic Based on Doppler Effect Used in Early Kick Detection for Deep Water Drilling. *International Conference on Communications, Circuits and Systems (ICCCAS)*, 488–491. https://doi.org/10.1109/ICCCAS.2013.6765389

17 Mud-Gas Separator Analysis

A Comprehensive Review of Sizing, Design Criteria, and Industry Standards

Daniel A. Tetteh, Saeed Salehi, Ali Ghalambor and Nabe Konate

ABBREVIATIONS

ABS	American Bureau of Shipping
AOBMGS	Atmospheric Open-bottom Mud Gas Separators
API	American Petroleum Institute
ASME	American Society of Mechanical Engineers
BOP	Blowout Preventer
BSEE	Bureau of Safety and Environmental Enforcement (BSEE)
CFD	Computational Fluid Dynamics
FDS	Functional Design Specifications
HPHT	High-Pressure, High-Temperature
IADC	International Association of Drilling Contractors
ID	Internal Diameter
MGS	Mud-Gas Separator(s)
MMSCF/D	Million Standard Cubic Feet
MPD	Managed Pressure Drilling
MR	Material Requirements
NACE	National Association of Corrosion Engineers
NORSOK	Norwegian Shelf's Competitive Position
NPS	Normal Pipe Size
NPT	Nonproductive Time
OBM	Oil-Based Mud
PRV	Pressure Regulating Valve
PSV	Pressure Safety Valves
QHSE	Quality Health Safety and Environment
R & D	Research and Development
RGH	Riser Gas Handling

DOI: 10.1201/9781003473770-17

RP	Recommended Practice
SCR	Slow Circulation Rates
SG	Specific Gravity
SGHMGS	Surface Gas Handling System and Mud Gas Separator Design
SPE	Society of Petroleum Engineers
SWP	Safe Work Practices (SWP)
TOGA	Total Gas Containment
UBD	Underbalanced Drilling
WBM	Water-based Mud
H_2S	Hydrogen Sulfide

17.1 INTRODUCTION

Gas kicks are one of the fundamental challenges associated with oil and gas drilling operations and occur when formation pressures exceed the pressure exerted by drilling fluids. When improperly handled, gas kicks can result in blowouts. Blowouts are catastrophic and can result in loss of lives and properties. Well control in oil and gas drilling operations involves all measures to mitigate the uncontrolled flow of formation fluids from the wellbore to maintain control of the well (Bourgoyne, 1997; Grace, 2017). Well control systems facilitate early detection and circulation of gas kicks to prevent blowouts. During well control operations, gas kicks are routed to the surface with the aid of drilling fluids while maintaining the pressure within the wellbore to prevent further influx of formation fluids. The gas-cut drilling fluids are subsequently directed to the mud-gas separator.

The mud-gas separator is an essential component in well control systems. It is sometimes referred to as the "gas buster" or "poor-boy degasser". It is the primary equipment for separating gases from gas-cut drilling fluids before reconditioning and recirculating the drilling fluids. In most instances, small quantities of gases remain in the drilling fluid after separation in the MGS and are removed by the vacuum degasser. Furthermore, the "poor-boy degasser" also plays a vital role in mitigating the challenges associated with having to shut down an entire well to prepare fresh mud for recirculation after circulating kicks from the wellbore. The MGS also facilitates trip gas handling, i.e., gas entering the wellbore due to the movement of the drill-string (Bland et al., 2006).

This work presents an extensive review of MGS sizing and design considerations. The study also presents findings from various consultations with industry stakeholders' engagement for design criteria used in their affiliated companies.

17.2 PRINCIPLE OF OPERATION

The MGS operates by a simple process. During kick circulation, gases in the gas-cut drilling fluid expand as gas bubbles and reach the surface due to decreased pressure. These bubbles form gas pockets at the surface, pushing the drilling fluids ahead of them and increasing the speed of the drilling fluid in the process. However, reducing the drilling fluids' velocity before they reach the mud pits is important to prevent accidents. Drilling fluids containing formation gas enter the separator via the choke-line and a separator inlet (MI Swaco, 2018). In the separator, the fluid is first routed toward an impingement plate perpendicular to the fluid flow direction. The impingement plates reduce the wear of the internals of the MGS due to erosion (MacDougall, 1991). From the impingement plates, gas-cut drilling fluids move toward a series of angled internal baffle plates. Fluids flow over these plates, and gases are separated from the liquids in the process. The separated gas is then vented through the vent line at an overhead position to a safe place on the rig. The resulting drilling fluid, on the other hand, is routed to the mud pits, where it is reconditioned. In conventional MGS designs, a float ball is on top of the mud and controls an inlet valve at the mud return line. On the one hand, when the mud level rises beyond a set threshold, the valve opens to allow some of the drilling fluid out of the separator to prevent it from being overfilled (MI Swaco, 2018). On the other hand, the valve closes when the mud level in the separator drops below a set threshold value, increasing the volume of drilling fluid within the separator. As shown in Figure 17.1a, most mud-gas separators have a U-tube mud leg design, generating a mud seal via its hydrostatic fluid head. The mud seal is sometimes called the MGS's liquid seal. Also, gases escaping through the vent line of the MGS generate a backpressure. For cases where a severe expansion of fluids entering the separator occurs, a backpressure valve assembly is used to dampen the pressure from the expansive gas to reduce its surge effects. Furthermore, the pressure in the mud seal must always be greater than the vent line friction pressure to prevent separator blow-through (sometimes called separator blowby or blowdown). Separator blow-through is an unwanted and fatal phenomenon where gaseous drilling fluid is "blown through" the separator to the shaker areas. Blow-through causes rig fire outbreaks and exposure of rig crew to hazardous gases like H_2S. Most conventional MGS designs have a siphon breaker line for preventing mud siphoning from the separator to the mud tanks. This line is located at the highest point of the mud return line and acts as a backup for the disposal of formation gas that passes through the U-tube mud seal. Typical challenges recorded from the industry on the usage of

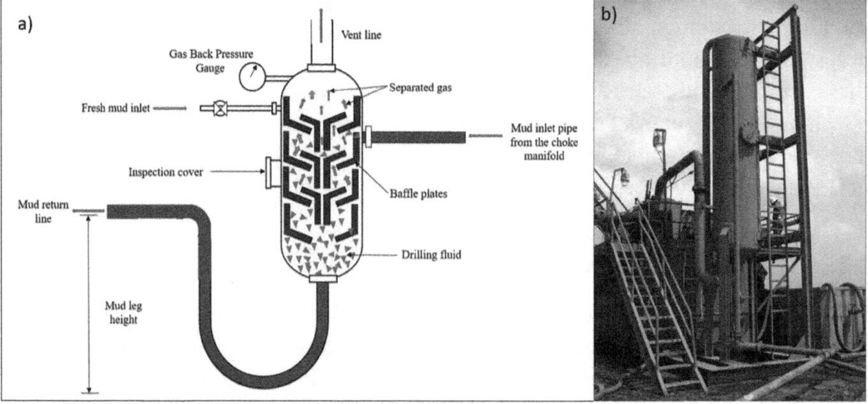

FIGURE 17.1 (a) Schematic representation of the mud-gas separator and (b) rig vertical MGS (Eby, 2008).

float-type MGS are the failure of the manual linkage between the float and the valve and frequent clogging of the mud return line by solids in the drilling fluid.

The ability of the MGS to efficiently separate gas from liquid is known as the separation capacity. It primarily depends on the number, size, and orientation of the baffle plates and the fluid dynamic properties of the separator (Lee, 2016). MGS's venting capacity, however, is its ability to vent gas safely and efficiently through its vent line when the separator is operating at its maximum pressure differential. The venting capacity of the MGS is a factor of the vent line backpressure and the hydrostatic pressure in the liquid seal. The venting capacity of the MGS is reduced when the vent line is long, and a significant number of bends exist in the vent line. Also, the venting capacity is reduced when high-density gas is present. The separator's liquid re-entrainment capacity is the measure of the ability of liquid droplets to suspend in the gas phase after being separated from the gas/liquid interface. At the gas/liquid interphase, momentum is transferred from the gas phase to the liquid phase due to changes in pressure. The agitation eventually causes the liquid droplets to separate from the liquid phase (Lee, 2016).

There are three major types of mud/gas separators, and they all have different operating procedures. The most common is the closed-bottom MGS, equipped with a U-shaped bend on the mud return line and a hydrostatic leg to generate mud seal. The fluid level in this separator can be controlled by adjusting the length of the U-shaped bend. The open-bottom MGS is another type of mud/gas separator. As the name suggests, the open-bottom MGS is opened at the bottom and entirely immersed in the mud. The mud level is also controlled by moving the separator vertically up or down. The height of the mud tank puts a restriction on the height of the mud level. This is an inherent limitation of the open-bottom mud-gas separator and, thus, is not very common. The final type of MGS is the float type. Unlike the closed and open-bottom MGSs, the float-type has a float valve configuration for manipulating the fluid level in the mud leg. Hence, the float-type MGS is not highly recommended.

The safe and effective separation of gases from gas-cut drilling fluids is a function of the MGS' separation, venting, and liquid re-entrainment capacities. These properties depend on MGS sizing considerations like vent line diameter and length, vessel diameter, hydrostatic head of the mud seal, vent line backpressure, and the throughput capacity of the MGS, among others. Moreover, there is no thorough understanding of these sizing considerations. Hence, there is an associated risk with the operation of the MGS.

17.3 SIZING AND DESIGN OF MUD-GAS SEPARATORS: A REVIEW

MGS sizing is vital in ensuring safety and efficiency in its operations (Oriji and Okechukwu, 2019), especially when high fluid flow rates are encountered (Patil et al., 2018a). Two major MGS designs are vertical and horizontal (Lee, 2016). Horizontal separators can handle larger fluid volumes, have larger surface areas, and can retain drilling fluids for longer periods. However, due to their large space requirements, horizontal separators are not preferable in rig settings with limited spaces. However, vertical separators (Figure 17.1(b)) are most suitable for rigs with limited space as they take less space and possess better separation efficiencies than horizontal separators. A common challenge for both separators, however, is hydrate plugging.

This section reviews and compares past and existing criteria and considerations for MGS sizing. The review scope encompassed both conventional and unconventional drilling techniques (i.e., underbalanced and managed pressure drilling), some special wellbore conditions, and the variability in formation pressures and temperatures.

One of the early designs of the MGS was described Louison et al. (1984) for an underbalanced drilling job. The process involved drilling a 9–7/8-inch hole to a depth of approximately 9,600 ft. The hole size was then reduced to 8 ½ -inch as the mud weight was raised from 9.6 to 11 ppg. According to the authors, after the Blowout Preventer (BOP) nipples up over the surface casing, a rotating head, a pressurized flow line, and a large MGS with flares were also installed. The MGS had a 5-ft internal diameter and a height of 14 ft. Furthermore, an important aspect of this well control set up is that the fluid from the well goes directly to the MGS without flowing through the choke manifold, which is the case for most conventional separators. The authors also mentioned the usage of a standard vacuum degasser in conjunction with the MGS. The schematic of the drilling and well control setup, including the MGS, is shown in Figure 17.2.

Furthermore, another early design of the MGS was described by Butchko et al. (1985). The authors described sizing considerations for the Atmospheric Open-Bottom Mud-Gas Separators (AOBMGSs) based on work done by the Canadian Petroleum Association. According to the authors, a key criterion considered for sizing the vessel is the maximum anticipated gas flow rate (the Absolute Open Flow). Besides the peak gas flow rate, the authors also mentioned MGS internal diameter, vent line sizing, vessel internal configurations, material type, and fabrications as other key elements considered in sizing. The authors described a rigorous and straightforward method

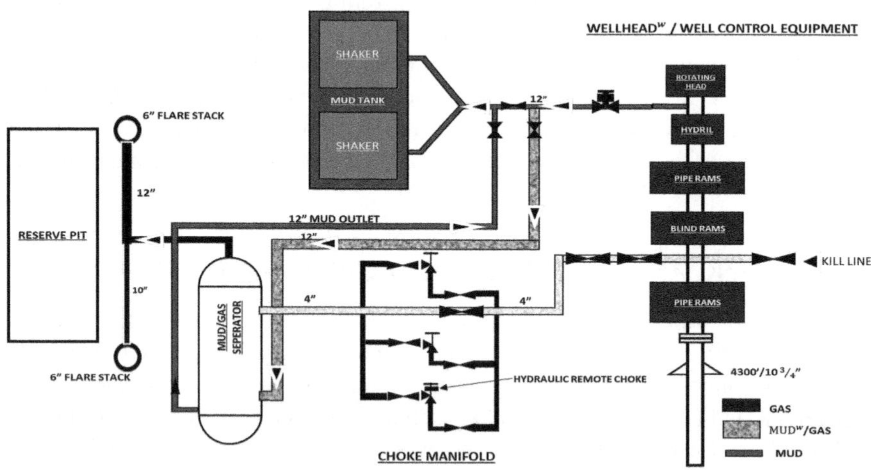

FIGURE 17.2 Schematic diagram of drilling and well control equipment – modified from Louison et al. (1984).

for determining peak gas flow rate. The former depends on the reservoir properties, while the latter assumes a constant density for gas and mud at the surface. The vessel diameter was determined on the basis of peak gas flow rate and fluid velocity using a set of equations and input data, as shown in Appendix A. Furthermore, based on these equations, the authors suggested the minimum height of the vessel to be 12.3 ft and required vessel diameter of 36 inches (3 ft) for a worst-case kick volume of 12.57 bbl and an unrealistically high circulation rate of 132 gal/minute. The proposed diameter by Butchko et al. (1985) is smaller than that of Louison et al. (1984) discussed earlier. Also, the minimum vent line length determined was 183 ft (55.8 m) according to the state of Alberta's regulations, which require the flare pit to be more than 50 m from the rig. Furthermore, the vessel should have impingement plates on its internals to prevent the baffle plates from wearing out and increase the surface area for gases to break out. The authors also recommended routine inspection of MGS components as erosional problems may occur due to high separator flow rates. The minimum required dimensions for the AOBMGS as proposed by Butchko et al. (1985) are shown in Figure 17.3.

Marsh and Altermann (1988) described the arrangement of surface gas handling equipment downstream of the choke and kill manifolds for handling marine riser gases in well control. The authors mentioned that this arrangement was suitable for moderate to deepwater drilling. The gas handling equipment described consisted of a twin-pump degasser, a mini-trip tank, a general-purpose vent line with a 12-inch diameter rising to the derrick's height, and a high-capacity MGS. The MGS was 10 ft tall and had a diameter of 48 inches (4 ft) (Figure 17.4(a)), which is bigger than that proposed by Butchko et al. (1985) for a worst-case kick volume and smaller than that proposed by Louison et al. (1984). Furthermore, the high-capacity MGS had a buffer tank and inlets opposite each other to mitigate erosion. The authors also described a gas handling system for the marine riser returns, which consists of telescoping joints, diverter packer seals, overboard diverter lines, a secondary flow line, a two-way 12-inch selector valve, and horizontal MGS which is 10 ft long with a diameter of 4 ft. The arrangement also consisted of two centrifugal fans for gas removal. During operations, the diverter packer is closed, and mud returns are routed toward the horizontal MGS or overboard downwind via the two-way flow selector valves, as shown in Figure 17.4(b). The choice of routing fluid toward the MGS or overboard downwind depends on the expected gas heading.

Riser unloading occurs when the gas in the riser is not detected early and adequately removed before it reaches the surface due to gas dissolution in oil-based muds. Riser gas unloading may be catastrophic as flammable gas may break out from the riser near the surface (Yuan et al., 2017). Common methods for handling gases in the riser include diverting gas flow overboard or directing the fluid flow with the gases to the shale shakers. Both of these methods have some notable flaws.

Collings and Stone (1989) proposed a method for handling the challenge of riser gas unloading due to drilling with oil-based muds. They introduced a two-stage mud-gas discharge system (Figure 17.5) for controlling drilling fluids where there is riser unloading. According to the authors, the proposed system can ensure the maintenance of the mud seal. The "first separator" in the two-stage MGS system is 15 ft tall and has a 48-inch diameter. It is equipped with a pneumatic float regulator for

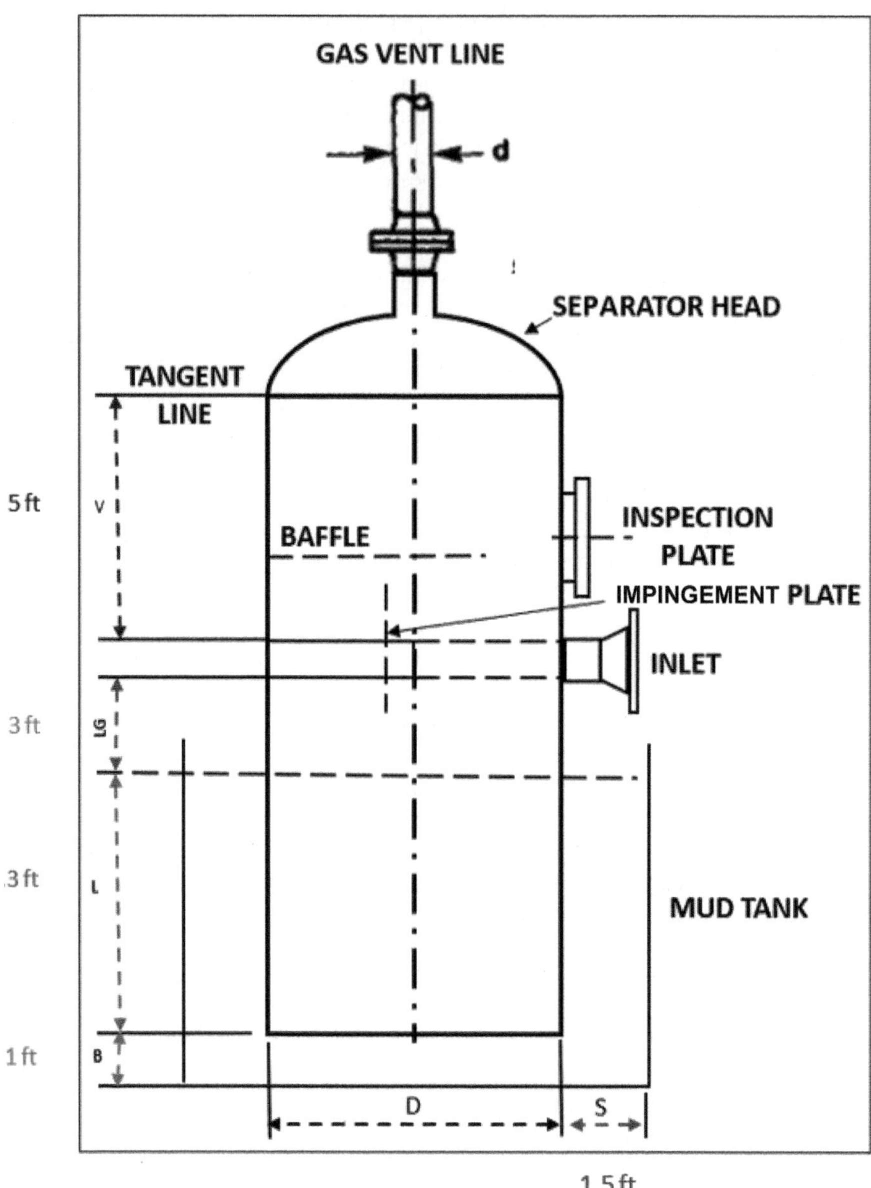

FIGURE 17.3 Minimum required dimensions for a typical open-bottom mud-gas separator. Note: B, open underflow; D, separator diameter; B, open-bottom underflow; L, liquid level; LG, Liquid–gas disengagement space; V, vapor space (Butchko et al., 1985).

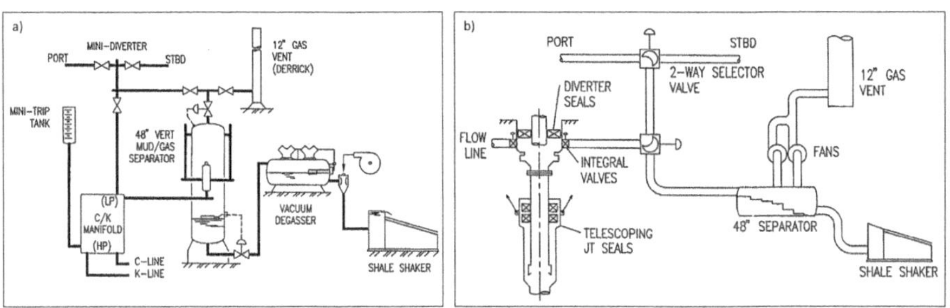

FIGURE 17.4 (a) MGS downstream the choke and (b) horizontal MGS dedicated to riser gas control (Marsh and Altermann, 1988).

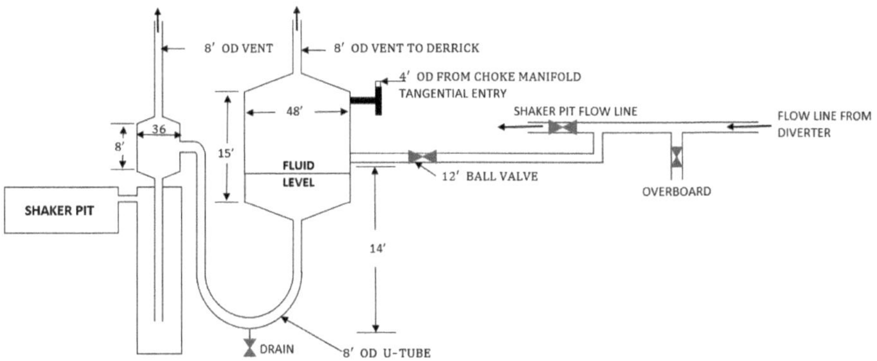

FIGURE 17.5 A two-stage mud-gas discharge system (Collings and Stone, 1989).

shutting down the mud discharge line when the mud seal is lost. In addition, its mud inlet line has a 12-inch internal diameter, whereas the fluid outlet and gas derrick vent lines had internal diameters of 8 inches. Also, the height of the mud seal for this separator was 14 ft. On the other hand, the "second separator" is downstream of the first, and drilling fluids are routed through it before reaching the mud tanks. It is 8 ft tall with a separator internal diameter of 36 inches and a vent line outer diameter of 8 inches. As per the authors' conclusions, including bleed valves between the annular and the flow line of the MGS ensures the safe discharge of trapped stack gas via the eight-inch outer diameter vent line. This reduces mud loss and pollution.

Hoopingarner et al. (1990) reported key MGS design and sizing recommendations proposed by a working committee of the International Association of Drilling Contractors (IADC) for drilling high-pressure, high-temperature (HPHT) (i.e., pressures ranging from 2,500 psi to 3,500 psi and temperatures ranging from 130°F and 160°F) wells. This followed concerns raised by the UK government on drilling such wells after the explosion of the "Ocean Odyssey", a semi-submersible rig, in 1988. According to the authors, the recommended vent line internal diameter for the

North Sea is 8 inches, although 10 inches was determined to be better. The proposed 8-inch internal diameter is similar to that proposed by Collings and Stone (1989). Furthermore, the committee recommended continuous maintenance of the liquid seal even during kill operations. Also, when dry gas is vented, it is recommended that the flow rate be lower than both the blowdown and separation capacity during the discharge of contaminated mud. A monitoring system and recording facility are also needed at the choke control station to keep track of the pressures on the separator, ensuring that the blowdown capacity is not exceeded.

An industry-renowned comprehensive work on MGS sizing and evaluation was done by G.R. MacDougall (MacDougall, 1991). The author presented a comprehensive approach for sizing the MGS as well as considerations for an economical upgrade. The author's methodology focused on preventing separator blowdown by optimizing certain key design parameters. According to MacDougall, separator blowdown occurs at peak gas flow rates and thus presented appropriate formulae based on kick data for determining peak gas flow rates, internal diameter of the vessel, and vent line friction pressure. The author also provided special MGS design considerations for oil-based muds, which may result in higher peak gas flow rates. The author further elaborated on a method for upgrading an existing MGS to facilitate rig bid analysis. The first step in the proposed method is to determine the MGS's internal diameter (ID). In instances where the ID is smaller than required to prevent a separator blow-through, the kill rate can be reduced to increase the mud/gas mixture retention time to improve the overall efficiency. Reducing the kill rate also reduces the peak gas flow rate and subsequently reduces vent line friction pressure such that it does not exceed hydrostatic pressure in the mud leg. The overall outcome of this process is an improvement in the separator's efficiency. The author analyzed a worst-case scenario kick size of 25 bbl and presented a summary of sizing dimensions of the MGS as follows: The inside diameter of 36 in is the same diameter as used by Butchko et al. (1985) for their worst-case scenario. The evaluated vessel height was 25 ft (7.6 m), which is greater than double the minimum required vessel height mentioned by Butchko et al. (1985). Furthermore, the recommended flare line diameter was 7-in, and the flare line length of the rig was 165 ft (50.2 m), indicating that MacDougall's standard is like that of Butchko et al. (1985).

Besides conventional drilling, MGS sizing is also very crucial in Managed Pressure Drilling, especially due to the comparatively higher fluid flow rates experienced.

MacDougall's method for determining the MGS' liquid and gas handling capacities was adapted and evaluated by Patil et al. (2018b) for applications in MPD. Their approach used a complex simulation model that considered gas compressibility, pressure, temperature, and density to estimate the frictional pressure loss in the vent line. Furthermore, the authors used a computational fluid dynamic (CFD) simulator to qualitatively examine the fluid velocities and gas separation from mud within the vessel to avoid separator blow-through. Figure 17.6 is a pictorial view of results from the CFD model indicating the velocity contours.

The authors also factored in the effects of gas deviation factors and gas gravity on the Atkinson-modified Darcy–Weisbach equation used by MacDougall (MacDougall, 1991); however, MacDougall did not consider these factors in his determination of

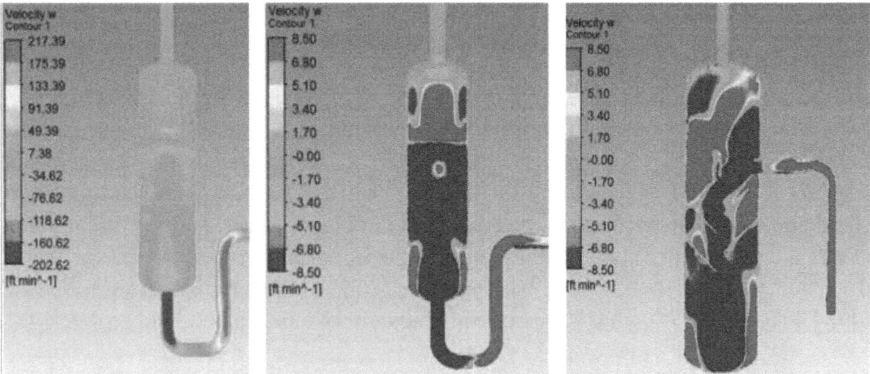

FIGURE 17.6 Pictorial view of results from the CFD model indicating the velocity contours (Patil et al., 2018b).

the vent line friction pressure. This was important in their approach as the critical time at which blow-through occurs at peak gas flow rates is dependent on the vent line friction pressure. Patil et al. (2018b) highlighted the importance of increasing the internal diameter of the MGS for MPD applications to reduce the velocity of the liquid–gas mixture entering the vessel. They also mentioned that gases entering the separator at high velocities and dispersing at the impingement plates caused very high velocities to reach the mud leg in the simulations; thus, there is the need for further studies on the geometry and orientation of the baffle and impingement plates to reduce the velocity contours inside the separator. This is important for the separator to be able to handle large gas rates. Finally, the authors also highlighted vessel height, internal diameters, and geometry as important factors on which the separation capacity of the vessel is dependent.

The use of a special MGS in drilling offshore wells in Abu Dhabi via MPD was reported by Al Hadidy et al. (2019). While many details were not given on the specific dimensions of the MGS used in this case study, the authors reported that the separator had a bigger diameter to accommodate the high fluid flow rates and hence handled gases separated from the mixtures faster.

17.4 MGS SIZING AND DESIGN CONSIDERATIONS FROM STAKEHOLDER ENGAGEMENT

As part of the stakeholder engagement exercise, the drilling manuals of some oil and gas operator companies were consulted for specifications on MGS sizing. Furthermore, target-specific industry personnel (i.e., MGS subject matter experts, well control specialists, and manufacturers of MGS) were also engaged via questionnaires and interviews on MGS design and sizing. Due to confidentiality reasons, the names of some of the businesses engaged are kept anonymous and represented with arbitrary letters, e.g., Company A and Company B.

17.4.1 MGS Design Specifications from Company A: An Oil and Gas Operator Company

In reference to the drilling manual of this operator company, three key factors need to be considered in MGS sizing: the vent line length, the vessel diameter, and the maximum expected gas flow rates when there is a severe kick. In addition, a bypass line to the vent line needs to be installed to cater for malfunctions, and measures should be put in place to reduce the erosion of impingement plates when drilling fluid enters the separators. The manual indicated the importance of using a 10-to-12-inch diameter vent line to minimize backpressure at the transition point from the vessel to the vent line. The manual also recommends the U-tube-shaped mud seal design.

17.4.2 MGS Design Specifications from Company A: An Oil and Gas Operator Company (Company B)

Table 17.1 summarizes the sizing considerations obtained from a national oil company. The vessel's internal diameter employed by this company in their design is less than that of MacDougall (1991) and Butchko et al. (1985). Furthermore, the specified vessel height is 19.68 ft, which is greater than the minimum required height of 12.3 ft by Butchko et al. (1985). Compared to MacDougall's specifications, the gas outlet line proposed by company B is six inches smaller. MacDougall proposed a diameter of seven inches.

17.5 ANALYSIS OF RESPONSES FROM ONLINE SURVEY AND STAKEHOLDER INTERVIEW

The questionnaires for the online surveys and stakeholder interviews inquired general information about MGS types, sizing and design, considerations, industry

TABLE 17.1
MGS Sizing Considerations of Company B

No.	*Parameter*	*Value*
1	Vessel Inside Diameter	39.37 in (1 m)
2	Vessel Height	19.68 ft (6 m)
3	Capacity	352 gpm (80 m^3/hr)
4	Inlet Line Diameter	4 (in)
5	Outlet Line for Gas	6 (in)
6	Outlet Line for Liquid	8 (in)
7	Working Pressure	150 psi (1 MPa)
8	Pressure Relief Valve	Yes
9	Distance to Burn Pit	1,312 ft (400 m)
10	Connections	Flanged

standards and regulations, and hazards and challenges associated with MGS usage for both onshore and offshore applications. This section discusses the major findings thereof.

The survey results indicated that the hydrostatic pressure within the mud leg and the vent line friction pressure are the most important parameters considered in MGS sizing using known industry methods. Besides, other factors like baffle plate orientation and measures put in place during vessel design to prevent hydrate plugging were among the least considered. Furthermore, drilling fluid type, formation pressure, temperature, and the presence of sour gases were also reported as factors considered in MGS sizing.

In addition, SPE publications, API, and NORSOK standards, according to the survey results, are the primary regulations that govern the sizing and design of MGS, besides others like the International Association of Drilling Contractors (IADC) and the American Society of Mechanical Engineers (ASME) standards.

Furthermore, the feedback from respondents indicated that the working pressure of the MGS depends on the specific well control application; thus, there is a need to conduct a thorough risk assessment, including worst-case scenarios of possible challenges, before deciding on MGS working pressures. Typical working pressures reported by respondents were between 20 and 30 psi. Some respondents also mentioned that although the working pressures of the MGS can be as high as 100 psi for high-risk wells, they do not exceed 264.7 psi (i.e., 250 psig), with both the mud leg and vent line operating at ambient pressures.

Additional information from the responses was that when there is excessive friction in the vent line, the mud leg blows dry, resulting in gas venting into the pit area. In such instances, the panic line is opened for the combined stream to be vented into the pit. These are, however, scarce systems.

Also, respondents indicated that the blowdown and separation capacities depend on the anticipated kick volumes and time required to stop the influx. However, a blowdown will occur when the MGS's working pressure (as determined by vent line friction pressure and pressure within the mud leg) is exceeded. Other responses indicated that not to exceed the blowdown capacity of the separator, the circulation (kill) rates used are relatively low, ranging from 20 to 40 strokes per minute. In other words, typical circulation rates of 150 gal/minute are expected for most MGS, although modern separators can handle up to about 1,000 gal/minute. Formation properties were also highlighted as key factors to consider in MGS sizing.

Regarding the internal configurations of the MGS, responses indicated that the standard industry design of baffle plates is acceptable in preventing vessel overpressurization. In contrast, the roughness of the internals of the vessel and changes in its diameter are not major sizing considerations. The typical height of the fluid leg is 10–15 ft for MGS sizing, and the fluid density depends on the mud weight and the quantity of gas being circulated.

According to the survey responses, the existing internal diameter for high-pressure, high-temperature wells is between 8 and 12 inches. The 12-inch vent lines are mainly for high-risk situations. It is crucial to ensure appropriate vent line internal diameter (ID) and vessel diameter sizing to prevent gas blowby.

17.6 EVALUATION OF RESULTS FROM INTERACTIONS WITH WELL CONTROL SPECIALISTS, SUBJECT MATTER EXPERTS, AND MANUFACTURERS

One important point discussed with some members of the IADC was the deliberations on updating the Petroleum industry guidelines for the design of the MGS. This will focus on maintaining a constant liquid flow rate through the MGS rather than building high-capacity separators. A survey respondent from a US-based drilling contractor reported that a standard MGS setup used in the company's industrial setting is the Total Gas Containment (TOGA) concept (Schlumberger, 2023). TOGA is a completely enclosed gas-separation system that combines the MGS and the degasser to separate and vent or flare gases, as shown in Figure 17.7(a). Also, the vent line is usually directed toward the flare stack where the separated gas is burnt. There are usually several bends in the vent line directing it toward the flare stack. The typical MGS design considerations for conventional drilling and well control operations are such that the pressure values range from 7 to 10 psi. As a result, the height of the fluid leg is usually between 10 and 12 ft and seldom over 15 ft because resizing the MGS is rarely done, as the standard industry practice is controlling pumping rates to stay within operational limits.

An oil and gas consulting company (Company D) respondent stated that sizing considerations for maximum vessel internal pressure consider a worst-case scenario for the hydrostatic pressure in the mud leg. A 5.77-ppg drilling fluid density is used to design maximum flow rates for such cases. Also mentioned by the respondent was the need for the proper estimation of separation and liquid entrainment capacities, as they are vital in sizing the MGS. Furthermore, the criteria for calculating the MGS separation capacity do not usually consider the baffle plates and other agitator devices within the separator.

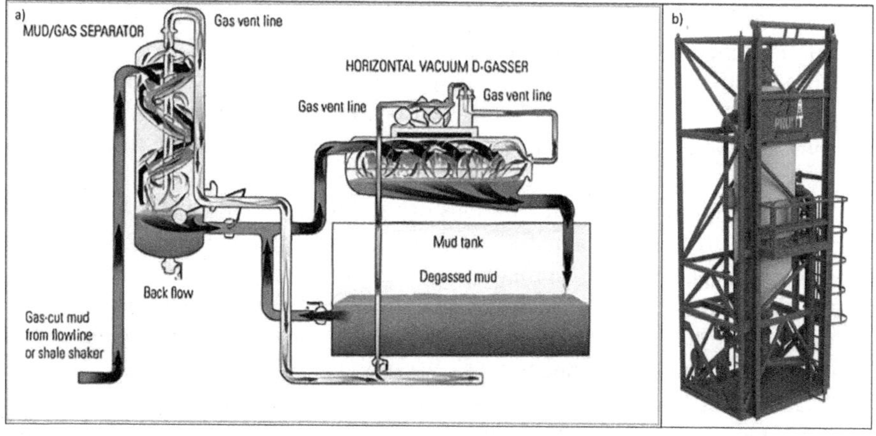

FIGURE 17.7 The Total Gas Containment (TOGA) system (Schlumberger, 2023) and (b) Drill Thru MGS for MPD (Pruitt, 2023a, 2023b).

The sizing criteria for a special MGS were also evaluated. An MPD equipment manufacturing company respondent introduced the Drill Thru MGS (Figure 17.7(b)) for MPD applications. This special MGS is designed to withstand a drilling rate of 1,000 gallons per minute. Its maximum gas rate is based on its ability to handle 10 MMSCF/D of gas with a mud weight of specific gravity of 1, where the vent line has a height of 180 ft and a diameter of 8 inches. The gas venting system of this separator has a vertical flare stack equipped with a 21-inch NFPA-68-compliant Flame flashback prevention device. Also, the product specifications mention an increase in its capabilities with an increase in mud weight. The Drill Thru MGS also has a centrifugal inlet device that helps to make gas separation from liquid more efficient. It has a mist extractor, alleviating the potential risks of condensate formation in the gas lines. It also has specially designed baffles with high angles and deliberate gaps between them, allowing the separated gas to rise. The bottom of the separator is conical, reducing the tendency to build drill cuttings. The MGS also has a hot loop that flushes the conical bottom to ensure the U-tube seal is clean, and the mud weight is always known. The vessel design conforms to the ASME VIII U stamp vessel rating code (i.e., ASME VIII Division 1-Rules for Construction of Pressure Vessels) and is equipped with a Pressure Regulation Valve (i.e., an over-pressure burst disc), which provides extra protection.

17.7 A REVIEW OF INDUSTRY STANDARDS, REGULATIONS, BEST PRACTICES, AND GUIDELINES ON THE MGS

In this section, a review of existing standards recommended practices, guidelines, and regulations on the design and operation of the mud-gas separator is conducted. Table 17.2 is a summary of the materials reviewed in these areas.

17.8 A COMPREHENSIVE EVALUATION OF REVIEWED REGULATIONS AND STANDARDS

The reviewed standards in Table 17.2 are evaluated in this section. As shown above, the NORSOK D-001 (1998) document, 2nd edition, discusses the degasser, which usually works in conjunction with the MGS. According to the document, the volumetric capacity of the degasser ranges from 320 to 450 m^3/h (i.e., approximately 1,410–1,980 gal/minute). These are high flow rates and require the flow rates of the MGS with which it works to be high as well to commensurate with the degasser. Furthermore, the document recommends that the vent line be 4 m from the derrick and have a minimum diameter of 10 inches. This is similar to the values discussed in the literature review section, although conventional vent line diameter values range between 8 and 12 inches, indicating that the 2nd edition of the NORSOK document is stricter. Also, the standard recommended that the atmospheric MGS be sufficient and should withstand wellbore fluids with S.G. of 2.2 to the top of the vent line. It also recommends a liquid seal height of 6 m, with which the liquid seal pressure (LSP) can be determined via Equation (17.1).

$$\text{LSP} = \text{hpg} = 2.2 \times 1000 \times 9.806 \times 6 = 129439 \text{ pa} = 18.76 \text{ psi} \tag{17.1}$$

TABLE 17.2

A Summary of Standards, Recommended Practices, Guidelines, and Regulations on MGS Sizing, Design and Operations

Reference	Area of Interest	Recommended Practices, Procedures, and Description
API RECOMMENDED PRACTICE (RP) 53: RECOMMENDED PRACTICES FOR BLOWOUT PREVENTION EQUIPMENT SYSTEMS FOR DRILLING WELLS. SUB-SECTION 15.9: MUD GAS SEPARATOR (API, 2012) **API RP 59: RECOMMENDED PRACTICE FOR WELL CONTROL OPERATIONS. SUB-SECTION 7.7.5: MUD GAS SEPARATORS** (API, 2006)	MGS operational capacity, Impingements plates, MGS vessel cleaning	Installation of a bypass line to the flare stack in case of any separator malfunction or if the separator's capacity is exceeded. Precautions must be taken to prevent erosion of impingement plates on the walls of the vessel. Provisions should be made to facilitate easy cleaning of the vessel and its lines where there is plugging.
API STANDARD 53: WELL CONTROL EQUIPMENT SYSTEMS FOR DRILLING WELLS. 5TH EDITION. SUB-SECTION 4.4.7: MUD GAS SEPARATORS; SUB-SECTION 4.6.2: EQUIPMENT VERIFICATION (API, 2018)	Recommended Practices for MGS Operation	The MGS shall not be used for well production and testing operations. Flow through the MGS shall always be controlled. Uncontrolled fluid flow (e.g., diverter flow shall not be directed to the MGS). The MGS vent line and discharge piping outlets shall be routed when handling H_2S gas to ensure the hazardous gas is appropriately discharged during well control. Block ells and tees of the choke and kill-lines should be targeted or have fluid cushions installed in the direction of flow or in both directions if a bidirectional flow is expected. Flare/vent lines should be as straight as possible to limit back-pressure. Remote kill-lines should be connected to the kill-line near the BOP stack and extended to an auxiliary high-pressure pump at a safe location. The kill-line should not be used as a fill-up line.

NORSOK D-001 (DRILLING FACILITIES, 2ND REVISED VERSION), SUB-SECTION 5.9.3.5: DEGASSER (NORSOK, 1998)	MGS Design	The design of the MGS should be such that it can handle all anticipated flow rates for the given well. It should safely separate the gases within the drilling fluid and safely vent the gas.
NORSOK D-001 (DRILLING FACILITIES, 2ND REVISED VERSION), SUB-SECTION 5.10.3.2: CHOKE AND KILL SYSTEM (NORSOK, 2012)	MGS Design and Operation	Venting of gases from the atmospheric mud-gas separator shall be done at the least distance of 4 m above the top of the derrick/hoisting structure or in a safe area. The design of the atmospheric mud-gas separator should be such that it can withstand pressures from drilling fluids with 2.2 specific gravity to the top of the vent line. The minimum diameter of the vent line shall be 10 inches (i.e., 254 mm). For high-pressure, high-temperature operations, the choke manifold shall be equipped with a 127 mm (5”) overboard vent line, and the MGS shall have a liquid seal of at least 6 m minimum.
NORSOK D-001 (DRILLING FACILITIES, 3RD REVISED VERSION), SUB-SECTION 6.36.1: GENERAL REQUIREMENTS FOR CHOKE AND KILL SYSTEMS (NORSOK, 2012)	General Choke and Kill System requirements in relation to MGS	The choke manifold shall have at least three chokes, two remotely controlled (at minimum) and the other manually operated. Remotely controlled chokes shall be operated at the driller’s panel, where information on MGS pressure monitoring, among others, is displayed. The MGS’ capacity to connect the kill and choke manifolds must be sufficient.

(*Continued*)

TABLE 17.2 (*Continued*)

A Summary of Standards, Recommended Practices, Guidelines, and Regulations on MGS Sizing, Design and Operations

Reference	Area of Interest	Recommended Practices, Procedures, and Description
NORSOK D-001 (DRILLING FACILITIES, 3RD REVISED VERSION), SUB-SECTION 6.37.1: GAS SEPARATION IN KILL AND CHOKE DOWNSTREAM ARRANGEMENTS; SUB-SECTION 6.43.4: FUNCTIONAL REQUIREMENTS OPERATIONS IN HIGH-TEMPERATURE, HIGH-PRESSURE WELLS (NORSOK, 2012)	MGS Design and Operational Requirements	Venting of gases from the atmospheric mud-gas separator shall be done at the least distance of 4 m above the top of the derrick/hoisting structure or in a safe area. The design of the atmospheric mud-gas separator should be such that it can withstand pressures from drilling fluids with 2.2 specific gravity to the top of the vent line. As a general design requirement, the minimum liquid seal height of the MGS shall be 3 m. The separator shall not be connected to the diverter. The discharge line of the MGS shall be directed upstream of the shakers to the header box, where gas detection can be done. MGS discharge and bleed-off lines from the manifold of the standpipe shall be directed to a flow line system before they branch to the trip tank system and then to the header box located upstream of the shale shakers, where gas detection is done. The material used in constructing the choke and kill manifold lines routed to the MGS shall be such that they have a low-temperature rating (about –40°C) and are well protected from freezing by means of an appropriate method. As a functional requirement, the capacity of the MGS shall be able to meet the needs of the well to be drilled at a minimum. Regardless, the throughput capacity of the MGS shall be at least 10 MMSCFD and not limited by blowdown capacity. From a functional requirement perspective, the minimum height of the liquid seal of the MGS shall be 6 m and shall be maintained by continuous circulation of fresh mud from the mud pits to the separator into the system with the aid of a dedicated bypass loop system. A pressure gauge with a 0–1.5 bar (21.75 psi) scale shall be installed at the top of the MGS with read-back to the choke control panel

		The essential parameters of the MGS shall be monitored from the driller's control panel. Also, it is recommended that appropriate arrangements be made to maintain a liquid seal during a well control situation through a "hot loop".
ASME BOILER AND PRESSURE VESSEL CODE SECTION VIII: DIVISION 1 (ASME, 2019a)	Design, fabrication, inspection, testing, and certification of pressure vessels working at internal and external pressures higher than 15 psig.	The divisions show the series types of stainless steel pipes, screw threads, series types of Quality Assurance Programs, and Pressure Relief Devices. References on these are obtained from API 579–1/ASME FFS-1 documents.
ASME BOILER AND PRESSURE VESSEL CODE SECTION VIII: DIVISION 2 (ASME, 2019B)	Requirements on the materials, design, and nondestructive examination. These requirements are quite more rigorous than those in Division 1.	
ASME BOILER AND PRESSURE VESSEL CODE SECTION VIII: DIVISION 3 (ASME, 2019c)	Requirements for pressure vessels operating at internal and external pressures greater than 10,000 psi.	
NACE STANDARD MR-01–75. (NACE, 2001)	MGS separator fabrication design	Specifies the requirements for materials used in vessel fabrications
API SPEC 12J (API, 2009)	Minimum requirements for oil field type separator design, shop-testing, and fabrications. Separators outside the scope of this specification include centrifugal separators, filter separators, and desanding separators.	

(Continued)

TABLE 17.2 (*Continued*)

A Summary of Standards, Recommended Practices, Guidelines, and Regulations on MGS Sizing, Design and Operations

Reference	Area of Interest	Recommended Practices, Procedures, and Description
API SPEC 12J: SECTION 5: FABRICATION, TESTING, AND PAINTING (API, 2009)		The separator must be shop-constructed, adequately tested, and stamped in line with the revised edition of the ASME codes. Upon agreement between the manufacturer and purchaser, additional testing may be conducted. Separators must be cleaned of scale, weld spatter, rust, and grease and coated with a commercial metal primer before they are shipped. In the event of a request for internal coatings by manufacturers, all internal nonrenewable attachments shall be seal-welded and prepared for coating based on purchasers' specifications. All foreign matter should be removed from the vessel before shipment. All openings shall also be protected with shipping covers and plugs.
API SPEC 12J: SECTION 6: MARKING (API, 2009).		Separators fabricated on the basis of API SPEC J specifications shall have a corrosion-resistant nameplate securely attached and shall have all items specified from 1 to 9, as shown in Figure 6.1 of Section 6 in the reference document. ASME code for allowable stamping of the vessel should be referenced for stamping procedures.
API RP 64: RECOMMENDED PRACTICE FOR DIVERTER SYSTEMS EQUIPMENT AND OPERATIONS. SUB-SECTION 6.3.1: SOUR GAS DRILLING OPERATIONS (API, 2001)	Operational Requirements of the MGS in sour gas drilling	Diverter systems should be used in conjunction with the BOP for sour gas drilling to reduce hazardous exposure to hydrogen sulfide. For such operations, the drilling fluid flow line should be designed to be directed by valves through the MGS and vented safely at an appreciable distance away from the rig.

API RP 96: DEEPWATER WELL DESIGN AND CONSTRUCTION. SUB-SECTION 10.2.2.5: POTENTIAL FOR GAS IN RISER (API, 2013)	Handling riser gas with reference to the MGS	The BOP shall be closed to seal the wellbore to prevent further formation of fluid influx when the diverter is activated. Flow shall then be directed through the overboard flowlines to a safe location. Flow can be directed through the MGS instead of the diverter system. This process interferes with the rig's safety and, thus, is only recommended for nonemergency situations and not during a well-control event. This is not a mandatory requirement. Also, appropriate permits must be obtained to conduct this operation.
AUSTRALIA QUEENSLAND DEPARTMENT OF NATURAL RESOURCES AND MINES: SURFACE GAS HANDLING SYSTEM AND MUD GAS SEPARATOR DESIGN (SGHMGS) (Lee, 2016)	Operational requirements of the MGS	The standard recommends using a liquid seal instead of a conventional backpressure regulator or liquid level control valve to control separator backpressure. The process prevents hydrate formation and plugging by solids as the Joule–Thompson effect is reduced. Where the delivery rate of reservoir fluids to the separator exceeds the separator capacity, flow can be diverted to overboard lines if shutting in the well is catastrophic. The MGS' operational procedures should include using glycol as a hydrate suppressant. Hydrate formation can also be prevented by heating the kick fluid before or during separation in the separator vessel. The instrumentation arrangement of the MGS should have a local pressure and temperature monitoring device. The differential pressure in the separator can be observed and monitored using a low-range pressure gauge installed and visible from the position of the choke. A mist extractor shall be installed on the siphon breaker (anti-siphon) line to remove entrained liquid droplets before venting to the atmosphere. This extractor's design shall depend on the operating conditions, gas flow rates, and liquid content. The MGS should have a pressure and temperature sensor to provide remote read-out on a panel in the driller's cabin.

(*Continued*)

TABLE 17.2 (*Continued*)

A Summary of Standards, Recommended Practices, Guidelines, and Regulations on MGS Sizing, Design and Operations

Reference	Area of Interest	Recommended Practices, Procedures, and Description
MICHIGAN ADMINISTRATION CODE: R.299.2358: MUD GAS SEPARATOR (LLI, 2004).	Handling of gas generated by the MGS	Cable-tool Drilling: All fluids from a kick shall be circulated through an MGS. The separated gas shall be appropriately transported to an adequately engineered incinerator of a flare line with a high enough discharge to the atmosphere for burning, unless the supervisor of the mineral well suggests an alternative method. Rotary Drilling: For gases routed to the incinerator or flare line, a permittee or a permittee's proxy must determine the H_2S content of the gas using an approved device for determining H_2S concentration as well as a valid method as determined by the supervisor of the mineral well.
CODE OF FEDERAL REGULATIONS: BUREAU OF SAFETY AND ENVIRONMENTAL ENFORCEMENT (BSEE) – DEPARTMENT OF INTERIOR: SUB-PART D SUB-SECTION 250.456 (CFR, 2012)	Safety in overall drilling concerning MGS.	It is incumbent to ensure that MGS and degasser are installed and in good operational condition before drilling begins.
CODE OF FEDERAL REGULATIONS: BUREAU OF SAFETY AND ENVIRONMENTAL ENFORCEMENT – DEPARTMENT OF INTERIOR: SUB-PART D SUB-SECTION 250.490 K (CFR, 2012).	Personnel safety and equipment concerning MGS operation.	In the event of well control, it is recommended that after the MGS separates H_2S and other gases from drilling fluids, drilling fluids must be treated to neutralize the H_2S to restore and maintain proper quality.

AMERICAN BUREAU OF SHIPPING (ABS): CLASSIFICATION OF DRILLING SYSTEMS: CHAPTER 3, SUB-SECTION 7.7 (ABS, 2021)	Design and operation of MGS	Separators should be manufactured following ASME Section VIII Boiler and Pressure Vessel Code and Chapter 2, Sub-Section 7.5 of the ABS Guide. In a situation where alternative design codes and standards are used, they shall be considered by the ABS via appropriate justifications.
AMERICAN BUREAU OF SHIPPING (ABS): CLASSIFICATION OF DRILLING SYSTEMS: CHAPTER 2, SUB-SECTION 7.5 (ABS, 2021).		MGSs used in drilling are to be submitted to the ABS for approval in accordance with ABS Mobile Offshore Unit rules. All pressure vessel (including the MGS) designs should be such that stresses associated with movement and installation and other external factors are considered. They should also be within acceptable design limits as specified by the design codes. The selection of materials for the design of pressure vessels (including MGS) must follow the specified design standards and should be selected for their intended use.
AMERICAN BUREAU OF SHIPPING (ABS): CLASSIFICATION OF DRILLING SYSTEMS: CHAPTER 3, SUB-SECTION 7.7 (ABS, 2021).	Operational Maintenance Practices for MGSs	Well control operation personnel should take precautions to prevent erosional wear at the drilling fluid's entry into MGS. The design pressure of the MGS shall be based on vent line pressure determined by the specified mud weight for the operation. When the capacity of the separator is exceeded, the choke manifold should be able to divert flow to other locations for an emergency discharge.
AMERICAN BUREAU OF SHIPPING (ABS): CLASSIFICATION OF DRILLING SYSTEMS: CHAPTERS 4, 5, AND 6 (ABS, 2021).		Discuss piping, materials, and nondestructive evaluation of the separator.
THE OIL AND GAS ACT: (C.C.S.M.C. 034), MANITOBA – REGULATIONS 111/94: PART 6 – SUB-SECTION 26 (Province of Manitoba, 1994)	MGS installation	The obtainer of a drilling license shall ensure that when using the MGS, it is connected to a line that ends in a flare pit and has a diameter of at least 25 mm larger than the inlet line.

(*Continued*)

TABLE 17.2 (*Continued*)

A Summary of Standards, Recommended Practices, Guidelines, and Regulations on MGS Sizing, Design and Operations

Reference	Area of Interest	Recommended Practices, Procedures, and Description
NABORS WELL CONTROL MANAGEMENT MANUAL (MGS BEST PRACTICE GUIDE)	Vent line dimensions	The vent line of the MGS routed to the flare should be at a minimum distance of 150 ft from an active well center in instances with space limitations. This distance may be reduced to 100 ft, but never less. It is also essential to ensure the flare is away from offices and buildings. This line should always be in place when the BOP is "nippled" up on the well after installing the surface casing. The line must also be as straight as it can be.
MUD GAS SEPARATOR SIZING AND EVALUATION (MacDougall, 1991)	General Considerations for MGS Sizing	A larger MGS vent line ID is preferable. The impingement plates shall be at right angles with the separator's inlet and be replaceable. Baffle plates should be positioned at the upper part of the MGS and may run downwards toward its lower parts. Baffles shall not obstruct fluid flow in the separator. Provision shall be made for an upper manway at the upper part of the separator for routine visual inspection of the MGS' interior. The manway should be large enough to make the replacement of the impingement plates possible. There shall be a sump at the bottom of the closed-bottom MGS, which shall prevent the plugging of the outlet of the mud return lines. The lower section of the separator vessel shall have a valved inlet to permit the pumping of mud into the separator. In some MGS settings and arrangements, the return line must be below the separator's elevation for mud flow to mud tanks. In such situations, an anti-siphon line may be required to prevent siphoning the mud from the separator to the mud tanks.

ASME BOILER AND PRESSURE VESSEL CODE SECTION VIII: DIVISION 1 (ASME, 2019a)	Vessel testing	All constructed vessels shall undergo a hydrostatic pressure test based on an agreement between the product manufacturer and the user for calculated pressures.
DRILLING MANUAL FROM COMPANY A	Recommended response procedures when MGS overloading occurs	Change to a slower kill rate. Switch the choke and kill manifold outlets to an overboard line with a high-pressure rating or to production facilities. Evacuate the gas from the annulus with the use of the volumetric methods.
API SPEC 16C: CHOKE AND KILL EQUIPMENT, 3RD EDITION, MARCH 2021	Requirements for performance, design, materials, testing, and inspection of subsea choke and kill equipment	Actuators shall be installed on the choke and be designed to prevent pressure build-up. The hydraulic circuit shall contain a pressure relief valve or pressure regulating valves. The well control hydraulic control system shall be designed with a backup operating system to open or close the well control after losing primary power. Choke and kill-lines shall be either manually actuated or equipped with a double-acting hydraulic actuator. Hydraulically actuated valves shall fail-in-place and have a visual position indication. There should be the use of active discharge control systems such as float system or actuated discharge control valve to increase the gas handling capacity of the MGS. The siphon breaker line shall be installed after the mud leg and should not be connected to any other lines. It is important to consider the mud-gas separator as part of the choke and kill system. In the event the MGS separation capacity is exceeded, provisions should be made to bypass the MGS and direct return flow directly to the flare system, safe location, or an overboard discharge. The bypass may be manual or an actuated control system.

Where h is the height of the fluid leg, "rho" is the mud density, and "g" is the gravitational constant. This minimum mud leg pressure requirement by the NORSOK D-001 (2nd edition) is sufficient for conventional drilling despite being inadequate for MPD operations. In the 3rd edition of the NORSDOK D-001 (2012) document, further recommendations were made on pressure monitoring at the MGS driller's panel and installing a scale bar ranging from 0 to 1.5 bar (i.e., one with a maximum pressure of 21.75 psi) on the top of the MGS that reads back to the choke control panel. While both new and later versions recommend a mud seal height of 6 m, the latter proposed using a "hot loop" system to maintain the mud seal. The vent line was still required to be at a minimum of 4 m above the derrick, and the 10 m vent line height requirement was also maintained. Furthermore, while the 2nd edition required discharge lines to be connected to the middle of the tank, the 3rd edition required the line to be connected to the header box located upstream of the shale shakers where gas detection takes place. The 3rd edition also prohibits connecting the MGS to a diverter system. Besides, an important functional requirement in both editions was the need for the throughput capacity of the separator to be greater or equal to 10 MMSCFD and not limited by blowdown capacity.

The ABS Guide (2021) for drilling systems specifies that the design pressures of the MGS need to be determined by a mud weight of 2.2 S.G. Also, the ABS guideline provided recommendations on vent line orientation, which is not present in any of the NORSOK D-001 (1998, 2012) standards; the vent line needs to be as straight as possible with no obstructions to mitigate backpressures. This specification on vent line orientation is analogous to provisions in the Nabors Well Control Management manual (MGS Best Practice Guide). The ABS also recommends that the vent line be at least 13 ft above the crown block and have a minimum diameter of 10-inch normal pipe size (NPS), similar to that of the NORSOK documents. However, ABS strongly recommends a 12-inch internal diameter vent line for HPHT wells.

Likewise, in their Well Control Management Manual, Nabors Industries recommended that the MGS vent line directed to a flare should be at a minimum distance of 150 ft from an active well center during normal rig operations and 100 ft where the rig has space limitations. They also recommended having the flare away from offices and buildings, which is not present in the NORSOK nor the ABS guidelines. The ABS guideline also recommended that the anti-siphon line be at a minimum distance of 10 m above the MGS and have an internal diameter of 4 inches.

The minimum liquid seal height specification by the ABS guideline was 3 m for conventional drilling, which is like the provisions in the 3rd edition of the NORSOK D-001 (2012) standard. However, the latter recommends a 6 m liquid seal height. This shows that for a given mud weight, the minimum required liquid seal pressure will be higher per NORSOK D-001 specifications compared to ABS specifications.

Specifications on vessel construction, material type, and testing were also reviewed. According to the ABS guidelines, MGS construction must comply with ASME Boiler and Pressure Vessel Code Section VIII Div.1 or Div. 2. The API SPEC 12J (2009) for Oil and Gas Separators supports this and adds that materials used must meet NACE Standard MR-01–75 specifications. NACE standard requirements for MGS construction materials were also present in the ABS guidelines but were not clearly specified in the NORSOK standards. On the one hand, API SPEC 12J (2009) provided guidelines for MGS construction and testing before usage, while on the other hand, the API RP standards focused on guidelines for safe and efficient operations. API RP 53 (2012) and API RP 59 (2006) are similar, and both reference MacDougall (1991) for MGS design and sizing. API RP 64 (2001) provided recommended practices for drilling formations with sour gas, and API RP 96 (2013) focused more on riser gas handling. API RP 54 (2019) and API RP 49 (2001) also provide specifications for handling sour gases. Handling riser gas is a critical issue; thus, the recommended practice provided by API RP 96 is non-exhaustive. However, it mentions that directing riser gas through the MGS interferes with rig safety and, thus, is not recommended if it does not interfere with the rig's safety.

17.9 STATE AND INTERNATIONAL-LEVEL REGULATIONS ON MGS OPERATIONS

State and international-level standards and best practices were also reviewed. The SGHMGS guidelines by the Queensland Department of Natural Resources and Mines (Lee, 2016) recommended using glycol to eliminate hydrates formed within the MGS. This stipulation is, however, not present in the Michigan Administration Code on Mud Gas Separators (LLI, 2004), the BSEE Code of Federal Regulations, sub-section 250.456 (CFR, 2012), and the Oil and Gas Act: (C.C.S.M.c. 034) as presented in Table 17.2.

The SGHMGS also recommends installing a pressure gauge visible from the position of the choke. Furthermore, temperature and pressure sensors should be installed on a panel in the driller's cabin. The Michigan Administration Code and the Oil and Gas Act (C.C.C.M.c. 034), Manitoba (Province of Manitoba, 1994), do not have these specifications.

Also, the C.C.C.M.c.034 recommends having a line whose diameter is 25 mm larger than the inlet line ending in the flare pit. The R.299.2358 standards from Michigan place MGS operation specifications under two headings: Cable-tool and rotary drilling and require a permittee for an MGS operation to determine the H_2S concentration of the formation. Although this may be present in other regulations, it is not reviewed at any state level in this material.

The BSEE code of federal regulations also discussed handling H_2S gas; however, unlike R.299.2358, it further recommends separating H_2S from drilling mud and reconditioning the mud for reuse. The Labrador Offshore Petroleum Board (C-NLOPB) of Canada Newfoundland, in their drilling and production guidelines, 2017, mentioned the importance of two valves, both remotely operated, in each choke and kill outlet where one is controlled hydraulically to provide a safe kill system (Vikse, 2008). Using color codes or other suitable means to distinguish clearly between choke and kill-lines is also important. The guideline also recommends having choke and kill-lines with appropriate sizes to ensure that pressure losses do not impede well control operations.

Similarly, the Offshore Well Control Inspection Guide published by the Health and Safety Executive of the United Kingdom also recommends specifying well control procedures for high-temperature wells and making them available to rig personnel prior to operations (HSE, 2021). The document recommended the specification of procedures for routine flushing of the choke and kill-lines, the MGS, and redundant standpipe manifolds. It also recommended that these components be flushed at least once every tour and covered in the pre-tour checklist. The guideline further endorses the need for specification of gas percentage at which drilling fluids will need to be circulated across the MGS as well as corrective actions in cases of MGS overloading. This is not present in most MGS guidelines. In addition, frequently draining the MGS was also recommended as a fingerprinting procedure in conventional drilling for early and correct detection of anomalies during well control procedures. Besides, the well control inspections guide also recommended the provision of up-to-date "as-rigged-up" Process and Instrumentation (P&ID) diagrams showing flow paths for circulating fluids through the MGS. This document should be made available to all important rig personnel. Furthermore, the specification of frictional pressure losses from the Rotating Control Device (RCD) to the shaker header box and from the RCD to the MGS was recommended as an additional fingerprinting procedure for MPD applications. Still, on MPD applications, the guidelines recommended updating the MPD matrix for the hole section being drilled with the latest information on the capacity of the MGS and ensuring that the inlet valves on the MGS are well positioned for proper handling of kicks. In some situations, influx may be circulated out through an MPD package. In such cases, the standpipe pressure (SPP) is maintained at a constant value so that the bottomhole pressure (BHP) does not fall below the new pore pressure. Also, the pressure relief system (PRV) should be adjusted according to the maximum expected pressure where the influx is close to the surface. The positions of the MGS inlet valves should be adjusted accordingly. Another key operational recommendation stipulated by the guideline is seen in the event of a leaking Rotation Control Device (RCD). The recommended practice in such an event or when the maximum allowable pressure of the MGS is being approached is to close the BOP immediately based on procedures for switching from MPD to conventional well control.

The oil and gas regulations on well control in the state of Alaska focused more on the diverter in handling fluids at the surface (AOGCC, 2023). The regulations recommended operating the diverter remotely with an annular pack-off device, a vent line valve, and a diverter vent line with a diameter of 16 inches. Furthermore, it was recommended to have the vent line within not less than 75 ft from any identified ignition source. Like other reviewed regulations, the AOGCC also recommends the choke and kill-lines be made of rigid steel pipe and positioned as straight as possible. Grace (2017) corroborated this recommendation and further mentioned that any possible bends or curves are most likely to cause erosion, making well control operations difficult. Furthermore, the authors reported the need to ensure firm anchorage of choke-lines to reduce fluid flow effect and the impact of drilling solids or vibrations. Also, they endorsed equipping the separator with a positive liquid level control.

In the case of MGS regulations stipulated by the Texas Railroad Commission as presented in Rule 36 Requirements, the MGS must be installed between the first and second mud tanks in well control. Also, flare lines should be parallel and perpendicular to the prevailing wind's direction and unrestricted to flow.

17.10 CONCLUSIONS

This chapter employs an extensive literature review and a robust stakeholder engagement approach to analyze the MGS comprehensively. The goal was to facilitate an understanding of the vessel's key sizing and design criteria. Furthermore, industry standards governing the MGS design process are presented. The chapter's outcomes will assist in developing improved standards and regulations to reduce well-control risks. The following conclusions were reached from the study.

1. The MGS operating principle is similar for most separators. However, the procedure for controlling and maintaining the mud leg differs.
2. The closed-bottom mud-gas separator is usually preferred to the open-bottom and float-type separators due to limitations associated with the latter two.
3. The sizing and designing of the MGS are specific to the well conditions, drilling and completion fluids, the expected maximum flow rate, and the drilling conditions.
4. Survey results from well control experts and MGS specialists indicated that most of the sizing and designing procedures used in their respective companies are based on the approach proposed by MacDougall (1991) in SPE-20430PA with the associated modifications required.
5. The MGS sizing and designing are more complex in underbalanced drilling (UBD), managed pressure drilling (MPD), and deep HPHT wells where a higher gas flow rate is usually expected.
6. A review of industry standards and regulations on MGS design highlighted the significance of installing safety relief equipment such as pressure relief valves, kill-lines, choke-lines, and diverters that can be operated in case an excessive is expected and the MGS is bypassed.

17.11 KEY TAKEAWAYS FROM THE INDUSTRIAL SURVEY

- The vertical MGS was identified as the most common mainly due to its compact nature. Also, its mud leg is comparatively easily maintained and is the most familiar among rig personnel.
- The backpressure from the vent line and the hydrostatic pressure in the mud leg were identified as the most important factors considered in sizing.
- SPE, IADC, NORSOK, API, and ASME standards were all represented in the survey results as standards that govern MGS sizing considerations.

ACKNOWLEDGMENTS

The authors would like to extend their gratitude to the Bureau of Safety and Environmental Enforcement (BSEE) under contract 140E0122Q0023 for funding the project. We would finally like to extend our appreciation to Nabors Industries and all other industry experts for providing valuable information for this project.

REFERENCES

Alaska Oil and Gas Conservation Commission (AOGCC), 2023. *20 AAC 25.033: Primary Well Control for Drilling*. www.akleg.gov/basis/aac.asp#20.25.033

Al Hadidy, K., Hamdy, I., Al Samahi, M., et al., 2019. Full-Automated Managed Pressure Drilling System Facilitates Unlocking Unconventional Gas Reservoir Potential in the United Arab Emirates. *Abu Dhabi International Petroleum Exhibition & Conference*, OnePetro, November 11–14. SPE-197831-MS. https://doi.org/10.2118/197831-MS

American Bureau of Shipping, 2021. *Guide for the Classification of Drilling Systems*. https://ww2.eagle.org/content/dam/eagle/rules-and-guides/current/offshore/57_Classification_of_Drilling_Systems_2021/cds-guide-feb21.pdf

American Petroleum Institute (API) Specifications 12J, 2009. *Specification for Oil and Gas Separators*. Washington, DC: API.

API RP 49, 2001. *Recommended Practice for Drilling and Well Servicing Operations Involving Hydrogen Sulfide*, 3rd edn. Washington, DC: API.

API RP 53, 2012. *Recommended Practices for Blowout Prevention Equipment Systems for Drilling Wells*. Washington, DC: API.

API RP 54, 2019. *Recommended Practice 54, Occupational Safety and Health for Oil and Gas Well Drilling and Servicing Operations*, 4th edn. Washington, DC: API.

API RP 59, 2006. *Recommended Practices for Blowout Prevention Equipment Systems for Drilling Wells*, 2nd edn. Washington, DC: API.

API RP 64, 2001. *Recommended Practices for Diverter Systems and Equipment Operations*, 2nd edn. Washington, DC: API.

API RP 96, 2013. *Deepwater Well Design and Construction*, 1st edn. Washington, DC: API.

API Specification 12J, 2009. *Specification for Oil and Gas Separators*. Washington, DC: API.

API Specification 16C, 2021. *Choke and Kill Equipment*, 3rd edn. Washington, DC: API.

API Standard 53, 2018. *Blowout Prevention Equipment Systems for Drilling Wells*, 5th edn. Washington, DC: API.

API Standard 521, 2007. *Pressure-Relieving and Depressurizing Systems*, pp. 129–141. Washington, DC: API.

ASME Boiler and Pressure Vessel Code, 2019a. *Rules for Construction of Pressure Vessels, Section VIII Div. 1, An International Code*. https://www.asme.org/codes-standards/find-codes-standards/bpvc-viii-1-bpvc-section-viii-rules-construction-pressure-vessels-division-1

ASME Boiler and Pressure Vessel Code, 2019b. *Rules for Construction of Pressure Vessels, Section VIII Div. 2, An International Code*. https://www.asme.org/codes-standards/find-codes-standards/bpvc-viii-2-bpvc-section-viii-rules-construction-pressure-vessels-division-2-alternative-rules

ASME Boiler and Pressure Vessel Code, 2019c. *Rules for Construction of Pressure Vessels, Section VIII Div. 3, An International Code*. https://asmedigitalcollection.asme.org/ebooks/book/243/chapter-abstract/25143235/Section-VIII-Division-3-Alternative-Rules-for?redirectedFrom=fulltext

Bland, R., Mullen, G., Gonzalez, Y., et al., 2006. HP/HT Drilling Fluid Challenges. *IADC/SPE Asia Pacific Drilling Technology Conference and Exhibition*, Bangkok, Thailand, November 13–15. SPE-103731-MS. https://doi.org/10.2118/103731-MS

Bourgoyne, A. T., 1997. Well Control Considerations for Underbalanced Drilling. *SPE Annual Technical Conference and Exhibition*, San Antonio, TX, October 5–8. SPE-38584-MS. https://doi.org/10.2118/38584-MS

Butchko, D., Davies, G. E., Fuchs, G. T., et al., 1985. Design of Atmospheric, Open-Bottom Mud/Gas Separators. *SPE/IADC Drilling Conference*, New Orleans, LA, March 5–8. SPE-13485-MS. https://doi.org/10.2118/13485-MS.

Canadian-NewfoundLand and Labor Offshore Petroleum Board, 2017. *Drilling and Production Guidelines*, August. ISBN: # 978-1-927098-76-9. https://www.cnlopb.ca/wp-content/uploads/guidelines/drill_prod_guide.pdf

Code of Federal Regulations (CFR), 2012. *Sub-Part D, Sub-Section 250.490, Hydrogen Sulfide*. Washington, DC: Bureau of Safety and Environmental Enforcement-Department of Interior.

Collings, B. J., & Stone, W. M., 1989. Drilling With Oil-Based Muds on Offshore Wells: Drilling Contractor's Viewpoint. *SPE/IADC Drilling Conference*, New Orleans, LA, February 28–March 3. https://doi.org/10.2118/18705-MS (Last Accessed in March 13, 2023).

Eby, D., 2008. Many Factors Are Involved to Properly Size Mud Gas Separator, or Gas Buster, Systems. *Drilling Contractor*, 64 (6).

Grace, R. D., 2017. *Blowout and Well Control Handbook*. Cambridge: Gulf Professional Publishing. ISBN: 978-0-12-812674-5.

Hoopingarner, J. B., Greif, C. V., Neme, E. E., et al., 1990. Rig Modifications Meet New UK High-Pressure Requirements. *IADC/SPE Drilling Conference*, Houston, TX, February 27–March 2. https://doi.org/10.2118/19976-MS

Lee, P., 2016. *Surface Gas Handling and Mud Gas Separator Design Principles*. Department of Natural Resources and Mines, State of Queensland. https://www.resources.qld.gov.au/__data/assets/pdf_file/0020/351317/sghs-mgsd-principles-for-drilling-operations.pdf

LII Legal Information Institute, 2004. *Well Drilling and Construction-Mud Gas Separators*. Michigan Administration Code R. 299.2358. https://www.law.cornell.edu/regulations/michigan/Mich-Admin-Code-R-299-2358

Louison, R. F., Reese, R. T., & Andrews, J. P., 1984. Case History: Underbalance Drilling the Midway and Navarro Formations Successfully in Hallettsville, TX. *SPE Annual Technical Conference and Exhibition*, Houston, TX, September 16–19. SPE-13112-MS. https://doi.org/10.2118/13112-MS

MacDougall, G. R., 1991. Mud/Gas Separator Sizing and Evaluation. *SPE Drilling Engineering*, 6 (4), 279–284. SPE-20430-PA. https://doi.org/10.2118/20430-PA

Marsh, G. L., & Altermann, J. A., 1988. Subsea and Surface Well Control Systems and Procedures the Zane Barnes. *Offshore Technology Conference*, Houston, TX, May 2–5. OTC-5627-MS. https://doi.org/10.4043/5627-MS

MISwaco,2018.*MudGasSeparatorforAssembly9620244.OperatingandServiceManual*.https://www.slb.com/products-and-services/innovating-in-oil-and-gas/well-construction/rigs-and-equipment/managed-pressure-drilling-equipment/mud-gas-separators

NACE Standard MR-01-75, 2001. *Sulfide Stress Cracking Resistant Metallic Material for Oil Field Equipment*. https://store.ampp.org/sulfide-stress-cracking-resistant-metallic-materials-for-oilfield-equipment-7

NORSOK D. 001, 1998. *NORSOK STANDARD, Drilling Facilities*, D-001, Rev 2. https://www.scribd.com/document/728675047/D-001-DRILLING-FACILITIES-1998

NORSOK D. 001, 2012. *NORSOK STANDARD, Drilling Facilities*, D-001, Rev 3. https://www.scribd.com/document/465719231/D-001

The Offshore Well Control Inspection Guide. *2021-OMAR/ED Offshore Inspectors*. www.hse.gov.uk/offshore/ed-well-control.pdf

Oriji, B. A., & Okechukwu, O. G., 2019. Optimal Sizing of Mud Gas Separators in Well Planning for Effective Well Control in Drilling Operations. *Petroleum & Coal*, 61 (1).

Patil, H., Deshpande, K., Dietrich, E., et al., 2018a. Functional Design Specification FDS for MPD Equipment Installation on Deepwater Drillships. *Offshore Technology Conference*, Houston, TX, April 30–May 3. OTC-28876-MS. https://doi.org/10.4043/28876-MS

Patil, H., Deshpande, K., Ponder, T. L. et al., 2018b. Advancing the Mud Gas Separator Sizing Calculation: The MPD Perspective. *SPE/IADC Managed Pressure Drilling and Underbalanced Operations Conference and Exhibition*, New Orleans, LA, April 17–18. SPE-190008-MS. https://doi.org/10.2118/190008-MS

Province of Manitoba, 1994. *The Oil and Gas Act: (C.C.S.M.c 034)*. Regulations: 111/94: Part 6, Sub-Section 26, Manitoba, Canada. https://www.gov.mb.ca/iem/petroleum/actsregs/dap5.html

Pruitt, 2023a. *Managed Pressure Drilling Mud Gas Separator*. https://rb.gy/qgi7ss (Last Accessed in February 1, 2023).

Pruitt, 2023b. *Managed Pressure Drilling Mud Gas Separator*. www.slb.com/drilling/rigs-and-equipment/managed-pressure-drilling-equipment/mud-gas-separators/toga-total-gas-containment-system (Last Accessed in March 13, 2023).

Schlumberger, 2023. *Total Gas Containment (TOGA) System*. www.slb.com/drilling/rigs-and-equipment/managed-pressure-drilling-equipment/mud-gas-separators/toga-total-gas-containment-system (Last Accessed in March 13, 2023).

Texas Railroad Commission. *Discussion of Rule 36 Requirements*. www.rrc.texas.gov/media/oxydxk2i/discussionrule36.pdf (Last Accessed in October 10, 2023).

Vikse, C. M. T. M., 2008. *Canada-Newfoundland and Labrador Offshore Petroleum Board CEAA Screening Report*. Health and Safety Executive. The Offshore Well Control Inspection Guide. Offshore Petroleum Regulator for Environment & Decommissioning. https://www.cnlopb.ca/wp-content/uploads/nhseis/nheascrpt.pdf

Yuan, Z., Morrell, D., Sonnemann, P., et al., 2017. Mitigating Gas-in-Riser Rapid Unloading for Deepwater-Well Control. *SPE Drilling & Completion*. SPE-185179-PA. https://doi.org/10.2118/185179-PA

17.A APPENDIX

Simplified Method for Determining Peak Gas Flow Rate Using a Simple Method from Butchko et al. (1985) (Sample Input Data and Results)

TABLE 17.A.1

Wells' Data	
Depth	16,404 ft
Pb Pressure	8,766 psia
Tb	711.9 °R
Zb	1.20
Pb	28.26 lbm/ft^3
Hole Size	8.5 inches
Drill-Collar Lengths	700 ft
Drill-Pipe Size	5.0 inches
Kick Circulation Rate	0.294 ft^3/sec
Input Data	
Pb (psi)	8,766
d (ft)	16,404
Gm (psi/ft)	0.5205
pg (psi)	54.7
Ts/Tb	0.81
hb (ft)	278.2
Vb (ft^3)	35.5

TABLE 17.A.1 (*Continued*)

Zs/Zb				0.833			
Ps (psia)				1023			
Qm (ft³/sec)				0.2943			
B (ft³/ft)				0.284			
MW (lbm/lbmole)				27.1896			
g (lbm/ft³)				0.0717			
Ts (°R)				574.5			
Zs				1.0			
Computations for Ps							
Iteration	Zs	Zb	Zs/Zb	Ps	Prs	Trs	Z's
1	1.0	1.20	0.83	1141.0	1.35	1.27	0.725
2	0.725	1.20	0.604	1005.0	1.19	1.27	0.760
3	0.76	1.20	0.633	1023.0	1.214	1.27	0.775
Output Data							
Ps (psi)				1023			
Vs (ft³)				158.6			
Zs				0.775			
Qg (MMSCFD)				2.713			

18 Riser Gas Modeling Challenge

Rahman Ashena

18.1 INTRODUCTION

In deepwater and ultra-deepwater wells (seawater > 600 m/1,969 ft) while drilling or after killing the well in case of well control events, a small gas influx may pass through Blowout Preventer (BOP) and enter the marine riser prior to its detection. The gas in the riser would expand and move upward the riser due to a pressure drop with height. Gas migration and expansion through the riser may lead to a gas-unloading event, causing blowout and potential explosions. Therefore, it is essential to anticipate such kick events leading to riser unloading in a time-bound manner. A popular way to escape encountering riser gas flow particularly in shallow and intermediate drilling depths is riserless drilling which is out of scope of this chapter.

In offshore drilling, with increasing water depth (longer risers) and high pressure and temperature conditions, the risk of gas in riser (GiR) going undetected gets bigger. The lack of a competent system to anticipate the controlling parameters at the choke below the BOP can make gas handling challenging or lead to possible blowouts. There are a number of blowouts due to riser gas unloading (RGU): The first reported one is the Zapata Lexington with four fatalities in the Gulf of Mexico in 1984 (Blake et al., 1988) and the last reported one being the recent Macondo incident on Deepwater Horizon (CSB Report, 2014). About 10% of riser volume can result in unloading of drilling mud at the rig floor and can exacerbate into hazardous or possible explosion at the rig floor if not dealt with appropriately (Kiran and Salehi, 2018). Controlling gas in riser is either via using diverters, which is extremely challenging and is not recommended, or via riser gas handlers (RGH) and managed pressure drilling (MPD). The latter applies a choke backpressure at the surface so that the flow can be maintained stable by keeping gas below its equilibrium pressure. A diverter gas influx will almost always go above the riser equilibrium depth/point, after which the model calculation is more challenging. Paul Sonnermann was a pioneer to use the term "equilibrium point/depth" and mention it as a depth above which there is a challenge of modeling gas behavior (IADC, 2018), but he did not present/publish any analytical models. Some CFD software packages may handle modeling this complex fluid flow problem. IADC (2023) gives an explanation of the importance of gas in riser flow to provides a riser gas management process flowchart for deepwater riser gas handling systems. Such a modeling is important if one wants to understand and assess the capability along with the risks associated with using riser handling system to handle gas in the riser. Aarsnes et al. (2016) presented some complex mathematical models to be used for estimating gas in riser expansion.

DOI: 10.1201/9781003473770-18

Therefore, to address the modeling challenges, first a Boyle's gas law model was used in a trial-and-error manner applied in this work to estimate the extent of gas expansion after a specified period of time, called as each stage, to find the top of gas bubble. The single bubble method is a simple analytical model to imitate the riser gas unloading event which is considered in this work. Mustafa (2023) used the Boyle's law method with some calculation errors which are corrected in this work. As the next method, an analytical model was used considering a real-gas law including gas compressibility factor/Z-factor to better predict gas expansion and migration in the riser. The performance of both methods is analyzed and compared in a case study.

18.2 LITERATURE REVIEW

Kiran and Salehi (2018) provided a mathematical modeling work to characterize effects of different parameters such as temperature, mud type, backpressure, and gas solubility in the mud on riser gas unloading.

18.3 CASE STUDY

Figure 18.1 shows riser schematic and dimensions including the gas bubble initially in the riser, with the wellbore being closed from the bottom by the BOP and the

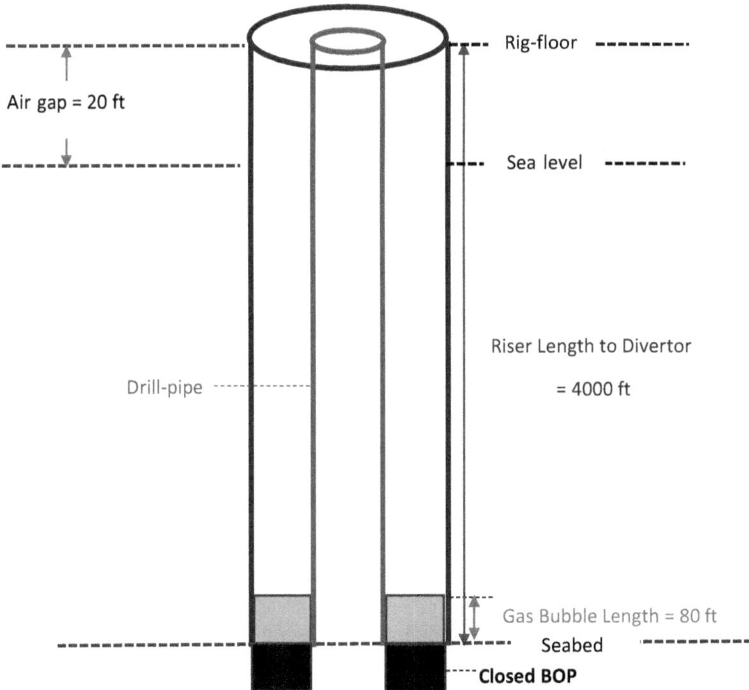

FIGURE 18.1 The riser schematic and dimensions including the initial gas length.

TABLE 18.1
Input Data for the Case Study

Riser Length to Diverter (slightly less than Seawater Depth)	4,000 ft
Mud Weight	12 ppg
Pressure of Gas Bubble (at the Top of Gas)	2446 psig/2460.7 psia
(Initial) Influx Size and Height in Riser	26.1 bbl 80 ft
Gas Type	Methane
	(P_c = 673 psi, T_c= 344 R)/Conductivity = 100 $\frac{\text{BTU}}{\text{hr.ft}^2.^\circ\text{F}}$
Migration Rate (of Gas Bottom)	100 ft/minute (assumption)
Pressure at Surface	14.7 psi (atmospheric)
Riser Size	21-in (OD)/19 ½-in (ID)
Drill-Pipe Size	6 5/8-in (OD)/5-in (ID)

diverter's packing element closed from the top. Table 18.1 shows the input data for the investigation of gas migration in a case study. The riser depth is 4,000 ft, the original mud weight in the riser is 12 ppg, and the gas influx height of 80 ft. The riser size is constant. For simplicity, a constant gas migration rate of 100 ft/minute was considered, but it should be kept in mind that this rate is not realistic.

18.4 GAS EXPANSION MODELING

First, the Boyle's gas law is applied which is a simpler form of ideal-gas law where the temperature is assumed constant. Thus, gas expansion is assumed to occur only due to pressure drop. Next, real-gas law is used as the analytical model. In the modeling, the following assumptions are made:

- A single gas bubble is in the riser and will rise like a single bubble.
- BOP is closed, and thus the riser is isolated from the wellbore/bottom.
- Gas is immiscible in the drilling fluid (i.e., the mud is water-based).
- The drilling mud is incompressible.
- Backpressure is considered in this work.

  Backpressure is included in this work's developed equations; however, their analysis was not performed. Therefore, the equations can take into account the diverter's pressure losses or the riser gas handling (RGH)'s backpressure. In case backpressure is ignored or just atmospheric pressure is considered, it is assumed that the diverter's pack-off is not closed which practically means that the crew have failed to detect the influx, or the operator failed, etc. This work's equations can be used for designing riser gas handling (RGH) systems.

It is possible to use a trial-and-error method to predict gas bubble expansion and migration in a riser. The next method is analytical and was carried out by using a real-gas law.

18.4.1 Boyle's Law Iterative Method

It is possible to use a trial-and-error method to predict gas bubble expansion and migration in a riser. Table 18.2 shows the iterative/trial-and-error methodology of finding the top of gas after five minutes. The same methodology applies to other depths until 28.8 minutes. It is noted that assuming the constant 100 ft/minute migration rate of gas bottom is not exact, and, therefore, it can cause errors for time-based analysis, indicating that in practice, the time might be shorter.

18.4.2 Real-Gas Analytical Method

Following the application of the iterative Boyle's law, the author became curious to investigate the problem analytically and identify the reason for the lack of convergence of the model from a specified depth upward. Thus, a second-order equation was found by which gas volume or length at each depth may be found. In this section, the real-gas law will be used to comprehensively consider the effects of gas Z-factor and temperature.

The key to solving the analytical problem is that the mud hydrostatic pressure above the gas bubble is the gas pressure required in the gas law. Assuming a single gas bubble with the length/height of L_g, its pressure P_g (applied on its top) is:

$$P_g = P_{atm} + 0.052 \times MW \times (D - L_g) \tag{18.1}$$

TABLE 18.2
The Iterative/Trial-and-Error Methodology of Finding Gas Length and the Top of Gas after Five Minutes

MW [ppg]	**12**			
Gas Bubble Info	Bottom [ft]	Length [ft]	Top	Pressure [psi]
	4,000	80	3,920	2460.78
Gas Migration Rate [ft/minute]	100			Temp [F]
P_{atm} [psi]	14.7			50
Time [minute]	5			
Gas BTM Depth [ft]	3500			
	1st Guess	2nd	3rd	4th
Gas Top Depth [ft]	3,300	3404.40	3407.33	3407.41
Gas Length [ft]	200	95.60	92.67	92.59
Pressure [psi]	2,059	2124	2126	2126
Updated Length [ft]	95.60	92.67	92.59	92.59
(Boyle's Law: $P_1V_1 = P_2V_2$				
Updated Gas Top [ft]	3404.40	3407.33	3407.41	3407.41
Difference	104.40	2.93	0.08	0.0022

where P_{atm} is the atmospheric pressure added for absolute pressure calculation and D is the true vertical depth (TVD) of the bottom of the gas bubble.

The modeling procedure and steps can be found in Appendix A at the end of this chapter. Equations (18.A.10), (18.A.13), (18.A.14), and (18.A.21) are the final equations which are useful for calculations and results given in Section 18.5.

18.4.3 Advanced Modeling

In advanced modeling, the gas influx is not considered a single bubble, and using finite elements, gas flow modeling is performed in a dynamic manner (dependent on time). This is a more robust and realistic modeling which is doable using software packages like Olga.

18.5 MODELING RESULTS AND DISCUSSION

Table 18.3 presents the derived equations for finding Lg and De.

For benchmarking of the analytical modeling, the results of the analytical model were compared with an iterative model based on Boyle's law, and the same results for L_g and D_e were obtained. As an example, at the depth of 1,200 ft, ignoring gas Z-factor and atmospheric pressure, L_g is found:

$$L_g = \frac{D - \sqrt{D^2 - 4D_{m,i}\,L_{gi}}}{2} = \frac{1200 - \sqrt{1200^2 - 4 \times 3920 \times 80}}{2} = 384.59\,\text{ft}$$

TABLE 18.3
Summary of Derived Equations

	Equations
Real Gas	$L_g = \frac{1}{2}\left[D' - \sqrt{D'^2 - \frac{4P_{gi}L_{gi}}{0.052MW}(\frac{ZT_g}{Z_iT_{gi}})}\right]$
	$D_e = 2\sqrt{\frac{P_{gi}L_{gi}}{0.052MW}\left(\frac{ZT_g}{Z_iT_{gi}}\right)} - \frac{P_b}{0.052\,MW}$
Boyle's Gas	$L_g = \frac{1}{2}\left[D' - \sqrt{D'^2 - \frac{4P_{gi}L_{gi}}{0.052MW}}\right]$
	$D_e = 2\sqrt{\frac{P_{gi}L_{gi}}{0.052MW}\left(\frac{ZT_g}{Z_iT_{gi}}\right)} - \frac{P_{atm}}{0.052\,MW}$
(Boyle's Gas + Ignore P_{atm})	$L_g = \frac{1}{2}\left[D - \sqrt{D^2 - \left(4D_{m,i}\,L_{gi}\right)}\right]$
	$D_e = 2\sqrt{D_{m,i} \times L_{gi}}$

Using Boyle's law and ignoring atmospheric pressure (Pg case), the analytical equation (18.A.21) gives the same D_e as by the iterative model:

$$D_e = 2\sqrt{D_{m,i} \times L_{gi}} = 2\sqrt{3920 \times 80} = 1120 \text{ ft}$$

However, using Boyle's law and considering atmospheric pressure (Pa case), D_e is:

$$D_e = 2\sqrt{\frac{P_{gi}L_{gi}}{0.052MW}\left(\frac{ZT_g}{Z_iT_{gi}}\right)} - \frac{P_{atm}}{0.052\ MW} = 2\sqrt{\frac{2460.78 \times 80}{0.052 \times 12}} - \frac{14.7}{0.052 \times 12} = 1099.8 \text{ ft}$$

For the real-gas law case, Equation (18.A.13) is used to find D_e while considering gas Z-factor of 0.9125 at the depth of 1,200 m (guessed to be nearest to the equilibrium depth).

$$D_e = 2\sqrt{\frac{P_{gi}L_{gi}}{0.052MW}\left(\frac{ZT_g}{Z_iT_{gi}}\right)} - \frac{P_{atm}}{0.052\ MW}$$

$$= 2\sqrt{\frac{2460.78 \times 80}{0.052 \times 12}\left(\frac{0.9125}{0.7733}\right)} - \frac{14.7}{0.052 \times 12} = 1{,}196.72 \text{ ft}$$

Therefore, as Figure 18.2 shows, D_e was estimated to be approximately 97 ft shallower when gas Z-factor was ignored. However, it was estimated to be approximately 77 ft shallower when both gas Z-factor and atmospheric pressure were ignored.

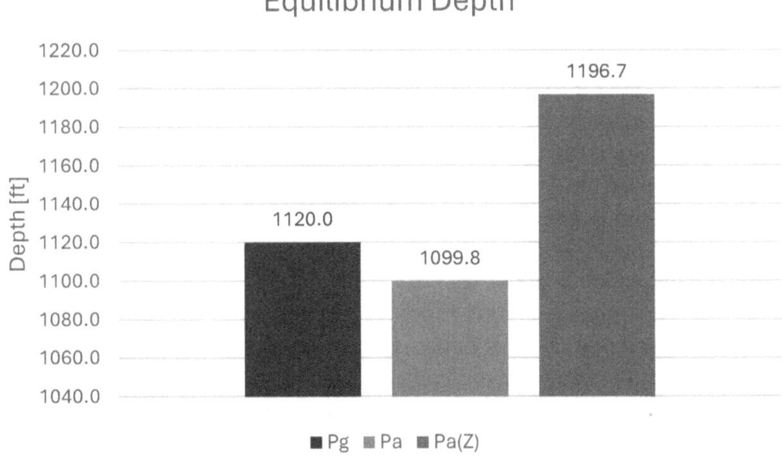

FIGURE 18.2 Equilibrium depth for three cases of real gas (Pa(Z)), ideal gas (Pa), and ideal gas ignoring atmospheric pressure (Pg).

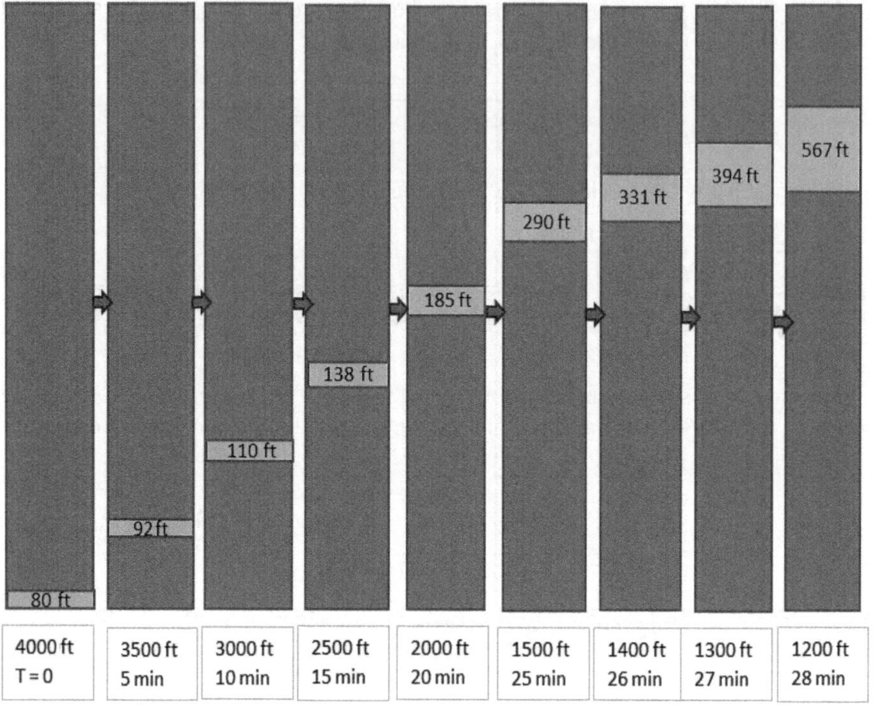

FIGURE 18.3 Gas migration process showing bottom depth and gas length at each step from time 0 to 28 minutes.

Figure 18.3 shows the gas migration process for the case study well. Gas bubble length/height increases steadily with time but then rises exponentially from 25 minutes onward until the depth of 1,200 ft (which is almost the equilibrium depth).

Figure 18.4a compares gas length versus depth for three different cases of ignoring atmospheric pressure effect (Pg), considering atmospheric pressure (Pa), and considering both atmospheric pressure and gas Z-factor (Pa(Z)). Figure 18.4b shows the errors of gas length calculation between the cases of Pg and Pa(Z). Briefly, ignoring atmospheric pressure causes a minor reduction of gas length prediction except near the equilibrium depth or say at low pressures. However, ignoring gas Z-factor causes an underestimation of gas length which becomes exponentially large near the equilibrium depth/at lower pressures.

Figure 18.5 shows gas migration curves for bottom and top depths and gas bubble lengths in the case study, and this figure is a complement to Figure 18.3. An interesting point is that at the equilibrium depth, the gas length equals the gas top depth of about 567 ft. From this depth above, the trial-and-error method does not converge, and that is the reason for not showing it.

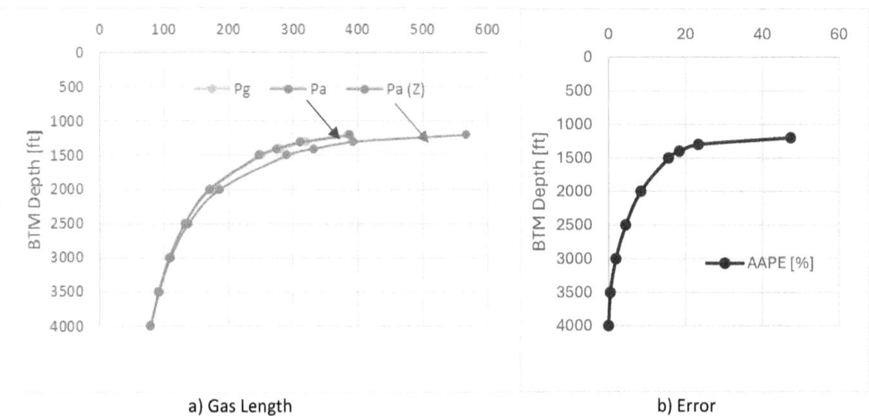

FIGURE 18.4 a) Gas length versus depth for three cases of 1) ideal-gas law while ignoring atmospheric pressure effect (Pg), 2) ideal-gas law with atmospheric pressure (Pa), and 3) real-gas law (Pa(Z)) and b) errors of gas length calculation between Pa(Z) and Pg.

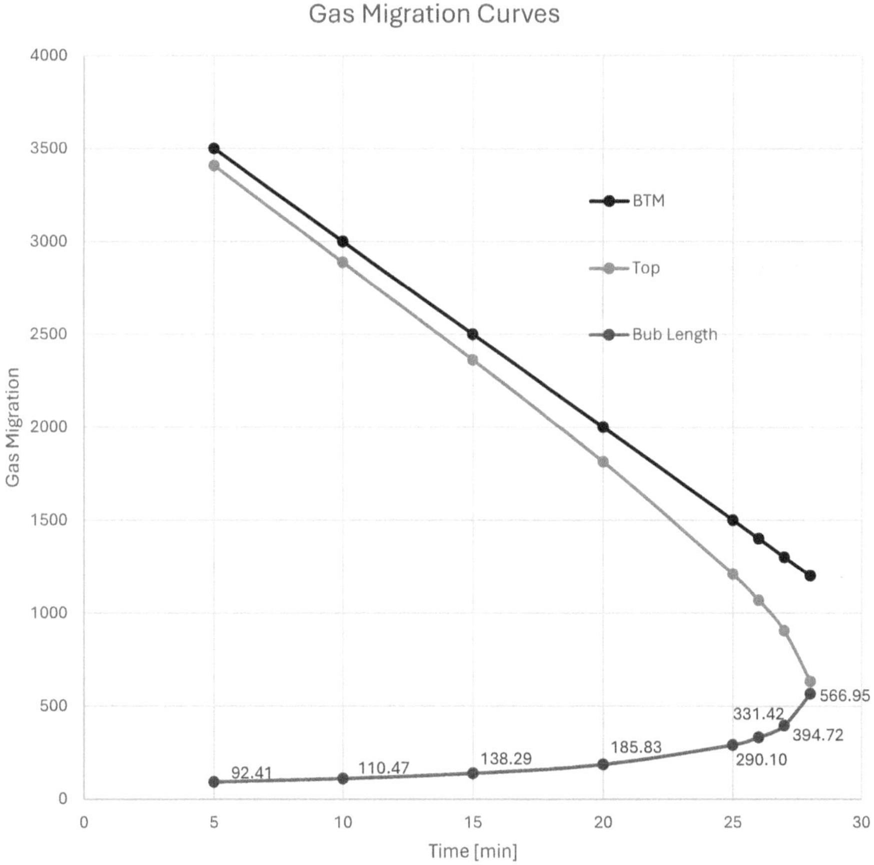

Gas Migration Curves

BTM
Top
Bub Length
Gas Migration
0 500 1000 1500 2000 2500 3000 3500 4000
92.41
110.47
138.29
185.83
290.10
331.42
394.72
566.95
0 5 10 15 20 25 30
Time [min]

FIGURE 18.5 Using real-gas law for finding gas bubble lengths as well as gas top and bottom depths for the case study.

18.6 CONCLUSIONS

1. Using real-gas law, an analytical model was found for estimating gas migration in riser versus depth and time. Since backpressure was considered in the developed equations, it can be used for riser gas handling (RGH) systems.
2. The performance of the analytical model was checked with another approach – iterative Boyle's law model. The nearer the gas to the surface, the greater the error in estimating gas length (up to ~50% average percent error).
3. A model was found for estimating the riser equilibrium depth. At this depth, the gas length equals the depth of top of gas.
4. The equilibrium depth was estimated to be approximately 97 ft shallower when gas Z-factor was ignored; it was estimated to be approximately 77 ft shallower when both gas Z-factor and atmospheric pressure were ignored.
5. Though the root cause of inefficiency and impracticality of gas laws above the equilibrium depth is not clear to authors, it may be attributed to the admission of gas into trans or super-sonic region whereby gas expands suddenly and extremely to reach the surface in a fraction of a second.
6. It is recommended to further study the root cause of the gas flow modeling problem from the equilibrium depth up using more sophisticated CFD models to provide a more reliable gas migration estimation. Further analysis is required to consider gas dissolution in oil-based muds also to find the temperature effect.

ACKNOWLEDGMENT

The author would like to thank Mr. Stan Christman for his consistent consultation and valuable discussions.

NOMENCLATURE

D	True vertical depth or bottom depth of gas	[ft]
D′	Dummy depth	[ft]
D_e	Equilibrium depth	[ft]
L_g	Length/height of gas	[ft]
MW	Mud weight/density	[ppg]
P_{atm}	Gas pressure (applied on its top)	[psi]
P_b	Backpressure applied	[psi]
P_g	Gas pressure	[psi]

D	True vertical depth or bottom depth of gas	[ft]
T_g	Gas temperature	[R]
$_{TVD}$	True vertical depth	[ft]
V_g	Gas volume	[ft^3]
Z	Gas Z-factor or compressibility factor	–

REFERENCES

Aarsnes, U. J., Hauge, E., & Godhavn, J., 2016. Mathematical Modeling of Gas in Riser. Paper presented at the *SPE Deepwater Drilling and Completions Conference*, Galveston, TX, September. https://doi.org/10.2118/180292-MS

Blake, K. W., Bourgeois, D. J., Howard, D. C., et al, 1988. *Investigation of September 1984 Blowout and Fire Lease OCS-G 5893*. OCS Report 86-0101. Gulf of Mexico: U. S. Department of the Interior/Mineral Management Service.

CSB Report, 2014. *Explosion and Fire at the Macondo Well. Investigation Report Overview*, Vol. 2. Report No. 2010-10-I-OS. U.S. Chemical Safety and Hazard Investigation Board. https://humanfactors101.com/incidents/macondo-deepwater-horizon/

IADC, 2018. *Sonnemann: New Approach May Be Needed in the Way Well Control Is Approached*. https://drillingcontractor.org/sonnemann-new-approach-may-needed-way-well-control-approached-49318 (Last Accessed in November 19, 2023).

IADC, 2023. *Deepwater Riser Gas Handling Guidelines*. https://store.iadc.org/product/riser-gas-handling-guidelines-rg-stand-alone-chapter-of-the-iadc-drilling-manual-12th-edition-revised (Last Accessed in November 19, 2023).

Kiran, R., & Salehi, S., 2018. Mathematical Modeling and Analysis of Riser Gas Unloading Problem. Vol. 8: Polar and Arctic Sciences and Technology, Petroleum Technology. *Proceedings of the ASME 2018 37th International Conference on Ocean, Offshore and Arctic Engineering*, Madrid, Spain, June 17–22. V008T11A063. https://doi.org/10.1115/OMAE2018-77719

Mustafa, A. S. E., 2023. *Mud Gas Separators Efficiency Improvement*. Bachelor's Thesis submitted in partial fulfillment of the requirement of the degree of BEng (Hons) in Petroleum Engineering, Asia Pacific University, Malaysia. Supervised by Dr. Rahman Ashena.

18.A APPENDIX

In this section, the real-gas law is used here to comprehensively consider the effects of pressure, gas Z-factor, and temperature for gas in riser flow.

The key to solving the analytical problem is that the mud hydrostatic pressure above the gas bubble is the gas pressure required in the gas law. Assuming a single gas bubble with the length/height of L_g, its pressure P_g (applied on its top) is:

$$P_g = P_{atm} + 0.052 \times MW \times \left(D - L_g\right) \tag{18.A.1}$$

where P_{atm} is the atmospheric pressure added for absolute pressure calculation; D is the true vertical depth (TVD) of the bottom of the gas bubble.

To simplify the equation form, we can embed the effect of P_{atm} in "D" parameter and express it as an updated TVD, called D':

$$D' = D + \frac{P_{atm}}{0.052\,MW} \tag{18.A.2}$$

Combining Equations (18.A.1) and (18.A.2):

$$P_g = 0.052 \times MW \times \left(D' - L_g\right) \tag{18.A.3}$$

The real-gas law comprehensively considers effects of all the known parameters on gas flow behavior. Using this law, we have:

$$\frac{P_{gi} \times V_{gi}}{Z_i nRT_{gi}} = \frac{P_g \times V_g}{ZnRT_g} \tag{18.A.4}$$

As the gas moles remain the same at any depth in the riser, equation becomes:

$$P_g = \frac{P_{gi}V_{gi}}{V_g}\frac{ZT_g}{Z_iT_{gi}} \tag{18.A.5}$$

Since the riser pipe size is constant from top to bottom, area (A) can be safely cancelled out from the numerator and denominator, and volume (V) may be safely replaced by length (L):

$$P_g = \frac{P_{gi}L_{gi}}{L_g}\frac{ZT_g}{Z_iT_{gi}} \tag{18.A.6}$$

Equaling Equations (18.A.3) and (18.A.6):

$$\frac{P_{gi}L_{gi}}{L_g}\frac{ZT_g}{Z_iT_{gi}} = 0.052 \times MW \times \left(D' - L_g\right) \tag{18.A.7}$$

Rearranging:

$$L_g^2 - D'L_g + \frac{P_{gi}L_{gi}}{0.052MW}\left(\frac{ZT_g}{Z_iT_{gi}}\right) = 0 \tag{18.A.8}$$

This is a quadratic equation with the following form, including parameter X and coefficients:

$$aX^2 + bX + C = 0$$

$$X = L_g,$$

$$a = 1, \quad b = -D' \ and \ c = \frac{P_{gi} L_{gi}}{0.052 MW}\left(\frac{ZT_g}{Z_i T_{gi}}\right)$$

If the discriminant $D = b^2 - 4ac > 0$, the solutions of the equation are:

$$X1 = \frac{-b + \sqrt{b^2 - 4ac}}{2a}, \ X2 = \frac{-b - \sqrt{b^2 - 4ac}}{2a}$$

Replacing coefficients a to c in the previous solutions gives:

$$L_g = \frac{D' +/- \sqrt{D'^2 - \frac{4 P_{gi} L_{gi}}{0.052 MW}\left(\frac{ZT_g}{Z_i T_{gi}}\right)}}{2} \tag{18.A.9}$$

For this problem, only ×2 (with the minus sign) applies in Equation (18.A.9):

$$\boldsymbol{L_g} = \frac{1}{2}\left[\boldsymbol{D'} - \sqrt{\boldsymbol{D'^2} - \frac{4 \boldsymbol{P_{gi} L_{gi}}}{0.052 \boldsymbol{MW}}\left(\frac{\boldsymbol{ZT_g}}{\boldsymbol{Z_i T_{gi}}}\right)}\right] \tag{18.A.10}$$

Equation (18.A.10) is not applicable if the discriminant gets negative $(D = (b^2 - 4ac) < 0)$ because the solution would not be valid any longer. Thus, the real-gas model/method cannot work/converge for low D', or specifically for gas depths above a specified depth. This depth may be called the equilibrium depth above where gas behavior abruptly changes, and thus real-gas law fails to predict gas flow behavior efficiently. The equilibrium depth (D_e) is found as follows:

$$D = D_e'^2 - \frac{4 P_{gi} L_{gi}}{0.052 MW}\left(\frac{ZT_g}{Z_i T_{gi}}\right) = 0 \tag{18.A.11}$$

Thus:

$$TVD_e' = 2\sqrt{\frac{P_{gi} L_{gi}}{0.052 MW}\left(\frac{ZT_g}{Z_i T_{gi}}\right)} \tag{18.A.12}$$

Combining Equations (18.A.2) and (18.A.12), the actual equilibrium depth D_e is found:

$$\boldsymbol{D_e = 2\sqrt{\frac{P_{gi}L_{gi}}{0.052MW}\left(\frac{ZT_g}{Z_iT_{gi}}\right)} - \frac{P_{atm}}{0.052\,MW}} \tag{18.A.13}$$

For riser handling systems, a backpressure (P_b) is applied. Thus, replacing P_{atm} with P_b:

$$\boldsymbol{D_e = 2\sqrt{\frac{P_{gi}L_{gi}}{0.052MW}\left(\frac{ZT_g}{Z_iT_{gi}}\right)} - \frac{P_b}{0.052\,MW}} \tag{18.A.14}$$

Simplifications:

a) *Ignoring Atmospheric Pressure:*
If "gauge pressure" is considered in the estimations which is not recommended (as it causes some slight error), D' in Equation (18.A.10) is replaced by D and P_{atm} drops in Equation (18.A.13). Thus:

$$L_g = \frac{1}{2}\left[D - \sqrt{D^2 - \frac{4P_{gi}L_{gi}}{0.052MW}\left(\frac{ZT_g}{Z_iT_{gi}}\right)}\right] \tag{18.A.15}$$

$$D_e = 2\sqrt{\frac{P_{gi}L_{gi}}{0.052MW}\left(\frac{ZT_g}{Z_iT_{gi}}\right)} \tag{18.A.16}$$

b) *Boyle' Law*
If Boyle's law is used, gas Z-factor is ignored ($Z = 1$) and $T_{gi} = T_g$. Thus, we have:

$$L_g = \frac{1}{2}\left[D' - \sqrt{D'^2 - \frac{4P_{gi}L_{gi}}{0.052MW}}\right] \tag{18.A.17}$$

$$D_e = 2\sqrt{\frac{P_{gi}L_{gi}}{0.052MW}} - \frac{P_b}{0.052\,MW} \tag{18.A.18}$$

c) *Boyles' Law + Ignoring Atmospheric Pressure:*

Both simplifications would apply here. In addition, P_{gi} in Equation (18.A.3) can be expressed as:

$$P_{gi} = 0.052MW \times D_{m,i} \tag{18.A.19}$$

where $D_{m,i}$ is the mud height (gas top TVD) at the initial state.

Applying both simplifications and replacing P_{gi} from Equation (18.A.19) would convert Equation (18.A.17) to:

$$L_g = \frac{1}{2}\left[D - \sqrt{D^2 - \left(4D_{m,i}\,L_{gi}\right)}\right] \tag{18.A.20}$$

$$\boldsymbol{D_e = 2\sqrt{D_{m,i} \times L_{gi}}} \tag{18.A.21}$$

19 Case Study
Root Cause of Blowout – Middle East, 2017

Rahman Ashena and Farzad Ghorbani

19.1 INTRODUCTION

Several incidents have occurred worldwide during oil and gas operations. Some of these incidents developed into accidents. Worldwide Offshore Accident Databank (WOAD) contains 6,183 records concerning near misses, incidents and accidents with their geographical distribution (Dhillon, 2016). During drilling, kick flows are among most serious drilling problems jeopardizing wellbore stability (Meng et al., 2019, 2020) and incidents which can occur. Kicks data can act as opportunities to elicit formation properties such as pressures (Miska et al., 1996; Baldino et al., 2019b, 2019a). However, kicks can be converted to blowouts with their catastrophic consequences. Etkin et al. (2017) listed 20 largest historical blowouts and considered the 1979 Ixtoc I and the 2020 Macondo as two examples of catastrophic ones. Blowouts have severe consequences in terms of fatalities, injuries, the detrimental environmental impacts (Lancaster, 2005), and the well control costs up to million (Grace, 2017). For many years, drilling engineers have made an effort directed toward better understanding of the blowout incidents as well as causes for their occurrence, with the aim of preventing kicks or at least controlling the kick flow to prevent blowouts (Editions Technip, 1981). Dobson (2009) and Bybee (2010) analyzed the kicks and blowouts occurred in the UK continental shelf and linked them to geological conditions at the wellsite and drilling conditions. The most efficient measure to prevent any blowouts is by using downhole early kick detection systems such as logging-while-drilling or seismic-while-drilling (see examples in Fraser et al., 2014; Jacobs, 2015; Shaker and Reynolds, 2020; Habib et al., 2020).

Statistically, many blowouts have occurred in development wells. Based on a ten-year (1979–1988) investigation of kicks and blowouts in conventional wells of the United States, frequency of blowouts in an exploratory well is 2.85 times that of a development well (Wylie and Visram, 1990). In unconventional high-pressure deepwater conditions (seawater > 600 m/1,969 ft), this frequency proportion is 5.14 (Etkin et al., 2017). This indicates that a large percentage of blowouts still occur in development wells. In addition, the same reference indicates that 51% of kicks and blowouts in development wells occur during tripping, which indicates the vivid role of human error role in the Reason's Swiss Cheese Model (Reason, 1997, 2000).

 DOI: 10.1201/9781003473770-19

Many researchers or practitioners have provided accounts of blowouts globally. Grace et al. (2000) discussed a gas blowout in Taiwan; US-CSB (2019) discussed how a blowout occurred in Pryor Trust well 1H-9 in the state of Oklahoma due to not filling in the hole during tripping in a horizontal well with oil-based mud. This case is similar to the blowout and explosion case discussed in this chapter. *St.* John (2016) discussed and compared the Macondo blowout (in Gulf of Mexico, GOM) and the Bardolino blowout (in North Sea) and how human error contributed to them. In GOM, regardless of the causes for the blowout, all the human factor barriers were absent. However, in Bardolino, one of the engagement factor barrier was present, and the rapid response could save the well (near miss). Skogdalen et al. (2011) presented information and indicators from the Risk Level Project in the Norwegian O&G industry related to safety climate, barriers, and undesired incidents and discussed the relevance for deepwater drilling based on experience from the Macondo blowout. Chen et al. (2021) developed a risk analysis model based on the bowtie model for the simulation of well blowout scenarios and updated the blowout risk based on operational field observations. Ashena et al. (2011) and Nabaei et al. (2011) discussed two blowouts in onshore wells of Iran, with the first happening due to lack of integrity in a production well and the second occurring due to too late kick detection while drilling. On July 23, 2013, a well operated by Walter Oil and Gas in the Gulf of Mexico (GOM) experienced a loss of well control event during well completion operations which escalated to an explosion and fire, causing damages. The primary factor causing the incident was an ineffective response to well control complications with both kick detection and well shut-in procedures that occurred (SEMS, 2014).

In 2017, a blowout occurred during drilling an onshore well in the Middle East. The loss of well control event happened following a sudden release of gas during tripping the drill-string to the surface. Tripping began to change the bit at a depth of 2,610 m (8,563.4 ft) when drilling the reservoir formation. The well was a development one with the objective of producing oil from the Northern sector of the oilfield. Due to the blowout and subsequent explosions, the drilling rig was burnt, and two crew members of the drilling contractor (the driller and derrickman) lost their lives. Some well capping methods were simultaneously applied to put out the blaze and finally cap the well. As the financial implication of the blowout, the operator and the drilling contract companies incurred huge economic loss due to this super-heavy land rig of nearly million investments being lost, in addition to the compensation and well capping expenses that had to be borne. As a result of this incident, the surrounding environment underwent some pollution. Historically, three blowouts (one exploration and two development wells) had already occurred during drilling wells in this hostile oilfield. In those occasions, the blowouts were controlled, except for one of the development wells where explosion occurred.

In this study, different aspects of the mentioned blowout were reviewed in detail with the purpose of determining the root causes of the incident. The lessons learned from this event are summarized, and recommendations to avoid similar incidents are presented. In this account, the pit gain/kick volume curve versus time is provided. Such an analysis of the root cause factors was rarely made and presented in the region, as was done in this work.

19.2 FIELD OF STUDY

The oilfield has recoverable reserves of 16.5 billion barrels of crude oil and 54.16 trillion cubic feet of natural gas in place. This field was discovered in 1964, and since then, production from its reservoir formation has started. The reservoir in this field is gas-cap drive. This field has a huge high-pressure gas-cap in the region, which makes drilling into its gas-cap challenging.

The subject well of this study is in the northern sector of the field where the oil zone is thinner than other sectors and estimated at 110 m (361 ft). The reservoir fluid properties including pressures, gas–oil-contact, and water–oil-contact are given in Table 19.1. This table will be used in the next sections as a reference to find the overbalance pressures during drilling and tipping. To prevent conning issues (especially gas conning), the well plan required horizontal drilling in the reservoir with subsequent cased-hole completion.

19.3 SEQUENCE OF EVENTS

The sequence of events that occurred prior and during the event of the kick and blowout in the field is divided into two phases of drilling (pre-kick) and tripping (when kick and blowout occur). In this section, a review of the events prior and during the incident is given to better understand the root causes of the event.

19.3.1 Past Kick and Blowout Events

Reviewing the previous well control events (kicks and blowouts) is essential to improve the learning curve and prevent or reduce the reoccurrence of such incidents in the future. The lessons learned are being incorporated in the updated companies' drilling policy handbooks and the drilling programs. The data collected for the purpose of this study are from the daily drilling reports of all the drilling wells

TABLE 19.1
Fluid Pressures and Contact Depths in the Oilfield. The Datum TVD Depth Was Also Found by Adding the Mean Seal Level (MSL) to the Rotary-Table-Elevation (RTE) of 122 m (~400 ft) of the Rig

Fluid	Density [psi/ft]	Pressure [psi]	Datum Depth (Mean Sea Level, from MSL)	Datum Depth (TVD Based)
Gas	0.057	3,200	−1,981 m (−6,500 ft)	2,103 m (−6,900 ft)
Oil	0.33	3,250	−2,316 m (7,599 ft)	2,438 m (7,999 ft)
Water	0.48	3,992	−2,743 m (9,000 ft)	2,865 m (9,400 ft)
Contacts				
Gas–Oil-Contact (GOC)	−2,290 m (−7,513.5 ft) below mean-sealevel, MSL			
Oil–Water-Contact (WOC)	−2400 m (−7,874.4 ft) below MSL			

in this field. To date, a total of 85 kick flows have been recorded during drilling the reservoir formation in this field. During drilling the first three exploratory wells in the reservoir formation, 30 kicks and one blowout were recorded. During drilling the development wells, 55 kick flows and 3 blowouts were recorded. The four recorded blowouts are listed and explained in Table 19.2. Reviewing the kicks and blowouts data in this field (Table 19.3 and Figure 19.1), a number of observations were made:

- In all the wells, the percentage of wells that experienced kicks are 57%. This percentage in the exploratory wells is 1,000% (10 kick per well), whereas it is 38.2% in development wells.
- In all the wells, the percentage of wells that experienced blowouts are 2.7%. This percentage in the exploratory wells is 33%, whereas it is 2.1% in development wells.
- In all the wells, the percentage of kicks that converted to blowouts are 4.7%. This percentage in the exploratory wells is 3.3%, whereas it is 5.5% in development wells.

Based on the recorded accounts and reports, the nature of the blowout in well #B was similar to the blowout studied in this work due to insufficient fluid in the wellbore during tripping. The well #B incident happened a few years prior to this blowout and it was a near miss, but the crew managed to save the well. Accounts of events in the literature such as the blast in BP Texas City Refinery disaster (Hopkins, 2008) shows the great importance of using past experiences to avoid future incidents. This requires that a reliable record of the kicks and blowouts be collected.

In the drilling well plan, it was explicitly stated that no gas-cap was expected along the drilling path. However, a gas-cap was illustrated in the schematic of the drilling plan. As a matter of fact, this discrepancy should have been detected by subsequent reviewers of the well plan and clarified.

19.3.2 Drilling the Reservoir Formation

In this well, 26-in, 17 ½-in, and 12 ¼-in holes were drilled with corresponding casing sizes of 18 5/8 in, 13 3/8 in and 9 5/8 in, respectively (Figure 19.2). A summary of the highlights of the drilling activities across the 8 ½-in section of the wellbore where the blowout occurred is given here.

1. The reservoir formation was drilled using the 8 ½-in drilling bit and 9 stands of 5-in heavy weight drill-pipes with 8.42 ppg mud weight from the measured depth MD of 2415.5 m (7925.3 ft), equivalent to TVD of 2348.1 m (7704.1 ft) to MD of 2,537 m (8323.9 ft), equivalent to TVD of 2424.7 m (7955.4 ft). The observed mud loss was 0–2 barrel per hour (BPH).
2. At the MD of 2,537 m (8,323.9 ft), TVD of 2,424.7 m (7,955.4 ft), an increase in mud loss rate, was observed. Attempts to cure the mud loss were made by pumping lost circulation material (LCM) pills periodically. Due to the lack of improvement, the mud weight was reduced to 8.15 ppg stepwise.

TABLE 19.2
Record of the Blowouts in the Oilfield, Prior to the Investigated Incident

Well	Type	When	Explosion	How Happened	Control Measures
#1	Exploratory	1964 (March 27)	No	Drilling in the reservoir formation was done to the depth of 2.258 m (7,408 ft). During running 7-inch liner to the depth of 1,343 m (4,406 ft), gas blowout occurred.	Liners were cut, and close blind rams were closed. Next, the well was bullheaded.
#A	Development	1975 (Oct. 31)	Yes	After drilling in the reservoir to the depth of 2,017 m (6,617 ft), following lost circulation and not filling the hole, kick and blowout occurred at the surface which was later exploded.	Relief well drilling and killing the well (after 85 days)
#B	Development	2013 (April 15)	No	Drilling in the reservoir was done to the depth of 2,403 m (7,884 ft). During pulling the drill-string out of the hole, the hole was not sufficiently filled up, thus the blowout occurred.	Shear rams were closed, but due to leakage in the surface facilities (tubing head spool) which was later sealed by the rig crew.
This Well	Development	2017	Yes	During pulling the drill-string out of the hole (with causes to be discussed in detail in this chapter)	Drilling a relief well and killing the well

TABLE 19.3
Statistics of Kicks and Blowouts in the Studied Oilfield

Well Type	Number of Kicks	Number of Blowouts	Kick Per Well [%]	Blowout Per Well [%]	Blowout Per Kick [%]
Exploratory (Including Appraisal) Wells	30	1	1,000	33	3.3
Development Wells	55	3	38.2	2.1	5.5
Total	85	4	57.8	2.7	4.7

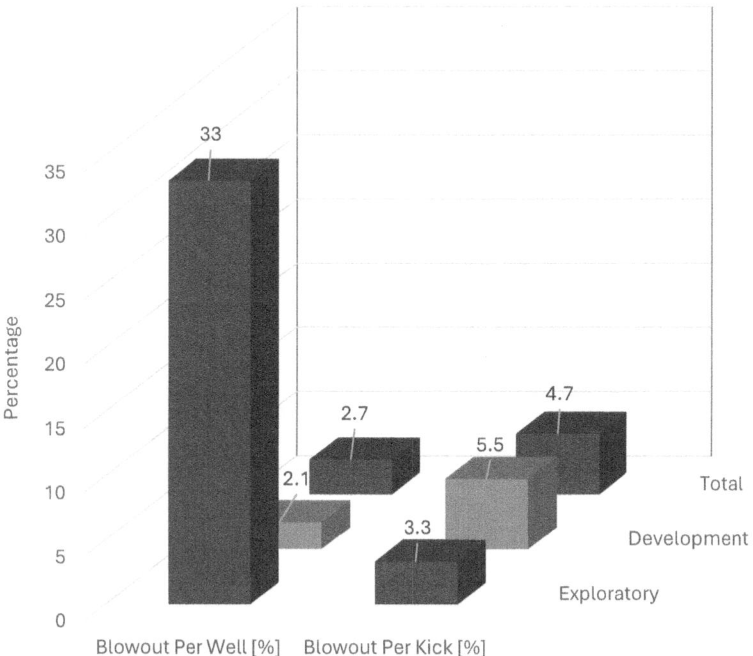

FIGURE 19.1 The percentages of blowouts per well and per kicks in exploratory, development, and all wells in the oilfield.

3. During directional drilling to MD of 2,540 m (8333.7 ft), TVD of 2,444 m (8,018.7 ft) with inclination angle of 82 degrees and azimuth of 161 degrees, the rate of mud loss increased to 80 BPH. Attempts were made to cure the loss by pumping periodic LCM pills, which were unsuccessful. Another decrease in mud weight was made to 8.02 ppg, the mud loss rate decreased to 40–60 BPH, which was still high and unacceptable. Therefore, to subsequently control the loss using a level-off magnesite plug (magnesium cement), the drill-string was pulled out. It is to be noted that as downhole motor was used in the bottomhole

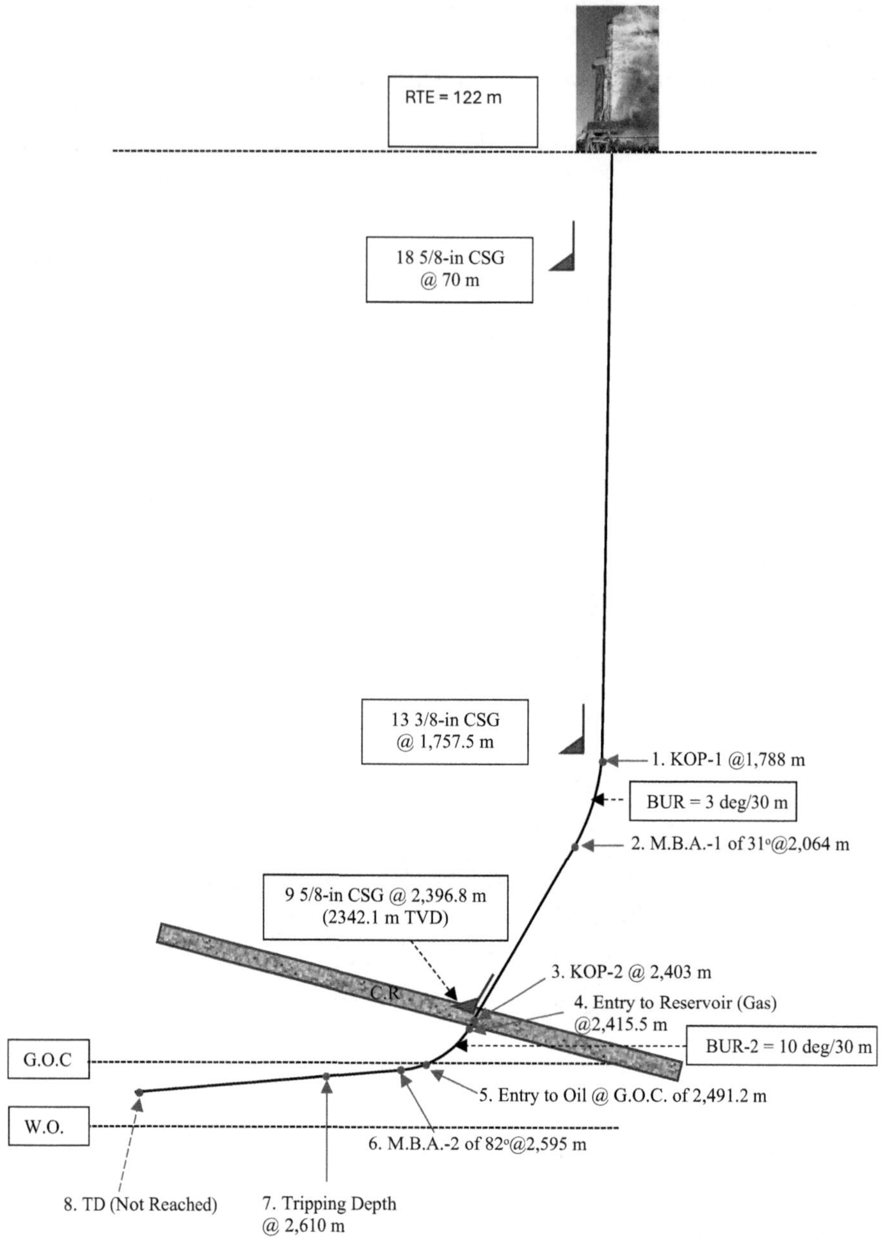

FIGURE 19.2 Schematic of the well trajectory. See Table 19.4 as the legend for this figure to obtain more information about the directional depths.
Abbreviations: CSG, casing; C.R., cap rock; G.O.C., gas oil contact; W.O.C., water oil contact; KOP, kick-off point; M.B.A., maximum build angle; BUR, build-up rate; TD, total depth.

TABLE 19.4
Well Schematic Details Corresponding to Figure 19.2

		MD		TVD		Inclination Angle	Azimuth
Point	**Description**	**[m]**	**[ft]**	**[m]**	**[ft]**	**[deg]**	**[deg]**
1. KOP-1	1st kick-off point	1,788	5,866	~1,788	~5,866	1.7	0
2. M.B.A.-1	1st maximum build angle point	2,064	6,772	2046.9	6,715.9	31	159
3. KOP-2	2nd kick-off point	2,403	7884.2	2,337	7667.7	31	159.5
4. Entry to reservoir (gas)	The point where the well entered the reservoir (gas-cap)	2415.5	7925.3	2348.1	7704.1	34	160.8
5. Entry to oil	The point where the well entered top of oil (at gas oil contact, GOC)	2491.2	8173.6	2,402	7880.9	54	159
6. M.B.A.-2	2nd maximum build angle point	2,595	8514.2	2,441	8008.9	82	161
7. Tripping depth	The depth where drilling stopped for tripping	2,610	8563.4	2443.1	8015.8	82	161
8. TD	Expected total depth (which never happened)	3,125	10,253	It was expected to hold the angle at 82 deg.			

assembly, the maximum possible LCM concentration in the pill was 5 pounds per barrel (PPB).

4. The drill-string was run into the hole to MD of 2,447 m (8028.6 ft), with TVD of 2372.5 m (7784.1 ft), and a 60-bbl of magnesium cement slurry was set at the depth of the loss zone (i.e., 2,540 m or 8333.7 ft). The well was observed static. The drill-string was pulled out of the hole.
5. The drill-string was run in the hole again. The magnesite plug was drilled out, and then the reservoir formation was drilled to 2,610 m (8563.4 ft) with the same mud weight and the mud loss range of 20–40 BPH. Complete loss occurred at two depths of a) MD of 2,542 m (8340.3 ft)/TVD of 2426.8 m (7,962 ft) and b) MD of 2,559 m (8,396 ft)/TVD of 2,432.6 m (7,981 ft). However, by periodically pumping LCM pills, it was reverted to the previous mud loss range. It is to be noted that, in order to prevent screening out of the LCM materials, drilling was done by by-passing the return flow from the shale shaker.
6. At MD of 2,610 m (8563.4 ft) and TVD of 2443.1 m (8015.8 ft), in order to change the bit and the adjustable kick-off (AKO) angle of the downhole motor, drilling was stopped at 3:18 am, and the crew became ready to pull the drill-string out of the hole.

19.3.3 Tripping Process with Kick and Blowout

In this section, the tripping process and the events which led to the incident (the kick and blowout) are discussed. These are sequentially explained and analyzed versus time in Table 19.5. The presented events and data are based on the field reports, sensor measurements, recordings of the mud engineer, and personal interviews with survivors. Figure 19.3 and Table 19.6 respectively show curves and values of changes in the actual pit volume totalizer (PVT_{act}) and the expected pit volume totalizer (PVT_{exp}) and pit gain/kick volume. The expected pit volume totalizer (PVT_{exp}) was found versus time as the total mud loss and pipe displacement volumes were removed from the well during pulling out of the hole. Due to sudden expansion of gas reaching near the surface, the pit gain/kick volume showed an extremely sudden rapid increase from 172 bbl at 12:53 o'clock to 341.5 bbl at 13:00 o'clock (when the blowout occurred). The trend of increase is exponential, indicative of gas expansion characteristic. It is to be noted that the involved companies did not use the trip tank for filling the hole during tripping, as part of their standard procedure. Instead, they used the rig pump to pump/inject fluid from the kill-line and back from the return line to establish a cross-wellhead circulation. It was required to level the tanks periodically and monitor the well condition to notice possible kick flow.

19.4 INVESTIGATION OF ROOT CAUSES

Based on the information provided in the previous section, it appears that the cause of the blowout is mainly due to the human error and some other technical aspects. We have classified the root causes of the blowouts into five categories of well planning, primary barrier issue, late kick detection, secondary barriers malfunctions, and organizational issues. These are discussed next.

19.4.1 Drilling Plan

19.4.1.1 Gas-Cap Drilling

In the drilling plan, the existence of the gas-cap with the inclined column of 75.7 m (248.4 ft), equivalent to the vertical column of 53.9 m (176.8 ft), was not mentioned. This is an important information which needs to be carefully considered for predicting the latent drilling risk to raise the crew awareness of the risks and to equip the rig with the gas detection sensors and necessary BOP component. Therefore, the drilling operation was in progress with a high risk as only one gas detection sensor was used at the shale shaker, while in the gas-cap drilling, three sensors are required to be placed at the mud return line and the suction line. This jeopardized the gas detection. In addition, no rotating head or BOP was requested to be installed; therefore, there was no well control equipment at the rig site.

19.4.1.2 Mud Type

Based on the drilling plan, the oil-based mud type was required and selected by the operator with multiple objectives which include improving the lubrication of the

TABLE 19.5

The Record of the Tripping Process from the Beginning to the Blowout and Explosion Time

Time and Tank Volumes				
From	**To**	**Operation/Event**	**SPM**	**CircSys.**
3:18	4:30	Mud circulation, hole cleaning and spotting 100 bbl LCM pill	8–10	
4:30	6:00	POOH from 2,610 to 2,255 m (8563.4 to 7398.6 ft)	8–10	
6:00	7:48	Hole cleaning & spotting 100 bbl LCM pill	8–10	
7:48	9:32	POOH from 2,255 to 1,860 m (7398.6 to 6102.6 ft), 10–12 BPH mud loss	8–10	
9:32	10:04	Mud circulation to check if the drill-string was not plugged by the LCM, and observing the well (for 13 minutes)	8–10	
10:04	10:20	POOH from 1,860 m to 1,773 m (6102.6 to 5817.2 ft) with 8 BPH mud loss. The continuous-fill method was applied to fill the hole during tripping.	8–10	
10:20 (S = 250) (M1 = 141) (M2 = 22) (D = 154) (T = 70) → PVT_{act} = 250	11:15 (S = 250) (M1 = 141) (M2 = 22) (D = 154) (T = 70) Trans = 0 → PVT_{act} (Exc Trans) = 250	POOH from 1,773 to 1,539 m (5817.2 to 5049.5 ft) At 10:20 am, the rate of mud pumped into the annulus was reduced and stopped. Therefore, almost zero stroke per minute (SPM) was recorded by the pump stroke-counter. As a continuous mud circulation was expected to be in progress using the rig pump, zero SPM indicates the pump had become off at that time. The active pit was only the suction pit (i.e., mud was pumped from the suction pit to the annulus and circulated back to the very suction). Next, it is expected that 18.3 bbl be reduced from the PVT due to pipe displacement and mud loss: Displaced closed-end drill-pipe vol. = $(5817-5049)\frac{5^2}{1029}$ =11 bbl Mud loss volume = 8 bbl/hr × 55/60 = 7.3 bbl Therefore, the expected PVT is: PVT_{exp} = 250–18.3 = 231.7 bbl The difference between PVT_{act} and PVT_{exp} is the kick influx volume: $Kick - Vol_1 = PVT_{act} - PVT_{exp} = 250 - 231.7 = 18.3$ bbl Assuming this 18.3 bbl kick influx entered at once, knowing the annular volume of 0.04892 bbl/ft, the mud level drop in the hole before influx was:	~ 0	Suction to Suction Tank

(continued)

TABLE 19.5 *(Continued)*

The Record of the Tripping Process from the Beginning to the Blowout and Explosion Time

Time and Tank Volumes				
From	**To**	**Operation/Event**	**SPM**	**CircSys.**
		Mud level drop = 18.3 bbl/0.04892 bbl/ft = 374.08 ft = 114 m The equivalent drop in the mud hydrostatic pressure was: $\Delta P_h = 0.052 \times 374.08 \times 8.02 = 156$ psi Knowing that the well was already minimally underbalanced (by making some calculations) at the gas-cap top, the underbalance pressure is at least 156 psi. Therefore, "MUD LOSS HAS CONVERTED TO KICK". The height of the kick influx in the hole is: $h_{in} = \frac{18.3 \text{ bbl}}{0.0732 \text{ bbl / ft}} = 250 \text{ ft}(76.2 \text{ m})$ The depth of the gas in the hole is MD = 2326.6 m/7,633 ft (TVD = 2271.9 m/7,454 ft). See Figure 19.4(a) to see schematic of gas kick influx at this time.		
11:15 (S = 250) (M1 = 141) (M2 = 22) (D = 154) (T = 70) → PVT_{act} = 413	11:53 (S = 255) (M1 = 295) (M2 = 22) (D = 16) (T = 70) Trans = 154 → PVT_{act} (Exc Trans) = 418	POOH from 1,546 to 1,393 m (5072.4 to 4570.4 ft) At 11:16 am, the crew began to transfer mud from the desander tank to the middle tank. To do this, the desander mud was pumped to the shale shaker box (also called possum-belly box) where it flowed together with the return mud back through the shale shaker, and then it flowed into the middle tank-1. Note that this type of flow means that the "shaker-out" system was reverted back to "shaker-in". It is expected that the 138 bbl transferred mud volume from the desander tank was to be added to the middle tank-1 ($\Delta M\text{-}1_{exp}$ = 138 bbl). However, the actual added volume to the middle-1 was 154 bbl ($\Delta M\text{-}1_{act}$). In addition, 5 bbl mud had been added to the suction tank (ΔS). This clearly indicates kick influx entry to the wellbore. Next, it is expected that 17.2 bbl was reduced from the PVT due to pipe displacement and mud loss: Displaced closed-end drill-pipe vol. = 12.2 bbl Mud loss volume = 8 bbl/hr × 38/60 = 5 bbl Therefore, the expected PVT (ignoring the mud transfer) is: $PVT_{exp} = 413 - 17.2 = 395.8$ bbl	~ 0	Suction to Middle Tanks

		The difference between PVT_{act} (excluding the transfer volume) and PVT_{exp} is the kick influx volume: $Kick - Vol_2 = PVT_{act} - PVT_{exp} = 418 - 395.8 = 22.2$ bbl Therefore, the total kick volume in the hole was: $Total\ Kick - Vol. = Kick - Vol_1 + Kick - Vol_2 = 18.3 + 22.2 = 40.5$ bbl	
<u>11:53</u> (S = 255) (M1 = 295) (M2 = 22) (D = 16) (T = 70) → PVT_{act} = 572	<u>12:53</u> (S = 243) (M1 = 297) (M2 = 175) (D = 6) (T = 32) Trans = 38 → PVT_{act} (Exc Trans) = 677	POOH from 1,393 to 1,160 m (4570.4 to 3,806 ft). See the pit gain in Figure 19.3 and the mud logging graph in Figure 19.5. At 11:54 am, mud from the middle-1 tank to the middle-2 tank was transferred. At 12:37, the transfer was stopped. However, still the middle-1 volume showed an increase in in the mud-log (Figure 19.5), which indicates a large kick was in progress. At 12:44 pm, the crew started to transfer 38 bbl of mud from the trip tank to the middle-1 tank. It is expected that this volume add up to the PVT. At 12:47 pm, an abrupt increase in the middle-1 tank can be detected in the same figure, which indicates the gas influx expansion. At 12:53 pm, the first explosion occurred which damaged and disconnected hydraulic connection line of the shear rams. At this time, the first gas slug reached the surface and ignited, but the explosion was just enough to disconnect the hydraulic connection line of the shear rams. Yet, the main gas influx had not reached the surface. It is expected that 26.5 bbl be reduced from the PVT due to pipe displacement and mud loss: Displaced closed-end drill-pipe vol. = 18.5 bbl Mud loss volume = 8 bbl/hr × 60/60 = 8 bbl Therefore, the expected PVT (ignoring the mud transfer) is: $PVT_{exp} = 572 - 26.5 = 545.5$ bbl The difference between PVT_{act} (excluding the mud transfer) and PVT_{exp} is the kick influx volume: $Kick - Vol_3 = PVT_{act} - PVT_{exp} = 677 - 545.5 = 131.5$ bbl Total Kick – Vol. $= Kick - Vol_1 + Kick - Vol_2 + Kick - Vol_3 = 18.3 + 22.2 + 131.5 = 172$ bbl	~ 0

(continued)

TABLE 19.5 *(Continued)*

The Record of the Tripping Process from the Beginning to the Blowout and Explosion Time

Time and Tank Volumes				
From	**To**	**Operation/Event**	**SPM**	**CircSys.**
12:53 (S = 243) (M1 = 297) (M2 = 175) (D = 6) (T = 32) → PVT_{act} = 715	13:00 (S = 310) (M1 = 394) (M2 = 177) (D = 12) (T = 32) Trans = 0 → PVT_{act} (Exc Trans) = 881	POOH from 1,160 to 1,131 m (3,806 to 3710.8 ft). See the pit gain graph in Figure 19.3 the mud logging graph in Figure 19.5. At 12:53 am, the kick flow was detected. The driller disconnected the last stand and immediately proceeded to lower the top drive system to connect to the box of the drill-string (to safe the drill-string). At around 12:59, the well was closed using the shear rams at the Hydraulic Control Unit (based on the interviews). A few minutes delay happened because the crew were trying to contact their superiors to ask for permission to close the well. Based on the standard, they needed no permission to do that, and they had to immediately proceed to close the well. At 12:59–13:00, three stages of flows were observed at the floor intermittently (with around five- second differences). The third flow which was gas rose up to the crown block and the explosion occurred. It is expected that 3.5 bbl be reduced from the PVT due to pipe displacement and mud loss: Displaced closed-end drill-pipe vol. = 2.3 bbl Mud loss volume = 8 bbl/hr × 9/60 = 1.2 bbl Therefore, the expected PVT is: $PVT_{exp} = 715 - 3.5 = 711.5$ bbl The difference between PVT_{act} and PVT_{exp} is the kick influx volume: $Kick - Vol_4 = PVT_{act} - PVT_{exp} = 881 - 711.5 = 169.5$ bbl Therefore, the total kick volume is: Total Kick – Vol. $= Kick - Vol_1 + Kick - Vol_2 + Kick - Vol_3 + Kick - Vol_4 = 18.3 + 22.2 + 129.5 + 169.5 = 339.5$ bbl The total mud volume in the hole (from the gas-cap top to the surface) is 490 bbl. Therefore, most of the hole volume was filled with gas, and it had percolated through the mud to the surface. Figure 19.4(b) shows the schematic of gas blowout and explosion at this time. Refer to Figure 19.6 to see a photo of the real blowout and explosion.	~ 0	Suction to Middle Tanks

Abbreviations: SPM, stroke per minute (pumping rate); Circ. Sys., circulation system; POOH, Pull out of the hole; LCM, lost circulation material; S, suction tank; M-1, =middle tank-1; M-2, middle tank-2; D, desander tank; T, trip tank; PVT, (actual) active pit volume; "act", actual; "exp", expected

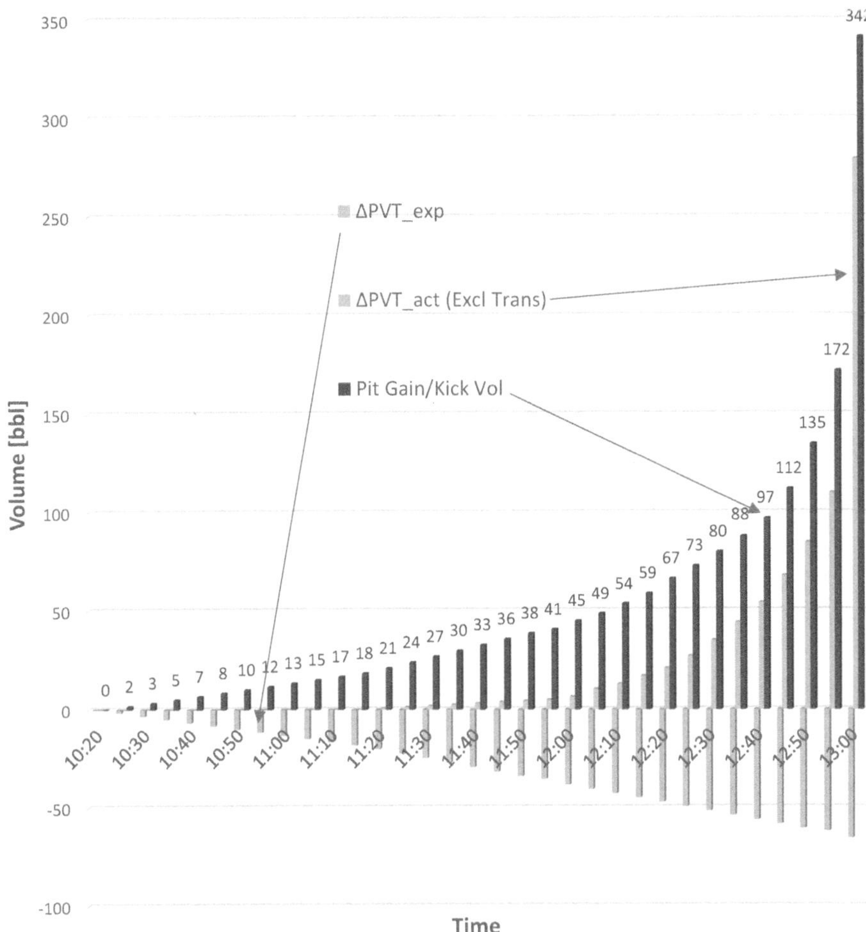

FIGURE 19.3 Graphs of changes in actual pit volume totalizer (ΔPVT_{act}), expected pit volume totalizer (ΔPVT_{exp}), and pit gain/kick volume versus time based on field data. ΔPVT_{act} was found using field sensors and mud engineer's measurements; PVT_{exp} was found as the total mud loss and pipe displacement volume removed from the well during pulling out of the hole. The pit gain/kick volume was found by subtracting ΔPVT_{exp} from ΔPVT_{exp}. Due to sudden gas expansion, the pit gain/kick volume shows a sudden exponential increase from 172 bbl at 12:53 o'clock to 341.5 bbl at 13:00 o'clock. Refer to Table 19.6 for the values and tabular presentation of the curves.

TABLE 19.6
Values of Changes in Actual Pit Volume Totalizer (ΔPVT_{act}), Expected Pit Volume Totalizer (ΔPVTexp), and Pit Gain/Kick Volume versus time, Based on Field Data. Plots of These Values Were Shown in Figure 19.3

Time	ΔPVT_{exp}	ΔPVT_{act} (Excluding Transits)	Pit Gain/ Kick-Vol
10:20	0	0	0.00
10:25	−1.66	0	1.66
10:30	−3.33	0	3.33
10:35	−4.99	0	4.99
10:40	−6.65	0	6.65
10:45	−8.32	0	8.32
10:50	−9.98	0	9.98
10:55	−11.65	0	11.65
11:00	−13.31	0	13.31
11:05	−14.97	0	14.97
11:10	−16.64	0	16.64
11:15	−18.30	0	18.30
11:20	−20.28	0.63	20.90
11:25	−22.58	1.25	23.83
11:30	−24.89	1.88	26.76
11:35	−27.20	2.50	29.70
11:40	−29.50	3.13	32.63
11:45	−31.81	3.75	35.56
11:50	−34.12	4.38	38.49
11:53	−35.50	5.00	40.50
12:00	−38.31	6.50	44.81
12:05	−40.55	10.50	48.50
12:10	−42.78	13.00	53.50
12:15	−45.02	17.00	59.00
12:20	−47.25	21.00	66.50
12:25	−49.49	27.00	73.00
12:30	−51.72	35.00	80.00
12:35	−53.96	44.00	88.00
12:40	−56.19	54.00	97.00
12:45	−58.42	68.00	112.00
12:50	−60.66	85.00	135.00
12:53	−62.00	110.00	172.00
13:00	−65.50	279.50	341.50

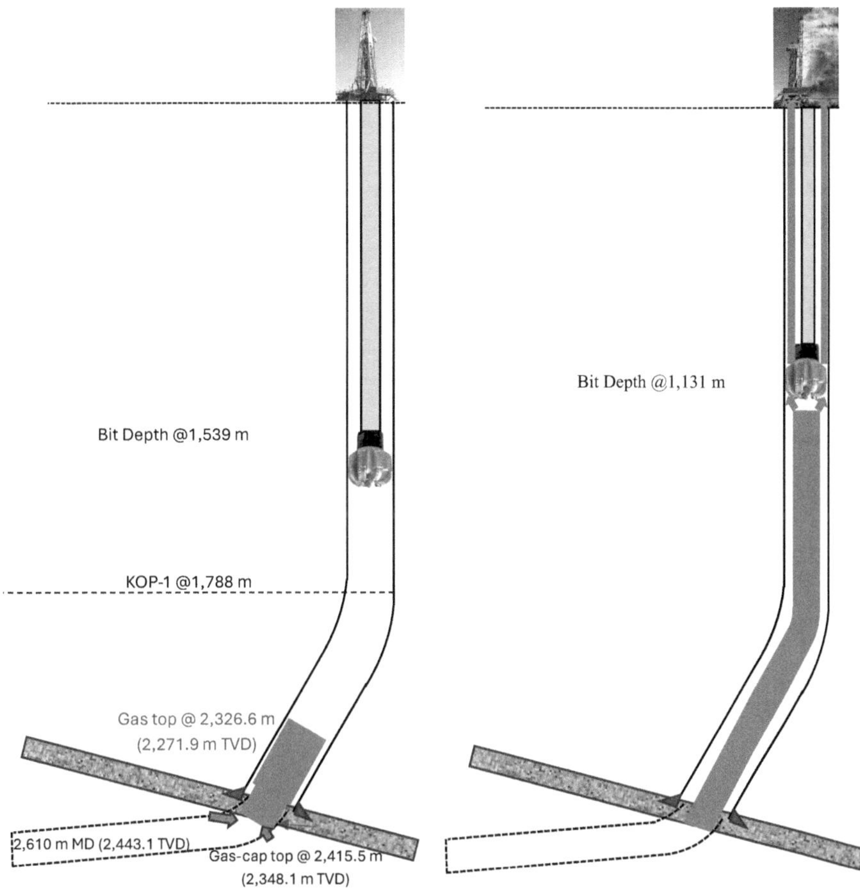

FIGURE 19.4 Schematics of gas influx in the wellbore at gas entry time and when the gas influx is at the surface in the studied incident (refer to Table 19.5 to see the related calculations and track of records). Note that the orange color (gray color in black and white print) shows the gas influx.

downhole tools and prevention of pipe stuck for drilling the horizontal wellbore, overcoming the challenge of drilling thin shale layers in the reservoir formation, and lowering the rate of mud loss because of lower differential pressure with the oil-based mud compared to a water-based mud. Due to the existence of a large gas-cap to be encountered in drilling of this well and the possibility of dissolution of gas in oil, an oil-based mud used for kick detection was not the best choice as a drilling fluid. Therefore, a water-based mud could have been potentially a better drilling fluid in this occasion. However, as the gas-cap had not been predicted in the drilling plan, an oil-based mud was selected.

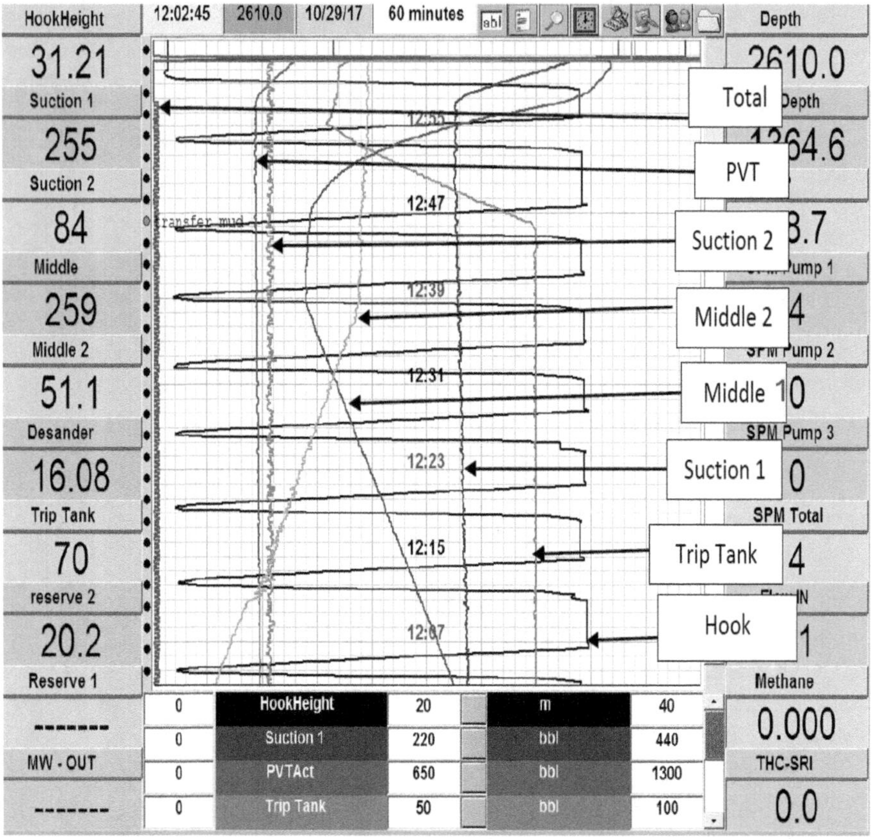

FIGURE 19.5 The mud logging graph from 12:07 to 13:00 pm during the explosion time.

FIGURE 19.6 Photo of the explosion following the gas blowout.

19.4.2 Primary Barrier Issues

19.4.2.1 Insufficient Mud Weight

While facing mud losses during drilling, reduction of the mud weight from 8.42 to 8.02 ppg was not a safe action and means that the operation was done at a very high risk. The most critical point of the kick occurrence was the top of the gas-cap at MD of 2415.5 m (7925.3 ft). The overbalance pressure at the time of drilling the gas-cap top was only 19 psi (using some calculations), which was an unsafe action to be taken. This was further jeopardized by further reducing the mud weight to 8.02 ppg at MD of 2,540 m (8333.7 ft) to control severe mud loss as mentioned in the previous sections. This has already put the well at the risk of an underbalance pressure of 27 psi at the gas-cap top. This indicates that for the subsequent tripping, there was a high risk of entry of gas kick influx entry, which was probably dissolved in the oil-based mud. As a standard procedure during tripping, the trip margin should be considered for possible swabbing effect, which was not considered in drilling this well.

As a matter of fact, the bottomhole (MD of 2,610 m/TVD of 2443.1 m) posed less risk to the kick than the gas-cap zone as it was in the oil zone section. The overbalance pressure at the bottomhole was 88 psi. The mud weight reduction to 8.02 ppg was against the company's policy which required 200 psi overbalance.

19.4.2.2 Insufficient Filling the Hole during Tripping

In order to fill the hole during the tripping operation, the trip tank should be used in order to prevent human error in displaced volumes' calculation. As the mud of the previous hole had been already transferred to this tank, the trip tank could not be used. In addition, the drilling contractor did not use the trip tank as part of their normal procedure.

At 10:20 am, the rate of mud pumped in the annulus reduced to 0–5 barrel per hour, which can be observed in the mud logging graph. However, this was not noticed by any of the drilling crew members until 12:57 pm (3 minutes before the blowout).

19.4.2.3 Late Kick Detection

The kick was not detected in this case until it was converted to a blowout. Even at about 12:42 pm when the first explosion occurred (with no gas blowout at the surface yet), the crew were not sure about the reason for the explosion. The blowout was detected and observed at 12:54 pm, only 8 minutes before the explosion. The crew were not alerted to detect the kick influx in time. The kick influx could be detected by leveling and measuring different mud tank levels and controlling the active pit volume. This was not done closely as per the standard procedure.

In addition, some essential mud logging sensors which could have contributed to the kick detection were not installed at the rig site. These include the outflow and gas detector sensors. The outflow sensor is installed at the mud return line and acts as the main primary kick indicator. Particularly, Coriolis flowmeters are recommended as an earlier kick detection system. At least three gas detection sensors are required to be installed at the return flow line, the shale shaker, and the suction pit. Installation of these sensors was even more important for this gas-cap drilling operation. Only one flammable gas detection sensor was installed on the shale shaker, but that was not effective due to the shaker-out condition of the return.

19.4.3 Secondary Barriers Malfunctions

19.4.3.1 Remote-Control Panels' Issue

A remote-control panel is commonly placed next to the tool pusher or drilling supervisor's offices so that they can operate it in emergency situations for shutting down and securing the well. It was not possible to operate the BOPs using the remote panel due to malfunctions which were not resolved in advance. Therefore, at 12:53 pm, the tool pusher decided to secure the well using the main accumulator control unit system. This is to be done while no drilling operations being allowed unless all the remote-control panels operate based on the standard procedure during the BOP tests.

19.4.3.2 Shear Rams Failure

The (pressure rating) Cameron shear rams were operated at the Hydraulic Control Unit at 12:59 pm (one minute before the incident). However, for two reasons, the shear rams did not operate properly. First, due to the first explosion at around 12:53 pm, one of the connection lines of the shear rams had come off, and, second, the bypass valve was not activated before proceeding to close the shear rams.

Subsequent investigations showed that the shear rams had just half-closed (Figure 19.7), and proper sealing was not obtained. This forces recall of similar global experiences such as the 2010 Macondo blowouts (Turley, 2014) where shear rams did not properly operate and resulted in the release of the gas to the surface. There are multiple blowouts like this in history. Another example was the Walter Oil and Gas blowout on Hercules-265 in 2013 where the BOP failed to close because the flow through the BOP became extremely high before the BOP was activated. The BOP would not close against the tremendous rushing flow.

After the incident, the shear rams were transported and tested in a workshop, and these showed to operate properly to cut the drill-pipe and close the well safely. This test confirmed the reasons stated for the connection line issue and not properly operating of the shear rams.

According to standard procedure in well control events/situations, first, the drill-string must be stabbed into the wellbore safely. To do this, an already pressure-tested safety valve must be installed on the drill-string, or alternatively the top drive is connected to the string. Next, the annulus is sealed by first closing the top pipe rams. As the companies adopt soft shut-in procedure, closing the rams must be preceded by opening the hydraulic control remote (HCR) 4-in valve. If it is not successful, then bottom pipe rams are closed. In emergency situations, shear rams are closed. This is done by activating the bypass valve in advance so that maximum accumulator pressure of around 3,000 psi can be applied to shear the drill-pipe. In this incident, the shut-in procedure was delayed for five to six minutes (until 12:58–12:59 o'clock) in order to receive permission from the superiors. It was expected to immediately take

FIGURE 19.7 Shear rams failed to seal the wellbore during the kick event in the studied well.

a decision to close the rams in such a critical situation. Next, since a hydraulic connection line of the shear rams had already come off (due to the first explosion), it was a more logical decision to proceed to shut down the well by closing the bottom pipe rams. Following that, by closing the BOP's hand wheels, it was possible to secure the well. However, making such wise decisions in such emergency situations is not an easy job, especially when the required training had not been provided to the crew.

19.4.4 Organizational Issues

Organizational issues fall into learning curve, miscommunications, and insufficient well control training.

19.4.4.1 Learning Curve

Also, to learn from past well control experiences and prevent similar mistakes, it is essential that all related data about kicks and blowouts be recorded and analyzed. This includes the causes of the previous kicks and the actions taken to control the well and a summary of the lessons learned. The occurrence of a kick must be explicitly reported in daily drilling reports, and the exact kick influx volumes must be reported in pit gains to expedite the company's learning rate. The lessons learned are expected to be added to the updated drilling policy. Unfortunately, the last update of the well control operation policy goes back to the 1990s in the case of this event. In addition, the use of the new standard early kick detection systems such as outflow sensors, particularly surface Coriolis flowmeters, must become mandatory.

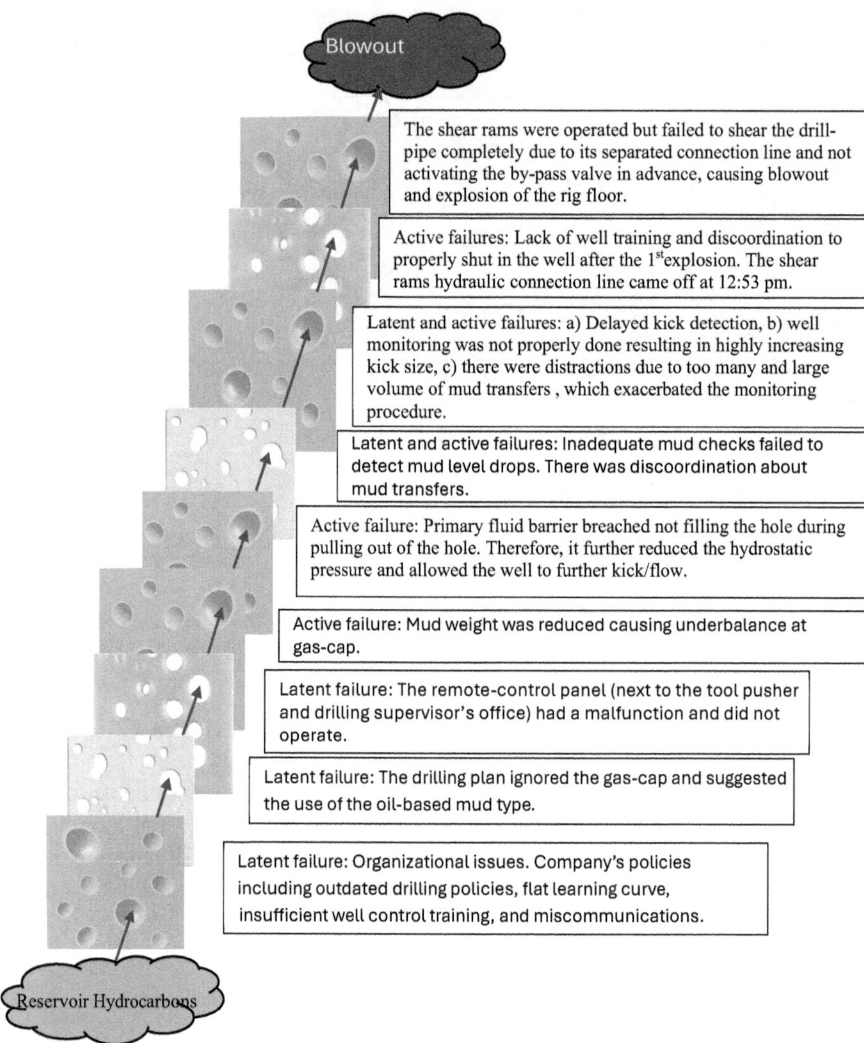

FIGURE 19.8 Reasons' Swiss Cheese Model for the investigated blowout incident.

19.4.4.2 Miscommunications

The mud transfer was in progress from 11:16 am onwards, but it was not closely monitored and not coordinated with the drilling supervisor. This could have a potential role in misleading the crew from the serious incident that was underway.

Based on our investigations, the rate of mud loss was reported as 6–8 BPH at 12:20 pm. A mud loss rate cannot be technically measured when the well is not filled. Therefore, this value was most likely not measured.

19.4.4.3 Insufficient Well Control Training

The drilling crew showed insufficient well control skills in kick detection and operation of the BOPs to secure the well. The evidence for this claim includes the very late kick detection and delayed operation of the rams (after five minutes), disregarding the fact that the connection line had come off from the shear rams before it was operated and failing to activate the bypass valve before closing the shear rams. Their poor performance can be attributed to insufficient theoretical and practical training. Therefore, as part of the enhanced readiness for well control operation, it is recommended that the involved drilling company mandates its drilling crew to regularly pass well control courses with equivalent international standards of the International Well Control Forum (IWCF) or the International Association for Drilling Contractors (IADC). Also, drills are important measures to check the practical preparation of the crew to operate in emergency and well control situations. Nowadays, the drilling companies have mandated frequent mock drills to enhance the awareness and experiences of the crew to act timely and confidently in case of real incidents.

Figure 19.8 summarizes the root causes of the incident as discussed here.

19.5 CONCLUSIONS AND RECOMMENDATIONS

Several conclusions were made from this work:

1. The review of the blowout incident showed that this catastrophic event occurred due to the following causal factors:
   - Absence of a critical review of the drilling program to resolve geological discrepancies.
   - Lack of effective organizational measures over operational practices, mud weight management, and management of changes in the basic well design parameters.
   - Operational complexities resulting from the decision to use OBM in a gas well subject to massive lost circulation problems.
   - Continuation of operations with defective safety critical equipment, sensors, and alarms.
   - Obscure routing of flows between tanks and lack of proper monitoring of the fluid transfers during tripping with losses obscuring an influx.
   - Lack of adherence to basic well control practices as evidenced by the elementary failure to employ the mud tank's level measurements to monitor hole fill.
2. It is essential that the companies improve their learning curve in terms of well control. To target that, the kick and blowout data were gathered for the whole field. The analysis of the kick volume data of the blowout well indicates that in the period of 2.66 hours, the kick converted to the blowout, while in the last 7 minutes, the kick volume reached from 172 to 341.5 bbl due to the sudden huge gas expansion.

3. Well planning should be carefully made to predict possible hazards in drilling a well. The part for rig selection and crew should be strictly performed by allocating competent rigs with experienced crew especially for critical cases. During operations, any mud weight change which compromises safety should *not* be allowed, particularly in gas-cap drilling. Management of *change* is strictly required.
4. Regular well-control training of the crews must never be compromised. Drills should be regularly practiced improving the crew's preparedness for emergency situations.
5. Finally, kick-related key-performance-indicators (KPIs) should be added for the rig and crew selection purposes. The required kick detection equipment should include accurate outflow sensors (preferentially Coriolis flowmeters) and early kick detection systems (EKDSs). Remote-control panels are required to always be operable. Otherwise, no permission should be given by no means to continue drilling.

REFERENCES

Ashena, R., Amani, M., Taei, B., et al., 2011. Management of the Disastrous Underground Blowout in South of Iran. Presented at *SPE Middle East Oil and Gas Show and Conference*, Manama, Bahrain, September 25–28. SPE 140781-MS. https://doi.org/10.2118/140781-MS

Baldino, S., Miska, S. Z., & Ozbayoglu, E. M., 2019a. A Novel Approach to Borehole-Breathing Investigation in Naturally Fractured Formations. *SPE Drilling & Completion*, 34, 27–45. https://doi.org/10.2118/189661-PA

Baldino, S., Miska, S. Z., Ozbayoglu, E. M., et al., 2019b. Borehole-Breathing/Kick Discriminator: Diagnostic Tool and Potential Resource for In-Situ Formation Characterization. *SPE Drilling & Completion*, 34, 248–267. https://doi.org/10.2118/195689-PA

Bybee, K., 2010. Kicks in Offshore UK Wells. *JPT SPE Journal Paper*. SPE 0110-0062-JPT. https://doi.org/10.2118/0110-0062-JPT

Chen, K., Wei, X., Li, H., et al., 2021. Operational Risk Analysis of Blowout Scenario in Offshore Drilling Operation. *Process Safety and Environmental Protection*, 149, 422–431. ISSN: 0957-5820. https://doi.org/10.1016/j.psep.2020.11.010

Dhillon, B. S., 2016. *Safety and Reliability in the Oil and Gas Industry*. Boca Raton: CRC Press, Taylor and Francis Group. ISBN: 13:978-1-4987-4656-4.

Dobson, J. D., 2009. Kicks in Offshore UK Wells—Where Are They Happening, and Why? Presented at the *SPE/IADC Drilling Conference and Exhibition*, Amsterdam, The Netherlands. SPE-119942-MS. https://doi.org/10.2118/119942-MS

Editions Technip, 1981. *Blowout Prevention and Well Control*. ISBN: 978-2710803973. https://www.abebooks.co.uk/9782710803973/Blowout-Prevention-Control-Technip-Editions-2710803976/plp

Etkin, D. S., French McCay, D., Horn, M., et al., 2017. Chapter 2-Quantification of Oil Spill Risk. In Fingas, M. (ed) *Oil Spill Science and Technology*, 2nd edn, pp. 71–183. Cambridge: Gulf Professional Publishing. ISBN: 9780128094136. https://doi.org/10.1016/B978-0-12-809413-6.00002-3

Fraser, D., Lindley, R., Moore, D. D., et al., 2014. Early Kick Detection Methods and Technologies. *Society of Petroleum Engineers*, October 27. https://doi.org/10.2118/170756-MS

Grace, R. D., 2017. *Blowout and Well Control Handbook*. Cambridge: Gulf Professional Publishing. ISBN: 978-0128126745.

Grace, R. D., Cudd, B., & Chen, J.-S., 2000. The Blowout at CHK-140W. *Society of Petroleum Engineers*. https://doi.org/10.2118/59120-MS

Habib, M. M., Imtiaz, S., Khan, F., et al., 2020. Early Detection and Estimation of Kick in Managed Pressure Drilling. *Society of Petroleum Engineers*. https://doi.org/10.2118/203819-PA

Hopkins, A., 2008. *Failure to Learn, the BP Texas City Refinery Disaster*, p. 186. Sydney: CCH Australia. ISBN: 978-1-921322-44-0.

Jacobs, T., 2015. Early Kick Detection: Testing New Concepts. *Society of Petroleum Engineers*. https://doi.org/10.2118/0815-0044-JPT

Lancaster, J., 2005. *Engineering Catastrophes-Causes and Effects of Major Accidents*, 3rd edn. ISBN: 978-1-84569-016-8.

Meng, M., Miska, S. Z., Yu, M., et al., 2020. Fully Coupled Modeling of Dynamic Loading of the Wellbore. *SPE Journal*, 25, 1462–1488. https://doi.org/10.2118/198914-PA

Meng, M., Zamanipour, Z., Miska, S., et al., 2019. Dynamic Wellbore Stability Analysis Under Tripping Operations. *Rock Mechanics and Rock Engineering*, 52, 3063–3083. https://doi.org/10.1007/s00603-019-01745-4

Miska, M., Samuel, G. R., & Azar, J. J., 1996. Modeling of Pressure Buildup on a Kicking Well and Its Practical Application. Presented at the *Permian Basin Oil and Gas Recovery Conference*, Midland, TX, March. https://doi.org/10.2118/35245-MS

Nabaei, M., Moazzeni, A. R., Ashena, R., et al., 2011. Complete Loss, Blowout and Explosion of Shallow Gas, Infelicitous Horoscope in Middle East. Presented at *SPE European Health, Safety and Environmental Conference in Oil and Gas Exploration and Production*, Vienna, Austria, February 22–24. SPE 139948-MS. https://doi.org/10.2118/139948-MS

Reason, J., 1997. *Managing the Risks of Organizational Accidents*. Aldershot, UK: Ashgate.

Reason, J., 2000. Human Error: Models and Management. *BMJ*, 320, 768–770. https://doi.org/10.1136/bmj.320.7237.768

SEMS, 2014. *Accident Investigation Report. Part 1 – Root Cause Investigation Results. In Re: South Timbalier Block 220, Well #A003ST01BP03. Walter Oil & Gas Corporation Blowout involving Hercules Offshore Incorporated Rig 265*. www.bsee.gov/sites/bsee.gov/files/panel-investigation/incident-and-investigations/sems-accident-investigation-report-walter-report-part-1.pdf (Last Accessed in February 17, 2021).

Shaker, S. S., & Reynolds, D. J., 2020. Kicks and Blowouts Prediction Before and During Drilling in the Over-Pressured Sediments. *Offshore Technology Conference*, Houston, TX. https://doi.org/10.4043/30711-MS

Skogdalen, J. E., Utne, I. B., & Vinnem, J. E., 2011. Developing Safety Indicators for Preventing Offshore Oil and Gas Deepwater Drilling Blowouts. *Safety Science*, 49 (8–9), 1187–1199. ISSN: 0925-7535. https://doi.org/10.1016/j.ssci.2011.03.012

St. John, M. F., 2016. Macondo and Bardolino: Two Case Studies of the Human Factors of Kick Detection Prior to a Blowout. *Offshore Technology Conference*, Houston, TX. https://doi.org/10.4043/26906-MS

Turley, J. A., 2014. An Engineering Look at the Cause of the 2010 Macondo Blowout. Presented at *IADC/SPE Drilling Conference and Exhibition*, Fort Worth, TX, March 4–6. SPE 167970-MS. https://doi.org/10.2118/167970-MS

US-CSB, 2019. *Investigation Report of Gas Well Blowout and Fire at Pryor Trust Well 1H-9*. www.csb.gov/pryor-trust-fatal-gas-well-blowout-and-fire/ (Last Accessed in January 1, 2020).

Wylie, W. W., & Visram, A. S., 1990. Drilling Kick Statistics. Presented at *IADC/SPE Drilling Conference*, Houston, TX, February 27–March 2. SPE 19914-MS. https://doi.org/10.2118/19914-MS

20 Case Study *Blowout in Australia Petrel No. 1 (SEDCO 135G), 1969*

Bryan Atchison

20.1 INTRODUCTION

This chapter presents an account of events leading to a blowout that happened in Australia in 1969.

20.2 DESCRIPTION

This is a key study of the Petrel-1 Well and what happens when well control goes catastrophically wrong. You will witness the sea on fire from footage never seen before.

I have spent my entire career involved with well operations, managing the risk of well control. This story is more than professional, rather, it is personal because my father was a pioneer in this industry. He was a mechanic on the SEDCO 135G, an eyewitness, and an active player in the blowout, its recovery, and the following tragedy. The information on this video is taken from family pictures, home movies and eyewitness accounts, and the report of the Commonwealth of Australia Department of National Development Bureau of Mineral Resources Geology and Geophysics.

In 1969, the Atlantic Richfield Company had contracted the SEDCO 135G to drill for oil off the coast of the Northern Territories of Australia in the Bonaparte Gulf. The 135G was the first-generation offshore drilling rig designed, built, and owned by the Southeastern Drilling Company SEDCO (see Figure 20.1). The triangular deck shape and milk bottle-shaped legs make the rig instantly recognizable. Situated on the deck were living and office accommodations, power generation system, deck cranes, substructure, derrick, and the equipment required for rotary drilling operations. The rig equipment list had all the subsea equipment required for marine drilling operations including one annular preventer and three ram-type preventers all rated to 5,000 psi (see Figure 20.2). SEDCO was one of the pioneers of the technology from marine drilling operations. The company mobilized crew who were resident in Darwin in the Northern Territories. Operations were going well with new crew learning to use the new technology while drilling the Petrel-1 well for the client.

On August 6, 1969, the rig was drilling the 8 3/8-inch hole around 13,000 feet with 10.5 pound per gallon density drilling mud (Figure 20.3) when the blowout happened (Figure 20.4). The events leading up to the blowout consisted of an increase in drilling rate for 5 feet at 13,052 feet. The drilling process was stopped, and the well

 DOI: 10.1201/9781003473770-20

FIGURE 20.1 The 135G drilling rig.

was observed to be flowing without pumping from the surface. The annular preventer was closed 5 minutes after the initial drilling rate increase to prevent any further influx from the formation into the well bore. The shut-in drill-pipe pressure was 750 psi, the Shut-In Casing Pressure was 1,800 psi, and the influx or kick volume was 190 barrels. I am a petroleum engineer with significant experience in the review and analysis of well control incidents.

The pressures in this incident are high, the global average for influx volumes is 25 barrels, and the excessive influx volume associated with this incident is very likely a function of poor drilling and well control practices which are typically rooted in human factor issues. There was obviously a very heavy reliance on human behaviors and competency which in this instance failed, resulting in an extremely large influx. The size of the influx is very important because the smaller the influx, the

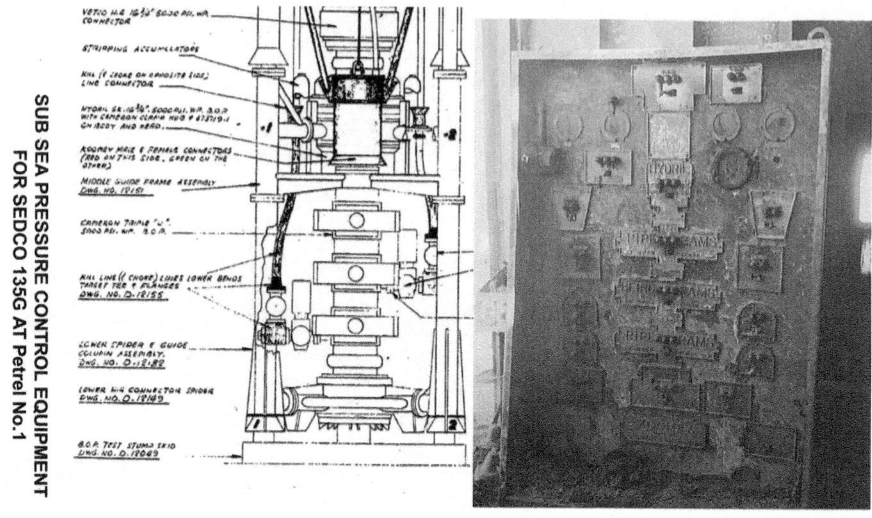

FIGURE 20.2 Schematic of the BOP stack including one annular preventer and three ram-type preventers.

FIGURE 20.3 Photo of the drilling crew prior to the blowout incident.

smaller the resulting well kill pressures experienced by the equipment. In all well control training, we stress to minimize the volume of the influx. We will analyze later why in this case the result was catastrophic. After the well was shut in on the 5,000-psi annular, the team followed the operator's procedure to make the situation safe. This involved pumping heavier fluid into the well, and this was to be done

FIGURE 20.4 Blowout in Petrel-1 well.

while periodically moving the drill-pipe with the well closed in on the large rubber annular. During this process, the annular failed which led ultimately to a complete loss of control on the well. The annular failed while the pipe was being moved by the driller, and a violent discharge of mud occurred at the drill floor knocking the driller off the brake handle. The lack of protection for the driller and proximity to the point of mud discharge meant the draw works were now out of control. The blocks were dropped, and the drilling line parted at the draw works' drum. Control of movement of the drill-pipe was no longer possible. Both sets of pipe rams were closed, and the upper pipe rams sealed. The BOP was unlikely to remain sealed because the drill-pipe was moving relative to the BOP rams, which was not designed for. The situation was critical; however, there was still a circulation path, as a result, heavy mud was continued to be pumped into the well. The annular pressure increased to such an extent that eventually the rams lost pressure integrity, and the gas was flowing from the well up to riser pipe to the rig.

In a final attempt to control the well, pumping of cement down the well started and continued for several hours. The flow from the well increased to a point where it was clear that the well control had been lost and the safety of the vessel took priority. The rig's anchor winches were used to move the vessel away from the well with the marine riser still connected to the BOP and well. The blind rams were shut; however, with drill-pipe across them, they would not seal the well. As a result, the gas had a path to the rig floor through the BOP's riser pipe and then onto the rig floor.

When the riser pipe connected to the BOP started to clash with the rig substructure, there was a failed attempt to disconnect the subsea equipment. Gas was flowing from the well through the riser to the drill floor. All personnel were evacuated from the rig. Fire fueled by the well gas broke out on the rig floor, under the rig floor, and in the office and accommodation block. As the rig continued to winch off location, the riser connected to the BOP parted just above the BOP which meant the gas-fed fire on the rig would cease. Several secondary fire incidences were still in place. Simultaneously, a gas plume from the seabed appeared in the water near the rig (see Figure 20.5). The gas plume was subsequently ignited to ensure that the gas plume and fire were recognizable to shipping. When the rig was a safe distance from the gas plume, the rig crew mobilized back onto the SEDCO 135G to put the fires out and address another critical vessel stability issue caused by the fire. With the fires extinguished and the rig stable, a damage assessment was conducted: Crew quarters and barge control room were completely gutted. Derek A-frame warped, all drill floor, and derrick equipment were badly damaged. All subsea equipment were on the seabed. Amazingly, there were no reported fatalities or injuries throughout this phase of the operation.

At this point, the rig was towed off location where the BOPs were sitting on the wellhead. A violent flow of hydrocarbon gas was flowing from the top of the BOP, forming a gas bloom which was then auto-igniting near the surface of the water. No oil slick was observed, following a detailed review of the situation with the regulator operator and drilling contractor. The quickest solution was to repair the rig in a shipyard, return it to location, and drill a relief well. A relief well was started on the 6th of February 1970 and was expected to conclude in May 1970. As a result, the sea was boiling from the discharge of the Petrel-1 well for around ten months. We can see from this home movie footage (see Figure 20.5), the seemingly supernatural image of the sea on fire. Tragically, while drilling the relief well, a marine accident resulted in the loss of nine personnel. The SEDCO, Helen work boat caught an anchor buoy in her propeller; the buoy impacted below the water line of the hull, quickly flooding and sinking the vessel.

The official report into this incident made several recommendations which have been adopted by the industry:

- Keep the drill-pipe stationary during well control operations.
- Provide equipment to shear the drill-pipe in the BOP.
- Provide two annular preventers in subsea-BOP.
- Maintain annular preventer elements.
- Provide a drill-pipe drop-in check valve.
- Provide a method to accurately measure return flow from the well.

A modern review of this incident with the limited information available highlights more issues. There was an unreasonable exposure to human factors' issues. Early well engineering standards and operational procedures fell far short of modern standards. The most striking of which is using the BOP equipment rated to 5,000 psi in a well that can produce pressures greater than 5,000 psi. While well engineering standards, procedures, and rig designs have improved dramatically in the ensuing

FIGURE 20.5 Gas plume visible at the surface of water at day and night times.

FIGURE 20.6 The rig floor after the blowout.

50 years, one aspect of well construction is the same: The recognition of an influx, decision-making, equipment control, and closing the BOP are still down to one man, i.e., the driller.

Blowouts are still happening on a regular basis with subsequent exposure to lives and the environment. The more notable recent events are in 2010, Deepwater Horizon with 11 fatalities, i2018 Prior Trust with 5 fatalities, and there are several more. They are predominantly caused by human factor issues just like on the SEDCO 135G on the Petrel-1 well over 50 years ago. Fortunately, a recent innovation is providing a tool to protect the driller personnel, the environment, assets, reputation, and against financial loss. Automated well control is a tool for the driller that automatically makes the well safe. The well construction industry owes a tremendous debt of gratitude to its pioneers. Their courage, innovation, and energetic enthusiasm paved the way for a cheap reliable energy for the world to benefit from. It may have taken over 50 years, but we finally have a solution for the constant industry threat of well control.

Index

L

M

N

O

P

R

S

T

U

V

W

For Product Safety Concerns and Information please contact our EU representative GPSR@taylorandfrancis.com Taylor & Francis Verlag GmbH, Kaufingerstraße 24, 80331 München, Germany

Batch number: 10397790

Printed by Printforce, the Netherlands